Grant Duncan
60 Chatham Drive, N.W
282-1915

PHYSICAL
CONSTANTS
COMMONLY
NEEDED IN
COMPUTATIONS

←

ELEMENTARY MODERN PHYSICS

ELEMENTARY
MODERN PHYSICS

RICHARD T. WEIDNER

PROFESSOR OF PHYSICS

RUTGERS UNIVERSITY

ROBERT L. SELLS

PROFESSOR OF PHYSICS

STATE UNIVERSITY COLLEGE

GENESEO, NEW YORK

ALLYN AND BACON, INC.   BOSTON

PRINTED IN THE UNITED STATES OF AMERICA

1st Printing . . . . March, 1960
2nd Printing . . . . November, 1960
3rd Printing . . . . August, 1961
4th Printing . . . . May, 1962
5th Printing . . . . May, 1963
6th Printing . . . . December, 1963
7th Printing . . . . May, 1964
8th Printing . . . . November, 1964
9th Printing . . . . June, 1965
10th Printing . . . . November, 1965

PREFACE

Our aim in this book has been to treat the fundamentals of modern physics fairly rigorously, but at an elementary level. The text is intended primarily for use in the concluding semester of, or immediately following, the general physics course for engineering and science students. The prerequisites are an understanding of elementary physics and introductory calculus.

The primary concern has been to give a logically coherent and sequential account of the basic principles of the relativity and quantum theories, of atomic and nuclear structure, and of a few topics in molecular and solid-state physics. All other considerations have been subordinated to this goal. We have tried to avoid two extreme approaches: one, so descriptive and non-mathematical as to be superficial (if not misleading); and another, a formal treatment, so abstract and analytical as to allow the student to evade the physics in his concern with strictly mathematical questions.

Insofar as the order of topics is concerned, the strategy has been to begin with a simple treatment of special relativity, followed by the basic quantum properties of electromagnetic radiation and of material particles. These form the foundations for the later chapters on atomic, nuclear, and solid-state physics. The theory of special relativity is discussed in some detail, not only because such relations as $E = mc^2$ and $m = m_0[1 - (v/c)^2]^{-1/2}$ are hardly self-evident and are really understood only in the light of Lorentz transformations, but also because of the considerable use to which relativity can be put in developing the properties of the photon. The quantum ideas are not introduced in the conventional fashion; namely, by a discussion of such a complicated many-body system as a blackbody. Rather, we use the simple photon-electron interactions as a basis for introducing the quantum concepts.

Each chapter (except Chapter 1) concludes with a summary of the most important concepts introduced in the chapter. References, principally to other texts at a comparable or somewhat higher level, are given at the ends of chapters, together with specific suggestions for collateral reading. The working of problems is, we believe, an indispensable pedagogical aid, and we have included many of them. Some problems are numerical (but we hope to have avoided the strictly plug-in variety), and others are in the nature of derivations and proofs. Problems of more than average difficulty are identified by asterisks, and answers to all odd-numbered numerical problems are given in the back of the book.

In its entirety this text provides enough material for a full one-semester course at the level of the sophomore or junior year. The arrangement of topics has been chosen, however, so that the book can be used for a shorter treatment of modern physics (perhaps 8 to 10 weeks) without serious discontinuities. In such a short course, the following chapters can safely be omitted: Chapter 6 (Many-Electron Atoms), Chapter 7 (X-Ray Spectra), Chapter 11 (Molecular and Solid-State Physics), and some portions of Chapter 8 (Instruments and Accelerating Machines Used in Nuclear Physics).

The rationalized mks system of units is used throughout. This choice is justified by its increasingly common use in physics texts, both elementary and advanced.

We are very grateful for the suggestions of our colleagues: Professor Herman Y. Carr, with whom the arguments in Appendix I were developed; Professor Ira M. Freeman, who used and criticized the preliminary text material; and Professor Henry C. Torrey, who reviewed one part of the manuscript. The criticisms of the manuscript by Professor Rolf M. Steffen of Purdue University have been most helpful. We have benefited greatly from the comments of our students, particularly Mr. Emory S. Fletcher. The manuscript was deciphered and typed with uncommon skill by Mrs. Patricia B. Kinder. It has been a pleasure working with Mr. Clifton A. Gaskill of Allyn and Bacon in the production of the book. Any errors of omission or commission are, of course, the authors' responsibility. Finally, we are indebted to our wives, without whose help, but with whose considerable forbearance, this book was possible.

Richard T. Weidner
Robert L. Sells

New Brunswick, New Jersey
January, 1960

CONTENTS

SEVEN
X-RAY SPECTRA

EIGHT
INSTRUMENTS AND ACCELERATING MACHINES USED IN NUCLEAR PHYSICS

NINE
NUCLEAR STRUCTURE

TEN
NUCLEAR REACTIONS

ELEVEN
MOLECULAR AND SOLID-STATE PHYSICS

TO JEAN AND PAT

ELEMENTARY MODERN PHYSICS

O N E

CLASSICAL PHYSICS

1-1 The program of physics The program of physics is to devise concepts and laws which can help us to understand the universe. Physical laws are human constructions, subject to all the limitations of human understanding. They are not necessarily fixed, immutable, or good for all time; and Nature is not compelled to obey them.

A law in physics is a statement, usually in the succinct and precise language of mathematics, of a relationship that has been found by repeated experiment to obtain among physical quantities, and which reflects persistent regularities in the behavior of the physical world. A "good" physical law has the greatest possible generality, simplicity, and precision. The final criterion of a successful law of physics is how accurately it can predict the results which experiments will yield. For example, our confidence in the essential correctness of the law of universal gravitation leads us to expect with (almost) complete certainty that, when the gravitational acceleration at the Moon's surface is measured there, it will be very close to 1.6 m/sec^2. We say that our certainty is *almost* complete inasmuch as extrapolating any law outside the range of its tested validity *may* predict results which come to be inconsistent with later experiment.

As physics developed, some early theories and laws were found to be inadequate to phenomena for which they had not been tested. These theories have been supplanted by more general, comprehensive theories and laws which more adequately describe phenomena in the new, as well as the old, regions. Figure 1-1 shows the various regions of applicability of *classical physics, relativity physics, quantum physics,* and *relativistic quantum physics.*

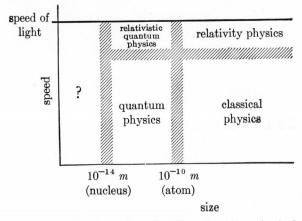

Figure 1-1. Regions of applicability of various physical theories.

Classical physics is the physics of ordinary-size objects moving at relatively low speeds; it embraces Newtonian mechanics (including heat—which can be described in terms of the random motion of molecules obeying Newton's laws) and electromagnetism (including the theory of light). At object-speeds approaching that of light, classical physics must be supplanted by relativity physics; for object-sizes of about 10^{-10} meter (approximately the size of an atom), classical physics must be supplanted by quantum physics. For subatomic dimensions and speeds approaching that of light, only relativistic quantum physics is adequate. The limits of the several physical theories are not sharply defined; in fact, they overlap. Relativistic quantum physics is the most comprehensive and complete theoretical structure in contemporary physics. At dimensions of about 10^{-14} meter (the approximate size of the atomic nucleus), different and perplexing phenomena appear; at the moment, they are only partly understood.

The foundations of our present understanding of atomic and nuclear structure lie in the two great ideas of modern physics—relativity physics and quantum physics. Both these theories had their origins early in this century, a period in which improved experimental techniques first made possible the study of phenomena in small dimensions, and at high speeds and high energies.

After reviewing some crucial aspects of classical physics, we shall study the theory of relativity and the quantum theory and then apply these theories to analysis of atomic and nuclear structure. We shall be concerned with situations in which some familiar notions in physics may be inapplicable, situations in which classical physics is downright wrong. Does this mean then, that all the time and effort spent in studying elementary classical physics is wasted, that one might better begin with relativity and quantum theory? Not at all! All results of experiment, however remote from our ordinary experience, must be expressed in classical terms; that is, in the classical concepts of momentum, energy, position, and time. We can most readily understand these quantities from classical physics. Furthermore, we shall see that many of the concepts and laws of classical physics are carried over into the new areas.

1-2   The conservation laws of physics   Of all the laws of physics the conservation laws are the most general and simple. Many of the conservation laws are carried over essentially unchanged into relativity and quantum physics.

A conservation law indicates the conditions under which some particular physical quantity remains unchanged in a system of objects; that is, under what circumstances the system is isolated from some particular external influence. The possibility of achieving this isolation is an idealization which can only be approximated but never perfectly realized, for any observation or measurement of a system necessarily interferes with it. For a simple example, consider what happens when the temperature of some thermally insulated liquid is measured with a simple mercury thermometer which is initially not at the temperature of the liquid. When the thermometer is placed in the liquid, the thermometer is either heated or cooled by the liquid, and the liquid has its temperature either lowered or raised. The final thermometer reading is *not* the actual temperature of the liquid before the measurement was made: it is the temperature to which the liquid has been brought by reason of insertion of the thermometer into it. Only if the liquid and the thermometer were at the same temperature before being brought into thermal contact with one another would the thermometer neither gain nor lose heat, and only then would the thermometer indicate the true temperature of the body. But this cannot be known in advance, and if it were, the measurement would be superfluous.† Interference with a system through the act of taking a measurement occurs not only for the example we have discussed; it happens in all measurements. In short, a completely isolated system is one that can never be studied or observed, and we cannot study a system without violating its isolation. In classical physics, however, it is always possible—by exercising experimental ingenuity—to reduce the disturbances to such an extent that the system may be regarded as *effectively* isolated.

Let us summarize briefly the conservation laws of classical physics: the conservation of (1) mass, (2) energy, (3) linear momentum, (4) angular momentum, and (5) electric charge.

(1) *The law of conservation of mass: the total mass of an isolated system is constant.* Despite changes that may occur in other quantities (e.g., energy, volume, or temperature) in a system, the total mass is unchanged. This law is also stated in the form: mass cannot be created or destroyed; or, mass cannot be produced or annihilated.

(2) *The law of conservation of energy: if no work is done on or by the system, and if no heat enters or leaves the system, the total energy of the system is constant.* In a collection of objects which make perfectly elastic collisions with one another, only purely mechanical energy need be considered; then the energy conservation law states that the sum of the kinetic and potential energies of the system is constant. Thermal energy is the mechanical energy of molecules or atoms in random motion on a scale so microscopic that the kinetic and potential energies of individual particles cannot be identified. The *first law of thermodynamics* is merely the law of conservation of energy expressed in its most comprehensive form.

(3) *The law of conservation of linear momentum: when a system is subject to no net external force, the total linear momentum of the system remains constant, both in magnitude and direction.* Thus, in a collision of two objects, free of external forces but interacting with one another during the collision, the total linear momentum vector before collision equals the total momentum vector after collision. This holds whether the collision is elastic or inelastic; that is, whether mechanical energy is conserved or dissipated into other forms. Here we see the particular utility of a conservation law. We need not be concerned with all the details of the interaction; we may in fact be ignorant of the details of the forces be-

---

† Of course, if one knew the heat capacities of the liquid and thermometer with complete precision, it would be possible to correct for the heat entering or leaving the thermometer. But the heat capacities become known only if the specific heats of the materials are measured in some prior experiment. This earlier experiment, however, if it is to give the specific heats without error, must involve a perfectly calibrated and corrected thermometer, and so on ad infinitum

tween the colliding objects. Nevertheless, the conservation of momentum law gives us to know that the momentum coming out of the collision is precisely the momentum going into the collision.

Newton's laws of motion are, of course, the foundation of classical mechanics, and it is useful to state these laws in the language of linear momentum.

(a) The momentum ($p = mv$) of a particle subject to no net external force is constant. (A vector quantity is indicated by boldface type, its magnitude, by ordinary type.)

(b) When a body is subject to a net external force, the force equals the time-rate-of-change of the linear momentum. In this form Newton's second law is sufficiently general to account for the motion of objects that accumulate or lose mass, as in the case of a snowball rolling on snow, or a falling raindrop losing water by evaporation. When the mass is unchanged, the force is simply the product of mass and acceleration. Thus,

$$F = \frac{d}{dt}(p) = \frac{d}{dt}(mv) = (v)\frac{dm}{dt} + m\frac{d}{dt}(v) = ma \qquad [1\text{-}1]$$

when $dm/dt = 0$.

This law has a profound consequence: when one knows the forces acting on a body and its initial position and velocity, it is possible in principle to predict in complete detail the future history of the body; that is, to project precisely its position and velocity for all future times.

In the rationalized meter-kilogram-second (mks) system of units, which will be used throughout this book, one newton is that force that acts on a one-kilogram mass to give it an acceleration of one meter per sec².

(c) When two bodies interact, the momentum imparted to one body during an infinitesimal time interval is equal but opposite to the momentum imparted to the second body during this time; therefore, the action and reaction forces are equal and opposite. The conservation of momentum law comes directly from Newton's third law.

(d) *The law of conservation of angular momentum: when a system is subject to no net external torque, the total angular momentum of the system remains constant, both in magnitude and direction.* Figure 1-2 shows the angular momentum, P_θ, with respect to the point O, *for a particle* of mass m and velocity v. The magnitude of the angular momentum P_θ is

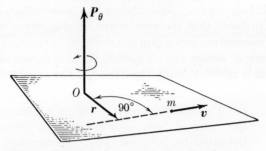

Figure 1-2. Angular momentum P_θ of a mass point m about the point O.

the product of the linear momentum mv and the perpendicular distance r from the point O to the line of action of the velocity vector. Thus,

$$P_\theta = (mv)r \qquad [1\text{-}2]$$

The direction of the angular momentum vector, as shown in Figure 1-2, is perpendicular to the plane of the velocity vector v and the vector r. Its direction is the same as the direction of advance of a right-hand screw obtained by rotating the vector r into the vector v through the smaller angle. For the special case of a particle which moves in a circle of radius r with an angular velocity ω, $v = \omega r$; and the magnitude of the angular momentum with respect to the axis of rotation is

$$P_\theta = (mr^2)\omega = I\omega$$

where I, the moment of inertia with respect to the axis of rotation, is defined as $I = mr^2$.

Figure 1-3 shows the quantities that enter into the definition of the

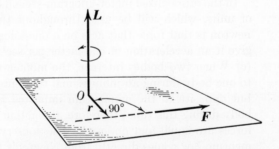

Figure 1-3. Torque L of a force F about the point O.

torque L. The magnitude of the torque L, with respect to the point O, is given by the product of the force F and its moment arm r; therefore, $L = Fr$. The direction of L is the direction of advance of a right-hand screw obtained by rotating the vector r into the vector F through the smaller angle.

In terms of torque and angular momentum, Newton's second law is written

$$L = \frac{d}{dt} (P_\theta) \qquad [1\text{-}3]$$

When a particle is subject to no net torque, its angular momentum P_θ is constant. It is also true that for a *system* of objects, subject to no net external torque, the *total* angular momentum of the system is of unchanged magnitude and direction. This is the law of conservation of angular momentum.

(5) *The conservation of electric charge: the total charge of an isolated electrical system is constant.* It is interesting to note that Kirchhoff's rules, which are the bases for analyzing electric circuits, are merely applications of the laws of conservation of charge and energy. Thus, the rule that the total current entering a junction at any point in a circuit equals the total current leaving the junction is, in effect, the law of conservation of charge; and the rule that the total potential drop around a closed loop is zero results from the conservation of energy.

1-3 Electromagnetic forces and fields   There are only two basic origins of force in classical physics, mass and electric charge; these give rise respectively to the universal gravitational force and to electromagnetic forces.

The gravitational force F_g between two masses m_1 and m_2 separated by a distance r is given by

$$F_g = Gm_1m_2/r^2$$

where G, the universal gravitational constant, is found by experiment to be 6.67×10^{-11} newton-m^2/kg^2.

Electromagnetic forces are of two basic types, the electric force and the magnetic force. Coulomb's law gives us the electric force F_e between two electric charges q_1 and q_2 (mutually attractive or repellent, according as they are of like or unlike signs) separated by a distance r:

$$F_e = q_1q_2/4\pi \, \epsilon_0 r^2 = kq_1q_2/r^2 \qquad [1\text{-}4]$$

Equation 1-4 applies for charges in free space, where ϵ_0, the electric permittivity of free space, is 8.85×10^{-12} coulomb^2/newton-m^2, and $k = (1/4\pi \, \epsilon_0)$ is 8.99×10^9 newton-m^2/coulomb^2.

The electric field intensity $\mathcal{E}$ is defined as the Coulomb force per unit positive charge. The magnitude of $\mathcal{E}$ is given by $\mathcal{E} = F_e/q$. The electric field can be represented by electric lines of force, which indicate by their density and direction the magnitude and direction of $\mathcal{E}$.

If we observe two electric charges (such as electrons) at rest with respect to us, we find that only the electric force and a negligibly small gravitational force act between them. But when the two charges are in motion with respect to us, there may be an additional force present, called the magnetic force F_m. As in the case of the electric force, where a field quantity is used, it is convenient to introduce a magnetic flux density B (in weber/m^2) which, for a point charge q_1 moving at a velocity v_1, has the magnitude

$$B = \frac{\mu_0}{4\pi} \frac{q_1 v_1 \sin \theta}{r^2} \qquad [1\text{-}5]$$

at the point P, which is a distance r from the charge q_1, and where θ is

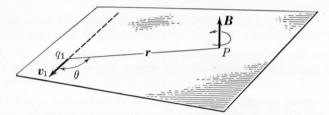

Figure 1-4. Magnetic flux density B at the point P produced by the moving charge q_1.

the angle between r and v_1. Equation 1-5 gives the flux density for free space, where μ_0, the permeability of free space, is given by $(\mu_0/4\pi) = 10^{-7}$ weber/amp-m. The direction of B in Figure 1-4 is given by the right-hand rule (turning v_1 into r).

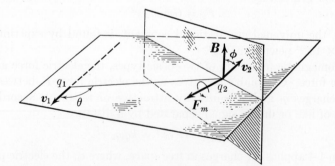

Figure 1-5. Magnetic force F_m on the moving charge q_2 produced by the moving charge q_1.

If a second charge q_2 is moving with a velocity v_2 in the vicinity of charge q_1, this second charge will experience a magnetic force F_m because of the magnetic flux density B, the magnitude of which is given by

$$F_m = q_2 v_2 B \sin \phi \qquad [1\text{-}6]$$

where ϕ is the angle between the vectors v_2 and B. The direction of the force is found by turning v_2 into B, as shown in Figure 1-5.

It is convenient to introduce the magnetic field intensity $\mathfrak{IC}$ which, for free space, is defined as

$$\mathfrak{IC} = \frac{B}{\mu_0}$$

and has the units of amp/m. Therefore, Equation 1-6 can be written

$$F_m = \mu_0 q v \, \mathfrak{IC} \sin \phi \qquad [1\text{-}7]$$

The magnetic intensity $\mathfrak{IC}$ can be represented in magnitude and direction by magnetic lines of force. Magnetic lines of force always form closed loops, whereas electrostatic lines of force originate on positive charges and terminate on negative charges.

Notice the utility of the field concept. When an electric force acts between two charges we can describe this interaction in the following way: the first charge produces an electric field $\mathcal{E}$ at the site of the second charge (as well as at other points in space), and the second charge experiences a force on it because it is present in this electric field. Similarly, we can describe the magnetic interaction of the two charges in terms of the magnetic field concept: the first moving charge produces a magnetic field $\mathfrak{IC}$ at the site of the second moving charge, and the second charge experiences a force because it is in motion within this magnetic field.

The electric and magnetic fields as portrayed by their respective lines of force certainly are valuable aids to visualizing electric and magnetic effects produced by electric charges at rest or in motion, but they have a far greater significance. We shall see that it is possible—indeed necessary—to ascribe energy, linear momentum, and angular momentum to electromagnetic fields.

Let us first consider the energy of an electric field, using as an example a parallel-plate capacitor, of capacitance C, area A, and plate separation d. The capacitance C is related to a charge q on one of the two plates and the potential difference V between the two plates by $C = q/V$. Work is required to charge the capacitor because, after the first small charge is placed on either plate, this charge repels other charges of like sign that are subsequently to be added to the plate. We wish to find the total work required to charge the capacitor to a final potential difference V_f with final charge Q on each plate. The total work done in bringing charges to the capacitor plates is the energy of the capacitor E_e. Because the potential difference increases as the capacitor is being charged, we must integrate from the initial zero charge to the final charge Q.

$$E_e = \int_0^Q V \, dq = \int_0^Q \frac{q}{C} \, dq = \frac{Q^2}{2C} = \tfrac{1}{2} C V_f^2 \qquad [1\text{-}8]$$

We can regard this energy E_e as being the potential energy of electric charges on the capacitor, or we can think of E_e as the energy to be attributed to the electric field $\mathcal{E}$ between the capacitor plates. For a parallel-plate capacitor in which the distance of separation between the plates is small, the electric field is uniform and confined entirely to the space between the two plates; therefore, $\mathcal{E} = V_f/d$ and $C = \epsilon_0 A/d$. Substituting for C and V_f in Equation 1-8 gives

$$E_e = \tfrac{1}{2}\epsilon_0 \mathcal{E}^2 (Ad)$$

We designate the electric energy density, or the energy of an electric field per unit volume, as (E_e/vol). Therefore, $(E_e/\text{vol}) = E_e/Ad$ and

$$\boxed{(E_e/\text{vol}) = \tfrac{1}{2}\epsilon_0 \mathcal{E}^2} \qquad [1\text{-}9]$$

This relation, Equation 1-9, although derived here for the special case of a parallel-plate capacitor, gives the correct energy density for *any* electric field.

To compute the energy density of a magnetic field, one might consider a toroidal (doughnut-shaped) solenoid through which there is a constant electric current. The magnetic field for this configuration is confined entirely to the interior of the solenoid. A derivation similar to the one just given for the capacitor shows that the energy density of the magnetic field of intensity $\mathcal{3C}$, which we designate by (E_m/vol), is

$$\boxed{(E_m/\text{vol}) = \tfrac{1}{2}\mu_0 \mathcal{3C}^2} \qquad [1\text{-}10]$$

Work is required to achieve a current through the solenoid because, as the current changes from zero to some final value, the magnitude of the magnetic field accompanying the current is changed, thus inducing a back emf determined by the self-inductance of the solenoid. The total work done in achieving a final current through the solenoid equals the energy residing in the magnetic field of the solenoid.

We have seen that it is possible to describe an electric or magnetic field not merely as a region of space in which a charged particle is subject to an electric or magnetic force, but also as a region of space containing electric and magnetic energy. Further, it is relatively easy to show not only that we *can* but that we *must* assign linear and angular momentum to electromagnetic fields.

Consider the situation shown in Figure 1-6 in which two positively charged particles marked A and B are moving at right angles to one another. There is an electric force of repulsion acting between the two particles, and this Coulomb force acts along the direction of the line connecting them. The electric force exerted on A by B equals the electric force exerted on B by A; the electric forces are equal, but opposite. Now consider the magnetic interaction between the moving charges A and B.

From our earlier discussion of magnetic fields we know that the magnetic field at charge B which derives from charge A is zero [noting that in Equation 1-5 $\sin \theta = 0$]; hence there is no magnetic force on B because of A. But now consider the magnetic field at charge A deriving from charge B. The magnetic field is into the paper at A (here, $\sin \theta = 1$), and therefore the magnetic force on A is to the right as indicated in Figure 1-6. A mag-

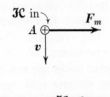

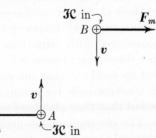

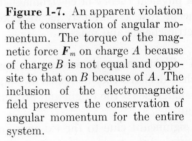

Figure 1-6. An apparent violation of Newton's third law. The magnetic force on A because of B is *not* equal and opposite to the magnetic force on B because of A. Inclusion of the electromagnetic field preserves conservation of linear momentum.

Figure 1-7. An apparent violation of the conservation of angular momentum. The torque of the magnetic force F_m on charge A because of charge B is not equal and opposite to that on B because of A. The inclusion of the electromagnetic field preserves the conservation of angular momentum for the entire system.

netic force acts on A, but not on B. At the instant when the charges are moving as shown in Figure 1-6, there is an apparent violation of Newton's third law! Inasmuch as the third law is the basis for the conservation of momentum principle, this isolated system, consisting *only* of the two moving charges, *violates* the fundamental law of conservation of linear momentum. The reason for this non-conservation of momentum is that our isolated system has *not* included the electromagnetic field associated with the particles. *An electromagnetic field carries linear momentum and is capable of exerting a force on electric charges.* If, in our present example, we include the momentum of the electromagnetic field, we preserve the conservation of momentum law.

We can also show that an electromagnetic field may carry angular momentum. Consider two charges moving as shown in Figure 1-7. Here again the electric forces are equal, but opposite. Moreover, the magnetic forces are equal and opposite, but they do *not* act along the line joining the charges. Therefore, without including the angular momentum of the electromagnetic field, there is an unbalanced torque acting on the two charges, which changes their angular momentum. If we are to preserve the conservation of angular momentum law for the *entire* system—the two charges *and* the associated electromagnetic field—it is necessary that the electromagnetic field have angular momentum. Again, the electromagnetic field can be shown to carry angular momentum whose value is always such as to conserve angular momentum for the entire system.

It may seem strange that we attribute to the electromagnetic field such properties as linear momentum and angular momentum since we usually regard these as strictly mechanical quantities. The criterion is *not* whether our "common sense" allows it, for our common sense is based mostly on our experience with situations in mechanics. The criterion *is* always whether there is agreement between theory and experiment. We are therefore led to conclude that the electric and magnetic fields are not merely useful constructions for mapping or visualizing electric and magnetic forces, but that they also have certain physical attributes in and of themselves. The electromagnetic field is, in fact, the most important element in a physical theory of electromagnetic effects, and we must take the field concept just as seriously as we take other concepts in physics, such as electric charge and mass.

1-4 Electromagnetic waves Inasmuch as electromagnetic radiation plays such a decisive role in relativity, quantum, atomic, and nuclear physics, we shall discuss the origin and properties of electromagnetic waves in some detail.

A significant clue to the existence of electromagnetic waves is found by examining the two constants ϵ_0 and μ_0 which appear in the basic relations for the electric and magnetic forces. If one considers the product $\epsilon_0 \mu_0$, and substitutes the experimentally determined values for the electric permittivity (ϵ_0) and magnetic permeability (μ_0), both for free space, one sees that

$$\epsilon_0 \mu_0 = (8.85 \times 10^{-12} \text{ coul}^2/\text{n-m}^2)(4\pi \times 10^{-7} \text{ weber/amp-m})$$

$$\epsilon_0 \mu_0 = \frac{1}{9 \times 10^{16} \text{ m}^2/\text{sec}^2} = \frac{1}{(3 \times 10^8 \text{ m/sec})^2}$$

This strongly suggests that light is an electromagnetic phenomenon inasmuch as its speed c through free space (3.00×10^8 m/sec) appears to be related to two constants of electromagnetism by the equation†

$$c = \frac{1}{\sqrt{\epsilon_0 \mu_0}} \qquad [1\text{-}11]$$

The rigorous, detailed theoretical prediction of the existence of electromagnetic waves was first made by James Clerk Maxwell in 1864. Maxwell

† Equation 1-11 is justified in Appendix I for a special situation, using a rather simple argument.

summarized all the known laws of electromagnetism in a set of equations, known as Maxwell's equations, which give the relations between electric and magnetic fields. The existence of electromagnetic waves moving at the speed $1/\sqrt{\epsilon_0\mu_0}$ is an inescapable logical consequence of Maxwell's equations. Heinrich Hertz (1887) first produced and detected high-frequency radio waves in the laboratory and thereby verified the theoretical predictions of Maxwell. These experiments showed that the waves were of electromagnetic origin and that they exhibited all the important wave properties of light: interference, diffraction, and polarization. To derive the properties of electromagnetic waves from Maxwell's equations requires a formal mathematical procedure which lies beyond the scope of this book. We shall instead present arguments based on simple non-mathematical considerations which make plausible certain important features of electromagnetic waves.

The production of electromagnetic waves depends crucially on the facts that (1) a changing magnetic field produces an electric field; and, reciprocally, (2) a changing electric field produces a magnetic field. We know from Faraday's law of electromagnetic induction (emf $= -d\phi/dt$) that a changing magnetic field induces an emf in a conductor; that is, an induced electric field acts on the free charges, causing an induced current. Even in the absence of a conductor, *a changing magnetic field produces an electric field*.

Now we consider the converse situation. The Oersted effect demonstrates that a current i in a conductor produces a magnetic field in the vicinity of the conductor. We wish to indicate how it is also possible for a changing electric field to produce a magnetic field in space. In Figure 1-8

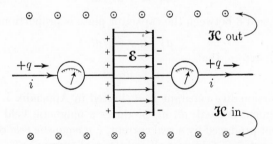

Figure 1-8. Magnetic field $\mathfrak{IC}$ around the capacitor plates is the same as that around the conducting wires, even though there is no *real* current across the gap of the capacitor.

is shown a parallel-plate capacitor which is being charged by a current i. In terms of the conventional current (flow of positive charge), positive charges are deposited on the left plate. Other positive charges depart the right plate, leaving that plate with a negative charge. At any instant of time the ammeters on the left and right sides both indicate a current, and the same current, but no actual charges flow through the space between the plates. Now, by the Oersted effect, a magnetic field is produced by the current in the left- and right-hand conducting wires, the magnetic lines of force going out of the paper above the conductors and into the paper below the conductors. If we were to examine the magnetic field in the circuit of Figure 1-8 we would find that magnetic lines of force surround the parallel-plate capacitor as well as the conducting wires. In fact, if one were to enclose the capacitor in some "black box," so that it could not be seen, and then examine the magnetic field surrounding the region of the box, it would be impossible to distinguish by the nature of the magnetic field alone the behavior of the real capacitor from the behavior of a simple current conductor replacing the capacitor. In short, the *charging* of a capacitor produces a magnetic field altogether equivalent to that produced by a real current; that is to say, the magnetic field surrounding the capacitor can be thought of as arising from a "fictitious" current in the space between the capacitor plates. This current i_d, called the *displacement current*, is proportional to the time-rate-of-change of the electric field between the plates. Thus, the displacement current i_d is equal to the real current i, and i is proportional to the rate of accumulation of charge q on the capacitor plate:

$$i_d \propto dq/dt$$

The potential difference V between the capacitor plates is proportional to the charge on the plates; hence

$$dq/dt \propto dV/dt$$

The electric field $\mathcal{E}$ between the capacitor plate is proportional to V; hence

$$dV/dt \propto d\mathcal{E}/dt$$

It follows, therefore, that

$$i_d \propto d\mathcal{E}/dt$$

The proportionality constant is evaluated in Appendix I. Inasmuch as this displacement current, $d\mathcal{E}/dt$, produces a magnetic field indistinguishable from that produced by a real current, i, *a magnetic field can be produced by a changing electric field*, as well as by a real current.

We have shown that a changing magnetic field, $d\mathcal{3C}/dt$, produces an electric field, and that a changing electric field, $d\mathcal{E}/dt$, produces a magnetic field. In both instances the fields are produced not at just one point in space; rather, the fields extend over regions of space. We have, then, the necessary elements for the production of a wave, since a wave consists in the propagation of a disturbance over a region. In the case of an electromagnetic wave, the disturbance consists of the changing electric and magnetic fields; the propagation is possible because changing electric and magnetic fields at one point produce changing magnetic and electric fields respectively at points more distant.

We can, however, follow the course of an electromagnetic wave in somewhat more detail by considering first the behavior of electric lines of force from a single accelerating positive charge. It is customary to regard the electric field lines surrounding a charge as rigidly attached to the charge, so that when the charge moves at a constant velocity, so do the lines of force. Now consider what happens when an electric charge is accelerated. Figure 1-9 shows a positive charge Q originally at rest at A, then accel-

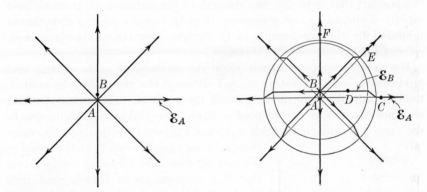

Figure 1-9. Electric field surrounding a charge which has been accelerated from A to B.

erated, and finally brought to rest at B. When the charge is at rest at A, electric lines extend outward in all directions in straight lines from A. One such line is shown as ε_A. Further, after Q has come to rest at B, the electric lines *near the charge* Q again extend outward, the corresponding line being shown as ε_B. But the far electric line ε_A and the near electric line ε_B are the same line of force! We know that this is true because the electromagnetic effects produced by an accelerated charge travel outward from the charge with a *finite* speed c. Thus, a small charge q_C at point C will not have "known" that Q has accelerated from A to B at the instant shown in Figure 1-9 because q_C is still experiencing the electric field ε_A. The kink in the electric line of force, that is the transverse component, is traveling toward q_C with the speed of light and reaches it at some later time. On the other hand, a charge q_D at point D has already experienced a transverse electric force arising from the acceleration of Q. A charge q_E at E will experience a smaller transverse electric force than q_C because the kink in the electric line is less pronounced than at C. Finally, a charge q_F at F, which lies along the direction of the acceleration (from A to B) of Q experiences *no* transverse force.

At a point such as E, both before and after the passage of the transverse kink, a small charge is acted upon only by a *radial* electric field, whose strength varies inversely as the square of the distance AE (or BE) [Figure 1-9] in accordance with Coulomb's law. During the passage of the kink there is, in addition, a *transverse* component of the electric field. The strength of the transverse electric field at E can be shown, from the geometry of Figure 1-9, to vary inversely as the distance AE, whereas the radial component of the electric field at E varies inversely as $(AE)^2$, following Coulomb's law. Therefore, at large distances from the charge Q, the radial component of the electric field is negligible as compared with the transverse component; that is to say, the strength of the ordinary electrostatic field rapidly decreases as one goes away from Q, but the electric disturbance produced by the acceleration of Q survives over the electrostatic field, although it too falls off with distance.

We have focused our attention on the disturbance in an electric field produced by an accelerating charge. Because the charge Q is in motion, there is, in addition, a disturbance in the magnetic field intensity, which also moves outward at the speed of light. The kinks, or disturbances, in the electric and magnetic fields represent a pulse of electromagnetic radiation traveling radially from Q at the speed c. We can see from the following consideration that this electromagnetic disturbance carries energy. Work is done on charge Q in accelerating it from rest at A. A fraction of this energy is recovered when Q is decelerated and finally brought to rest at B, the remaining energy being radiated and carried away by the electromagnetic wave. When this wave or kink reaches some distant point, the electric and magnetic fields can exert forces on charges and thus the electromagnetic wave is capable of doing work on a charge placed at this point. In this way, a portion of the electromagnetic energy can be transferred to a distant charge.

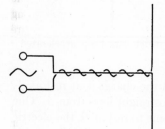

Figure 1-10. Electric-dipole oscillator.

We have discussed a pulse of electromagnetic radiation arising from a single accelerated charge. Let us now turn to the more interesting and useful situation in which a continuous electromagnetic wave, having a well-defined frequency and wavelength, is produced. A common source of this type of radiation is an electric-dipole oscillator, schematically represented in Figure 1-10. The applied alternating voltage produces an alternating current in the two vertical straight-wire conductors (the antenna). At the end of one wire the charge is alternately positive and negative; at the other end, the charge similarly alternates, but is instantaneously 180°

out-of-phase with the first end. We can think of this behavior in terms of the motion of two equal but opposite charges undergoing simple harmonic motion along a vertical axis and at the same frequency as that of the alternating voltage source. Figure 1-11 shows the positions of the charges at

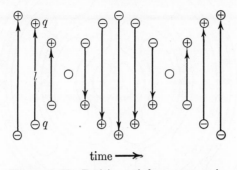

Figure 1-11. Positions of charges comprising an oscillating electric dipole as a function of time.

various times. Such a pair of equal but opposite charges is known as an *electric dipole*, whose electric-dipole moment, ql, is given as the product of one charge q and the distance of separation l. In our example, the charges oscillate, changing the distance between them; thus, the electric-dipole moment of the pair changes sinusoidally with time, and it is appropriate to describe the electric oscillator as an *oscillating electric dipole*. Because charges in sinuosoidal oscillation are continuously accelerated, we know that they produce electromagnetic radiation.

Figure 1-12 is a representation of the electric and magnetic fields produced by an electric-dipole oscillator at one instant of time. We see in this figure the following important features:

(1) At short distances from the dipole, electric lines of force extend from the positive to the negative charge of the dipole. At large distances, the electric field lines have become detached from the dipole and form closed loops. These detached electric-field loops are formed in space because, at great distances from the dipole charges, the electric lines do not have enough time—because the speed of light is finite—to collapse into the electric-dipole oscillator as the electric-dipole moment alternates. To put it another way, the changing electric and magnetic fields near the oscillating electric dipole respectively produce alternating magnetic and electric fields at more distant points, which in turn generate alternating electric and magnetic fields still farther from the accelerating charges.

(2) The electric field ε is always perpendicular to the magnetic field $\mathcal{K}$, and both field vectors are in turn perpendicular to the

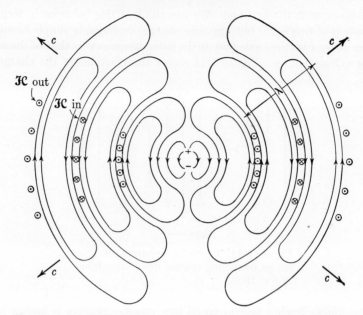

Figure 1-12. Radiation pattern showing the electric and magnetic fields surrounding an electric-dipole oscillator at one instant of time.

direction of propagation of the electromagnetic wave. Thus, electromagnetic waves are transverse.

(3) The wave fronts, each consisting of the locus of points of constant phase, are spherical.

(4) The wave is *linearly* polarized; that is, the electric field at any point in space oscillates along a fixed straight line, this direction being called the direction of polarization of the wave.

(5) The radiation pattern is symmetrical about the axis of the electric dipole. The power radiated through a unit area which lies on a wave front is greatest in the plane perpendicular to the dipole axis, and is zero along the dipole axis. (Note that the same behavior was seen in Figure 1-9.)

(6) The wavelength λ of the electromagnetic wave is the distance between two adjacent spherical wave fronts which are in the same phase. When each charge of the dipole oscillator completes one cycle of oscillation, a wave front travels outward a distance λ. Therefore, the frequency of the electric-dipole oscillator is precisely equal to the frequency ν of the electromagnetic wave. The frequency ν, wavelength λ and velocity c of the wave in free space are related by the equation

$$c = \lambda\nu$$ [1-12]

The radiation pattern shown in Figure 1-12 applies only when λ is much larger than the maximum separation of the dipole charges.

At very great distances from the radiation source, small portions of the spherical wave fronts become planes, and the radiation over this region can be regarded as a monochromatic, plane, linearly-polarized, electromagnetic wave. The electric and magnetic fields of a plane wave are shown in Figure 1-13, where we note that at any instant of time and point

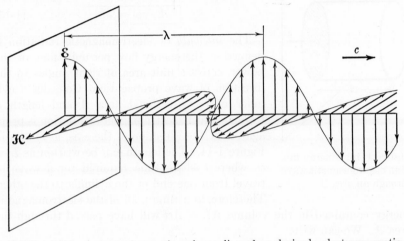

Figure 1-13. A monochromatic, plane, linearly-polarized, electromagnetic wave.

in space, $\mathcal{E}$ and $\mathcal{H}$ are in phase, and at right angles to one another and to the direction of propagation. The direction of energy transport is the same as the direction of advance of a right-hand screw, and is obtained by rotating the $\mathcal{E}$ vector into the $\mathcal{H}$ vector through 90°. We can represent the electric and magnetic fields in the plane, monochromatic wave for any time t and at any point x along the direction of propagation by

$$\mathcal{E} = \mathcal{E}_0 \sin 2\pi \left(\frac{x}{\lambda} - \nu t \right)$$

$$\text{and} \quad \mathcal{H} = \mathcal{H}_0 \sin 2\pi \left(\frac{x}{\lambda} - \nu t \right) \qquad [1\text{-}13]$$

It can be shown (Appendix I) that the magnitudes of the magnetic and electric fields of the wave are related by

$$\mathcal{H} = \sqrt{\frac{\epsilon_0}{\mu_0}}\, \mathcal{E} \qquad [1\text{-}14]$$

The total energy density (E/vol) of an electromagnetic wave is the sum of the electric and magnetic energy densities, (E_e/vol) and (E_m/vol), respectively. From Equations 1-9 and 1-10,

$$(E/\text{vol}) = (E_e/\text{vol}) + (E_m/\text{vol})$$

$$E/\text{vol} = \tfrac{1}{2}(\epsilon_0 \mathcal{E}^2) + \tfrac{1}{2}(\mu_0 \mathcal{H}^2)$$

Using Equation 1-14 to eliminate $\mathcal{H}$ gives

$$\boxed{E/\text{vol} = \epsilon_0 \mathcal{E}^2} \qquad [1\text{-}15]$$

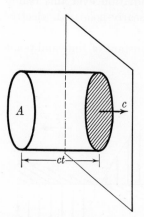

Figure 1-14. Energy flux of an electromagnetic wave through an area A.

The *intensity* of electromagnetic radiation is defined as the energy flow per unit time, or the power, across a unit area at right angles to the direction of wave propagation. Consider a cylinder of cross-sectional area A and length L through which electromagnetic radiation is propagated in the direction of the axis, as shown in Figure 1-14. The length can be written as $L = ct$, where t is the time required for a wave to travel from one end of the cylinder to the other. Therefore, in a time t, all of the electromagnetic energy contained in the volume $AL = Act$ will have passed through the area A. We can write

$$\text{intensity} = \frac{\text{energy}}{\text{area} \times \text{time}} = \frac{(\text{energy/volume}) \times \text{volume}}{\text{area} \times \text{time}}$$

Therefore,

$$I = \frac{(\epsilon_0 \mathcal{E}^2) Act}{At}.$$

and

$$\boxed{I = \epsilon_0 \mathcal{E}^2 c} \qquad [1\text{-}16]$$

Using Equations 1-11 and 1-14, Equation 1-16 can be written in the form

$$I = \mathcal{E}\mathcal{H}$$

When one assigns a direction to the intensity, that of the propagation of the electromagnetic energy, the vector intensity is sometimes known as the *Poynting vector*.

The intensity I of an electromagnetic wave at a distance r from a point source of the radiation is governed by the *inverse-square law*. This follows from Equation 1-16, which shows that $I \propto \mathcal{E}^2$, and from the fact that the electric field $\mathcal{E}$ is proportional to $1/r$ (see page 18). Therefore,

$$I \propto 1/r^2$$

We will summarize here those aspects of the classical theory of electromagnetic radiation which we will have occasion to re-examine later in the light of relativity and quantum physics. It must be emphasized that all of the following properties have been experimentally verified for the long wavelength electromagnetic waves known as radio waves.

(1) Electromagnetic radiation is produced when an electric charge is accelerated. The accelerated charge loses energy, which is carried away in the electromagnetic field.

(2) If an electric charge is accelerated in simple harmonic motion, the frequency of the radiated electromagnetic waves is precisely equal to the frequency of oscillation of the electric charge.

(3) Electromagnetic waves travel at the speed of light c through a vacuum.

(4) Electromagnetic waves consist of oscillating electric and magnetic fields, perpendicular to each other, and both at right angles to the direction of propagation of the electromagnetic energy.

(5) That electromagnetic radiation consists of waves is shown by the fact that it exhibits the phenomena of interference and dif-

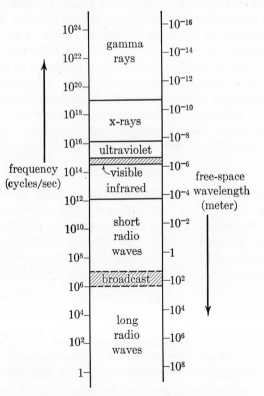

Figure 1-15. The electromagnetic spectrum.

fraction. That the waves are transverse is demonstrated by the fact that electromagnetic waves exhibit the phenomenon of polarization.

(6) The intensity of an electromagnetic wave is proportional to the square of the intensity of the electric (or magnetic) field.

(7) Electromagnetic fields, or electromagnetic waves, can carry linear momentum and angular momentum. The linear momentum and the angular momentum of electromagnetic waves can be demonstrated for visible light in experiments of great delicacy. Formidable experimental difficulties prevent easy direct observation of the linear and angular momentum of long-wavelength radio waves.

At the time of Hertz's experiments, which confirmed Maxwell's classical theory of electromagnetism, only two types of electromagnetic radiation were recognized—visible light and radio waves. Now it is known that a whole electromagnetic spectrum exists, as shown in Figure 1-15. The several regions differ according to the frequency, or wavelength, of the radiation and also according to the origin of the radiation and its effects on materials. The waves of lowest frequency (longest wavelength) are radio waves, which are produced by oscillating electric circuits. Infrared radiation (heat radiation) is produced by heated solids, or by the molecular vibrations of liquids and gases. Visible light, which lies in the very narrow region to which the human eye is sensitive, is produced by rearrangements of outer electrons in atoms, as is ultra-violet light. X-rays can be produced by the rearrangement of the innermost electrons of atoms, and gamma rays have their origin in the atomic nucleus. The boundaries between adjoining regions are not sharply defined, and we shall in later chapters explore the characteristics of the various types of electromagnetic radiation in detail.

1-5 The ideal gas law and the kinetic theory Many gases at room temperature are found to "obey" a simple law which relates three large-scale (macroscopic) characteristics of a gas: the pressure p, volume V, and absolute (Kelvin) temperature T. The equation summarizing the experimental relation found to obtain among these three quantities is the *ideal gas equation*, and any gas which follows this law is said to be an ideal gas.

$$pV = nRT \qquad [1\text{-}17]$$

In the mks system of units, p is given in newton/m^2, V is in m^3, T is in degrees Kelvin (K), n is the number of moles of gas, and the experimentally determined universal gas constant $R = 8.31$ joule/mole-K$°$. The ideal gas law gives no information concerning the particle-like structure of a gas and, in fact, suggests that a gas may be regarded as a continuous compressible fluid. As the density of a gas increases, the experimental results agree less and less with Equation 1-17; therefore, this "law" gives a correct description of the behavior of a gas at low density only.

The *kinetic theory of gases* marks an attempt—an extraordinarily successful one—to bring the behavior of gases within the realm of classical mechanics; that is, to derive the ideal gas law from a *mechanical model*. The development of the kinetic theory by J. C. Maxwell and L. Boltzmann in the middle of the nineteenth century ranks as one of the great triumphs of theoretical physics.

The assumptions of the kinetic theory are these:

(1) A gas consists of very small particles called molecules.

(2) The molecules are in constant motion, at various speeds, making perfectly elastic collisions with the container walls and with other molecules. The size of a molecule is assumed to be very small compared to the distance between molecules.

(3) The motion of the molecules is assumed to be random. We abandon all hope of describing in detail the motion of each and every molecule, though to do this is, of course, possible in principle; instead, it is assumed that there are so many molecules that a statistical treatment of the mechanics of molecular motion may be applied.

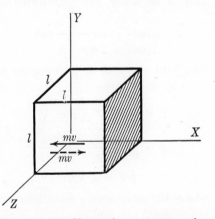

Figure 1-16. Change in momentum of a molecule resulting from a collision with the container wall.

Although the speeds of the molecules vary from zero to extremely high values and their directions of motion are randomly distributed, we shall, for the purpose of simplifying our derivation, make the following additional assumptions. We consider a cube of edge l and volume V containing N identical molecules each of mass m. All particles travel with the same speed v. $N/3$ particles move in each of the three mutually perpendicular directions of the cube. The density of the gas is so small that we can neglect any collisions between the particles.

With these assumptions as the basis for our model of a gas, we will relate the microscopic properties (N, m, v) of the gas to the measurable macroscopic quantity, the pressure p. We focus our attention on the $N/3$ molecules moving back and forth parallel with the X-axis and colliding elastically with the container walls in the YZ-plane. The average pressure on the right-hand, shaded wall (see Figure 1-16), because of this molecular bombardment, is the total average force per unit area. This total average force is $N/3$ times the average force F_{av} of a single, typical molecule. By Newton's second law,

$$F_{av} = \frac{\Delta(mv)}{\Delta t} \qquad [1\text{-}18]$$

where $\Delta(mv)$ is the change in momentum of a molecule occurring in the time interval Δt. We choose Δt to be the time required for any molecule to make one complete trip between the end walls, traveling a distance $2l$ at a speed v; therefore,

$$\Delta t = 2l/v \qquad [1\text{-}19]$$

During this interval Δt, the molecule has made a single collision with the right-hand wall, and in so doing has transferred a momentum $\Delta(mv)$ given by

$$\Delta(mv) = mv - (-mv)$$
$$\Delta(mv) = 2\,mv \qquad [1\text{-}20]$$

Using Equations 1-18, 1-19, and 1-20 we have for the pressure on the right-hand wall

$$p = \frac{N}{3}\frac{F_{av}}{l^2} = \frac{N}{3l^2} \cdot \frac{\Delta(mv)}{\Delta t}$$

$$p = \frac{N}{3l^2}\frac{(2mv)}{(2l/v)} = \tfrac{1}{3}\frac{Nm}{l^3}v^2$$

Or,
$$pV = \tfrac{1}{3}Nmv^2 \qquad [1\text{-}21]$$

This equation relates the macroscopic quantities p and V to the microscopic molecular properties N, m, and v.

The density of a gas, ρ, is the total mass Nm divided by the volume V; therefore, Equation 1-21 can be written in the form

$$p = \tfrac{1}{3}\rho v^2 \qquad [1\text{-}22]$$

Avogadro's number, $N_o = 6.02 \times 10^{23}$ molecules/gm-mole, is defined as the number of molecules in 1 gram-mole of gas, where one mole contains one gram-molecular-weight of gas. Therefore, $n = N/N_o$, and we can re-**write** Equation 1-21 in the form

$$pV = n(\tfrac{2}{3}N_o)(\tfrac{1}{2}mv^2) \qquad [1\text{-}23]$$

Comparing this with the ideal gas equation

[1-17]
$$pV = nRT$$

we find
$$\tfrac{1}{2}mv^2 = \tfrac{3}{2}\frac{R}{N_o}T \qquad\qquad [1\text{-}24]$$

The ratio of the universal gas constant R to Avogadro's number N_o is called the *Boltzmann constant* k.

Boltzmann constant $= k = \dfrac{R}{N_o} = 1.38 \times 10^{-23}$ joule/molecule-K°.

Equation 1-24 can finally be written

$$\boxed{\text{average kinetic energy of a gas molecule} = \tfrac{1}{2}mv^2 = \tfrac{3}{2}kT} \qquad [1\text{-}25]$$

We see from Equation 1-25 that the kinetic theory gives us a mechanical interpretation of the concept of "temperature," since, for a large collection of gas molecules, the temperature is merely a measure of the average kinetic energy per molecule.

At room temperature ($T = 300°$K) the average kinetic energy per molecule is

$$\tfrac{3}{2}(1.38 \times 10^{-23} \text{ joules/molecule-K°})(300°\text{K}) = 6.21 \times 10^{-21} \text{ joule/molecule}$$

A commonly used energy unit for molecules or atoms is the *electron volt* (ev), which is defined as the energy acquired by a particle of charge e, the charge on the electron (1.60×10^{-19} coulomb), when it is accelerated through a potential difference of 1 volt.

Therefore,
$$1 \text{ ev} = (1.60 \times 10^{-19} \text{ coulomb})(1 \text{ joule/coulomb})$$

$$\boxed{1 \text{ ev} = 1.60 \times 10^{-19} \text{ joule}} \qquad [1\text{-}26]$$

In the units of electron-volt, the average kinetic energy of any molecule at room temperature is

$$\boxed{\text{average kinetic energy of }any\text{ molecule at room temperature} = \tfrac{3}{2}kT = \tfrac{1}{25}\text{ev}} \qquad [1\text{-}27]$$

At any fixed temperature, all molecules have the same average kinetic energy; therefore, in a mixture of gases the average speed of a light molecule must be greater than that of heavier molecules.

We have assumed in our treatment that all molecules in a gas move at the same average speed v. Of course, the speeds of the molecules are distributed over a tremendous range of values as shown in Figure 1-17. The ordinate in this graph is the number of molecules per unit speed with speed v. This *distribution function*, derived theoretically by Maxwell and Boltzmann, agrees completely with the distribution of speeds determined experimentally (Section 11-4). Actually, if we had not assumed that all the molecules have the same speed, but rather that the molecular speeds were distributed as shown in Figure 1-17, then we would have obtained

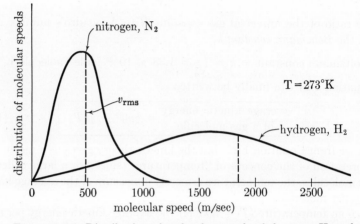

Figure 1-17. Distribution of molecular speeds of the gases H_2 and N_2 at 273°K.

the same final equations, where the speed v appearing in these equations would be the square *root* of the average (or *mean*) of the *square* of the speeds. This root-mean-square, or rms, speed, v_{rms}, is very close to the average speed. Further, it must be remembered that the derivation given here applies only for gases at low density. The finite size of molecules (approximately 10^{-10} m diameter) and the finite interaction between the molecules must be taken into account to give a satisfactory description of high-density gases, liquids, and solids.

Let us note finally some salient features of the kinetic theory of gases which will be of concern to us in later chapters.

(1) An apparently continuous gas is, in actuality, composed of small discrete molecules, the "volume" of a gas being mostly empty space. The molecules are considered to be impenetrable spheres. But molecules have an internal atomic and nuclear structure; therefore the hard-sphere model of the molecule in the kinetic theory is valid only because the energy transferred in molecular collisions occurring at room temperature is too small to alter this internal structure. In the kinetic theory a molecule is considered to behave *as if* it were an inert particle.

(2) The laws of classical mechanics can be applied to objects as small as molecules with apparently complete success. Such ideas as "heat" and "temperature" can be interpreted in terms of the mechanical attributes of molecules. It is no wonder that many physicists of the 19th

century confidently expected that all microscopic phenomena could ultimately be described in classical mechanical terms.

(3) Despite the fact that classical mechanics is capable of describing perfectly the motion of a single particle, the kinetic theory, taking advantage of the enormous number of molecules involved, employs *statistical* methods to describe the *average* behavior of many molecules rather than to detail the behavior of any single molecule. The mechanics of the kinetic theory is one part of that branch of physics called *statistical mechanics*.

1-6   The correspondence principle   Any theory or law in physics is, to a greater or lesser degree, tentative and approximate. This is true because applying a physical law to situations in which it has not been experimentally tested *may* show it to be incomplete or even incorrect. Thus, when we extrapolate a theory to untested situations, we cannot be sure that the theory will hold. However, if a new, more general theory is proposed, there is a completely reliable guide for relating this more general theory to the older restricted theory. This guide is the *correspondence principle*, first applied to the theory of atomic structure by the Danish physicist Niels Bohr in 1923. We shall find it helpful to apply this principle in a broadened sense, using it to great advantage in relativity physics as well as in quantum physics.

THE CORRESPONDENCE PRINCIPLE: *We know in advance that any new theory in physics—whatever its character or details—must reduce to the well-established classical theory to which it corresponds when the new theory is applied to the circumstances for which the less general theory is known to hold.*

Consider a simple, familiar situation that illustrates the correspondence principle. When we have a problem in projectile motion of relatively small range, the following assumptions are made: (1) the weight of the projectile is constant in magnitude, given by the mass times a *constant* gravitational acceleration, and (2) the Earth can be represented by a plane surface and (3) the weight of the projectile is constant in direction, vertically downward. With these assumptions, a parabolic path is predicted and one gets perfectly satisfactory results provided that the projectile motion extends over only relatively short distances. But if we try to describe the motion of an Earth satellite with the same assumptions, *very* serious errors will be made. To discuss the satellite motion one must assume, instead, that (1) the weight of the body is *not* constant but varies inversely with the square of its distance from the Earth's center, (2) the Earth's surface is round, not flat, and (3) the direction of the weight is *not* constant but always points toward the Earth's center. *These* assumptions lead to a prediction of an elliptical path, and to a proper description of satellite motion. Now, if we apply the second, more general theory to the motion of a body traveling a distance small compared to the Earth's radius at the surface of the Earth, notice what happens: the weight appears to be constant both in magnitude and direction, the Earth appears flat, and the elliptical path becomes parabolic. This is precisely what the correspondence principles requires!

The correspondence principle asserts that when the conditions of the new and old theories correspond, the predictions will also correspond. We have then an infallible guide with which to test any new theory or law: it must reduce to the theory which it supplants. Any new proposed theory which fails to meet this test is clearly defective in so fundamental a way that it cannot possibly be accepted. In the next section we shall see another familiar example of the correspondence principle.

1-7 Ray optics and wave optics There are two ways of describing the propagation of light: ray optics (geometrical optics), and wave optics (physical optics). We know that only wave optics is capable of explaining such phenomena as interference and diffraction. On the other hand, ray optics, although incapable of describing interference and diffraction, is a perfectly satisfactory theory for describing such phenomena as the rectilinear propagation, reflection, and refraction of light. Wave optics is also able to explain these phenomena; therefore, wave optics is a comprehensive theory of light, and ray optics is an adequate theory only in certain restricted situations.

The correspondence principle requires that the comprehensive theory reduce to the restricted theory in the correspondence limit. Thus, wave optics must become, in effect, ray optics under those conditions for which such distinctively wave phenomena as diffraction and interference are unimportant. But we know that interference and diffraction are clearly discernible only if the dimensions d of obstacles or apertures that the light encounters are comparable to the wavelength λ of the light. When $\lambda \ll d$, the wave treatment gives the same results as the ray treatment. Symbolically we can write:

$$\underset{\lambda/d \to 0}{\text{Limit}}\ (\text{wave optics}) = (\text{ray optics})$$

Figure 1-18 illustrates the transition from conditions in which wave optics is required to the simpler conditions in which both wave and ray optics are adequate. The figure shows the diffraction pattern for monochromatic light passing through a single parallel slit for (a) the wavelength λ comparable to the slit width d, (b) the wavelength less than the slit width, and (c) the wavelength *much* less than the slit width. In Figure 1-18a, the wavelength is comparable to the slit width and we notice that the wave disturbance is spread far beyond the geometrical shadow and that there are alternate light and dark bands. In Figure 1-18b, the diffraction pattern is less pronounced, and the light is concentrated mainly in the region of the geometrical shadow predicted by ray optics. In Figure 1-18c, where the wavelength is *much* less than the slit opening, we notice that the intensity pattern is indistinguishable from that predicted by ray optics.

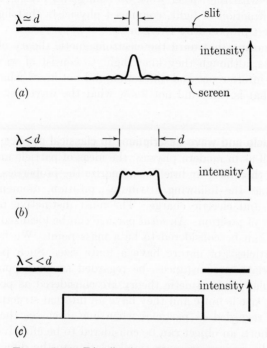

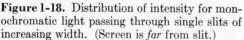

Figure 1-18. Distribution of intensity for monochromatic light passing through single slits of increasing width. (Screen is *far* from slit.)

Ray optics is concerned solely with the path of light, which can be represented as rays showing the direction of light propagation. This suggests that a possible model for describing the character of light is the *particle model.* In such a model, light is assumed to consist of small, essentially weightless particles or corpuscles. Such a particle model of ray optics is consistent with the observed facts that (1) in free space, light follows a straight-line path, in the manner of a stream of particles, (2) upon reflection, light behaves like particles making elastic collisions with the surface, and (3) upon refraction in a transparent material, such as glass, the light particles can be imagined to have their direction abruptly changed at the interface.

The most celebrated advocate of the particle theory of light was Sir

Isaac Newton. It was he who showed that, according to the particle concept, the speed of light in a refracting medium should be *greater* than the speed of light in a vacuum. The wave theory, of course, predicts a slower speed in a refracting medium. But Foucault experimentally found the speed of light through water to be *less* than through air. This experiment, together with the earlier work of Young and Fresnel on the interference and diffraction of light, convinced physicists that light consisted of waves, as first proposed by Huygens.

It is worth noting that until the electromagnetic theory of Maxwell in 1864, physicists, although they knew light to consist of waves and thus could describe interference and diffraction, did not know what it was that was waving; that is, they did not know what the wave disturbance consisted of.

1-8   The particle and wave descriptions in classical physics   In classical physics, as well as in modern physics, the ideas of particle and wave play a central role; therefore, we briefly summarize the properties of each.

A particle has the following attributes: position, momentum, kinetic energy, mass, and electric charge. The most distinctive property of a particle is that of position. An *ideal* particle can be localized with infinite precision and can be considered to be a mass point. We find, however, that the "particles" of nature have a finite size. Such particles may, under appropriate circumstances, be regarded as mass points. For example, molecules in the kinetic theory are considered as point particles, although their size is finite and they have an internal structure; similarly, stars may be regarded as particles when one discusses the behavior of galaxies. In short, an object can be considered to be effectively a particle whenever its dimensions are very small relative to the dimensions of the system of which it is a part and the internal structure of the particle is unimportant for the problem under consideration.

We make some further observations that will be of interest to us later in atomic structure. Our usual visualization of a particle is that of a small, impenetrable sphere having no internal structure. This leads us to an obvious assertion: two such particles cannot occupy the same point in space at the same time. Finally, let us remember that the greatest achievement of classical mechanics is its ability, through the Newtonian laws of motion, to determine completely the path of a particle; that is, to give its position and velocity at any instant of time, provided that the initial position and velocity of the particle, along with the forces acting on it, are known.

Next, we list the important attributes of a wave: frequency, wavelength, velocity, amplitude of the disturbance, energy, and momentum. The simplest form of a wave is one which shows a strictly sinusoidal wave form. Consider the electric field ε of a perfectly monochromatic electromagnetic wave of amplitude $\mathcal{E}_0$, frequency ν, and wavelength λ traveling along the positive X-direction with a speed v, as given by Equation 1-13.

[1-13] $$\mathcal{E} = \mathcal{E}_0 \sin 2\pi\left(\frac{x}{\lambda} - \nu t\right)$$

Such a wave shows a sinusoidal variation in $\mathcal{E}$ in space for any fixed time; conversely, at any point x in space the electric field $\mathcal{E}$ varies sinusoidally in time. The frequency of the wave disturbance is precisely specified, and Equation 1-13 requires that the electric disturbance extend over all possible values of x for all possible instants of time t; that is, if the frequency is known with complete precision, the wave *must* have an infinite extension in space. This sort of behavior is found for any sinusoidal wave disturbance; for example, a monochromatic sound wave is represented by Equation 1-13 when we replace the electric field intensity $\mathcal{E}$ by the pressure p

$$p = p_0 \sin 2\pi\left(\frac{x}{\lambda} - \nu t\right)$$

where again the velocity of propagation of the sound wave is given by $v = \nu\lambda$.

The most distinctive attribute of a wave is its frequency, or its wavelength. If we are to have an ideal wave, one whose wavelength is known with infinite precision, the wave cannot be confined to any restricted region of space, but must rather have an infinite extension along the direction in which it is propagated. This property of a monochromatic wave can be developed mathematically, and the reader is referred to Appendix II for details of this treatment. Here, however, we shall indicate, by a hypothetical experiment, that if one is to measure and know the frequency of a wave with complete accuracy it is necessary to have an infinite wave.

Suppose that we have a standard clock for measuring the number of wave crests that pass any fixed point per unit time. For simplicity, we imagine that the standard clock is an oscillator which produces waves whose frequency we wish to compare with that of some incoming wave. Let us see what we must do in order to be able to state with complete assurance that the frequency of the incoming wave is precisely the same as the frequency of the wave generated by our standard clock.

A beat arises in consequence of differing frequencies of two waves. We will allow the two waves to interfere with one another to produce beats; the number of beats per second will be equal to the difference in frequencies of the two waves. If the two waves are of precisely the same frequency, then we will detect no beats whatsoever. If we observe the resultant amplitude of the two interfering waves over some limited period of time, we *may* find that there is no appreciable change in this amplitude. But we can *not*, on the basis of measurement through a limited period of time only, assert that the two frequencies are alike, because at some later time we might find the combined amplitude of the two waves to be decreasing or increasing (see Figure 1-19), thus indicating an incipient beat, or a difference in the frequencies. To be absolutely sure that no beats occur—that

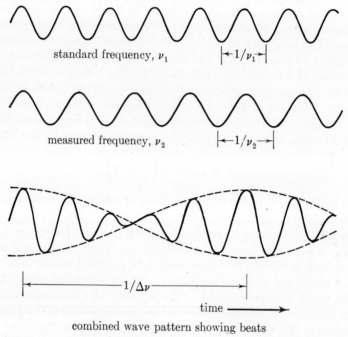

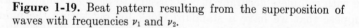

Figure 1-19. Beat pattern resulting from the superposition of waves with frequencies ν_1 and ν_2.

the frequencies of the two waves are precisely the same—we must wait for an infinite period of time. But if, to measure the frequency with no uncertainty, we wait for an infinite time, then the wave we measure will have traveled for an infinite time and have an infinite extension in space.

We now wish to determine how great is the uncertainty in the unknown frequency ν_2 when the beats it produces with a standard clock of frequency ν_1 are observed over a finite time Δt. If ν_1 and ν_2 differ by an amount $\Delta \nu$, then $\Delta \nu$ beats are observed per second; the time required to observe one beat is $1/\Delta \nu$ sec. To be confident of observing one beat we must make a measurement over a period of time Δt equal to at least the time for the occurrence of one beat; therefore $\Delta t \geq 1/\Delta \nu$, or

$$\Delta t \, \Delta \nu \geq 1$$ [1-28]

If the measured frequency differs from the standard clock's frequency by an amount $\Delta\nu$, then $\Delta\nu$ is a measure of our uncertainty in the frequency ν_2. It follows from Equation 1-28 that, if the frequency of a wave is measured over a very short time, the uncertainty in this frequency is large, and conversely. To have $\Delta\nu = 0$, Δt must be infinite, as we have just argued.

A corresponding relationship for the uncertainty in the wavelength can easily be deduced from Equation 1-28. If the wave has been observed only over the finite time interval Δt, then during this time the wave will have traveled a distance $\Delta x = v\,\Delta t$, where v is the speed of the wave; therefore, we will have observed the wave only over the distance Δx.

$$\Delta x = v\,\Delta t$$

and from Equation 1-28, $\Delta x \geq v/\Delta\nu$ [1-29]

But $\nu = v/\lambda$

therefore, in magnitude $\Delta\nu = v\,\Delta\lambda/\lambda^2$ [1-30]

Substituting Equation 1-30 into Equation 1-29 gives

$$\boxed{\Delta x\,\Delta\lambda \geq \lambda^2}$$ [1-31]

Therefore, if a wave is observed so that its extension is certain for an amount Δx, the wavelength will be uncertain by at least an amount $\Delta\lambda \geq \lambda^2/\Delta x$. Equation 1-31 shows that $\Delta\lambda = 0$ only if $\Delta x = \infty$.

Our discussion on waves has, to this point, been concerned only with monochromatic sinusoidal waves. We know that wave pulses can also be propagated. Any pulse—a wave disturbance confined at a given time to some limited region of space—can be shown mathematically to be equivalent to a number of superimposed sinusoidal waves of different frequencies. It is for this reason that our analysis was carried out for a simple monochromatic sinusoidal wave. If we compute the number of waves of different frequencies that must be added together to give a completely sharp pulse, we find that *all* frequencies from zero to infinity must be included (see Appendix II for details). This agrees completely with what we have already found: if a wave pulse is confined to an infinitesimally small region of space, then we can give no information as to what its wavelength is. One cannot speak of the "frequency" of a pulse. This is equivalent to saying that if we observe a sharp pulse, a single sharp wave crest, and no others, we will be altogether uncertain as to the wavelength or frequency of the pulse because we will never know where the next adjacent crest is, nor how long we will have to wait until the next one arrives.

A simple procedure for following the course of an advancing wave is *Huygens' principle*, which states that every point on a given wave front may be regarded as a new point source of a Huygens wavelet, and that a wave front for some future time is found by drawing the envelope of these

Huygens wavelets. This method of wave tracing is shown in Figure 1-20, where the corresponding rays, giving the direction of energy flow, are shown as lines which are perpendicular to the wave fronts. Just as it is possible, in principle, to predict in complete detail the motion of the particle, using Newton's laws, so it is possible to determine with complete certainty the future wave front of a traveling wave if one knows the speed of the wave at all points in space and the geometry of the region through which it travels; that is, the location of apertures and obstacles that the wave may encounter.

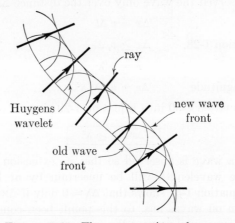

Figure 1-20. The construction of wave fronts according to Huygens' principle.

Waves and particles play such an important role in physics because they represent the *only two* modes of energy transport. We can transport energy from one point in space to a second point only by sending a particle from the first to the second site, or by sending a wave from the first to the second site. Particles and waves are the only means of communicating between two points. For example, we can signal another person by throwing an object at him (particle), calling to him (sound waves), motioning to him (light waves), telephoning him (electric waves in conductors), or radioing him (electromagnetic waves).

Only three interactions, or modes of energy transfer, are possible between particles and waves: (1) the interaction between two particles, (2) the interaction between a particle and a wave, and (3) the interaction between two waves. Two particles interact when they collide. We have already discussed an example of the interaction between a particle and a wave when we showed (see Figure 1-9 and Figure 1-12) an electric charge generating an electromagnetic wave. This wave can in turn interact, and give energy to, a second charged particle. There is no interaction between two waves, and their combined effects at any point in space are governed by the *principle of superposition,* which states that one can superimpose two or more wave disturbances to find the resultant disturbance. Everyone has seen this: if two water waves travel, say, at right angles to one another on a pond, they interfere with one another at the point where and time when they cross, and then travel onward *as if* each wave were completely oblivious of the existence of the other. This behavior is in contrast to the behavior of two small, impenetrable particles, which cannot occupy the same spot at the same time. The superposition principle is, of course, the basis for treating all problems in interference and diffraction.

REFERENCES

Born, M., (trans. W. M. Deans) *The Restless Universe.* New York: Dover Publications, Inc., 1951. One of the most attractive features of this paperbound book is the set of "moving pictures." Flick from page 121 through 239 to see the electromagnetic radiation from an electric-dipole oscillator, and flick from page 120 to page 2 to see the behavior of gas molecules and a method of measuring molecular speeds.

Fowler, R. G., and D. I. Meyer, *Physics for Engineers and Scientists.* Boston: Allyn and Bacon, Inc., 1958. Chapters 19 through 21 give the fundamentals of wave motion, including some impressive three-dimensional wave pictures.

Holton, G., and D.H.D. Roller, *Foundations of Modern Physical Science.* Reading, Massachusetts: Addison-Wesley Publishing Company, Inc., 1958. Chapters 8, 13, 14, and 15 discuss critically, but at an elementary level, "The Nature of Scientific Theory," "The Nature of Concepts," "Duality and the Growth of Science," and "Scientific Discovery."

Kennard, E. H., *Kinetic Theory of Gases.* New York: McGraw-Hill Book Company, Inc., 1938.

Lindsay, R. B., and H. Margenau, *Foundations of Physics.* New York: Dover Publications, Inc., 1957. "The Meaning of a Physical Theory" is discussed in Chapter 1 at a sophisticated philosophical level.

Skilling, H. H., *Fundamentals of Electric Waves.* New York: John Wiley & Sons, Inc., 1942. Starting from elementary principles, Maxwell's equations of electromagnetism and the associated mathematical tools are clearly and rigorously developed.

PROBLEMS

1-1   Consider what disturbance of the observed quantity arises when
the following measurements are made: (a) potential difference and
current in an electric circuit, (b) length of an object by a meter
stick.

1-2 *   A hard sphere makes a *perfectly elastic collision* with a second
identical sphere initially at rest. For *all*, except head-on, collisions
show that the angle between the velocities after the collision must
be 90°.

1-3   Formulate Newton's laws of motion in the language of angular
momentum and torque.

1-4   Two identical spheres approach one another along parallel lines
at equal speeds and make an inelastic collision (not necessarily
head-on). Show that the particles leave the site of the collision
in opposite directions and equal speeds.

1-5   A spinning ice-skater pulls in his arms, thereby reducing his
moment of inertia. (a) Is angular momentum conserved? (b)
Is mechanical kinetic energy conserved? (c) Is total energy conserved? Explain.

1-6   Prove that the angular momentum of a planet, moving in an
elliptical orbit, subject to the gravitational attraction of the Sun,
is constant. Assume that the Sun is fixed. (*Hint:* consider the
torque acting on the planet.)

1-7 *   The angular momentum of a spinning top is the product of the
moment of inertia of the top about its axis of rotation and its angular velocity. When under the influence of a gravitational torque,
the magnitude of the top's angular velocity does *not* change, but
the top *precesses* about its point of support. Explain this motion
qualitatively in terms of the vector properties of angular momentum and torque.

1-8   Show that a constant magnetic field acting on a moving charge
cannot do work.

1-9   Compute the ratio of the Coulomb to the gravitational force between two electrons. The electron mass is 9.11×10^{-31} kg; the
electronic charge, 1.60×10^{-19} coulomb.

1-10   A charged particle moves along the Z-axis with a speed v. A unit
electric field (1 newton/coulomb) is applied along the X-axis; a
unit magnetic field (1 amp/m) is applied along the Y-axis. Show
that the ratio of the magnetic force to the electric force on the
particle is $(v/c) \sqrt{\mu_0/\epsilon_0}$.

1-11   Two protons are moving parallel to one another with equal speeds as shown in Figure 1-21. Show that the ratio of the magnetic (attractive) force to the electric (repulsive) force on one proton because of the other is equal to $(v/c)^2$.

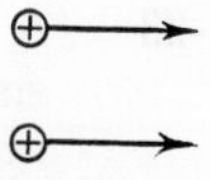

⊕——→

⊕——→

Figure 1-21.

1-12   Show that the momentum of a particle of mass m, speed v, and electric charge q moving at right angles to a magnetic flux density B is given by $(mv) = qBr$, where r is the radius of the circle in which the particle moves.

1-13 *   Show that the energy density of a magnetic field (E_m/vol) is given by $\frac{1}{2}\mu_0\mathcal{K}^2$.

1-14   Show that $\frac{1}{2}\epsilon_0\mathcal{E}^2$ and $\frac{1}{2}\mu_0\mathcal{K}^2$, the energy densities of an electric and magnetic field respectively, have the units of joules/m^3.

1-15 *   Describe the operation of an electric oscillator, consisting of a capacitor and inductor, in terms of changes that occur in the energies of the electric and magnetic fields.

1-16   Using Equations 1-5 and 1-6, show that the magnetic force per unit length between two long parallel current-carrying conductors a distance a apart, and carrying currents i_1 and i_2, is given by

$$F/l = \mu_0 i_1 i_2/2\pi a$$

1-17   An electron moves at a speed of 3.00×10^4 m/sec to the North in a region where an electric field of 1.00×10^6 volt/m is applied to the East. What is the magnitude and direction of the magnetic flux density which must be applied so that the electron can continue in an undeviating path?

1-18   Show that $1/\sqrt{\epsilon_0\mu_0}$ gives the correct units for the speed of light.

1-19   State in words the meaning of a "monochromatic, linearly-polarized, plane, electromagnetic wave."

1-20 *   Show that the ratio of the magnetic force to the electric force acting on a charged particle, subject to the influence of an electromagnetic wave, is equal to the ratio of the particle's speed to the speed of light. Assume the particle moves at right angles to the magnetic field.

1-21 *   Prove that the intensity of a light beam passing through two Polaroid sheets varies as $\cos^2 \theta$, where θ is the angle between the directions of polarization of the two Polaroid sheets.

1-22 *   A beam of light having an intensity of 1 watt/m^2 strikes a mirror which reflects the light completely back along the incident direction. (a) What is the total intensity crossing an area at right angles to the beam? (b) What is the *average* energy density in the region containing the incident and reflected beams?

1-23   A sensitive radio receiver can detect an electromagnetic signal as small as 1.0×10^{-3} microvolt per meter. What must the minimum power output of a radio transmitter in a satellite 1000 miles "above" the point of reception be so that its signal may be detected? Assume that the radiation pattern is isotropic (uniform intensity in all directions).

1-24   What are the maximum electric and magnetic field intensities at a distance of 10 m from a 100 watt lamp bulb? (Assume 50 watts of monochromatic, isotropic electromagnetic radiation.)

1-25 *   (a) What is the path traced out by an electric field on a wave front in an electromagnetic wave produced by two linearly-polarized, plane, electromagnetic waves traveling in the same direction, with the same frequency and intensity, but 90° out-of-phase with one another? (b) Show that such a circularly-polarized electromagnetic wave carries angular momentum in (or opposite to) the direction of propagation.

1-26   The tail of a comet, consisting of fine particles, is always directed away from the Sun and is therefore not necessarily "behind" the comet as it travels. What property of the electromagnetic field might this illustrate?

1-27   Derive Avogadro's law—equal volumes of gases at the same pressure and temperature contain the same number of molecules—from the kinetic theory of gases.

1-28   Show that the kinetic theory predicts the law of partial pressures: in a mixture of gases in thermal equilibrium, the total pressure is equal to the sum of the partial pressure exerted by each gas in the mixture.

1-29   The velocity of escape from the Earth's gravitational field is 7 mi/sec. Show from the kinetic theory that the Earth's atmosphere contains essentially no hydrogen gas.

1-30   Show that the average kinetic energy of a molecule at room temperature is 1/25 ev.

1-31   A neutron has a mass of 1.67×10^{-27} kg. Compute the speed of a thermal neutron which, by definition, has an average kinetic energy of $(3/2)kT$, where $T = 300°K$.

1-32 *   The operation of the thermal diffusion method of separating uranium 235 (atomic weight 235) from uranium 238 depends upon the fact that the average speed of the gas UF_6, uranium hexafluoride, differs for the two isotopes. Compute the ratio of the rms speeds of the gases at room temperature. (Atomic weight of fluorine, 19.)

1-33   Compute the average kinetic energy per particle (both in joules and in ev) in the interior of the Sun, where the temperature is approximately 20 million °K.

1-34 *   The weight of a vessel containing a gas exceeds the weight of the same vessel when evacuated. Explain why a weighing balance indicates a greater weight when the vessel is filled with gas than when it is empty *in terms of the motion of the gas molecules.*

1-35   A 50-ev electron enters a uniform electric field of intensity 2500 n/coul. If the electron moves in the same direction as the electric lines of force, how far does it move in this field before coming to rest?

1-36   A 50-ev electron enters a uniform magnetic field of strength 0.50 weber/m². If the electron enters at right angles to the magnetic flux lines, what is its energy after it has traveled a distance of 1.00 mm?

1-37   A doubly-ionized oxygen molecule is accelerated from rest through a potential difference of 1000 volt. Compute its final kinetic energy in joules and in electron volts.

1-38   Using Equations 1-28 and 1-31, show that $\Delta\nu/\nu = -\Delta\lambda/\lambda$.

1-39   Light from an atom is radiated typically in a time of 10^{-8} sec. (a) What is the fractional uncertainty in the frequency and in the wavelength for visible light of wavelength 5000 Å (1 Ångstrom $= 1\overset{\circ}{A} = 10^{-10}$ m)? (b) What is the spatial extension of this electromagnetic wave? (c) How many wavelengths are contained within this distance?

1-40   The National Bureau of Standards radio station WWV transmits standard time signals of precisely 10 megacycles per second. Over what period of time must the WWV signal be compared with a laboratory signal so that the error in calibration of the laboratory oscillator is no greater than 10^{-4} per cent? Assume that the only error arises from the finite period of observation.

T W O

THE THEORY OF SPECIAL RELATIVITY

2-1   The principle of relativity   The theory of special relativity, set forth by Albert Einstein in 1905, is fundamental to all modern physics and one of the greatest achievements of the human intellect. Despite the fact that this theory is often regarded as being esoteric and recondite, we shall find that its principal features arise in a natural way from basic physics and two fundamental postulates of relativity. The first postulate, *the principle of relativity*, is also basic to classical, or Newtonian, mechanics; the second postulate, *the constancy of the speed of light*, is at seeming variance with both classical physics and Postulate I, if the classical concepts of space and time are adhered to. It was the brilliant work of Einstein which reconciled these two postulates to yield a self-consistent theory of the physical universe which is in many fundamental respects quite different from that presented through classical physics. The "theory" of relativity is not hypothetical or conjectural inasmuch as a variety of experiments have firmly established its essential correctness.

Since relativity theory plays a major role in the study of atomic and nuclear physics, we shall discuss it in some detail. We first explore the meaning and consequences of Postulate I in the light of classical physics.

POSTULATE I: *The Principle of Relativity. The laws of physics are the same, or invariant, in all inertial systems; that is, the mathematical form of a physical law remains the same.*

An *inertial system* is defined as a coordinate frame of reference within which the law of inertia, Newton's *first* law, obtains. A body which is subject to no net external force will move with a constant velocity if it is in an inertial system. A simple test of whether an observer is within an inertial system can be made by having him throw an object and then notice whether this object travels in an undeviating path at a constant speed. This will occur only in a true inertial system, and such a system can exist only in empty space, far from any mass. A reference, or coordinate, system attached to the Earth's surface can, however, be regarded as an approximate inertial system when the gravitational force on a body is balanced out by a second force. Thus, an object sliding on a frictionless flat plane on the Earth would move in a nearly† straight line with a nearly constant speed. The first postulate of relativity physics states that all inertial systems are equivalent; that is, no one inertial system can be distinguished by any experiment in physics from any other inertial system since the laws of physics are the same for all inertial systems.

To examine the full significance of Postulate I, we must find the relationships between the spatial and temporal coordinates of one inertial system and the spatial and temporal coordinates of a second inertial system moving relative to the first.

2-2 Galilean transformations The equations in classical physics which relate the space and time coordinates of two coordinate systems moving at constant velocity relative to one another are called the *Galilean* (or Newtonian) *transformations.*

Consider two observers, 1 and 2, located in two separate coordinate systems S_1 and S_2, respectively. System S_2 travels with a constant velocity v to the right with respect to S_1; conversely, S_1 moves to the left with a velocity v with respect to S_2.‡ We can speak only of the *relative* motions of S_1 and S_2. Each observer carries a meter stick and a clock to measure the location and time of a particle or object relative to his own system. By specifying the location and time of some physical phenomenon, the observer describes an *event*. The space and time coordinates of an event at point P (Figure 2-1), as described by observer 1, are (x_1, y_1, z_1, t_1), and the coordinates of the *same* event as described by observer 2 are (x_2, y_2, z_2, t_2). The space coordinates, x_1, y_1, and z_1, give the distance from the origin in

---

† The body will deviate from its straight-line motion because of the Earth's rotation about its axis.

‡ We choose the direction of relative motion along the X-axes solely for convenience.

the X-, Y-, and Z-directions as measured by the meter stick of observer 1, and t_1 gives the time that observer 1 reads on his clock. We assume that the clocks are synchronized and the meter sticks in S_1 and S_2 are compared when both observers are temporarily at rest with respect to one another. System S_2 is then set in motion with respect to S_1, the clocks being set such that when the origin of S_2 passes the origin of S_1 both clocks read zero. When $t_1 = 0$, then $t_2 = 0$, and at this instant $x_1 = x_2$. It is further assumed that the y- and z-axes of the two coordinate systems are always respectively parallel.

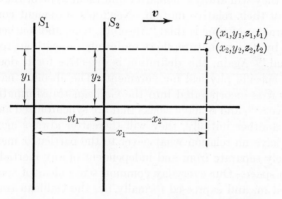

Figure 2-1. Space and time coordinates of an event as measured by two observers moving at a constant relative velocity v.

From Figure 2-1 we can immediately write down the Galilean coordinate transformations, expressing the coordinates as measured by observer 2 in terms of the coordinates as measured by observer 1:

Galilean
Coordinate
Transformations

$$\begin{aligned} x_2 &= x_1 - vt_1 \\ y_2 &= y_1 \\ z_2 &= z_1 \\ t_2 &= t_1 \end{aligned}$$

[2-1]

That $y_2 = y_1$ and $z_2 = z_1$ follows from the fact that the relative motion between S_2 and S_1 is at right angles to these coordinates. To obtain the coordinates of 1 in terms of 2, we merely interchange subscripts and change v to $-v$; this is proper because the labels 1 and 2 are purely arbitrary. Saying that S_2 moves with velocity v with respect to S_1 is equivalent to saying that S_1 moves with a velocity $-v$ with respect to S_2.

These classical transformation equations may seem completely axiomatic and self-evident, but it is crucial for us to appreciate the profound assumptions implicit in them. These assumptions are that space is absolute and that time is absolute. Newton described the absolute character of space as follows: "Absolute motion is the translation of a body from one absolute place to another absolute place." This implies that it is possible and meaningful to speak of an "absolute place," unreferred to anything external. Newton's definition really gives no method of measuring such a place; and from the point of view of the Galilean transformations, Newton's assumption of absolute space means to us that if observers S_1 and S_2 compare their meter sticks at one and the same time and find them to be of the same length, then they will always thereafter find them to be of the same length, irrespective of their relative motion. Newton's statement concerning the absolute character of time is this: "Absolute, true, and mathematical time of itself and by its own nature, flows uniformly on, without regard to anything external." Again, this definition is defective for it does not admit time to any specific physical measurement. The absolute nature of time intervals, as it is incorporated into the Galilean transformations, implies that if observers S_1 and S_2 have clocks that are synchronized and calibrated against one another initially, they will thereafter always agree. Time is assumed to have no relation whatsoever to the particular inertial system; it is absolutely separate from and independent of any inertial system and any point in space. Our everyday common sense ideas of space and time are contained in, and expressed formally by, the Galilean transformation equations.

The velocity and acceleration transformations follow directly from Equation 2-1 by differentiation with respect to time. We define the x-component of the velocity as measured by observer 1 as dx_1/dt_1, and for convenience designate this velocity component as $\dot{x}_1$, the dot above the coordinate implying the first derivative with respect to time. Similarly, we write the y and z velocity components as $\dot{y}_1 = dy_1/dt_1$ and $\dot{z}_1 = dz_1/dt_1$. The velocity components in system S_2 are $\dot{x}_2 = dx_2/dt_2$, etc. The acceleration is given by $\ddot{x}_2 = d^2x_2/dt_2^2$, etc., for observer 2. It is important to emphasize the exact meaning of the concept of velocity. We define dx_1/dt_1 as the limit of the distance traversed in the X-direction, dx_1, measured by the meter stick of observer 1, divided by the time interval, dt_1, as measured by the clock of the same observer 1, both as the time interval approaches zero. It is meaningless to speak of dx_1/dt_2, etc., since the measurement of length and time to determine velocity must be made with respect to a *single* coordinate system. For the Galilean transformations this careful definition of velocity may not appear important—since time is regarded as absolute, $dt_1 = dt_2$, and therefore, $dx_1/dt_2 = dx_1/dt_1$. We shall find later, however, that the coordinate transformations which satisfy the postulates of the theory of special relativity do not have such simplicity.

By differentiating Equation 2-1, we immediately obtain the velocity transformations, and the derivatives of these give the acceleration transformations.

$$\begin{aligned} \textit{Galilean} && \dot{x}_2 &= \dot{x}_1 - v \\ \textit{Velocity} && \dot{y}_2 &= \dot{y}_1 \\ \textit{Transformations} && \dot{z}_2 &= \dot{z}_1 \end{aligned} \qquad \text{[2-2]}$$

$$\begin{aligned} \textit{Galilean} && \ddot{x}_2 &= \ddot{x}_1 \\ \textit{Acceleration} && \ddot{y}_2 &= \ddot{y}_1 \\ \textit{Transformations} && \ddot{z}_2 &= \ddot{z}_1 \end{aligned} \qquad \text{[2-3]}$$

Equation 2-2 shows that the velocity of a particle as measured by S_2 is equal to the velocity of the same particle as measured by S_1 minus the velocity, v, of S_2 relative to S_1. Thus, velocities can be combined following the usual rules for vector addition. From Equation 2-3 we see that the corresponding acceleration components in two inertial systems, moving with respect to one another at a *constant* velocity, are equal.

2-3 Invariance of classical mechanics under Galilean transformations

To see more clearly the significance of Postulate I, the principle of relativity (Section 2-1), let us consider two well-known physical laws of mechanics under a Galilean transformation: (1) conservation of momentum, and (2) conservation of energy.

CONSERVATION OF LINEAR MOMENTUM   An observer in S_2 watches a head-on collision between two particles of respective masses m and M, as shown in Figure 2-2a. Figure 2-2b shows the same collision as seen by an observer in S_1. As before, S_2 moves to the right of S_1 with a velocity v, and the velocities as measured by the two observers are related by the Galilean velocity transformations, Equation 2-2.

The notation used in Figure 2-2 is this: (a) the small letters and large letters refer to particles m and M, respectively; (b) the subscripts 1 and 2 refer to the two observers, respectively; (c) the unprimed and primed velocities refer to velocities measured before and after the collision, respectively.

We now ask, "Is the law of conservation of momentum a 'good' physical law in that a 'good' physical law satisfies Postulate I of relativity theory and, therefore, is invariant under a Galilean transformation?" To answer this question we must see whether observers S_1 and S_2 will both find the *same mathematical form* for the statement of the conservation of momentum law as each of the two observers watches the same head-on collision of the masses m and M, each measuring the velocities with his own meter stick and clock.

For the observer in the inertial system S_2, the conservation of momentum law is written:

momentum before collision = momentum after collision

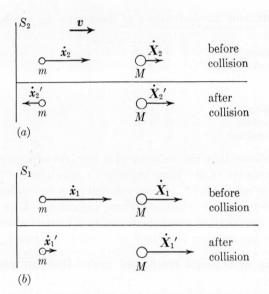

Figure 2-2. Collision of two particles as viewed by two observers moving at a constant relative velocity v.

Therefore,
$$m\dot{x}_2 + M\dot{X}_2 = m\dot{x}_2' + M\dot{X}_2' \qquad [2\text{-}4]$$

Using the Galilean velocity transformations, Equation 2-2, we can rewrite Equation 2-4 in terms of the velocities measured by the observer in inertial system S_1.

$$m(\dot{x}_1 - v) + M(\dot{X}_1 - v) = m(\dot{x}_1' - v) + M(\dot{X}_1' - v)$$

which reduces to
$$m\dot{x}_1 + M\dot{X}_1 = m\dot{x}_1' + M\dot{X}_1' \qquad [2\text{-}5]$$

Equations 2-4 and 2-5 are of identical mathematical form; that is, they differ only in the subscripts, 1 and 2. Therefore, an observer S_1 in some one inertial system and an observer S_2 in *any* other system which moves with a constant velocity with respect to S_1 both agree that *momentum is conserved*. In short, the conservation of momentum law *is* a "good" law of classical mechanics. It must be noted, however, that the total linear momentum before (or after) the collision in S_1, $m\dot{x}_1 + M\dot{X}_1$, is *not* the same as the total linear momentum before (or after) the collision in system S_2, $m\dot{x}_2 + M\dot{X}_2$, the total momentum being greater in S_1 in this example.

Let us note that Newton's second law of motion is invariant under a Galilean transformation. Consider two bodies, of respective masses m and M, which interact with one another as a result of some force, such as the gravitational force. If no net external force is applied to the system (the two bodies), the system is isolated and the total linear momentum of the system must remain constant. For simplicity, we assume that both masses lie on the X-axis and move along this axis. Observer S_1 states this in the form

$$m\dot{x}_1 + M\dot{X}_1 = \text{constant}$$

Taking the time derivative of this equation gives

$$d/dt_1(m\dot{x}_1) = -d/dt_1(M\dot{X}_1) \qquad [2\text{-}6]$$

Because the force acting on a body is defined in terms of the rate of change of that body's momentum, the left-hand side of Equation 2-6 is the force f_1 on m because of M as measured in S_1, and the right-hand side is the force F_1 on M because of m, again as measured in S_1. Therefore,

$$f_1 = -F_1$$

which is Newton's third law.

Now considering the force acting on the body m alone, we have

$$f_1 = d/dt_1(m\dot{x}_1) = m\ddot{x}_1 \qquad [2\text{-}7]$$

where it is assumed that the mass m is the same in S_1 and S_2. In a similar fashion, the observer in S_2 would write

$$f_2 = d/dt_2(m\dot{x}_2) = m\ddot{x}_2 \qquad [2\text{-}8]$$

Equation 2-3 shows that $\ddot{x}_1 = \ddot{x}_2$, and therefore, from Equations 2-7 and 2-8, $f_1 = f_2$. From the invariance of Newton's second law and from the fact that the forces and acceleration are unchanged, it immediately follows that, if S_1 is an inertial system, then S_2, which represents *any* coordinate system moving with respect to S_1 at a constant velocity v, is also an inertial system. From the point of view of Newton's second law, all inertial systems, of which there are an infinite number, are equivalent and indistinguishable. Clearly, any coordinate frame of reference F which is *accelerated* with respect to some inertial system cannot itself be an inertial system because no longer does $\ddot{x}_1 = \ddot{x}_F$.

We will restrict our considerations to the *special* case of inertial systems moving with a constant velocity with respect to one another, and will not discuss the more *general* case in which one system is accelerated with respect to another. Thus our discussion will be confined to the theory of *special relativity*. The more general case of accelerated systems is treated in the theory of *general relativity*.

CONSERVATION OF ENERGY   To examine the invariance of the energy-conservation law under a Galilean transformation, consider again the collision in Figure 2-2, assuming it to be *elastic*. From the viewpoint of *observer 2*, the conservation of energy law is written:

kinetic energy before collision = kinetic energy after collision

$$\tfrac{1}{2} m(\dot{x}_2)^2 + \tfrac{1}{2} M(\dot{X}_2)^2 = \tfrac{1}{2} m(\dot{x}_2')^2 + \tfrac{1}{2} M(\dot{X}_2')^2 \qquad [2\text{-}9]$$

This equation can be rewritten in terms of the velocities as measured by *observer 1* by using Equation 2-2, the Galilean velocity transformation,

$$\tfrac{1}{2} m(\dot{x}_1)^2 - m\dot{x}_1 v + \tfrac{1}{2} mv^2 + \tfrac{1}{2} M(\dot{X}_1)^2 - M\dot{X}_1 v + \tfrac{1}{2} Mv^2$$
$$= \tfrac{1}{2} m(\dot{x}_1')^2 - m\dot{x}_1' v + \tfrac{1}{2} mv^2 + \tfrac{1}{2} M(\dot{X}_1')^2 - M\dot{X}_1' v + \tfrac{1}{2} Mv^2$$

Using Equation 2-5, the invariance of the conservation of momentum law, we can cancel the terms involving v; and because the terms in v^2 also cancel, there remains

$$\tfrac{1}{2} m(\dot{x}_1)^2 + \tfrac{1}{2} M(\dot{X}_1)^2 = \tfrac{1}{2} m(\dot{x}_1')^2 + \tfrac{1}{2} M(\dot{X}_1')^2 \qquad [2\text{-}10]$$

Equations 2-9 and 2-10 are of identical mathematical form, differing only in the subscripts 1 and 2; therefore, the conservation of energy law is valid for all inertial systems.

We have found that *the laws of classical mechanics—the conservation of momentum, Newton's laws of motion, and the conservation of energy—are all invariant under a Galilean transformation.* Thus, all inertial systems are equivalent for classical mechanics and it is impossible, by means of any experiment in mechanics, to distinguish one inertial system from any other. The invariance of the laws of mechanics, which has been formally proved here, is implicitly assumed in all elementary physics. For example, we are confident that an experimenter performing experiments in mechanics in a train moving at a constant velocity will arrive at results which are equally valid and which involve the same concepts and laws as those obtained by some other experimenter in a "non-moving" laboratory. We can draw another conclusion from the invariance of the laws of classical mechanics under a Galilean transformation; namely, that the basic transformation equations confirm our assumption of the absolute character of space and time. Our analysis has also contained the assumption that the mass of a body is a constant and is completely independent of its motion with respect to an observer; so that it can be said that *the Galilean transformations and classical mechanics imply that length, time, and mass,* the three basic quantities in physical measurements, *are all independent of the relative motion of an observer.* But we shall see that the relativity physics of Einstein drastically revises this notion.

2-4   The failure of Galilean transformations   Our discussion thus far has been limited to mechanics. One might well ask whether the other laws of physics, particularly those of electromagnetism and optics, are invariant under a Galilean transformation. Inasmuch as Postulate I requires that *all* the laws of physics be invariant, the Galilean transformations are universally valid only if the laws of electromagnetism can also be shown to be invariant. To examine this question we will restrict our discussion to the propagation of electromagnetic waves, because this alone will enable us to analyze the invariance of classical electromagnetism.

First, consider an example from mechanics. Assume that a *sound* pulse is traveling to the right with respect to the medium transmitting the pulse. The medium is assumed to be at rest in S_1, and the velocity of the pulse, as measured in S_1, is $\dot{x}_1$. By the velocity of a sound wave is ordinarily meant the velocity of propagation with respect to the medium—in this example, air. For an observer in S_2 moving at a velocity v with respect to S_1, the measured (apparent) velocity is, by Equation 2-2, $\dot{x}_2 = \dot{x}_1 - v$ (Figure 2-3). Therefore, an observer in S_2 measures a different velocity for the pulse than does the observer in S_1.

We illustrate this by an example. A sound pulse from a cannon travels at 1100 ft/sec ($\dot{x}_1$) in still air with respect to the Earth (S_1). An observer in an airplane (S_2), moving at 400 ft/sec (v) away from the source of the pulse measures a velocity $\dot{x}_2 = \dot{x}_1 - v = 1100 - 400 = 700$ ft/sec. This is the velocity of the pulse with respect to *his* inertial frame (S_2). On the other hand, if the airplane approaches the source, $v = -400$ ft/sec, $\dot{x}_2 = 1100 + 400$ ft/sec, and the observer in S_2 measures the speed of the pulse to be 1500 ft/sec. It follows that, in general, the measured speed of the sound pulse depends on the relative speed obtaining between the observer and the medium through which the pulse travels, and only when the observer is at rest with respect to the medium (here, air) will he find the measured speed to be 1100 ft/sec. *This result is confirmed by experiment for sound waves.* The measured velocity of the sound pulse in the system S_1, in which the air is at rest, *does not*, however, *depend on the velocity of the source* producing the sound. Of course, if a source of sound generates sinusoidal variations in the air pressure, rather than a pulse, the frequency and wavelength of the sound will, according to the Doppler effect, depend on the relative motion of the source and the medium—but the velocity of propagation of the disturbance will be independent of the source's relative motion.

We now turn to the completely analogous situation for light. A pulse of *light* travels to the right with respect to the medium through which it is propagated at a speed $\dot{x}_1 \equiv c$ (Figure 2-3). This medium for light propagation was historically given the name, the *ether*. Because nineteenth-century physicists were so firmly convinced that all physical phenomena were ultimately mechanical in origin, it was, for them, unthinkable that an electromagnetic disturbance could be propagated in empty space. Thus, the *ether* concept was invented. The only conspicuous property attributed to ether was that it "carried" electromagnetic disturbances, and that in the inertial system in which the ether was at rest, and in this system alone, was the speed of light equal to c.

In terms of the Galilean transformations, an observer at rest in S_1

would measure the speed of the light pulse to be c, in the X_1 direction, as well as in *any other direction* in which the light pulse might travel. Any other observer at rest in another system S_2 measures the speed to be $\dot{x}_2 = \dot{x}_1 - v = c - v$ when S_2 travels to the right; if S_2 moves to the left, he measures a different speed, $c + v$. This implies that the measured speed of light for any observer, except S_1, depends on the velocity of the coordinate system with respect to the medium, the ether, through which the pulse of light is propagated. Therefore, the speed of light is certainly *not* invariant under a Galilean transformation. In fact, if these transformations apply to light, then there exists in nature a unique inertial system in which the ether is at rest; in this one system, and in this system alone, is the measured speed of light exactly c.

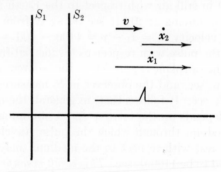

Figure 2-3. Observation of a pulse by two observers moving at a constant relative velocity v.

The essence of an experiment to find and confirm the existence of the ether is simple: to measure the speed of light in a variety of inertial systems, noting whether the measured speed is different in the different systems, and most especially, noting whether there is evidence for a single, unique inertial system—the ether—in which the speed of light is c. To perform such an experiment, however, is a far more difficult task than an experiment with sound waves because of the very high speed of light. In one of the most celebrated experiments of all time, Michelson and Morley, in 1887, sought to find this unique inertial system. The experiment was simply that of measuring whether there was a change in the measured speed of light as the Earth drifted through a conjectured ether in its axial rotation and its revolution about the Sun.

Let us analyze how the Michelson-Morley experiment attempted to find the unique inertial system. It is assumed that the unique inertial system is the system S_1. The experimenter has no prior knowledge of whether he is at rest in S_1; hence, he must assume that he is, in general, in *any* system S_2 which moves with a velocity v with respect to S_1. If he is, at some moment, at rest in S_1, then S_2 is S_1, and the speed of light is measured to be c. But six months later, when the Earth is moving in the opposite direction in its motion around the Sun, S_2 will surely be in motion with respect to S_1, and the measured speed of light now will be different.

Because of the extremely large magnitude of c, a measurement of the speed of light on Earth must, for practical reasons, be based on the measurement of the time interval required for a light beam to travel a known distance from some starting point to a reflecting mirror and back again. Therefore, to detect a change in the speed of light, one must detect a change in the time interval for a round trip by the light beam.

Consider a cylinder of length l which is at rest in S_2 and aligned along the X_2 axis, the direction of relative motion of S_1 and S_2 (Figure 2-4d, e, and f). As before, S_2 moves to the right with a velocity v relative to the unique inertial system, or ether, S_1. While a light pulse travels to the right, S_2 measures the speed of the pulse to be $c - v$, and the time required for the pulse to reach the right-end plate is $l/(c - v)$; after reflection, the pulse travels to the left with a speed $c + v$ relative to S_2 and reaches the left-end plate in a time $l/(c + v)$. Therefore, the time interval Δt_x for the light pulse to travel a complete round trip is

$$\Delta t_x = l/(c - v) + l/(c + v)$$

$$\Delta t_x = \frac{2\,l/c}{1 - (v/c)^2} \qquad [2\text{-}11]$$

The sequence of events as seen by observer S_1 is illustrated in Figure 2-4a, b, and c.

Now consider the situation when observer S_2 aligns the same cylinder along his Y_2 axis. The time interval required for the pulse to travel a round trip between what are now the bottom- and top-end plates is designated by Δt_y. The sequence of events as seen by observer S_1 is shown in Figure 2-5a, b, and c; and Figure 2-5d, e, and f shows the events as seen by S_2. From the point of view of S_1, only a light pulse which leaves the origin of S_1 in the particular direction θ, necessarily traveling at the speed c, will reach the center of the top-end plate at A, as shown in Figure 2-5b. The pulse goes from point O to A at a speed c in a time $\Delta t_y/2$, while the cylinder goes from O to B with a speed v in the same time. Therefore,

$$OA = c\,\Delta t_y/2, \qquad OB = v\,\Delta t_y/2, \qquad \text{and } AB = l$$

But
$$OA^2 = OB^2 + AB^2$$

and by substitution
$$(c\,\Delta t_y/2)^2 = (v\,\Delta t_y/2)^2 + l^2$$

Solving for Δt_y gives, finally,

$$\Delta t_y = \frac{2\,l/c}{\sqrt{1 - (v/c)^2}} \qquad [2\text{-}12]$$

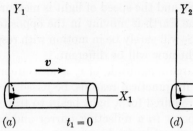

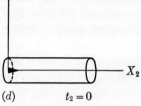

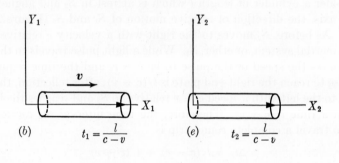

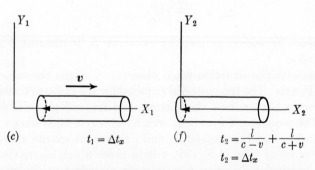

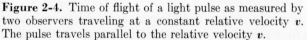

Figure 2-4. Time of flight of a light pulse as measured by two observers traveling at a constant relative velocity v. The pulse travels parallel to the relative velocity v.

Comparing Equation 2-11 with Equation 2-12 shows that $\Delta t_x \neq \Delta t_y$; that is, the time interval for a round trip by the light pulse is *not* the same for the two perpendicular orientations. Of course, if the cylinder were at rest in S_1, v would be zero and

$$\Delta t_x = 2\,l/c, \qquad \Delta t_y = 2\,l/c$$

Therefore, $\Delta t_x - \Delta t_y = 0$ when $v = 0$. When S_2 is in motion with respect to S_1,

$$\Delta t_x - \Delta t_y = (2\,l/c)\{[1 - (v/c)^2]^{-1} - [1 - (v/c)^2]^{-1/2}\}$$

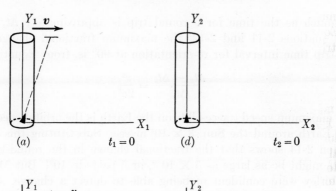

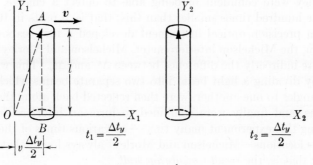

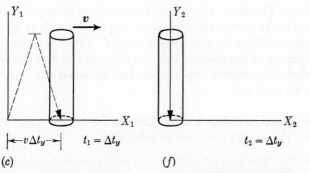

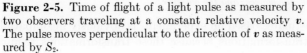

Figure 2-5. Time of flight of a light pulse as measured by two observers traveling at a constant relative velocity v. The pulse moves perpendicular to the direction of v as measured by S_2.

It can be assumed that $v/c \ll 1$. The binomial expansion can be used (neglecting higher order terms) to yield

$$\Delta t_x - \Delta t_y = (2\,l/c)[1 + (v/c)^2 - 1 - \tfrac{1}{2}(v/c)^2]$$
$$\Delta t_x - \Delta t_y = (2\,l/c)(v^2/2c^2) \qquad\qquad [2\text{-}13]$$

Inasmuch as the time for a round trip is approximately $\Delta t_x = 2\, l/c$, from Equations 2-11 and 2-12, the maximum fractional change in the round-trip time interval for re-orientation at $90°$ is, from Equation 2-13,

$$\frac{\Delta t_x - \Delta t_y}{\Delta t_x} = \frac{v^2}{2c^2} \qquad [2\text{-}14]$$

The maximum speed v accessible on the Earth is the orbital speed of the planet Earth around the Sun, 3×10^4 m/sec. Substituting this value in Equation 2-14 shows that the fractional change in the round-trip time interval might be as large as 5×10^{-9}, or 5 parts in 10^9! But Michelson and Morley were confident of being able to detect a change, should it occur, one hundred times smaller than this; that is, 5 parts in 10^{11}.

Using a precision optical instrument developed a few years earlier by Michelson, the Michelson interferometer, Michelson and Morley were able to measure indirectly the difference between Δt_x and Δt_y. This was accomplished by dividing a light beam into two separate beams which traveled at right angles to one another, were then reflected back along their respective paths, and finally were combined to form an interference pattern. Performing this experiment many times—at various times of the year and in various locations—Michelson and Morley† always found that $\Delta t_x - \Delta t_y$ was zero; that is, the *result was always null.*

This can have only one meaning—*any inertial system S_2 behaves as if it were the unique inertial system S_1*; or, the measured speed of light in *every* inertial system is found to be the same, namely c, for all directions and for all observers. Therefore, there is no experimental evidence to indicate the existence of a unique inertial system, or ether, inasmuch as *all inertial systems are equivalent for the propagation of light.* This fundamental assertion of the constancy of the speed of light for all observers is supported not only by the Michelson-Morley experiment, but by a variety of other experiments, as we shall see.

2-5   The second postulate and the Lorentz transformations   We can now write down the second postulate of relativity.

POSTULATE II: *The speed of light in a vacuum is a constant, independent of the inertial system, the source, and the observer.*

This postulate, based on experiment, is obviously incompatible with the Galilean transformations because these transformations require that the

---

† A number of workers have conducted similar experiments. A particularly interesting ether-drift experiment is that of L. Essen, Nature *175*, 793 (1955). This experiment is the microwave radio analog of the optical Michelson-Morley experiment in that short-wavelength radio beams replace the light beams, and beats between two resonant cavities at right angles to one another are observed rather than interference between two light beams.

measured speed of light depend on the motion of the observer. Einstein observed the inconsistency between Postulate II and the Galilean transformations. Postulate II could not be relinquished, for it was an experimental fact. The Galilean transformations, despite their apparent success in classical mechanics and their obvious appeal to our common experience, had to be supplanted by less restrictive transformation equations which would reduce to the Galilean transformations under appropriate conditions. How drastic such a change would be is demonstrated by the fact that our very ideas of space and time, and their apparently absolute character, are contained in the Galilean transformations.

Having seen that the Galilean transformations were fundamentally defective, Einstein sought the coordinate transformation equations which would be in harmony with Postulates I and II, the invariance of physical laws under coordinate transformations and the invariance of the speed of light. These two postulates, which have been discussed separately to this time, can be regarded as a single postulate, in which the invariance of the speed of light is regarded as a fundamental physical law.

We will find, as did Einstein in 1905, the transformation equations that satisfy the relativity requirements. These transformation equations are known as the *Lorentz transformations* because they were originated by H. A. Lorentz in 1903 in his theory of electromagnetism.

We assume the transformations satisfying relativity Postulates I and II can be written in the form

$$\left.\begin{aligned}
x_2 &= A(x_1 - vt_1) \\
y_2 &= y_1 \\
z_2 &= z_1 \\
t_2 &= A'(t_1 - Bx_1)
\end{aligned}\right\} \quad [2\text{-}15]$$

where A, A', and B are constants to be determined. It is assumed again that when $t_1 = t_2 = 0$ the origins of the coordinate systems coincide. As before, subscripts 1 and 2 refer to the meter sticks and clocks of observers 1 and 2 in inertial systems S_1 and S_2, respectively. Again S_2 moves to the right with a speed v relative to S_1. The equations must be written as *linear* equations rather than, say, quadratic equations, for only then will some one event (x_1, y_1, z_1, t_1) in S_1 correspond to a *single* event (x_2, y_2, z_2, t_2) in S_2, and conversely. Our equation for x_2 anticipates the reduction to the Galilean form (see Equation 2-1). The equation for t_2 now includes a term for x_1, a possibility which cannot be precluded a priori. In order to solve for A, A', and B, we impose the requirement of invariance of the speed of light c for any two observers S_1 and S_2.

Imagine that at the time $t_1 = t_2 = 0$, a spherical pulse of light leaves the common origin of S_1 and S_2. By the invariance of c, each observer, through measurements with his own meter stick and clock, sees a spherical "bubble" expand outward at the speed c in his own system. It follows then that for S_1,

$$c = \frac{\sqrt{x_1^2 + y_1^2 + z_1^2}}{t_1}$$

and similarly for S_2,

$$c = \frac{\sqrt{x_2^2 + y_2^2 + z_2^2}}{t_2}$$

Consequently,

$$x_1^2 + y_1^2 + z_1^2 - c^2t_1^2 = x_2^2 + y_2^2 + z_2^2 - c^2t_2^2$$

Using Equation 2-15 to cancel the y's and z's and to substitute for x_2 and t_2, we have

$$x_1^2 - c^2t_1^2 = A^2(x_1^2 - 2vx_1t_1 + v^2t_1^2) - c^2A'^2(t_1^2 - 2Bx_1t_1 + B^2x_1^2)$$

Since x_1 and t_1 are independent coordinates, the coefficients of x_1^2 on each side of the equation must be equal; hence,

$$1 = A^2 - c^2B^2A'^2 \qquad [2\text{-}16]$$

Similarly for the coefficients of x_1t_1 and t_1^2

$$0 = -2A^2v + 2c^2A'^2B \qquad [2\text{-}17]$$

and

$$-c^2 = A^2v^2 - c^2A'^2 \qquad [2\text{-}18]$$

We wish to find A, A', and B in terms of v and c. Solving Equation 2-17 for A^2 gives

$$A^2 = c^2A'^2B/v \qquad [2\text{-}19]$$

Substituting Equation 2-19 into Equation 2-16 yields

$$1 = A'^2Bc^2[(1/v) - B] \qquad [2\text{-}20]$$

and substituting Equation 2-19 into Equation 2-18 yields

$$-c^2 = A'^2c^2(vB - 1) \qquad [2\text{-}21]$$

Multiplying Equation 2-20 by $-v$ gives

$$-v = A'^2Bc^2(vB - 1) \qquad [2\text{-}22]$$

and dividing Equation 2-22 by Equation 2-21 yields finally

$$B = v/c^2 \qquad [2\text{-}23]$$

Substituting Equation 2-23 into Equation 2-21 and solving for A' gives

$$A' = \frac{1}{\sqrt{1 - (v/c)^2}} \qquad [2\text{-}24]$$

and using Equation 2-23 in Equation 2-19 shows that

$$A = A' = \frac{1}{\sqrt{1 - (v/c)^2}} \qquad [2\text{-}25]$$

Substituting the values for A, A', and B from Equations 2-23 and 2-25 into Equation 2-15 gives the Lorentz transformations:

*Lorentz
Coordinate
Transformations*

$$x_2 = \frac{x_1 - vt_1}{\sqrt{1 - (v/c)^2}}$$

$$y_2 = y_1$$

$$z_2 = z_1$$

$$t_2 = \frac{t_1 - (v/c^2)x_1}{\sqrt{1 - (v/c)^2}}$$

[2-26]

As before, to get the inverse transformation equations, giving x_1, y_1, z_1, and t_1, in terms of x_2, y_2, z_2, and t_2, one merely interchanges the subscripts 1 and 2 and replaces v by $-v$ in Equation 2-26.

The Lorentz transformations are the logical consequence of the experimentally established fact that the speed of light c is a true constant of nature, independent of the motion of the source or observer, and independent of the inertial system.

We have arrived at the unique transformations which meet the requirements of the two relativity postulates and which, therefore, supplant the Galilean transformations. By the correspondence principle (Section 1-6), we know that the Lorentz transformations *must* reduce to the Galilean transformations in that range in which the Galilean transformations are known to be essentially correct. By comparing Equation 2-26 with Equation 2-1, we see that these two transformations become identical when $v/c \to 0$. Thus, when $v \ll c$, the Galilean transformations are an excellent approximation to the universally valid Lorentz transformations. The Galilean transformations are adequate to describe all low-speed phenomena. Mathematically, putting $v/c \ll 1$ is equivalent to letting $c \to \infty$; therefore, we can regard the Galilean transformations as being the correct coordinate transformations in an hypothetical universe in which the speed of light is infinite. We can write symbolically:

Limit (Lorentz transformations) = (Galilean transformations)
$\scriptstyle c \to \infty$

Conversely, when the speed of an object is close to the speed of light, only the Lorentz transformations will apply.

We note two results of the Lorentz transformations: (1) the upper speed limit, c, for all particles, and (2) the relative character of space and time. From Equations 2-26, it is seen that the space and time coordinates of S_2 will be real, rather than imaginary, only if the speed v of S_2 with respect to S_1 is less than c. If $v < c$, then $[1 - (v/c)^2]^{1/2}$ is real. If $v > c$, $[1 - (v/c)^2]^{1/2}$ is imaginary; but this is impossible because one can measure only real physical quantities. According to the Lorentz transformation for the time coordinate, Equation 2-26, the time coordinate (t_2) is no longer absolute; that is, it is no longer independent of the space coordinate (x_1). The clocks (t_1 and t_2) of observers in two different coordinate systems (S_1 and S_2) may not always agree ($t_1 \neq t_2$). We shall discuss this matter further in Section 2-7.

2-6   The Lorentz velocity transformations   In order to find the proper transformation equations for the velocity satisfying the relativity requirements, we follow the same procedure that was used in Section 2-2 for the Galilean velocity transformation equations.

Taking the differential of each coordinate in Equation 2-26 gives

$$dx_2 = \frac{dx_1 - v\,dt_1}{\sqrt{1 - (v/c)^2}}$$

$$dy_2 = dy_1$$

$$dz_2 = dz_1$$

$$dt_2 = \frac{dt_1 - (v/c^2)\,dx_1}{\sqrt{1 - (v/c)^2}}$$

Dividing dx_2, dy_2, and dz_2 in turn by dt_2 gives the velocity components in S_2.

$$\dot{x}_2 \equiv dx_2/dt_2 = \frac{dx_1 - v\,dt_1}{dt_1 - (v/c^2)\,dx_1}$$

and

$$\dot{y}_2 \equiv dy_2/dt_2 = \frac{dy_1\sqrt{1 - (v/c)^2}}{dt_1 - (v/c^2)\,dx_1}$$

The relation for $\dot{z}_2$ is similar to that for $\dot{y}_2$. Dividing the numerator and denominator above by dt_1 gives

Lorentz
Velocity
Transformations

$$\dot{x}_2 = \frac{\dot{x}_1 - v}{1 - (v/c^2)\dot{x}_1}$$

$$\dot{y}_2 = \frac{\dot{y}_1\sqrt{1 - (v/c)^2}}{1 - (v/c^2)\dot{x}_1}$$

$$\dot{z}_2 = \frac{\dot{z}_1\sqrt{1 - (v/c)^2}}{1 - (v/c^2)\dot{x}_1}$$

[2-27]

One surprising result of the Lorentz velocity transformations, Equation 2-27, is that the Y- and Z-components of the velocity of a particle as measured in S_2 depend on the X-component of velocity as measured in S_1! As before, to obtain the velocity components as measured in S_1 in terms of those as measured in S_2, we simply interchange subscripts and replace v by $-v$ in Equations 2-27. Applying the correspondence principle to the

Lorentz velocity transformations, we obtain the Galilean velocity transformations, Equation 2-2:

Limit (Lorentz velocity transformations)
$c \to \infty$

= (Galilean velocity transformations)

As an example of the Lorentz velocity transformations we consider the following problem.

A fast car moves at a speed $c/3$ with respect to a man holding a lantern. A passenger in the car measures the speed of light reaching him from the lantern with his own meter stick and clock; by the relativity principle, he must measure the light's speed to be c no matter what the direction of his velocity with respect to the lantern. Can we confirm this by using the Lorentz velocity transformations, Equations 2-27?

We will test the invariance of c for the simple case in which the car directly approaches the lantern. Assume the car to be at rest in system S_2 and to be traveling toward the lantern in S_1 with a velocity $v = -(c/3)$ as shown in Figure 2-6.

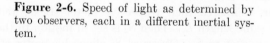

Figure 2-6. Speed of light as determined by two observers, each in a different inertial system.

Since the light travels at a speed c with respect to the lantern, $\dot{x}_1 = c$; and using Equations 2-27 to find $\dot{x}_2$, we have

$$\dot{x}_2 = \frac{\dot{x}_1 - v}{1 - (v/c^2)\dot{x}_1} = \frac{c - (-c/3)}{1 - (-c/3c^2)c} = c$$

We notice immediately from this example, as well as from the general Equations 2-27, that the velocity of a particle as measured by S_1 is no longer the vector sum of the velocity as measured by S_2 and the relative velocity of S_2 with respect to S_1. For the example above, the Galilean velocity transformations, Equation 2-2, would give

$$\dot{x}_2 = \dot{x}_1 - v = c - (-c/3) = (4/3)c$$

which *does* obey vector addition,

$$\overrightarrow{c} + \overrightarrow{c/3} = \overrightarrow{(4/3)c}$$

but is inconsistent with experimental results. Symbolically, under the Lorentz transformations, we have for the propagation of light in the above example

$$\overrightarrow{c} + \overrightarrow{c/3} = \overrightarrow{c}$$

It must be remembered, however, that in any one inertial system, such as S_1, the resultant velocity v_1 of a particle traveling in any direction is still obtained by the vector addition of the components in S_1, the speed being given by

$$v_1 = \sqrt{\dot{x}_1^2 + \dot{y}_1^2 + \dot{z}_1^2}$$

The Lorentz transformations assure us that the speed of light—and only this speed—will be the same for all inertial coordinate systems. *The laws of optics and electromagnetism are invariant under a Lorentz transformation, and all inertial systems are equivalent for these laws.*

2-7   Length, time, and mass in relativity physics

SPACE CONTRACTION   In classical physics, the length of an object, for instance a meter stick, is the same for all observers, whatever their velocities with respect to the meter stick. Let us now examine the meaning of length in relativity theory according to the Lorentz transformations, remembering that only these transformation equations give a complete description of space and time.

Suppose that at some time observers S_1 and S_2 were at rest with respect to each other, compared their respective meter sticks, and agreed that both had the same length L_0. Then S_2 is set in motion to the right with a speed v with respect to S_1. Observer S_1 aligns his meter stick along the X axis, with its left end at x_1 and its right end at x_1'; therefore, from the point of view of S_1, $L_0 = x_1' - x_1$. Similarly, S_2 aligns his meter stick along the X axis, with the left and right ends of his meter stick at the points x_2 and x_2'; therefore, from the point of view of S_2, $L_0 = x_2' - x_2$. Each observer sees *his own* meter stick as having length L_0, which, of course, must follow from the fact that all inertial systems are equivalent and indistinguishable.

Now we ask, "What is the length of S_2's meter stick as measured by S_1?" Using the Lorentz coordinate transformations, Equation 2-26, we can write

$$(x_2' - x_2) = \frac{(x_1' - x_1) - v(t_1' - t_1)}{\sqrt{1 - (v/c)^2}} \qquad [\text{2-28}]$$

Inasmuch as the meter stick in S_2, the length of which S_1 wishes to measure, is moving with respect to S_1, it is essential that S_1 mark simultaneously the point at which each end of S_2's meter stick falls; that is, $t_1 = t_1'$. The length of the meter stick of S_2, *as measured by* S_2, is $x_2' - x_2 = L_0$; the length *as measured by* S_1 is $x_1' - x_1 \equiv L$. Therefore, Equation 2-28 becomes

$$L = L_0 \sqrt{1 - (v/c)^2} \qquad [2\text{-}29]$$

Thus, when measuring the length of the moving meter stick in S_2, S_1 finds it to be contracted by a factor $[1 - (v/c)^2]^{1/2}$ along the direction of the relative motion. This contraction is reciprocal; that is, S_2 finds meter stick S_1 to be $L_0[1 - (v/c)^2]^{1/2}$. Because the contraction is *not* in consequence of any physical disturbance (cooling, compression, etc.) but rather reflects the properties of space and time as contained in the Lorentz transformations, the phenomenon is known as *space contraction*. Since $y_1 = y_2$ and $z_1 = z_2$, there is no space contraction in a direction perpendicular to that of the relative motion. Clearly, the length of an object is *not* absolute but depends upon that object's relative motion with respect to the observer. It is a maximum when at rest in the observer's inertial system. We are accustomed to experiencing phenomena for which $v \ll c$, and therefore, from Equation 2-29, $L \simeq L_0$.

TIME DILATATION   Relativity physics shows that time as well as length is not absolute, but rather, depends on the relative motion of the observer.

Two observers S_1 and S_2, while at rest with respect to one another, synchronize their respective clocks, and both agree that the interval between two events as measured on their respective clocks is the same interval T_0. Now we imagine S_2 to be moving at a speed v to the right with respect to S_1. Observer S_1 keeps his clock at rest at a particular point x_1 in his system, and measures the time interval T_0 to be that time interval elapsed between the instants t_1 and t_1'. Similarly, S_2 keeps his clock at the fixed position x_2, and measures the time interval T_0 to be that interval from the instants t_2 to t_2'. Thus each observer measures the time interval on his own clock to be T_0. We wish to find out how the time intervals registered on clocks S_1 and S_2 compare when *both* intervals are measured by observer S_2. From Equation 2-26 we have

$$(t_2' - t_2) = \frac{(t_1' - t_1) - (v/c^2)(x_1' - x_1)}{\sqrt{1 - (v/c)^2}} \qquad [2\text{-}30]$$

Because observer S_2 sees S_1 in motion with respect to him, it is essential that the clock in S_1 remain fixed in inertial system S_1; therefore, $x_1 = x_1'$. The time interval that S_1 measures for his own clock is $T_0 = t_1' - t_1$. According to S_2's measurement, the time elapsed on S_1's clock is $T \equiv t_2' - t_2$. Substituting in Equation 2-30 gives

$$T = \frac{T_0}{\sqrt{1 - (v/c)^2}} \qquad [2\text{-}31]$$

This relation shows the phenomenon of *time dilatation*. As an example of the consequences of time dilatation, suppose that $v = 0.98c$, then

$[1 - (v/c)^2]^{1/2} = 1/5$, and $T = 5T_0$. Therefore, a clock which registers an interval T_0 when at rest with respect to an observer will register the interval $5T_0$ when in motion past the observer at 98 per cent the speed of light. When an observer compares two previously synchronized clocks, one at rest and the other in motion with respect to him, he finds that the moving clock runs slower than the fixed clock. An observer finds time in a moving system to be expanded, or dilated, as compared to his time. Again, this effect is reciprocal in that each observer finds the other moving observer's clock to run slow. For $v \ll c$ the relative time of relativity physics reduces to the absolute time of classical physics. Despite the fact that space contraction and time dilatation strike us as being extraordinarily bizarre, there is direct and inescapable experimental evidence of time dilatation in the decay of high-speed meson particles (Section 10-9).

RELATIVISTIC MASS   In Section 2-3 we found that the laws of classical mechanics—conservation of momentum, Newton's laws of motion, and conservation of energy—were invariant under the Galilean transformations, assuming the invariance of mass. Clearly, these laws in their classical forms *cannot* be invariant under the Lorentz transformations. Therefore, we must reinvestigate the concepts and laws of mechanics in the light of these new transformation equations. Experimental evidence indicates that the laws of the *conservation of momentum and of energy*, in modified form, *are still valid in relativistic mechanics*. We first wish to explore the consequences of the invariance of the law of conservation of momentum under a Lorentz transformation.

Consider a situation in which two observers, S_1 and S_2, are observing a perfectly elastic collision between two smooth, hard spheres A and B, whose masses are *equal when compared at rest* in the same inertial system.

Observer 1 throws particle A in his $+Y_1$ direction with a speed $(\dot{y}_1)_A$ so that it collides with particle B and returns in the $-Y_1$ direction, as shown in Figure 2-7a (in which the X_1 and X_2 axes are displaced for clarity). Observer 2 throws particle B in his $-Y_2$ direction with a speed $(\dot{y}_2)_B$ so that it collides with A and returns, as shown in Figure 2-7b. For simplicity, the circumstances of the collision are chosen such that *the behavior of particle A is to observer 1 exactly the same as is the behavior of particle B to observer 2*.

Observer S_1 sees particle A travel out a distance $(y_A/2)$ at a speed $(\dot{y}_1)_A$ and return with the same speed, $(\dot{y}_1)_A$, the total time elapsed being T_0 according to S_1's clock. Therefore, $(\dot{y}_1)_A = y_A/T_0$. Similarly, observer S_2 sees particle B travel out a distance $y_B/2$ at a speed $(\dot{y}_2)_B$ and return with the same speed, $(\dot{y}_2)_B$, in a total time T_0 according to S_2's clock. Hence, $(\dot{y}_2)_B = y_B/T_0$. Inasmuch as *there is no spatial contraction in a direction perpendicular to the relative velocity* of the two inertial systems, $y_A = y_B \equiv y$; therefore,

$$(\dot{y}_1)_A = (\dot{y}_2)_B = y/T_0 \qquad\qquad [2\text{-}32]$$

and both observers agree that the collision takes place midway between the two X axes; that is, at $y/2$.

An observer in inertial system S_1 now applies the conservation of momentum in the Y direction to the collision between the particles A and B. Recalling that the Y component of the velocity of each particle merely changes direction upon collision, observer S_1 writes, considering magnitudes only,

$$m_A(\dot{y}_1)_A = m_B(\dot{y}_1)_B \qquad\qquad [2\text{-}33]$$

where, as in classical mechanics, *the momentum of a particle is defined as the product of the mass and the velocity of the particle*. It must be remembered that *all* masses and velocities appearing in Equation 2-33 are measured by

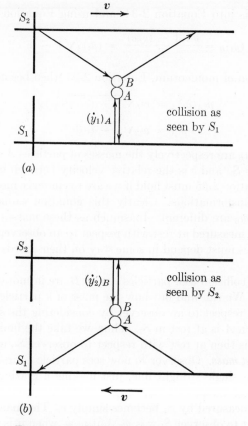

(a)

(b)

Figure 2-7. Collision of two particles as observed by two observers, each in a different inertial system.

the observer in S_1. The magnitude of the Y component of the velocity of particle B as measured by S_1 is simply y/T, where T is the total time measured by S_1 from the instant S_2 throws particle B until the instant when S_2 catches B.

Thus,

$$(\dot{y}_1)_B = y/T \qquad [2\text{-}34]$$

We have already seen that both S_1 and S_2 agree on the distance y. However, these observers *disagree* on the time interval between the throwing of particle B and its return to observer S_2; observer S_1 measures the time interval T, whereas observer S_2 measures the time interval T_0. These time intervals are related by the time-dilatation formula, Equation 2-31

$$T = \frac{T_0}{\sqrt{1 - (v/c)^2}}$$

Substituting this into Equation 2-34, and using Equation 2-32, we have

$$(\dot{y}_1)_B = \frac{y\sqrt{1 - (v/c)^2}}{T_0} = (\dot{y}_1)_A \sqrt{1 - (v/c)^2}$$

The conservation of momentum, Equation 2-33, then becomes

$$m_A(\dot{y}_1)_A = m_B\sqrt{1 - (v/c)^2}(\dot{y}_1)_A$$

or,
$$m_A = m_B\sqrt{1 - (v/c)^2} \qquad [2\text{-}35]$$

where m_A and m_B are respectively the masses of particles A and B as measured by *observer* S_1, and v is the relative velocity between inertial systems S_1 and S_2. Equation 2-35 must hold if we are to conserve momentum under the Lorentz transformations. Clearly this equation cannot be satisfied unless m_A and m_B are different. Inasmuch as these masses were identical when they were measured at rest with respect to an observer, it is apparent that their masses must depend in some way on their speeds relative to the observer.

In the above collision *both* particles A and B are in motion with respect to observer S_1. We can find out how the mass of a particle depends upon its motion with respect to an observer by considering the special collision that occurs when A is at rest in S_1. Thus, we take the limit as $(\dot{y}_1)_A \rightarrow 0$. The particle A is then at rest with respect to observer S_1, and we label its mass m_0, the *rest mass*. Observer S_1 now sees particle B first approach and then recede in a single straight line, just making a grazing collision with particle A. Using Equation 2-32, $(\dot{y}_2)_B \rightarrow 0$ and then the speed of B (see Figure 2-8) as measured by S_1 becomes simply v. The mass of B when at rest with respect to observer S_1 was m_0, but now, when it is in motion with a speed v with respect to observer S_1, it is different and we label it m. For this grazing collision, Equation 2-35 becomes

$$m_0 = m\sqrt{1 - (v/c)^2}$$

or $$\boxed{m = \frac{m_0}{\sqrt{1 - (v/c)^2}}}$$ [2-36]

By the same token, observer S_2 sees his own particle B at rest with mass m_0 and the second particle A with a mass m in motion at a speed v. Any observer finds that if a particle has a rest mass m_0 when measured at rest with respect to him, then the mass of this particle when moving at a speed v with respect to him is $m_0/\sqrt{1 - (v/c)^2}$.

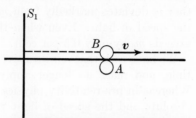

The relativistic mass, Equation 2-36, follows from the invariance of the conservation of the quantity mv, which we still define as momentum p in relativity physics. The *relativistic mass* m depends on the speed v and, therefore is not invariant; on the other hand, the rest mass m_0 is still an

Figure 2-8. Grazing collision of a moving particle B with a particle A at rest, as viewed by an observer in inertial system S_1.

invariant quantity. We must relinquish a classical conservation law, the conservation of mass; however, we shall soon find a more general, and in some ways more simple, conservation law which supplants the classical law of mass conservation. Of course, in the low-speed limit, $v/c \ll 1$, and hence, $m = m_0$, as required by the correspondence principle.

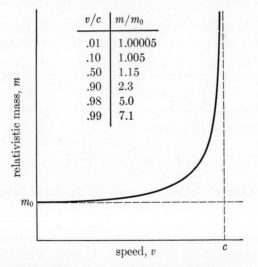

v/c	m/m_0
.01	1.00005
.10	1.005
.50	1.15
.90	2.3
.98	5.0
.99	7.1

Figure 2-9. Variation of relativistic mass with speed.

Experiments with high-speed particles (electrons, etc.) verify Equation 2-36 for they show that the relativistic mass of a particle is not invariant in nature, but is indeed a function of the speed of the particle with respect to the observer. Of course, an observer moving with the particle cannot detect any change in mass.

In Figure 2-9 is shown the relativistic mass as a function of the particle's speed with respect to the observer, as given by Equation 2-36. It is clear that m deviates markedly from m_0 only when the speed is comparable to the speed of light. Even when the speed is 10 per cent that of light, m exceeds m_0 by only 0.5 per cent.

We have seen that our intuitive notions as to the absoluteness of length, time, and mass no longer apply, except for speeds much less than c. Whereas in pre-relativity physics space, time, and mass were regarded as absolute and the speed of light was regarded as relative to some unique inertial frame, the relativity physics of Einstein requires that, because the speed of light is absolute for all inertial systems, space, time, and mass must be relative and depend upon the motion of an observer.

2-8 Relativistic mechanics In this section we shall develop the relativistic forms for the mechanical quantities: momentum, force, and energy.

We have already pointed out that the relativistic momentum p is defined to be mv, where from Equation 2-36,

$$p = mv = \frac{m_0 v}{\sqrt{1 - (v/c)^2}}$$ [2-37]

Experiment shows that the force F on a body is the time-rate of change of the relativistic momentum. Therefore, Newton's second law must be written in the form:

$$F = \frac{dp}{dt} = \frac{d}{dt}(mv) = \frac{d}{dt}\left[\frac{m_0 v}{\sqrt{1 - (v/c)^2}}\right]$$ [2-38]

Equation 2-38 can also be written

$$F = m\frac{dv}{dt} + v\frac{dm}{dt}$$ [2-39]

In classical mechanics, when the mass of the body remains constant, $dm/dt = 0$; and the two forms for Newton's second law, $F = (d/dt)\cdot(mv)$ and $F = m\,dv/dt = ma$, are equivalent. In relativistic mechanics, however, these two forms are not equivalent; this is so because the mass varies with the speed of the particle and, therefore, with time if the *speed* changes with time. It is interesting to note that Newton wrote his second law in the form, $F = (d/dt)\cdot(mv)$ not because he foresaw the relativistic variation in mass, but rather to take into account the loss or accumulation of material mass from the system; a common example of this occurs in the propulsion of rockets, where mass is ejected from the rocket, thereby decreasing its mass m_0.

There is one very important situation in which the speed of a body is constant while the velocity changes; that is, of course, the motion of a particle in a circle under the influence of a centripetal force. Because the centripetal force is always normal to the velocity, no work is done on the particle; thus it moves at a constant speed, but with an ever-changing direction of its velocity.

Consider a particle of relativistic mass m and electric charge q which moves with a velocity v at right angles to lines of uniform magnetic flux density B. The magnetic force, which causes the charged particle to move in a circle, is given by Equation 1-6

$$F = qvB$$

Because the particle moves with a constant speed, its relativistic mass is constant, and hence $dm/dt = 0$. Equation 2-39 then becomes $F = m\, dv/dt = ma$, where a is the centripetal acceleration with a magnitude v^2/r, r being the radius of the circular path (see Figure 2-10). Using this relation for the magnetic force, we have

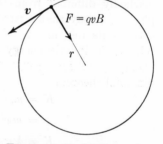

$$qvB = mv^2/r,$$

or

$$\boxed{p = mv = q(Br)} \qquad [2\text{-}40]$$

It must be remembered that, although Equation 2-40 is of exactly the same form as that obtained by classical mechanics, the mass m appearing in Equation 2-40 is the relativistic mass, not the rest mass.

Figure 2-10. Motion of a charged particle in a transverse magnetic field.

Equation 2-40 is the basis of a simple method for determining the relativistic momentum of a charged particle. In this method, q is known, B and r are measured, and the momentum p is computed from Equation 2-40. Inasmuch as the speed v can be measured by using crossed electric and magnetic fields (Section 8-8), as in J. J. Thomson's e/m experiment, the relativistic mass m can be computed from Equation 2-40. This was essentially the method used by A. H. Bucherer (1909) and others to verify the variation of relativistic mass with speed predicted by relativity theory.

We now have expressions for relativistic mass, Equation 2-36, relativistic momentum, Equation 2-37, and the correct form for Newton's second law of motion, Equation 2-38. We next ask, "What is the relativistic kinetic energy K?" To find this, we define, as in classical physics, the kinetic energy K to be the total work done in bringing a particle from rest to the final speed v under a constant force F.

$$K = \int_0^s F \, ds = \int_0^s \frac{d}{dt}(mv) \, ds = \int_0^t \frac{d}{dt}(mv) \, v \, dt$$

$$= \int_0^{mv} v \, d(mv) = \int_0^v v \, d\left(\frac{m_0 v}{\sqrt{1 - (v/c)^2}}\right)$$

$$= m_0 \int_0^v \left\{\frac{v}{[1 - (v/c)^2]^{1/2}} + \frac{(v^3/c^2)}{[1 - (v/c)^2]^{3/2}}\right\} dv$$

$$= m_0 \int_0^v \frac{v \, dv}{[1 - (v/c)^2]^{3/2}} = m_0 c^2 \left[\frac{1}{\sqrt{1 - (v/c)^2}}\right]_0^v$$

$$= m_0 c^2 \left(\frac{1}{\sqrt{1 - (v/c)^2}} - 1\right)$$

$$K = mc^2 - m_0 c^2 = (m - m_0)c^2 \qquad [2\text{-}41]$$

Thus, the relativistic kinetic energy is the increase in mass, arising from its motion, multiplied by c^2. We see that the relativistic kinetic energy is markedly different from the classical form. Furthermore, in relativity physics, to say that a particle has kinetic energy is to say that its mass exceeds its rest mass. Of course, by the correspondence principle, the relativistic kinetic energy must reduce to the familiar classical kinetic energy, $\frac{1}{2} m_0 v^2$ for $v/c \ll 1$. To show this, we expand Equation 2-41 by the binomial theorem

$$K = m_0 c^2 \{[1 - (v/c)^2]^{-1/2} - 1\}$$

$$= m_0 c^2 [1 + \tfrac{1}{2}(v/c)^2 + \tfrac{3}{8}(v/c)^4 + \ldots - 1]$$

$$K = \tfrac{1}{2} m_0 v^2 + \tfrac{3}{8} m_0 v^4 / c^2 + \ldots$$

and $\qquad \underset{c \to \infty}{\text{Lim }} K = \underset{c \to \infty}{\text{Lim }} (m - m_0)c^2 = \tfrac{1}{2} m_0 v^2$

It is important to recognize that the relativistic kinetic energy is *not* given by $\frac{1}{2} mv^2$, where m is the relativistic mass.

Analysis shows that an increase in the potential energy of a system of particles corresponds to an increase in the mass of the system, just as a gain in kinetic energy does. Thus, when the total energy of a particle, or a system of particles, increases by virtue of an increase in either its kinetic energy or its potential energy, its mass also increases.

For a constant potential energy, the kinetic energy represents the increase in the total energy of the particle.

$$\boxed{K = E - E_0 = mc^2 - m_0 c^2} \qquad [2\text{-}42]$$

where the *rest energy* E_0 is the energy of the particle when at rest, and the *total energy* E is the energy of the particle when in motion with kinetic energy K. Whereas in classical physics we are free to choose E_0 to be zero, in relativity physics it is convenient and in fact necessary, as more sophisticated analysis shows, to choose the rest energy E_0 as

$$E_0 = m_0c^2 \qquad [2\text{-}43]$$

and, therefore, from Equation 2-42

$$\boxed{E = mc^2} \qquad [2\text{-}44]$$

Equation 2-44 is the famous Einstein relation, $E = mc^2$, which shows the equivalence of energy and mass, each being a different manifestation of the same physical entity. A particle at rest with respect to an observer has a rest mass m_0 and a rest energy m_0c^2. Because mass and energy are equivalent and interchangeable, we no longer have the separate conservation laws of energy and mass; instead, relativity physics combines these into a single, simple conservation law, the conservation of mass-energy. The conservation of mass-energy is invariant under a Lorentz transformation. It is interesting to note that in relativistic mechanics mass and energy are inextricably united, just as in relativistic kinematics space and time are inseparably related.

In physics, the momentum is, in general, a more useful concept than the velocity (e.g., we have the conservation of momentum, but not the conservation of velocity). Therefore, it is often convenient to express the energy E in terms of p rather than v.

We can eliminate v in the following manner: squaring Equation 2-36 and multiplying both sides by $c^4[1 - (v/c)^2]$ gives

$$m^2c^4 - m^2v^2c^2 = m_0^2c^4$$

Substituting Equations 2-37, 2-43, and 2-44 into the above equation for mv, mc^2 and m_0c^2 immediately gives the desired relation between E and p

$$\boxed{E^2 = (pc)^2 + E_0^2} \qquad [2\text{-}45]$$

It is interesting to examine the relativistic equations under two limiting conditions:

(a) $v \ll c$. This is the classical region in which Newtonian mechanics is adequate, and the relativistic quantities reduce to their familiar classical forms, namely

$$m \simeq m_0$$
$$p \simeq m_0v$$
$$K \simeq \tfrac{1}{2}\, m_0v^2$$

In this region the kinetic energy is much less than the rest energy; that is,

$$K \ll E_0$$

because

$$\frac{K}{E_0} = \frac{\tfrac{1}{2}\, m_0v^2}{m_0c^2} = \tfrac{1}{2}\left(\frac{v}{c}\right)^2 \ll 1$$

(b) $v \simeq c$. This represents the extreme relativistic region. Therefore, Equations 2-36, 2-42, and 2-45 become:

$$m \gg m_0$$
$$E \gg E_0$$
$$p \simeq E/c$$
$$K \simeq E$$

If a particle exists in nature which has energy and momentum, but has zero *rest* mass—a possibility which makes no sense from a classical viewpoint, but is admissible in relativity theory—then the above equations are *exactly true*, and a zero-rest-mass particle must necessarily travel with the speed of light, c. That is, for a zero-rest-mass particle:

$$m_0 = 0; \quad E = pc, \quad K = E, \quad \text{and } v = c \qquad [2\text{-}46]$$

Conversely, if a particle with a non-zero energy travels with the speed of light, it *must* have a zero rest mass, since from Equations 2-36 and 2-44,

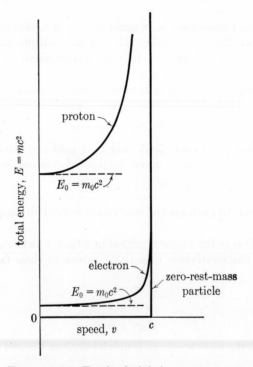

Figure 2-11. Total relativistic energy versus speed of three particles having different rest masses. (The rest energies are *not* to scale relative to one another.)

$$m = \frac{E}{c^2} = \frac{m_0}{[1 - (v/c)^2]^{1/2}} \qquad [2\text{-}47]$$

If $v = c$, the denominator equals zero, and for a non-zero rest mass the energy of the particle would be infinite, an impossibility. However, if the rest mass is zero, the right-hand side of Equation 2-47 becomes $0/0$, an indeterminate. The relativistic mass m can be determined from Einstein's equation $m = E/c^2$, and is finite even though m_0 is zero. The energy (and therefore mass, because $E = mc^2$) variation as a function of speed is shown in Figure 2-11 for a proton, an electron, and a particle of zero rest mass. For the zero-rest-mass particle the relativistic energy E has a non-zero value only if the particle moves at the single speed c; but for this speed, its energy (and mass) can have any value from zero to infinity. A zero-rest-mass particle has zero rest energy; therefore, its total energy is purely kinetic, as shown in Equation 2-46.

It is useful to examine the dependence of energy on momentum. The relativistic total energy E is shown as a function of the relativistic momentum p in Figure 2-12, following Equation 2-45:

$$E^2 = (pc)^2 + E_0^2$$

For low speeds $v \ll c$, $m \simeq m_0$, and $p \simeq p_0 = m_0 v$. Therefore, in this, the

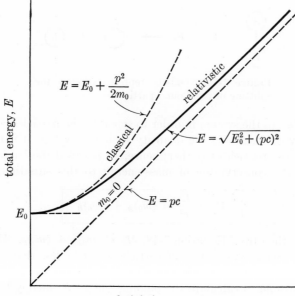

Figure 2-12. Total relativistic energy versus relativistic momentum of a particle.

classical region, the total energy is given by

$$E \simeq p_0^2/2m + E_0 = \tfrac{1}{2}\,m_0 v^2 + E_0$$

On the other hand, for very high speeds, $E \simeq pc$ as shown in Equation 2-46. In the classical region the kinetic energy $(E - E_0)$ varies as the *square* of the momentum, whereas in the extreme relativistic region the kinetic energy $(\simeq E)$ varies *linearly* as the momentum. It follows from Equation 2-45, that the slope, dE/dp, of Figure 2-12 is equal to the speed v of the particle.

2-9 Mass-energy equivalence and binding energy To illustrate the significance of mass-energy equivalence, one of the most important consequences of Einstein's relativity theory, consider two particles, A and B, bound together by some attractive force and forming a single system. To break the system into its separate components requires work; that is, the adding of energy to the system. Let the rest mass of the bound system be M_0 and the rest masses of the individual particles be $(m_0)_A$ and $(m_0)_B$, when A and B are completely separated and the attractive force is zero. The bound system (before it is separated into its component particles) and the two particles after separation are all assumed to be at rest.

This is shown symbolically in Figure 2-13, where E_b is the energy that

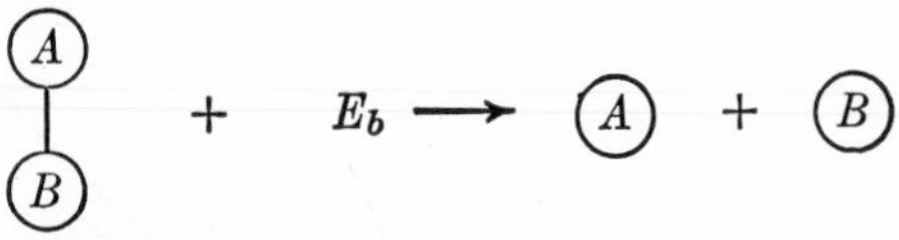

Figure 2-13. Symbolic representation of the splitting of two bound particles.

must be added to the system in order to separate the particles completely. Another way of describing the quantity E_b is to say that it is the energy which binds the particles together; hence, E_b is called the *binding energy*.

Applying the conservation of mass-energy to this situation, we have

$$\boxed{M_0 + \frac{E_b}{c^2} = (m_0)_A + (m_0)_B}\qquad\qquad [2\text{-}48]$$

If $E_b > 0$, then from Equation 2-48, $M_0 < (m_0)_A + (m_0)_B$. That is, the rest mass of a bound aggregate of particles must be less than the sum of the rest masses of the individual particles when separated. In principle, it is then possible to calculate the binding energy E_b merely by knowing the rest mass of the system as a whole and the rest masses of its constituents. Only for the case of nuclear forces is the binding energy between particles large enough to produce a measurable mass difference.

EXAMPLE   One kg (approximately 1 quart) of water at 100°C is converted into vapor at 100°C. (a) What is the total binding energy between the water molecules in the liquid state (assuming no binding between the molecules in the gaseous state)? (b) What is the average binding energy per molecule? (c) What is the fractional gain in rest mass?

(a) The latent heat of vaporization for water is

$$540 \text{ cal/gm} = 5.4 \times 10^5 \text{ cal/kg}$$

Therefore, $E_b = 5.4 \times 10^5 \text{ cal/kg} = 2.3 \times 10^6 \text{ joule/kg}$

(b) Inasmuch as 18 gm (1 mole) of water contains 6.0×10^{23} molecules, there are $(6.0 \times 10^{23} \times 1000)/(18) = 3.3 \times 10^{25}$ molecules in one kg of water. Consequently, the average binding energy per molecule is

$$(E_b)/\text{molecule} = \frac{(2.3 \times 10^6 \text{ joule/kg})}{(3.3 \times 10^{25} \text{ molecules/kg})}$$

$$= 7.0 \times 10^{-20} \text{ joule/molecule} = 0.44 \text{ ev/molecule}$$

(c) By Equation 2-48, the mass of the vapor is greater than the mass of the water by

$$\frac{E_b}{c^2} = \frac{(23 \times 10^5)}{(3 \times 10^8)^2} = 2.6 \times 10^{-11} \text{ kg}$$

and the fractional gain in rest mass is

$$\frac{(2.6 \times 10^{-11}) \text{ kg}}{1 \text{ kg}} = 2.6 \times 10^{-9} \text{ per cent!}$$

We are free to choose the zero of *total kinetic and potential energy* of a

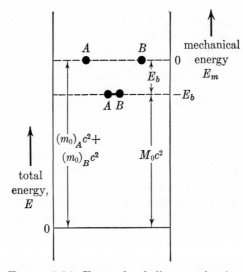

Figure 2-14. Energy-level diagram showing the energy of two particles when bound together, and when completely separated from one another and at rest.

system of particles at any convenient value. We call this the total *mechanical energy*, E_m. The most convenient choice for particles that attract one another is to make E_m of the system zero when the particles are all infinitely separated and at rest.

When particles are *bound* to one another, the energy E_m of the bound system is then *negative* because energy must be added to the bound system to separate the particles completely and to bring the energy of the system up to zero. Figure 2-14 shows the relationships among rest masses, total relativistic energy, binding energy, and mechanical energy.

EXAMPLE   The binding energy of an electron and a proton together to form a stable hydrogen atom is known experimentally to be 13.58 ev. This binding energy is also called the ionization energy since it represents the energy that must be added to a hydrogen atom to separate it into two oppositely charged particles. Inasmuch as the energy of a separated electron and proton, both at rest, is taken to be zero, when a stable hydrogen atom exists, its total mechanical energy is -13.58 ev. Using Equation 2-48 it is possible to compute the difference between the mass of the hydrogen atom, $(M_0)_H = 1.67 \times 10^{-27}$ kg, and the combined rest masses of the separated electron $(m_0)_e$ and proton $(m_0)_p$.

$$(m_0)_e + (m_0)_p - (M_0)_H = E_b/c^2 = \frac{13.58 \text{ ev}}{c^2}$$

The fractional change in the mass is

$$(E_b/c^2)/M_0 = \frac{(13.58 \text{ ev})(1.6 \times 10^{-19} \text{ joules/ev})}{(1.67 \times 10^{-27} \text{ kg})(3.0 \times 10^8 \text{ m/sec})^2}$$

$$E_b/(M_0 c^2) = 1.43 \times 10^{-8}$$

This fractional mass difference of slightly more than one part in 10^8 is much smaller than the experimental fractional error in the measurement of the masses of the hydrogen atom, proton, and electron, which is at best one part in 10^5. Therefore, in a reaction for which the binding energy is *several electron volts*—and all *chemical* reactions are of this order of magnitude—it is impossible to detect a change in the total mass of the system. A mass change can, however, be detected in *nuclear* reactions, for which the binding energy is typically *several million electron volts*.

We can illustrate the reduction of the general mass-energy conservation law into the separate classical conservation laws of mass and energy in a simple way. Again consider the example of two bound particles, A and B, which when separated and at rest have masses $(m_0)_A$ and $(m_0)_B$, and when bound have an aggregate rest mass M_0. When an amount of energy equal to the binding energy E_b is added to the bound system, the system separates into its components, A and B, both particles being at rest when separated. See Equation 2-48 and Figure 2-14.

We now add an amount of energy W to the bound system so that the particles A and B are not only separated but also have acquired kinetic energies, K_A and K_B. Clearly for this to occur, W must exceed the binding energy E_b. Then, the conservation of mass-energy gives

$$W/c^2 + M_0 = (m_0)_A + (m_0)_B + K_A/c^2 + K_B/c^2 \qquad [2\text{-}49]$$

We note that when $K_A = K_B = 0$, $W = E_b$, by comparing Equations 2-49 and 2-48. Figure 2-15 shows an energy diagram of this situation,

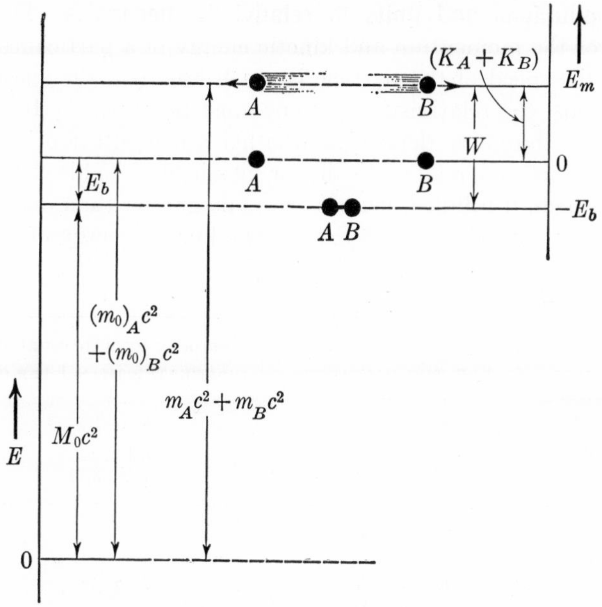

Figure 2-15. An energy-level diagram of two particles (a) bound together; (b) completely separated and at rest; and (c) completely separated and moving.

where E represents the *total* energy of the particles, including the rest-mass energy, and E_m is the *mechanical* energy, which is taken to be zero when the particles A and B are separated and at rest.

When the binding energy E_b is very much less than M_0c^2, as is the case for all except nuclear interactions, we see from the left-hand side of Figure 2-15 that

$$M_0 \simeq (m_0)_A + (m_0)_B$$

—the classical conservation-of-mass law! We recall that the fractional mass difference for the hydrogen atom was only one part in 10^8 and that this is typical of the fractional mass change in all *chemical* reactions. Thus it is hardly surprising that all experiments in chemistry show no perceptible deviations from the conservation-of-mass law.

The right-hand side of Figure 2-15 shows that

$$W = E_b + K_A + K_B$$

This is simply the classical conservation-of-energy law!

Thus, as we expect from the correspondence principle, the general mass-energy conservation law reduces in a natural way to the separate conservation laws of mass and of energy in the classical region.

2-10 Computations and units in relativistic mechanics The classical equations for the momentum and kinetic energy of a particle can be used only when the speed of the particle is much less than the speed of light; for high speeds, the relativistic relations must be invoked. It is useful to have a rule-of-thumb to determine whether a computation in a problem can safely be treated relativistically or classically. Table 2-1 shows the conditions which, if fulfilled, lead to errors no greater than one per cent in the computed momentum or energy. If the kinetic energy of a particle is

Table 2-1

	FOR THE CONDITION	ERROR IN RELATION BELOW IS NO GREATER THAN 1 PER CENT
Classical region	$K/E_0 < \frac{1}{100}$ or $v/c < \frac{1}{10}$	$K \simeq \frac{1}{2} m_0 v^2$ and $p \simeq m_0 v$
Extreme relativistic region	$E/E_0 > 7$ or $K/E_0 > 6$ or $v/c > 0.99$	$E \simeq pc$

a very small fraction of its rest energy, classical mechanics applies; on the other hand, if the total energy, or the kinetic energy, greatly exceeds a particle's rest energy, then the extreme relativistic relation $E = pc$ (which holds strictly only for $m_0 = 0$) can be applied.

A typical unit for specifying the kinetic energy, rest energy, or total energy of a particle in atomic and nuclear physics is the *electron volt* (ev), 1.60×10^{-19} joule, the energy gained by a particle having a charge e when accelerated by a potential difference of one volt. Other units related to the electron volt are:

$$\text{Kilo-electron volts} \quad = 1 \text{ Kev} = 10^3 \text{ ev}$$
$$\text{Million electron volts} = 1 \text{ Mev} = 10^6 \text{ ev}$$
$$\text{Billion electron volts} = 1 \text{ Bev} = 10^9 \text{ ev}$$

The masses of particles in atomic physics are frequently given in units of the *atomic mass unit*, or amu. *One atomic mass unit is defined as 1/16 the mass of a neutral oxygen atom (isotope 16).* Avogadro's number, 6.025×10^{23}, gives the number of atoms in 16 grams of atomic oxygen. Therefore,

$$1 \text{ amu} = \frac{1}{16} \left(\frac{16 \text{ gm}}{6.025 \times 10^{23}} \right) = 1.660 \times 10^{-27} \text{ kg}$$

The relation between the amu mass unit and the Mev energy unit is particularly useful. We find this from the general mass-energy relation, $E = mc^2$.

$$E = mc^2 = (1 \text{ amu})c^2$$

$$E = \frac{(1.66 \times 10^{-27} \text{ kg})(3.00 \times 10^8 \text{ m/sec})^2}{(1.60 \times 10^{-19} \text{ joule/ev})(10^6 \text{ ev/Mev})}$$

$$E = 931 \text{ Mev}$$

Therefore,

1 atomic mass unit = 1 amu = 931 Mev

This relation, 1 amu = 931 Mev, is worth memorizing because it is used frequently. It can be regarded as a conversion factor between the mass and energy units. The masses in amu and in Mev of the electron and proton are:

electron rest mass = 0.00055 amu = 0.51098 Mev
proton rest mass   = 1.00759 amu = 938.21 Mev

It is worth memorizing that the rest energy of an electron is approximately $\frac{1}{2}$ Mev, and that the rest energy of a proton is approximately 1 Bev. By convention, when a particle is described as, say, a 3.0 Mev particle, this means that the *kinetic* energy, not the total energy E, is 3.0 Mev.

EXAMPLE   What is the speed of a 2.0 Mev electron? The kinetic energy $K = 2.0$ Mev, and the rest energy $E_0 = 0.51$ Mev. Because $K > E_0/100$, a relativistic calculation must be made. Equation 2-36 shows that

$$mc^2 = \frac{m_0 c^2}{\sqrt{1 - (v/c)^2}}$$

or

$$E = \frac{E_0}{\sqrt{1 - (v/c)^2}} \qquad [2\text{-}50]$$

Equation 2-50 is often useful in relativistic computations. Solving for v/c gives

$$v/c = \sqrt{1 - E_0^2/E^2} = \sqrt{1 - E_0^2/(K + E_0)^2}$$

$$v/c = \sqrt{1 - (0.51/2.51)^2} = 0.98$$

EXAMPLE   What is the momentum of a 20.0 Mev electron? Because $K/E_0 = 20/0.51 \simeq 40$, Table 2-1 shows that we can use $E = pc$ with an error of less than 1 per cent; therefore,

$$p = E/c = (K + E_0)/c = (20.0 + 0.5) \text{ Mev}/c = 20.5 \text{ Mev}/c$$

The momenta of high-speed particles are frequently given in units of Mev/c.

2-11   Summary   Classical mechanics is invariant under the Galilean transformations and is in agreement with experiment for $v \ll c$.

The correct description of the physical universe is based on the principle of relativity: all the laws of physics, including that for the speed of light in a vacuum, are invariant in all inertial systems under the Lorentz transformations. The special theory of relativity was introduced by Einstein in 1905.

The Lorentz transformations:

$$x_2 = \frac{x_1 - vt_1}{\sqrt{1 - (v/c)^2}}$$

$$y_2 = y_1$$
$$z_2 = z_1$$

$$t_2 = \frac{t_1 - (v/c^2)x_1}{\sqrt{1 - (v/c)^2}}$$

No signal or particle can move at a speed exceeding the speed of light in vacuum.

Relativity physics is supported by the following experimental observations: (1) Michelson-Morley (and related) experiments—the invariance of c. (2) Mass variation with speed. (3) Direct observation of time dilatation in meson decay. (4) The equivalence of mass-energy.

The transition from relativity physics to classical physics can be written symbolically as:

$$\underset{v/c \to 0}{\text{Limit}} \text{ (relativity physics)} = \text{(classical physics)}$$

	RELATIVISTIC FORM	CLASSICAL FORM
Transformation equations	Lorentz transformations	Galilean transformations
Length	$L = L_0\sqrt{1 - (v/c)^2}$	$L = L_0$
Time	$T = \dfrac{T_0}{\sqrt{1 - (v/c)^2}}$	$T = T_0$
Mass-energy	$\dfrac{m}{m_0} = \dfrac{E}{E_0} = \dfrac{1}{\sqrt{1 - (v/c)^2}}$	$m = m_0$
Momentum	$p = mv$	$p_0 = m_0v$
Kinetic energy	$K = E - E_0 = (m - m_0)c^2$	$K = \frac{1}{2} m_0 v^2$
Newton's second law	$F = (d/dt)(mv)$	$F = (d/dt)(m_0v) = m_0a$

For a particle with a zero rest mass ($m_0 = 0$), but with a finite energy and momentum,

$$v = c, \quad \text{and} \quad K = E = pc = mc^2$$

The rest mass of a system of bound particles is less than the sum of the rest masses of the separated particles by E_b/c^2, where E_b is the total binding energy of the system.

$$1 \text{ amu} = 931 \text{ Mev}$$
$$\text{electron rest energy} \simeq \tfrac{1}{2} \text{ Mev}$$
$$\text{proton rest energy} \simeq 1 \text{ Bev}$$

REFERENCES

French, A. P., *Principles of Modern Physics*. New York: John Wiley & Sons, Inc., 1958. A detailed discussion of the Michelson-Morley and other related experiments, together with a discussion of relativistic geometry, is given in Chapter 6.

Gamow, G., *Mr. Tompkins in Wonderland; or, Stories of c, G, and h.* New York: The Macmillan Co., 1940. The remarkable experience of Mr. Cyril George Henry Tompkins in a universe in which the speed of light is 20 miles per hour.

Goldstein, H., *Classical Mechanics*. Reading, Massachusetts: Addison-Wesley Publishing Co., Inc., 1950. A thoroughly rigorous treatment of relativistic mechanics at a somewhat advanced level is given in Chapter 6.

Lieber, L. R., *The Einstein Theory of Relativity*. New York: Rinehart & Company, Inc., 1945. An elementary and lucid account; arguments are clearly stated and easy to follow.

PROBLEMS

2-1 An oxygen molecule (molecular weight, 32) initially traveling in the positive X direction with a speed of 1.00×10^3 m/sec makes a *perfectly elastic collision* with another oxygen molecule at rest. All velocities are with respect to an observer in the laboratory. After the collision, the first molecule is observed to be moving in the direction 45° above the positive X axis. Therefore, the second molecule must move in the direction 45° below the positive X axis (see Problem 1-2). (a) In this laboratory system compute the numerical values of the conserved quantities—total momentum (X and Y components) and total energy. (b) Using the Galilean transformations calculate the numerical values of these conserved quantities as measured by an observer moving in the positive X direction with a speed of 0.50×10^3 m/sec.

2-2 A plane flying horizontally at an altitude of 5000 ft and speed of 400 ft/sec releases a bomb with zero initial velocity with respect to the plane. (a) Write down the equations of motion (the equations giving the position and velocity as functions of the time) of the bomb relative to an observer in the plane. (b) By means of the Galilean transformations derive the equations of motion of the bomb relative to an observer on the ground.

2-3 A river 3 miles wide flows with a constant speed of 4 mi/hr. Three men leave simultaneously from the same point on the river bank. The first man rows 3 miles directly upstream and returns. His speed in still water is 5 mi/hr. The second man, who also rows 5 mi/hr in still water, crosses the river in such a direction that he always moves at right angles to the bank and upon returning he arrives at the starting point. The third man, walking at 5 mi/hr, travels 3 miles upstream along the bank and returns. (a) Find the total traveling time for each man. (b) Compare the time difference between the first two men as found above and as given by Equation 2-13. (c) Why are these different?

2-4 Prove that Newton's second law of motion is *not* invariant under a transformation between an inertial system and a second coordinate frame which has a constant acceleration with respect to the inertial system.

2-5 Show that if the length of a cylinder in Figure 2-4 and Figure 2-5 is contracted by a factor $[1 - (v/c)^2]^{1/2}$ along the direction of its motion with respect to the "ether," Δt_x of Equation 2-11 is equal Δt_y of Equation 2-12. (In 1893 G. F. Fitzgerald proposed this contraction as an *ad hoc* explanation of the null result of the Michelson-Morley experiment. A similar contraction appeared in the electromagnetic theory of the electron by H. A. Lorentz in 1895. Hence, this contraction became known as the *Lorentz-Fitzgerald contraction*.)

2-6 Solve algebraically for x_1, y_1, z_1, and t_1 in terms of x_2, y_2, z_2, and t_2 in Equation 2-26. This proves the statement that the inverse Lorentz transformations result from changing v to $-v$ and interchanging subscripts.

2-7 A man standing on the platform of a railroad station observes two trains approaching him from opposite directions at speeds of 0.9 c and 0.8 c, respectively. At what speed does one train move with respect to the other? (*Hint:* consider the railroad platform and one of the trains as the two inertial systems.)

2-8 * An inertial system S_2 is moving to the right with a speed $(c/2)$ with respect to another inertial system S_1. An observer in S_1 sees a particle moving in the positive Y direction with a speed $(c/\sqrt{3})$. (a) Calculate the speed and direction of the particle as seen by an observer in S_2. (b) If the rest mass of the particle is 2.0 kg, what is its mass as measured by each of the two observers above?

2-9 * (a) Show that if a particle is traveling at a speed c in some direction θ_1 with respect to the positive X_1 axis in the inertial system S_1, then its speed with respect to *any* other inertial system S_2 is also c. (b) Show that the direction θ_2 of the particle as measured in S_2 is given by

$$\tan \theta_2 = [1 - (v/c)^2]^{1/2}\{1 - [v/(c \cos \theta_1)]\}^{-1} \tan \theta_1$$

2-10 Using the Lorentz velocity transformations show that if $c^2 = \dot{x}_2^2 + \dot{y}_2^2 + \dot{z}_2^2$ is true in the inertial system S_2, then $c^2 = \dot{x}_1^2 + \dot{y}_1^2 + \dot{z}_1^2$ in the inertial system S_1.

2-11 An observer in S_2 holds a 1.00 m stick at an angle of 45° with respect to the positive X axis. If S_2 is moving to the right with a velocity of 0.98 c with respect to S_1, what is the length and angle of the meter stick as measured by S_1?

2-12 A student is given an examination to be completed in 50 minutes (by the professor's clock, always). The student and professor are moving at 0.98 c with respect to each other. How much time has elapsed on the professor's clock, *as measured by the student*, when the professor says, "Time is up"?

2-13 * Two observers approach each other with a relative velocity of 0.98 c. According to one of the observers the initial distance of separation of the two observers is 10 meters. (a) What is the distance of separation as measured by the second observer? (b) Calculate the time elapsed before the two observers meet, as measured by *each* observer.

2-14 How much energy (ev) is required to accelerate an electron from: (a) 0.500 c to 0.900 c? (b) 0.900 c to 0.990 c? (c) 0.990 c to 0.999 c?

2-15 (a) At what speed must a proton approach the Earth so that the Earth appears, to an observer at rest with respect to the proton, like a lima bean whose diameter is five times its thickness? (b) What is the proton's momentum with respect to the Earth?

2-16 Show that the ratio (v/c) of a particle having a rest energy E_0 and a total energy E is given by

$$(v/c) = [1 - (E_0/E)^2]^{1/2}$$

2-17 An electron and a proton are each accelerated from rest by a total potential difference of 5.0×10^8 volt. (a) What is the increase in mass of each particle? (b) What is the *fractional* increase in mass of each? (c) What is the final speed of each?

2-18 It is now possible to accelerate electrons to energies of approximately 1.0 Bev by an electron synchrotron. (a) What is the ratio of the electron's mass at 1.0 Bev to its rest mass? (b) What is the percentage difference between the speed of light and the speed of the electron?

2-19 * A high-energy accelerator accelerates a beam of electrons through a total potential difference of 1.0 billion volts. These electrons are then brought to rest by a metal target. If the average number of electrons striking the target is 6×10^{12} per sec, (a) What is the average electron current? (b) What is the average force exerted on the target by the electron beam?

2-20 At what fraction of the speed of light must a particle move so that its kinetic energy just equals its rest-mass energy?

2-21 What is the ratio of the relativistic mass to the rest mass of an electron when the electron's relativistic mass equals the rest mass of a proton?

2-22 * (a) What is the maximum speed a particle can move so that its kinetic energy can be written as $\frac{1}{2} m_0 v^2$ with an error in the kinetic energy no greater than one per cent? (b) What is the kinetic energy of an electron moving at this speed? (c) What is the kinetic energy of a proton moving at this speed?

2-23 * (a) What is the minimum speed of a particle such that its kinetic energy can be written as its total energy E, and therefore as pc, with an error in the total energy no greater than one per cent? Under these conditions, what is the kinetic energy of (b) an electron? (c) a proton?

2-24 With what energy, in Bev, must a proton approach the Earth at the Equator so that it is deflected in a path whose radius of curvature equals the Earth's radius? For simplicity, assume that the Earth's magnetic flux density B is constant and equal to 5.0×10^{-5} weber/m^2. (Protons of this energy are a primary part of the cosmic rays striking the Earth.)

2-25 * (a) What is the radius of curvature of a 400 Kev electron which enters a uniform magnetic field (0.24 weber/m^2) at right angles to the magnetic lines of force? (b) Calculate the mass of this electron in units of its rest mass.

2-26 (a) Show that the momentum of a particle can be written as $p = (1/c)(K^2 + 2E_0 K)^{1/2}$. (b) Show that this reduces to $(m_0 v)$ in the classical limit, and to (E/c) in the extreme relativistic region.

2-27 Prove that the slope of Figure 2-12 is the speed of the particle; that is, $dE/dp = v$.

2-28 (a) Derive the space contraction relation, Equation 2-29, assuming that observer S_2, rather than observer S_1, measures the lengths. (b) Derive the time dilatation relation, Equation 2-31, as measured by observer S_1, rather than observer S_2.

2-29 * An electron has a momentum of 4.0 Mev/c. (a) What is the kinetic energy of a proton having the same momentum? (b) What is the total energy of a zero-rest-mass particle having the same momentum?

2-30   Show that when a force F acts on a particle in the same direction as the velocity, Newton's second law becomes

$$F = \frac{m_0(dv/dt)}{[1 - (v/c)^2]^{3/2}}$$

2-31 * Show that a particle of rest-energy E_0, kinetic energy K, and electric charge q, will, when moving at right angles to a uniform magnetic field of flux density B, travel in a circle of radius $r = [K^2 + 2E_0K]^{1/2}/qcB$.

2-32   It is observed that electrons move in a circle of radius 0.40 m in a uniform magnetic field of flux density 0.80 weber/m². What is (a) the electron's momentum and (b) its kinetic energy?

2-33 * A 5.0 Mev electron makes a head-on elastic collision with a proton initially at rest. Show that: (a) The proton recoils with a speed approximately equal to $(2E_e/E_p)c$; (b) The fractional energy transferred from the electron to the proton is $(2E_e/E_p)$, where E_e is the *total* energy of the electron and E_p is the *rest* energy of the proton. (Hint: (a) since the electron's energy is much greater than its rest energy, it can be treated as an extreme relativistic particle, and (b) since the rest energy of the proton is much greater than the *total* energy of the electron, the proton can be treated classically.)

2-34   The total intensity of radiation from the Sun at the Earth's surface is 8.0 joule/cm²-min. Calculate the loss in the Sun's mass per second and the fractional loss in the Sun's mass in 10^9 years (approximately one-tenth of the age of the universe) resulting from its radiation. The distance from the Sun to the Earth is 93×10^6 mi and the Sun's mass (at present) is 2.0×10^{30} kg.

2-35   A rocket ship having a final pay-load rest mass of 2000 kg is accelerated to a speed of 0.98 c. (a) What minimum energy is required to accelerate the rocket ship to this speed? (b) How much equivalent mass does this represent? (c) What amount of nuclear fuel (assume one per cent conversion of mass to energy) would be needed to achieve this?

2-36   To separate a carbon monoxide molecule, CO, into carbon and oxygen atoms requires 11.0 ev. (a) What is the fractional change in mass of a CO molecule when it is broken into the atoms C and O? (b) What is the binding energy (ev) per molecule?

2-37   The latent heat of fusion of water is 80 cal/gm. Compute the average binding energy (ev) per molecule between the molecules of water in the solid state at 0°C as compared with that between molecules of water in the liquid state at 0°C.

2-38   Compute the difference in the average binding energy (ev) per molecule between ice molecules at $-40°C$ and water vapor molecules at $100°C$. The specific heats of ice and water are 0.5 and 1.0 cal/gm-C°, respectively, and the latent heats of fusion and vaporization are 80 and 540 cal/gm, respectively.

2-39   Twin ships, one powered by oil (4×10^6 cal/lb) and the other powered by nuclear fuel (1.5×10^{26} Mev/kg), travel an equal distance. What is the ratio of the fuel used by the two ships?

2-40   Use the binomial expansion to show that the expression $\frac{1}{2} mv^2$ does *not* give the relativistic kinetic energy ($E - E_0$), where m is the relativistic mass.

T H R E E

QUANTUM EFFECTS: THE PARTICLE ASPECTS OF ELECTROMAGNETIC RADIATION

3-1 Quantization in classical physics The theory of relativity and the quantum theory constitute the two great theoretical foundations of twentieth-century physics. Just as the theory of relativity leads to new insights into the nature of space and time and to profound consequences in mechanics and electromagnetism, so too does the quantum theory lead to drastically new modes of thought which form the basis of understanding atomic and nuclear structure. Some aspects of the quantum description of nature are not, however, totally new, and indeed are to be found in classical physics.

In the study of the physical world we find two general kinds of physical quantities: (1) quantities which have a continuum of values; and (2) quantities which are *quantized*—that is, quantities which are restricted to certain discrete values.

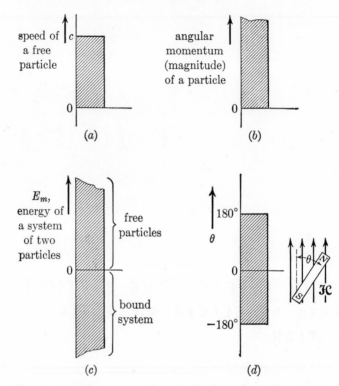

Figure 3-1. Some examples of classical physical quantities having a continuum of allowed values.

Figure 3-1 shows several examples of continuous, or non-quantized, physical quantities:

(a) The speed of a free particle, which can range from zero through any value up to the speed of light.

(b) The magnitude of the angular momentum of a particle, which can take on any value from zero to infinity.

(c) The mechanical energy of a system of two particles, which can assume any *negative* value when the particles are *bound* together ($E_m < 0$), and any *positive* value when the particles are *free* ($E_m > 0$).

(d) The angle between the direction of a magnet and an external magnetic field, which may vary between $-180°$ and $+180°$.

In Figure 3-2 are shown examples of physical quantities which are quantized.

(a) The observed rest masses of atoms, which do not occur in a continuous range but rather display an atomicity, or quanti-

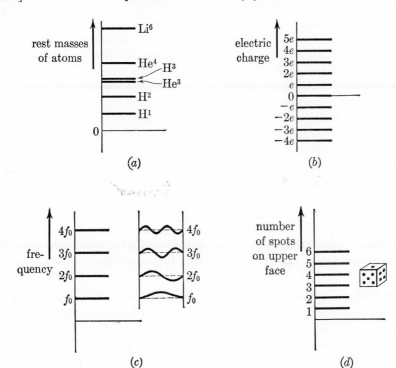

Figure 3-2. Some examples of classical physical quantities having quantized values.

zation. This was first perceived in the fundamental studies on chemical combination which led to the atomic theory of Dalton. The masses of atoms occurring in nature are now known with great precision, and it is interesting to note that, whereas the atomic masses are *nearly* in the ratio of integers, they are *not precisely* in integral ratios. One of the principal tasks of nuclear physics, we shall see, is to explain on some fundamental basis these departures from integral ratios.

(b) The atomicity of electric charge shows that the total charge of any body is precisely an integral multiple, positive or negative, of the fundamental electric charge in nature, the charge of an electron, e. This quantization of charge is clearly revealed in the chemical idea of valence and in the laws of electrolysis, and it was most directly demonstrated in the oil-drop experiments of R. A. Millikan, in which the charge of the electron was measured.

(c) A particularly interesting manifestation of quantization in classical physics is to be found in situations involving standing waves and resonance. Thus, the frequency of oscillation of a resonating vibrating string, fixed at both ends, can be only integral multiples of the lowest, or fundamental, frequency of oscillation, f_0. The fundamental frequency f_0 is determined in turn by the physical properties of the string and its length. The wave on the string is repeatedly reflected from the boundaries, or fixed ends, and (so to speak) constructively interferes with itself to produce standing waves. The standing-wave condition can be achieved, however, only if the distance between the end points is precisely an integral multiple of half-wavelengths. It was argued in Section 1-8 that the frequency of a wave is precisely determined only when the wave has an infinite extension in space. This argument is still valid, and the frequency of a resonant standing-wave pattern is precisely determined, because the wave is folded on itself an infinite number of times.

(d) There are many everyday examples of quantized quantities. One familar illustration is a pair of dice, either one of which when rolled can have only 1, 2, 3, 4, 5, or 6 spots appear on the upper face.

The quantum theory is based in large measure on the discovery that certain quantities which in classical physics had been regarded as continuous are, in fact, quantized. Historically, the quantum theory had its origins in the theoretical interpretation of electromagnetic radiation from a black body (a perfect absorber and radiator). Near the end of the nineteenth century it was found that the experimentally observed variation in wavelength of the intensity of electromagnetic radiation from a black body was in disagreement with the theoretical expectations of classical electromagnetism. Max Planck, formulator of the quantum theory, showed in 1900 that a revision of classical ideas through the introduction of quantization led to satisfactory agreement between experiment and theory. Because a detailed analysis of black-body radiation (Section 11-6) involves rather sophisticated arguments, we shall introduce the quantum concepts through the much simpler, and in many ways more compelling, considerations that arise in the phenomenon of the photoelectric effect.

3-2   The photoelectric effect   The photoelectric effect was discovered by Heinrich Hertz (1887) during the course of experiments the primary intent of which was confirmation of Maxwell's theoretical prediction (1864) of the existence of electromagnetic waves produced by oscillating electric currents.

The photoelectric effect is one of several processes for removing electrons from a metal surface. Electrons are released from metals in the following ways:

Thermionic emission—heating the metal, thereby giving thermal energy to the electrons;

Secondary emission—transfer of the kinetic energy from particles which strike the surface to the electrons in the metal;

Field emission—extraction of electrons from the metal by a strong electric field; and

Photoelectric emission—with which we are concerned here.

The photoelectric effect occurs when light shines on a clean metal surface, and electrons are released from this surface. Electrons are free to move about within the surface of a metal but are bound to the metal as a whole. One can most simply interpret the photoelectric effect, therefore, as arising when a light beam supplies any electron with an amount of energy which equals or exceeds the energy with which the electron is bound to the surface, thus allowing that electron to escape. A more detailed study of the photoelectric effect, however, requires a knowledge, based on experiment, of how the several variables involved in photoelectric emission are related to one another. These variables are: the frequency of the monochromatic light, ν; the intensity of the light beam, I; the photoelectric current, i; the kinetic energy of the emitted photoelectrons, $\frac{1}{2} m_0 v^2$ (we shall see shortly that the use of the classical kinetic-energy formula is justified); and the chemical identity of the surface from which photoelectrons emerge.

Figure 3-3 shows a schematic diagram of an experimental arrangement for studying important aspects of the photoelectric effect. Monochromatic light shines on the metal surface, which is the *anode*, enclosed in a vacuum

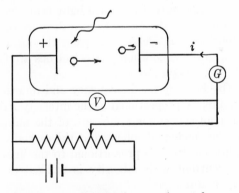

Figure 3-3. Schematic experimental arrangement for studying the photoelectric effect.

tube. An evacuated tube is used so that collisions between photoelectrons and gas molecules are essentially eliminated. When photoelectrons are emitted, some travel toward the cathode and, upon reaching it, comprise the current flowing in the circuit (conventional current, as shown in

Figure 3-3). The negatively charged cathode repels the photoelectrons; and, when the work done on an electron by the retarding electrostatic field of potential difference V equals the initial kinetic energy of a photoelectron, the photoelectron is brought to rest just in front of the cathode. Thus, $eV = \frac{1}{2} m_0 v^2$, where v is the speed of the photoelectron as it leaves the anode surface, and V is the potential difference that stops the electron of rest-mass m_0 and charge e. Photoelectrons leave the anode (the photo-surface) with a variety of kinetic energies. When the most energetic electrons, having a speed v_{max}, are brought to rest before reaching the cathode by a sufficiently large potential difference V_0, all photoelectrons are stopped, no photocurrent flows, and $i = 0$. Therefore,

$$eV_0 = \frac{1}{2} m_0 v_{max}^2 \qquad [3\text{-}1]$$

For still higher retarding potential differences, all electrons are turned back before reaching the cathode.

We first list below the results of experiment. We shall then give the results that might be expected on the basis of the classical theory of electromagnetism; it will be seen that the experimental results strongly disagree with the classical expectations. Finally, we shall see how the photoelectric effect can be understood on the basis of a quantum interpretation.

EXPERIMENTAL RESULTS   The results of experiment for the photoelectric effect are summarized in Figure 3-4.

(a) When photoelectrons are emitted, the photocurrent begins almost instantaneously, even for a light beam having an intensity as small as 10^{-10} watts/m^2 (the intensity at a distance of 200 miles from a 100-watt light source). The delay time from the instant that the light beam is turned on to the time when photoelectrons are first emitted is no greater than 10^{-9} sec. (Figure 3-4a.)

(b) For any fixed frequency and retarding potential, the photocurrent i is directly proportional to the intensity I of the light beam. Inasmuch as the photocurrent is a measure of the number of photoelectrons released at the anode and collected at the cathode, this implies that the number of photoelectrons is proportional to the light intensity. (The variation in photocurrent with intensity is utilized in practical photoelectric devices.) (Figure 3-4b.)

(c) For a constant frequency v and light intensity I, the photocurrent decreases with increasing V and finally reaches zero when the retarding potential is equal to V_0. With a small retarding potential, the low-speed, low-energy photoelectrons are brought to rest and no longer contribute to the photocurrent; when the retarding potential is equal to V_0, even the most energetic photoelectrons have been brought to rest, following Equation 3-1, and $i = 0$. (Figure 3-4c.)

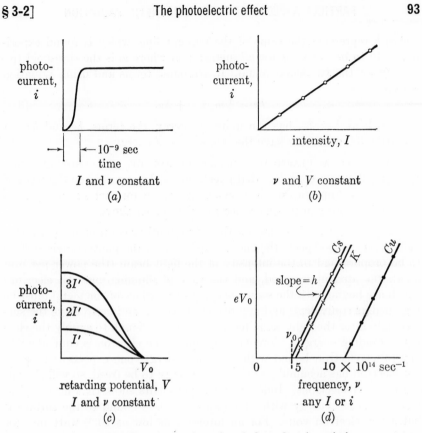

Figure 3-4. Experimental results of photoelectric emission.

(d) For any particular surface, the value of the stopping potential V_0 depends on the frequency of the light, but V_0 is independent of the light intensity and, therefore from (b), independent also of the photocurrent. Figure 3-4d shows the results for the three metals: cesium, potassium, and copper. For each material there is a well-defined frequency, ν_0, the *threshold frequency*, which must be exceeded for photoemission to occur at all; that is, no photoelectrons are produced, however great the light intensity, unless $\nu > \nu_0$. For most metals the threshold frequency for photoemission lies in the region of ultraviolet light. A typical stopping potential is several volts. The photoelectrons have energies of several electron volts for visible or ultraviolet light; therefore, we are justified in using the classical kinetic energy formula for the photoelectrons.

For any one particular metal, the experimental results of Figure 3-4d can be represented by the equation for a straight line

$$eV_0 = h\nu - h\nu_0$$

where h represents the slope of the straight line, which is found experimentally to be the *same* for *all* metals, and where ν_0 is the threshold frequency for the particular metal. Rearranging terms and using Equation 3-1 gives

$$h\nu = \tfrac{1}{2} m_0 v_{max}^2 + h\nu_0 \qquad [3\text{-}2]$$

Inasmuch as $\tfrac{1}{2} m_0 v_{max}^2$ has the units of energy, the terms, $h\nu$ and $h\nu_0$, in Equation 3-2 must also have the dimensions of energy.

ATTEMPTS AT A CLASSICAL INTERPRETATION OF THE PHOTOELECTRIC EFFECT   We now consider what effects should be expected on the basis of the classical properties of electromagnetic waves for each of the four experimental results in the photoelectric effect given above.

(a) Because of the (apparently) continuous nature of light waves, one expects the energy absorbed on the photoelectric surface to be proportional to the intensity of the light beam (the power per unit area), the area illuminated, and the time of illumination. All electrons which are bound with the same energy in the surface of the metal must be regarded as equivalent, and any one electron will be free to leave the surface only after the light beam has been on long enough to supply the electron's binding energy. Moreover, since any one electron is equivalent to any other electron bound with the same energy, one expects that when one electron has accumulated sufficient energy to be freed, so will a number of other electrons. Independent experiments show that in a typical metal the least energy with which an electron is bound to the surface is but a few electron volts. For an intensity as low as 10^{-10} watt/m^2—for which delay times no longer than 10^{-9} sec have been observed—a conservative calculation (see Problem 3-2) shows that no photoemission could be expected until at least several hundred hours have elapsed! Clearly, the classical theory is impotent to account for the essentially instantaneous photoelectric emission. (Figure 3-4a.)

(b) Classical theory predicts that as the light intensity is increased, so is the energy absorbed by electrons in the surface. Hence, the number of photoelectrons emitted, or the photocurrent, is expected to increase proportionately with light intensity. The classical theory agrees with the experimental result. (Figure 3-4b.)

(c) The results of these observations show first that there is a distribution in the speeds, or energies, of the emitted photoelectrons; this distribution in photoelectron energies is, in itself, not incompatible with classical theory because the distribution of energies can be attributed to the varying degrees of binding of electrons in the surface, or to the varying amounts of energy extracted by electrons from the incident light beam. The fact, however, that there is a very well defined stopping potential V_0 for a given frequency, independent of the intensity, indicates that the maximum energy of released electrons is in no way dependent on the total amount of energy reaching the surface per unit time. Classical theory predicts no such effect. (Figure 3-4c.)

(d) The existence of a threshold frequency for a given metal, below which no photoemission occurs, however great the light intensity, is completely inexplicable in classical terms. From the classical point of view, the primary circumstance that determines whether or not photoemission will occur is the energy reaching the surface per unit time (or the intensity), but *not* the frequency. Further, the appearance of a single constant h that relates the maximum energy of photoelectrons to the frequency, through Equation 3-2, for any material cannot be understood in terms of any constants of classical electromagnetism. (Figure 3-4d.)

In short, *classical electromagnetism cannot give a reasonable basis for understanding the experimental results* (a), (c), *and* (d).

QUANTUM INTERPRETATION OF THE PHOTOELECTRIC EFFECT An adequate understanding of the photoelectric effect is to be found only by the quantum theory. Albert Einstein first applied the quantum theory to the nature of electromagnetic radiation in 1905, and this led to a satisfactory explanation of the photoelectric effect. According to the quantum theory, the apparently continuous electromagnetic waves are quantized and consist of discrete *quanta*, called *photons*. Each photon has an energy E which depends only on the frequency (or wavelength) and is given by

$$E = h\nu = h(c/\lambda)$$ [3-3]

The constant h is, in fact, the h appearing in Equation 3-2, which summarizes the results of experiments on the photoelectric effect. This fundamental constant of the quantum theory is called *Planck's constant* because its value was first determined and its significance first appreciated by Planck in 1900 in the interpretation of blackbody radiation. The value of Planck's constant is

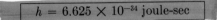

$$h = 6.625 \times 10^{-34} \text{ joule-sec}$$

Planks Constant

According to the quantum theory, a beam of light of frequency ν consists of particle-like photons, each having an energy $h\nu$. A single photon can interact only with a single electron in the metal surface of a photoemitter; it cannot share its energy among several electrons. Inasmuch as photons travel with the speed of light, they must, on the basis of relativity theory, have zero rest-mass and an energy which is entirely kinetic. When a particle with a zero rest-mass ceases to move with a speed c, it ceases to exist; as long as it exists, it moves at the speed of light. Thus, when a photon strikes an electron bound in a metal and no longer moves at the speed c, it relinquishes its entire energy $h\nu$ to the single electron it strikes. If the energy the electron gains from the photon exceeds the binding energy of the electron to the metal surface, the excess energy becomes the kinetic energy of the photoelectron.

We are now prepared to interpret, on the basis of the quantum theory, the experimental results of the photoelectric effect, which we now take in reverse order for convenience.

(d) The terms in Equation 3-2 now have a simple meaning in terms of the energies of the photon and photoelectron.

[3-2]
$$h\nu = \tfrac{1}{2}\, m_0 v^2_{max} + h\nu_0$$

The left expression in this equation shows the energy supplied to any electron. Those electrons which are least tightly-bound leave the surface with a maximum kinetic energy. On the right side of Equation 3-2 is the energy gained by the electron from the photon; namely, the kinetic energy and the binding energy. The binding energy for the least tightly-bound electrons to the metal surface is often represented by ϕ and called the *work function*. Therefore,

$$\phi = h\nu_0 \qquad\qquad\qquad\qquad [3\text{-}4]$$

and Equation 3-2 can be written in the form

$$h\nu = \tfrac{1}{2}\, m_0 v^2_{max} + \phi \qquad\qquad [3\text{-}5]$$

The value of ϕ for a particular material, determined from the photoelectric effect, agrees with the value of the work function obtained through other experiments based on different physical principles.

An electron bound with an energy ϕ can be released only if a single photon supplies at least this much energy; that is, if $h\nu > \phi = h\nu_0$, or if $\nu > \nu_0$. Figure 3-4d takes on new meaning in that the ordinate can now be interpreted in terms of the photon energy, as shown in Figure 3-5. (Figure 3-5 should be compared with the right-hand side of Figure 2-15, where for the photoelectric effect, $W = h\nu$ and $E_b = \phi$.)

(c) A well-defined maximum kinetic energy for photoelectrons exists for any given frequency because the frequency of the electromagnetic radiation determines precisely the photon energy ($E = h\nu$).

(b) The intensity of a monochromatic electromagnetic wave takes on a new meaning; it is, from the quantum point of view, the energy of one photon times the number of photons crossing a unit area per unit time. An increase in the intensity of a light beam means, therefore, a proportionate increase in the number of photons striking the metal surface. It is expected then that the number of photoelectrons, or the photocurrent, i, will be proportional to I.

(a) Photoemission occurs with no appreciable delay because even for the smallest intensity whether an electron is released de-

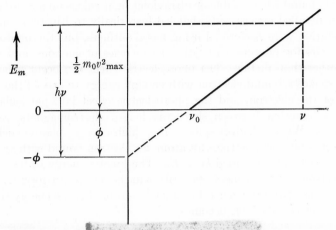

Figure 3-5. Maximum kinetic energy of photoelectrons as a function of the frequency of the incident photons for a particular material.

pends not upon its accumulating energy (which can be spread among many electrons) but simply upon the fact of being hit by a photon which, upon stopping, gives all of its energy to the single, struck electron.

Table 3-1 summarizes the results of experiment, the classical interpretation, and the quantum interpretation for each of the four effects shown in Figure 3-4.

Our discussion of the photoelectric effect has thus far been in terms of

Table 3-1

EFFECT	EXPERIMENT	CLASSICAL ELECTROMAGNETISM	QUANTUM THEORY
(a) (Figure 3-4a)	Essentially instantaneous photoemission $(10^{-9}$ sec)	Emission only after several hundred *hours* $(10^6$ sec)	A single photon gives its energy to a single electron essentially instantaneously
(b) (Figure 3-4b)	$I \propto i$	Energy/area/time $\propto i$	$I \propto$ number of photons $\propto i$
(c) (Figure 3-4c)	A well-defined $\frac{1}{2} m_0 v_{max}^2$, dependent only on ν	Inexplicable	A photon gives all its energy to a single electron
(d) (Figure 3-4d)	A threshold for photoemission, independent of I and i $h\nu = \frac{1}{2} m_0 v_{max}^2 + h\nu_0$	Inexplicable	Photon energy $= h\nu$ Work function $= \phi = h\nu_0$

the effects found when visible or ultraviolet light shines on a *metal* surface. The first detailed experiments that led historically to Einstein's quantum interpretation were performed using metal surfaces, but the photoelectric effect occurs for photons with other frequencies or energies, and for materials other than metals. The photoelectric effect can occur whenever a photon strikes a bound electron with enough energy to exceed the binding energy of the electron, and the photoelectric effect is a particularly important interaction between short-wavelength electromagnetic radiation and atoms. When a high-frequency (i.e., high-energy) photon such as an x-ray or a gamma ray, strikes an atom, an electron bound with an energy E_b can be released, provided $h\nu > E_b$. The kinetic energy of the released photoelectron must, in general, be written in the relativistic form $(E - E_0)$, or $(mc^2 - m_0c^2)$; and the general form of Equation 3-2, the energy equation of the photoelectric effect, becomes

$$h\nu = (E - E_0) + E_b$$
$$h\nu = (mc^2 - m_0c^2) + E_b$$

[3-6]

The photoelectric effect thus provides an indirect method for measuring the energy of a photon; for if the photoelectron's kinetic energy $(mc^2 - m_0c^2)$ is measured and the binding energy E_b is known on some other basis, $h\nu$ can be computed from Equation 3-6. Conversely, if $h\nu$ and $(mc^2 - m_0c^2)$ are measured, then E_b can be determined.

It is worth noting here that the photoelectric effect is only one of several ways in which photons can be removed from a beam of electromagnetic radiation; the photoelectric effect can occur simultaneously with and compete with the processes of the Compton effect and pair production, which will be discussed in detail in Sections 3-4 and 3-5.

The fundamental insight into the nature of electromagnetic radiation which the photoelectric effect provides is the quantization of electromagnetic waves, or the existence of photons. One can properly speak of the quantization of electromagnetic waves because the radiation can be regarded as a collection of particle-like photons, each of energy $h\nu$. When the frequency of the radiation is specified to be ν, the photon can have but one energy, $h\nu$.

The total energy of a beam of monochromatic electromagnetic radiation is always precisely an integral multiple of the energy $h\nu$ of a single photon, as is shown in Figure 3-6.

This granularity, or atomicity, of electromagnetic radiation is not conspicuous in ordinary observations because of the very small energy of any one photon, and because the number of photons in a light beam of moderate intensity is enormous. The situation here is rather like that occurring in the molecular theory: the molecules are so small, and their numbers so great, that the molecular structure of all matter is disclosed only in observations of considerable subtlety.

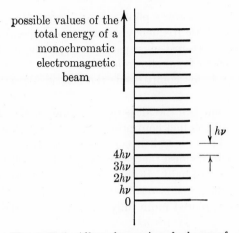

possible values of the
total energy of a
monochromatic
electromagnetic
beam

$h\nu$

$4h\nu$
$3h\nu$
$2h\nu$
$h\nu$
0

Figure 3-6. Allowed energies of a beam of
monochromatic electromagnetic radiation.

Whereas in Figure 1-15 the electromagnetic spectrum was shown simply
as a function of frequency, we now see that, from the point of view of the
quantum theory, this spectrum also represents the energy per photon,

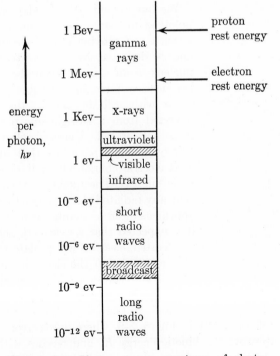

energy
per
photon,
$h\nu$

1 Bev —

1 Mev —

1 Kev —

1 ev —

10^{-3} ev —

10^{-6} ev —

10^{-9} ev —

10^{-12} ev —

gamma
rays

X-rays

ultraviolet

visible

infrared

short
radio
waves

broadcast

long
radio
waves

proton
rest energy

electron
rest energy

Figure 3-7. Photon energy spectrum of electromagnetic radiation.

which is smallest for radiowave photons and largest for gamma-ray photons. The electromagnetic frequency spectrum corresponds exactly to the energy spectrum of a zero-rest-mass particle, or photon, whose energy can extend from zero to infinity, as shown in Figure 2-11. Figure 3-7 shows the photon spectrum and, for comparison, the rest energies of the electron and proton.

As we saw in Section 1-8, the ideas of wave and particle are apparently mutually incompatible, in fact contradictory. The fact that, in the photoelectric effect, light behaves as if it consisted of particles or photons does not mean that we can dismiss the incontrovertible experimental evidence for the wave properties of light. Both descriptions must be accepted. The way in which this dilemma is resolved will be postponed (Section 4-4) until after we have explored more fully the quantum attributes of light.

3-3 X-ray production and Bremsstrahlung

In the photoelectric effect a photon transfers all of its electromagnetic energy to a bound electron; the photon's energy appears as the binding energy and kinetic energy of the photoelectron. The inverse effect is that in which an electron loses kinetic energy and, in so doing, creates a photon. This process is most clearly illustrated in the production of x-rays.

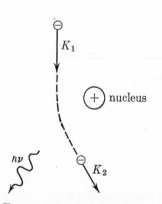

Figure 3-8. *Bremsstrahlung* collision between an electron and a positively charged nucleus with the emission of a single photon.

We first consider the fundamental process involved in x-ray production, which occurs when a fast moving electron comes close to and is deflected by the positively charged nucleus of an atom. When the electron experiences a force in consequence of its collision with a heavy atom, the electron is deviated from a straight-line path; that is, it is accelerated. Classical electromagnetic theory predicts (page 23) that any accelerated electric charge will radiate electromagnetic energy. The quantum theory predicts that any radiated electromagnetic energy will appear as discrete quanta, or photons. Thus, it is expected that a deflected, and therefore accelerated, electron will radiate one or more photons, and that the electron will leave the site of the collision with a smaller kinetic energy. The radiation produced in such a collision is often referred to as *Bremsstrahlung* ("braking radiation" in German). A *Bremsstrahlung* collision is shown schematically in Figure 3-8, where an electron approaches the deflecting atom with a kinetic energy K_1 and recedes with a kinetic energy K_2 after having produced a single photon of energy $h\nu$. The conservation-of-energy law requires that

$$\boxed{K_1 - K_2 = h\nu}$$ [3-7]

Because the mass of an atom is at least 2000 times greater than the electron's mass, we have neglected the very small energy of the recoiling atom. It is important to note that whereas classical electromagnetic theory predicts continuous radiation throughout the time that the electron is accelerated, the quantum theory requires the radiation of single, discrete photons. The *Bremsstrahlung* process is clearly illustrated in the production of x-ray photons.

X-rays were discovered and first investigated by Wilhelm Roentgen in 1895, who assigned this name because the true nature of the x-radiation was at first unknown. X-rays are now known to consist of electromagnetic waves, or photons, having wavelengths centered around 10^{-10} m = 1 Ångstrom unit = 1 Å (see Figure 1-15). It has been experimentally confirmed that x-rays exhibit the wave phenomena of interference, diffraction, and polarization; because x-rays pass readily through many materials that are opaque to visible light, and because a typical x-ray wavelength is far shorter than the wavelengths of visible light, these experiments require considerable ingenuity. We shall postpone discussion of the measurement of x-ray absorption and intensity to Section 3-7, and the measurement of x-ray wavelengths to Section 4-3; our chief concern here shall be with the energy characteristics of x-ray production.

The essential parts of a simple x-ray tube are shown in Figure 3-9. Electric current through the filament F heats the cathode C, and the

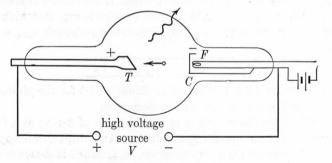

Figure 3-9. Essential parts of an x-ray tube.

electrons in the cathode are supplied with enough kinetic energy to overcome their binding to the cathode surface and thus are emitted by thermionic emission. The electrons are then accelerated through a vacuum by a large electrostatic potential difference V, typically several thousand volts, and strike the target T, which is the anode. In going from the cathode to the target each electron gains a kinetic energy K, just before striking the target, which is given by

$$K = Ve$$

where e is the electron charge. We have neglected the electron's kinetic energy at the cathode because it is typically much less than Ve. When the electron enters the target, it acquires an additional energy ϕ, the binding energy of the electron to the target surface; because ϕ is always a *few* electron volts, whereas K is at least several *thousand* electron volts, we can also properly ignore the binding energy ϕ.

Upon striking the target, the electrons are decelerated and essentially brought to rest. Each electron loses its kinetic energy $K = Ve$ because of its impact with the target. Most (around 98 per cent) of this energy appears as heat in the target, but there is in addition the production of electromagnetic radiation through the *Bremsstrahlung* process. Any electron striking the target may make a number of *Bremsstrahlung* collisions with atoms in the target, thereby producing a number of photons. The *most* energetic photon is produced by an electron whose *entire* kinetic energy is converted into the electromagnetic energy of a *single* photon when the electron is brought to rest in a single collision. Thus, $K_1 = Ve$ and $K_2 = 0$, and Equation 3-7 becomes

$$Ve = K = h\nu_{max}$$

where ν_{max} is the maximum frequency of x-ray photons produced. More often, electrons lose their energy at the target by heating it or by producing two or more photons, the sum of whose frequencies will then be less than ν_{max}. Thus a distribution in photon energies is expected with a well-defined maximum frequency ν_{max}, or minimum wavelength $\lambda_{min} = c/\nu_{max}$, given by

$$\boxed{K = h\nu_{max} = hc/\lambda_{min} = Ve} \qquad [3\text{-}8]$$

Note that Equation 3-8 is equivalent to Equation 3-5 for the photoelectric effect when the binding energy term is neglected.

Figure 3-10 shows the variation in the intensity of x-rays as a function of frequency for typical operating conditions. An abrupt cutoff appears at the limit of the *continuous* x-ray spectrum, ν_{max}, which is determined only by the accelerating potential V of the x-ray tube. The value of hc/e can be determined with considerable precision, using Equation 3-8, by simultaneous measurements of λ_{min} and V; the value obtained for Planck's constant h agrees completely with values deduced from the photoelectric effect and other experiments. Superimposed on the continuous spectrum are sharp increases in the intensity, or peaks, whose wavelengths are

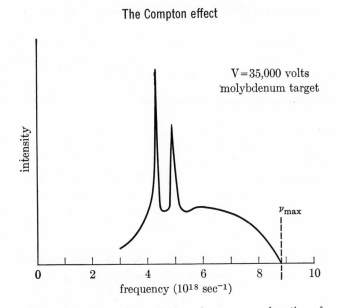

Figure 3-10. Intensity variation of x-rays as a function of frequency.

characteristic of the target material; the explanation of these *characteristic x-ray lines* is to be found in the quantum description of the atomic structure of the target atoms, and will be given in Chapter 7. When the accelerating voltage V is changed, but not the target material, the limit of continuous x-ray spectrum changes but the characteristic x-ray frequencies remain unchanged; conversely, when the target material is changed, but not the accelerating voltage, the characteristic x-ray spectrum changes, but the limit of the continuous x-ray spectrum remains unchanged.

It is found that appreciable x-ray production occurs only if the accelerating potential V is of the order of 10,000 volts or higher. Equation 3-8 shows that for $V = 1.0 \times 10^4$ volts, $\lambda_{min} = 1.2$ Å. Even at 10 kilovolts, somewhat less than 1 per cent of the total energy appears as electromagnetic radiation, the remainder appearing as heat in the target.

3-4 The Compton effect In the photoelectric effect a photon gives all of its energy to an electron. It is also possible for a photon to give only part of its energy to a charged particle. This type of interaction between electromagnetic waves and a material substance is the *scattering* of the waves by the charged particles of the substance. The quantum theory of the scattering of electromagnetic waves is known as the *Compton effect*. We shall first review briefly, however, the classical theory of the scattering of electromagnetic waves by charged particles.

When a monochromatic electromagnetic wave impinges upon a free, charged particle whose size is much less than the wavelength of the radiation, the charged particle will be acted upon principally by the electric field of the wave. Under the influence of this changing electric force, the particle will oscillate in simple harmonic motion at the same frequency as that of the incident radiation (see Figure 3-11). We have seen (Section 1-4)

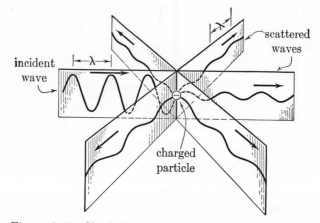

Figure 3-11. Classical scattering of electromagnetic radiation by a charged particle.

that the classical theory of an oscillating charged particle predicts: an accelerated charged particle radiates electromagnetic energy in all directions, the intensity being greatest in the plane perpendicular to the direction of motion of the oscillating charge and zero along the line of oscillation. Furthermore, the frequency of the radiated electromagnetic waves is the same as the frequency of oscillation of the charged particle. The classical theory predicts, therefore, that the scattered radiation will have the *same* frequency as that of the incident radiation. The charged particle plays the role of transfer agent, absorbing energy from the incident beam and re-radiating this energy at the same frequency (or wavelength), but scattering it in all directions. The net energy change of the scattering particle is zero, since it reradiates the electromagnetic energy at the same rate as it absorbs energy from the incident beam. The classical theory of scattering is in agreement with experiment for wavelengths of visible light and all other longer wavelength radiation. A simple example of the unchanged frequency of coherent scattered radiation is this: light reflected from a mirror (a collection of scatterers) undergoes *no* apparent change in frequency.

Now we consider scattering from the point of view of the quantum theory. Utilizing Einstein's successful photon interpretation of the photoelectric effect, Arthur H. Compton in 1922 applied the particle-like, quantum nature of electromagnetic radiation to explain the scattering of x-rays From the quantum theory, electromagnetic radiation consists of photons, each with an energy given by $E = h\nu$. Because a photon can be regarded as a zero-rest-mass particle moving at speed c, Equation 2-46 shows that the magnitude of the corresponding linear momentum p is given by E/c. Thus,

$$p = \frac{E}{c} = \frac{h\nu}{c} = \frac{h}{\lambda} \qquad\qquad [3\text{-}9]$$

Each photon in a beam of monochromatic electromagnetic radiation of wavelength λ has a momentum equal to h/λ. Equation 3-9 shows that the momentum of a photon is precisely specified when the wavelength, the frequency, or the energy of the photon is known. The direction of $\boldsymbol{p}$ is along the direction of propagation of the wave.

We can derive Equation 3-9 in a slightly different way, noting that the momentum of a photon p must be the product of its relativistic mass m and its speed c, where m is given by E/c^2.

$$p = mc = \left(\frac{E}{c^2}\right)c = \frac{E}{c} = \frac{h\nu}{c} = \frac{h}{\lambda}$$

Just as the energy of a photon increases as its frequency increases, so too the momentum of a photon increases with frequency. Therefore, the momentum of a high-frequency and high-energy photon, such as a gamma ray, will be greater than the momentum of a low-frequency and low-energy photon, such as a radio photon.

Using this particle interpretation of monochromatic electromagnetic radiation—a beam of photons having precisely defined energy and mo-

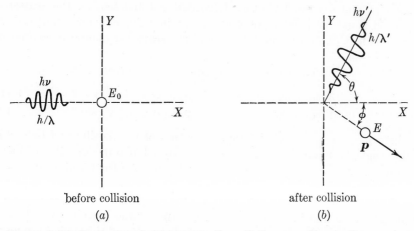

before collision

(a)

after collision

(b)

Figure 3-12. Scattering of electromagnetic radiation according to the quantum theory.

mentum—we see that the problem of the scattering of electromagnetic radiation becomes, in effect, a problem involving the elastic collision of a photon with a charged particle. It is assumed that the charged particle is initially at rest with a rest mass m_0 and rest energy $E_0 = m_0c^2$. As in classical physics, such a collision, if free from external influences, is solved by applying the conservation laws of energy and momentum. Figure 3-12 shows the photon and particle before and after the collision. In applying the conservation laws we notice that we need not be concerned with the details of the interaction during the collision, but only with the energy and momentum of the particles going into and coming out of the collision.

Unlike the classical scattering of electromagnetic waves, in which the scattering particle is assumed to gain no energy, the quantum treatment requires a partial transfer of electromagnetic energy; and because the kinetic energy of the charged particle may be large after the collision, we must treat the recoil particle relativistically.

Applying the conservation of energy to the collision of Figure 3-12 gives

$$\boxed{h\nu + E_0 = h\nu' + E} \qquad [3\text{-}10]$$

The symbol E refers to the energy of the scattering *particle*; $h\nu$ and $h\nu'$ give the energies of the incident and scattered photons, respectively. Since the final energy (rest energy plus kinetic energy) of the recoil particle, $E = mc^2$, is greater than its initial energy, E_0, we immediately see from the conservation of energy, Equation 3-10, that $h\nu' < h\nu$. Consequently, the scattered photon has *less* energy, a lower frequency, and a longer wavelength than the incident photon. This disagrees with the classical prediction of no frequency change upon scattering. Because the incident and scattered photons have different frequencies, the scattered photon is *not* the incident photon moving in a different direction; rather, the incident photon is annihilated and the scattered photon is created in the collision.

The conservation of momentum requires that (Figure 3-12):

$$X\text{-components: } h/\lambda = \quad p\cos\phi + (h/\lambda')\cos\theta \qquad [3\text{-}11]$$
$$Y\text{-components: } \quad 0 = -p\sin\phi + (h/\lambda')\sin\theta \qquad [3\text{-}12]$$

where $p = mv$ is the relativistic momentum of the recoiling particle, θ is the angle between directions of the scattered photon and the incident photon, and ϕ is the angle between the direction of the recoil particle and the incident photon.

We wish to solve for the change in wavelength, $\lambda' - \lambda = \Delta\lambda$, between the scattered and incident photons in terms of the angle θ. We do this because the wavelength of the scattered photon λ' and the angle θ are more easily measured than the kinetic energy $(E - E_0)$ and the angle ϕ of the recoil particle. Therefore, we wish to eliminate ϕ, E, and p from Equations 3-10, 3-11, and 3-12.

We first eliminate the angle ϕ by squaring the two momentum equations, Equations 3-11 and 3-12, and adding $p^2 \sin^2 \phi + p^2 \cos^2 \phi$, which equals p^2. This yields

$$p^2 = \frac{h^2}{\lambda^2 \lambda'^2} (\lambda^2 + \lambda'^2 - 2\lambda\lambda' \cos \theta) \qquad [3\text{-}13]$$

The energy equation, Equation 3-10, when solved for E in terms of $\lambda = c/\nu$ and $\lambda' = c/\nu'$, becomes

$$E = \frac{hc}{\lambda\lambda'} (\lambda' - \lambda) + E_0$$

Squaring this equation, we have

$$E^2 = \frac{h^2 c^2}{\lambda^2 \lambda'^2} (\lambda^2 + \lambda'^2) - \frac{2h^2 c^2}{\lambda\lambda'} + \frac{2hcE_0}{\lambda\lambda'} (\lambda' - \lambda) + E_0^2 \qquad [3\text{-}14]$$

Solving Equation 3-13 for $(h^2/\lambda^2\lambda'^2)(\lambda^2 + \lambda'^2)$, replacing E_0 by $m_0 c^2$, and recalling Equation 2-45 from relativity theory, $E^2 = (pc)^2 + E_0^2$, we obtain for Equation 3-14

$$0 = \frac{2h^2 c^2}{\lambda\lambda'} \cos \theta - \frac{2h^2 c^2}{\lambda\lambda'} + \frac{2hcm_0 c^2}{\lambda\lambda'} (\lambda' - \lambda)$$

Solving finally for the increase in wavelength $\Delta\lambda$, we have

$$\boxed{\Delta\lambda = \lambda' - \lambda = \frac{h}{m_0 c} (1 - \cos \theta)} \qquad [3\text{-}15]$$

The *Compton effect equation*, Equation 3-15, gives the increase $\Delta\lambda$ in the wavelength of the scattered photon over that of the incident photon. We see that $\Delta\lambda$ depends only on the rest mass of the scattering particle m_0, Planck's constant h, the speed of light c, and the angle of scattering θ. It is perhaps surprising to find that $\Delta\lambda$ is *independent* of the incident photon's wavelength, λ. The quantity $h/m_0 c$, appearing on the right-hand side of Equation 3-15 and having the dimensions of length, is known as the *Compton wavelength*. It is to be noted that knowing θ enables us to determine $\Delta\lambda$ unambiguously; but, for a single collision, we cannot predict in advance the angle at which the scattered photon will emerge.

If the scattering particle is a free electron within the scattering material, $m_0 = 9.108 \times 10^{-31}$ kg, and $(h/m_0 c) = 0.02426$ Å. For a scattered photon emerging at, for example, $\theta = 90°$ with respect to the incident photon direction, the wavelength change by Equation 3-15 is 0.024 Å. For $\theta = 180°$, with the incident photon making a head-on collision with the electron, $\Delta\lambda$ is a maximum and is equal to 0.048 Å; in such a collision, the electron recoils in the forward direction with the maximum possible kinetic energy. For the 90°-scattering by a free electron of incident radiation in the visible region—say, 4000 Å—the fractional increase in wavelength, $\Delta\lambda/\lambda$ is only 0.006 per cent. The Compton shift in wavelength is completely masked for visible light by the fact that the electrons in an ordinary scattering material are not at rest but are in thermal motion. An observable shift, say, $\Delta\lambda/\lambda = 2$ per cent, can be obtained by using incident radiation of wavelength $\lambda = 1$ Å ($\Delta\lambda = 0.024$ Å). Thus, there is an observable shift for x-ray and shorter wavelength photons. For longer wavelength photons, the *fractional* wavelength change $\Delta\lambda/\lambda$ becomes very small, and the scattered radiation has essentially the same wavelength or frequency as the incident radiation. Classically, the incident and scattered wavelengths are essentially equal; hence, Compton scattering agrees with classical scattering in the region where $\Delta\lambda/\lambda \ll 1$. We see here an example of the correspondence principle as applied to quantum effects, since, by Equation 3-15,

$$\underset{\substack{m_0 \to \infty \\ \text{or } h \to 0}}{\text{Limit}} (\Delta\lambda/\lambda) = \underset{\substack{m_0 \to \infty \\ \text{or } h \to 0}}{\text{Limit}} (h/m_0 c\lambda) = 0$$

A. H. Compton showed that the scattering of x-rays agrees with the photon model rather than the classical model, which predicted no wavelength change. Figure 3-13 shows a schematic experimental arrangement for x-rays incident on a target of carbon, a substance having many "free" electrons. For any fixed angle θ, the detector (see Sections 4-3 and 8-5)

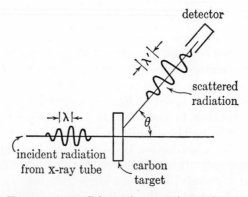

Figure 3-13. Schematic experimental arrangement of the Compton effect.

can measure the scattered radiation intensity as a function of wavelength. (Compare Figure 3-13 with Figure 3-11, where $\lambda = \lambda'$ and $\Delta\lambda = 0$.) Figure 3-14 shows the intensity versus the scattered wavelength for three fixed angles θ.

For any fixed angle θ, two predominant wavelengths are present in the

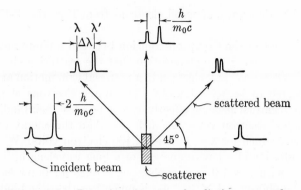

Figure 3-14. Intensity of scattered radiation versus the wavelength of scattered radiation for three different angles θ.

scattered radiation; one of the same wavelength λ as the incident beam (the *unmodified wave*), and a second longer wavelength λ' (the *modified wave*), given by the Compton equation, Equation 3-15. The unmodified wavelength results from the coherent scattering of the incident radiation by the inner electrons of atoms; these electrons are so tightly bound to the atom that an incident photon cannot strike the electron without moving, at the same time, the entire atom. The mass of these inner, tightly bound electrons is then effectively the mass of the atom, M_0. Therefore, in a Compton collision with a tightly bound electron, $\Delta\lambda = (h/M_0 c)(1 - \cos\theta) \simeq 0$, because M_0 is always thousands of times larger than the electron mass m_0.

The Compton effect provides a simple method for determining the energy of a photon by measuring the kinetic energy K of the recoil electron. From Equation 3-10 we have

[3-10] $$K = E - E_0 = h\nu - h\nu'$$

Because $\nu = c/\lambda$ and $\nu' = c/\lambda' = c/(\lambda + \Delta\lambda)$,

$$K = h\nu \frac{\Delta\lambda}{\lambda + \Delta\lambda}$$ [3-16]

where $\Delta\lambda$ depends on the scattering angle θ and is given by Equation 3-15. The kinetic energy of the recoil electron is a maximum $K_{\max}$ when a head-on collision occurs, with the electron recoiling in the forward direction and the scattered photon traveling in the backward direction. For this collision, $\theta = 180°$, $\Delta\lambda = 2h/m_0 c$, and Equation 3-16 becomes

$$K_{\max} = h\nu \left[\frac{(2h\nu/m_0 c^2)}{1 + (2h\nu/m_0 c^2)} \right]$$ [3-17]

Therefore, if the energy of a photon $h\nu$ is known, one can compute the maximum kinetic energy of recoil electrons from Equation 3-17, and conversely.

Our treatment of the Compton collision between a photon and an electron has been based on the assumption that the scattering electron is *free* and at rest. Of course, any electron in a material is in motion and is bound to some degree to its parent atom. The outer electrons of atoms can, however, be regarded as being effectively free because their binding energy, typically a few electron volts, is much less than the energy of a typical x-ray photon, which for $\lambda = 1$ Å, is 12,400 electron volts. When electromagnetic radiation of a low frequency, or a long wavelength, for example, radio waves with $\nu = 1.0$ megacycles/sec, strikes an outer electron, the energy of the photon, 4.1×10^{-9} ev, is much less than the binding energy of an outer electron; therefore, the m_0 in the Compton equation is effectively the mass of the atom, $\Delta\lambda \ll \lambda$, and the scattered radio radiation has essentially the frequency of the incident radiation.

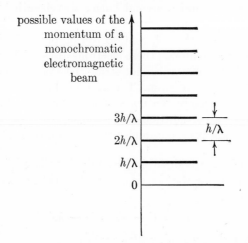

Figure 3-15. Allowed values of the linear momentum of a beam of monochromatic electromagnetic radiation.

The Compton effect shows clearly the particle-like aspects of electromagnetic radiation, for not only can a precise energy, $h\nu$, be assigned a photon, but also, a precise momentum h/λ. The total momentum of a monochromatic electromagnetic beam can then assume *not* any value, but only an exact integral multiple of the linear momentum of a single photon as shown in Figure 3-15. In this sense, the momentum, as well as the energy, of electromagnetic radiation is quantized.

3-5   Pair production and annihilation   The photoelectric effect, *Bremsstrahlung* (the inverse photoelectric effect), and the Compton effect are all examples of the conversion of the electromagnetic energy of photons into the kinetic energy and potential energy of material particles, and vice versa. It is natural to ask whether it is possible to convert a photon's energy into *rest* mass—that is, to create pure matter from pure energy— or, conversely to convert rest mass into electromagnetic energy. The answer is "Yes," provided such conversions do not violate the conservation laws of energy, momentum, and electric charge.

PAIR PRODUCTION   Let us first consider the minimum energy required to create a single material particle. Inasmuch as the electron has the smallest non-zero rest mass of all known observable particles, it requires the least energy for its creation. The photon has zero electric charge; thus the conservation of charge precludes the creation of a *single* electron from a photon. The creation of an electron pair consisting of two particles, having opposite electric charges, is possible. The positively charged particle is called a *positron* and is said to be the *anti-particle* of the electron. The electron and positron are similar in all ways except for the sign of their charges, $-e$ and $+e$ (and the effects because of this sign difference). The minimum energy $h\nu_{\min}$ to create an electron-positron pair with rest masses m_0^- and m_0^+ is, by the conservation of energy,

$$h\nu_{\min} = (m_0^-)c^2 + (m_0^+)c^2 = 2m_0c^2$$

Since the rest energy, m_0c^2, for an electron (or a positron) is 0.51 Mev, the threshold energy, $2m_0c^2$, for pair production is 1.02 Mev. The photon wavelength corresponding to this threshold energy is 0.012 Å; hence, pairs can be produced only by gamma-ray photons or very short wavelength x-ray photons. This creation of matter from electromagnetic radiation is called *pair production*, because a particle and its anti-particle must always be created together to satisfy the conservation laws. The pair production effect is a most emphatic demonstration of the interconvertibility of mass and energy.

If the photon has more energy than the threshold energy, the excess appears as kinetic energy of the created pair. The application of the conservation-of-energy to pair production gives

$$h\nu = m^+c^2 + m^-c^2 = (m_0^+c^2 + K^+) + (m_0^-c^2 + K^-)$$

$$\boxed{h\nu = 2m_0c^2 + (K^+ + K^-)} \qquad [3\text{-}18]$$

where ν is the frequency of the incident photon, and K^+ and K^- are the kinetic energies of the created particles. The minimum energy, $h\nu_{\min}$, just large enough to produce the pair is obtained by setting the kinetic energies of the created particles equal to zero, $(K^+ + K^-) = 0$,

Pair production cannot occur in empty space; it is easy to prove that energy and momentum cannot be simultaneously conserved in particle-anti-particle production unless the photon is near the nucleus of a heavy particle. We can see that the presence of a heavy particle is essential by considering Figure 3-16, which gives a plot of the relativistic energy versus

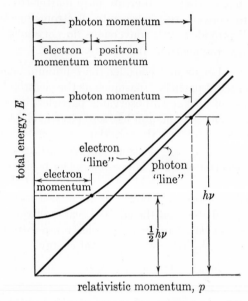

Figure 3-16. An illustration of the impossibility of pair production in empty space.

the relativistic momentum, as discussed earlier for Figure 2-12. First assume that there is no heavy particle present in the pair production process and that, for simplicity, the energy of each of the created particles is exactly half of the photon energy. Examination of Figure 3-16 shows that the sum of the momenta of the two particles is *less* than the momentum of the photon. Thus, the total momentum after the collision will be less than the momentum of the photon unless a heavy particle is present to carry away some of the photon's momentum. We can be assured that the energy carried away by the recoiling heavy particle is negligibly small, and need not be included in Equation 3-18, because its mass is very large as compared to the mass of the electron or positron.

Figure 3-17 is a schematic drawing of pair production, and Figure 3-18 is a cloud-chamber photograph showing the creation of electron-positron pairs. In Figure 3-18 high-energy gamma-ray photons entering from the top of the photograph came close to lead nuclei, were annihilated, and electron-positron pairs were created. The paths of the charged particles are visible because of the ionization effects they produce as they travel through the gas; the trajectories of the oppositely charged particles (with approximately equal kinetic energies†) show opposite curvatures because the charged particles were deflected into oppositely directed circular orbits by a uniform magnetic field.

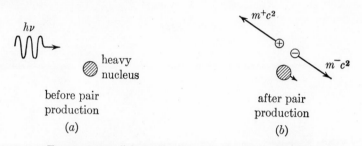

Figure 3-17. Schematic diagram of pair production.

The energy of a photon which produces an electron-positron pair can be computed, using Equation 3-18, if the kinetic energies of the electron and positron are measured. These energies can be determined from a photograph, such as that of Figure 3-18, if the magnetic flux density B and the radius of curvature r of the trajectories are measured. The relativistic momentum p of each particle is given, using Equation 2-40, by

$$p = mv = qBr$$

and the total energy E or kinetic energy $(E - E_0)$ of the particle can be computed using Equation 2-45, $E^2 = E_0^2 + (pc)^2$.

The existence of positrons—the anti-particles of electrons—was predicted theoretically by P.A.M. Dirac in 1928. Four years later, C. D. Anderson observed and identified a positron during his studies of cosmic radiation. Shortly thereafter, electron-positron pairs were produced in the laboratory by means of particle accelerators operating at a few Mev of energy. Electron-positron pair production is now a commonly observed phenomenon in cosmic rays and in interactions of high-energy photons produced in accelerators. Proton-anti-proton and neutron-anti-neutron pairs were first created in the laboratory in 1955. The threshold energies for these particles are a few Bev (the proton and neutron masses are approximately equal to 1 Bev), and therefore require very-high-energy accelerating machines.

PAIR ANNIHILATION   Pair annihilation and the creation of photons is the inverse of pair production. Consider the annihilation of pure matter and

† To be precise, the positron always has, on the average, a larger kinetic energy than the electron because the positron is repelled and the electron is attracted by the positively charged nucleus.

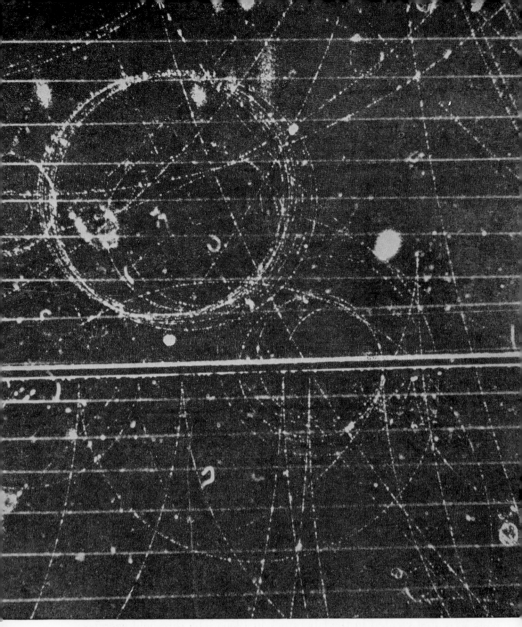

Figure 3-18. Cloud-chamber photograph showing the creation of electron-positron pairs. Photons of approximately 200 Mev, producing no tracks, enter from the top. Some photons are annihilated and electron-positron pairs are created in the thin horizontal lead foil; above the foil and to the right, one sees a pair produced by the collision of a photon with a gas molecule. The external magnetic field of flux density 1 weber/m² bends the electrons and positrons into paths of opposite curvature. (From *Cloud Chamber Photographs of the Cosmic Radiation*, C. D. Rochester and J. G. Wilson, Pergamon Press, Ltd., 1952. Courtesy of Pergamon Press, Ltd.)

the creation of electromagnetic energy that occurs when an electron and positron are close together and essentially at rest. The total linear momentum is initially zero; therefore, a *single* photon cannot be created if these two particles unite and are annihilated, because this would violate the conservation of momentum. Momentum can, however, be conserved if two photons, moving in opposite directions with equal momenta, are created. Such a pair of photons would have equal frequencies and energies (Figure 3-19). (The creation of three photons can occur, but with a *much*

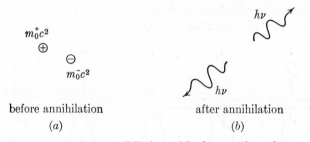

before annihilation   after annihilation

(a)   (b)

Figure 3-19. Pair annihilation with the creation of two photons.

smaller probability than for two-photon creation.) The conservation of energy requires

$$m_0^+ c^2 + m_0^- c^2 = h\nu_1 + h\nu_2$$

where the electron and positron are assumed to be at rest initially. But $m_0^+ = m_0^-$, and by momentum conservation, $\nu_1 = \nu_2 = \nu_{min}$; hence,

$$2h\nu_{min} = 2m_0 c^2$$
$$h\nu_{min} = m_0 c^2 \qquad\qquad [3\text{-}19]$$

Since $h\nu = m_0 c^2 = 0.51$ Mev, the minimum energy of a created photon is 0.51 Mev, and the maximum wavelength 0.024 Å, for electron-positron annihilation.

Annihilation is the ultimate fate of positrons. When a high-energy positron appears, as in pair production, it loses its kinetic energy in collisions as it passes through matter, finally moving at a low speed. It then combines with an electron forming a bound system (called *positronium*) which decays very quickly (10^{-10} sec) into two photons of equal energy. Thus, the death of a positron is signalled by the appearance of two $\frac{1}{2}$ Mev annihilation quanta, or photons. Positrons have a transitory existence, not because they are intrinsically unstable, but because of the high risk of their collision and subsequent annihilation with electrons.

In our part of the universe there is a preponderance of electrons, protons, and neutrons; the anti-particles associated with these particles when created will quickly combine with their respective particles in annihilation processes. It is conceivable, although presently purely conjectural, that there may exist a part of the universe consisting of anti-matter in which positrons, anti-protons, and anti-neutrons predominate.

Pair production and annihilation are particularly striking examples of mass-energy equivalence. These phenomena provide irrefutable confirmation of the theory of relativity.

3-6   Photon-electron interactions   Figure 3-20 summarizes the important photon-electron interactions, or collisions, that we have discussed in this chapter: (a) the photoelectric effect, (b) the Compton effect, (c) pair

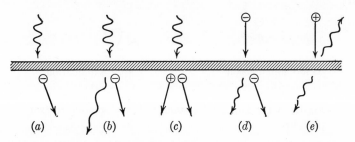

Figure 3-20. Photon-electron interactions: (a) photoelectric effect; (b) Compton effect; (c) pair production; (d) *Bremstrahlung*; and (e) pair annihilation.

production, (d) *Bremsstrahlung*, and (e) pair annihilation. In each instance, a photon, electron, or positron approaches a slab of material, a collision occurs, and one or more particles emerge.

We will briefly summarize the salient features of each of these interactions.

(a) The photoelectric effect: a photon disappears and a bound electron is dislodged. (Figure 3-20a.)

(b) The Compton effect: a photon collides with a free electron, thereby producing a second photon of lower energy and a recoiling electron. (Figure 3-20b.)

(c) Pair production: a photon is annihilated in the vicinity of a heavy nucleus and an electron-positron pair is created. (Figure 3-20c.)

(d) *Bremsstrahlung:* an electron is deflected and a photon is created. (Figure 3-20d.)

(e) Pair annihilation: A positron combines with an electron to produce a pair of photons. (Figure 3-20e.)

It is interesting to note that the principal features of these photon-electron collisions were derived simply by applying the conservation laws of energy, momentum, and electric charge, and by assuming the existence of photons of energy $h\nu$ and momentum h/λ. In no case did we concern ourselves with the details of the interaction. Further, we have not calculated the probability for the occurrence of any of these processes. For example, in the Compton effect, we are able to predict the wavelength of a photon scattered in any particular direction, but we are not able to predict in advance what will be the direction of any one scattered photon. The probabilities for the occurrence of these various photon-electron interactions can be calculated with high precision by the methods of *quantum electrodynamics*.

3-7 Absorption of photons   Three important processes that can remove photons from a beam of electromagnetic radiation are the photoelectric effect, the Compton effect, and pair production. These are shown as the photon-electron interactions (a), (b), and (c) of Figure 3-20. In each of these three processes a photon is removed from the forward direction of the beam and an electron appears. Furthermore, each process occurs only when atoms are present with which the incoming photons can collide and interact; these atoms provide bound electrons for the photoelectric effect, nearly free electrons for the Compton effect, and heavy nuclei for pair production. Consequently, the intensity of the photon beam is reduced only to the extent that the photons encounter and interact with atoms.

In each of these three interactions an electron with kinetic energy appears as one of the outgoing particles. This fact can be utilized in detecting photons. The fast moving electrons produce ionization which can be electrically measured; thus, the intensity of high-frequency photons to which the eye is not sensitive, such as x-rays and gamma rays, can be measured by ionization effects. Ionization measurements will be discussed in Chapter 8. Our concern here shall be with the absorption of electromagnetic radiation in a material.

The intensity of electromagnetic radiation is defined as the energy per unit time passing through a unit area at right angles to the direction of propagation. In terms of a monochromatic photon beam, the intensity I is the energy of a single photon $h\nu$ times the number of photons per unit time crossing a unit area placed perpendicular to the direction of the beam; that is,

$$\text{intensity of a photon beam} = \frac{\text{energy}}{\text{photon}} \times \frac{\text{number of photons}}{\text{area} \times \text{time}}$$

We define the *photon flux* of a beam of monochromatic electromagnetic radiation to be the number of photons crossing a unit area per unit time, and we represent this photon flux by N. Therefore,

$$\boxed{I = (h\nu)N} \qquad\qquad [3\text{-}20]$$

If a photon beam is absorbed in a material, the photon flux N is reduced because photons are removed or scattered from the forward direction. The absorption of photons by a material of thickness dx is shown schematically in Figure 3-21. Clearly, the probability that a photon will be

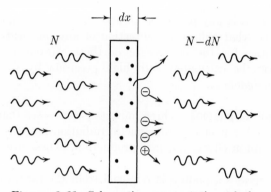

Figure 3-21. Schematic representation of the absorption of photons by a material substance.

removed from the forward beam increases as the number of atoms increases; therefore, the probability for photon removal increases as the thickness of the absorber is increased. In Figure 3-21 photons with a flux N are incident on a very thin absorber of thickness dx, and flux $(N - dN)$ emerges from the absorber in the forward direction; the number of photons removed per unit time by a unit area of the absorber is then dN. When the number of photons in the incident beam is increased, the number of

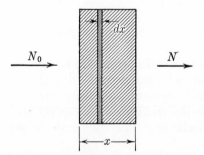

Figure 3-22. Change in photon flux in passing through an absorber.

photons removed by encounters with atoms in the absorber is increased proportionately; that is, dN is proportional to N. Also, inasmuch as the number of atoms "seen" by the incident photon beam is directly proportional to the thickness of the thin absorber, dN is also proportional to dx. Therefore,

$$dN = -\mu N dx$$

where the proportionality constant is μ, the so-called *absorption coefficient*. The minus sign appears because N decreases as x increases. Rearranging terms and integrating x from zero thickness to a finite thickness x (Figure 3-22), and integrating the flux from N_0, the flux incident on the absorber, to N, the flux emerging from an absorber of thickness x, we have

$$\int_{N_0}^{N} \frac{dN}{N} = -\mu \int_0^x dx$$

$$\ln (N/N_0) = -\mu x$$

$$N = N_0 e^{-\mu x} \qquad [3\text{-}21]$$

Using Equation 3-20, we can write Equation 3-21 as

$$\boxed{I = I_0 e^{-\mu x}} \qquad [3\text{-}22]$$

where $I_0 = (h\nu)N_0$ is the intensity incident on the absorber, and $I = (h\nu)N$ is the intensity at a distance x from the front surface. Equation 3-22 shows that the intensity of monochromatic electromagnetic radiation falls off exponentially through an absorber. The absorption increases as the absorber is made thicker (x increases), or as the absorption coefficient μ increases. We also see from Equation 3-22 that when $\mu x = 1$, or $x = 1/\mu$, then $I = I_0/e$; therefore, the quantity $1/\mu$ represents that absorber thickness at which the beam intensity I is $(1/e)^{\text{th}}$ of the incident intensity I_0.

For a particular photon energy and for a particular absorbing material, the absorption coefficient μ is a constant having the units of $(\text{meter})^{-1}$. The value of the absorption coefficient does, however, change from one material to another, and μ depends as well, for a given absorbing material, on the energy (or frequency) of the electromagnetic radiation. Figure 3-23 shows the absorption coefficient for aluminum and for lead as a func-

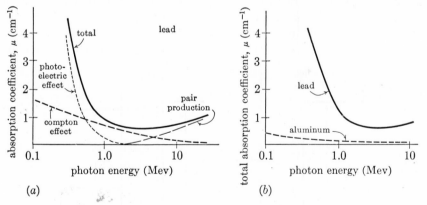

Figure 3-23. Absorption coefficients for aluminum and lead as a function of photon energy.

tion of the photon energy (plotted on a logarithmic scale). It is to be noted that μ is large for small-energy photons, where photon removal occurs principally through the process of the photoelectric effect. At intermediate energies the absorption coefficient is reduced, and here the Compton collisions are most effective in absorbing photons. The threshold for pair

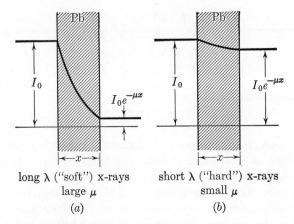

long λ ("soft") x-rays
large μ

(a)

short λ ("hard") x-rays
small μ

(b)

Figure 3-24. Exponential absorption of (a) soft and (b) hard x-rays by a lead absorber.

production occurs for photons of 0.012 Å wavelength or 1.02 Mev energy, and for very high energies this process predominates. We also notice that the absorption coefficient reaches a minimum in the vicinity of a few Mev and then increases with photon energy.

Figures 3-24 and 3-25 show how the absorption of electromagnetic radiation varies with the identity of the absorbing material and with the energy of the photon radiation. In these figures, the intensity is shown for the incident beam, for the beam at various penetration depths within the absorber, and for the emergent beam. Figure 3-24a shows the absorption of long wavelength x-rays, which are appreciably absorbed in moderate

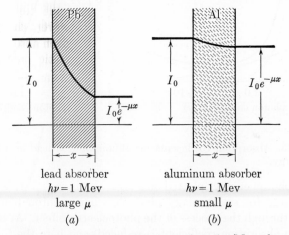

lead absorber
$h\nu = 1$ Mev
large μ

(a)

aluminum absorber
$h\nu = 1$ Mev
small μ

(b)

Figure 3-25. Exponential absorption of 1 Mev photons by (a) lead and (b) aluminum absorbers.

thicknesses of lead, and hence are appropriately called *soft* x-rays. In Figure 3-24b is shown the absorption of short wavelength, or *hard* x-rays, which are only slightly attenuated through the same thickness of lead. This illustrates that for a particular absorbing material and for relatively low-energy photons (of less than 4 Mev) the absorption coefficient decreases as the photon energy increases. Figure 3-25 shows the absorption of the same wavelength x-rays through lead and through aluminum absorbers of the same thickness. Clearly, lead is a much more effective x-ray absorber than aluminum and therefore has a significantly larger absorption coefficient for the same photon energy. It is generally true that heavy materials, such as lead, are more effective absorbers than lighter materials. This fact is the basis of x-ray photography, in which the darkening of the photographic emulsion is a measure of the x-ray intensity and for which the relative absorption properties of light and heavy materials are the basis of a recognizable image.

3-8 Summary Two kinds of physical quantities can be distinguished: those with a continuous range of values, and those with a discrete or quantized set of values. The quantum theory, formulated by Max Planck in 1900, has shown that many quantities which appear to have continuous values actually have only discrete values.

Monochromatic electromagnetic radiation, when interacting with matter, must be considered to consist of photons, each photon having a discrete energy and momentum:

$$E = h\nu \qquad \text{and} \qquad p = h/\lambda$$

[A useful relation between the energy (in Mev) and the wavelength (in Å) of a photon is: $E = 0.0124/\lambda$.] This is exemplified in the following photon-particle interactions:

(1) Photoelectric effect: the *complete* transfer of a photon's energy to a bound electron

$$h\nu = \phi + (E - E_0)$$

where ϕ is the binding energy (work function) of an electron to a surface or to an atom.

(2) *Bremsstrahlung* (x-ray production): *partial* or *complete* transfer of a particle's *kinetic energy* to *electromagnetic energy*. The maximum frequency (minimum wavelength) photons are produced when an electron is brought to rest in a single collision.

$$Ve = K = h\nu_{\max} = hc/\lambda_{\min}$$

(3) Compton effect: the *partial* transfer of electromagnetic energy to the *kinetic energy* of a particle. When a photon of wave-

length λ interacts with a (nearly) free particle essentially at rest, a scattered photon emerges at an angle θ and the particle recoils with kinetic energy K.

$$\Delta\lambda = \lambda' - \lambda = \frac{h}{m_0 c}(1 - \cos\theta)$$

$$K = (h\nu)\frac{\Delta\lambda}{(\lambda + \Delta\lambda)}$$

where m_0 is the rest mass of the recoil particle.

(4) Pair production and annihilation: *complete* conversion of electromagnetic energy into the *rest energy* and *kinetic energy* of the created particles, and the converse reaction.

Pair production: $h\nu = 2m_0 c^2 + (K^+ + K^-)$
Pair annihilation: $(m^+ + m^-)c^2 = 2h\nu$

The intensity of a monochromatic photon beam is the product of the photon energy and the photon flux N

$$I = (h\nu)N$$

Absorption in materials of thickness x of monochromatic electromagnetic radiation (by the processes of the photoelectric effect, the Compton effect, and pair production) follows the relation

$$I = I_0 e^{-\mu x}$$

where μ is the absorption coefficient. The value of μ is dependent on the nature of the absorber and the photon energy.

REFERENCES

Handbook of Chemistry and Physics. Cleveland, Ohio: Chemical Rubber Publishing Company. This handbook, issued annually, contains comprehensive tables of work functions, absorption coefficients, etc., for various materials.

Kaplan, I., *Nuclear Physics*. Reading, Massachusetts: Addison-Wesley Publishing Company, 1955. Chapter 15 gives a comprehensive treatment of gamma-ray absorption and the experimental details for measuring gamma-ray energies.

Richtmeyer, F. K., E. H. Kennard, and T. Lauritsen, *Introduction to Modern Physics*. New York: McGraw-Hill Book Company, Inc., 1955. Chapters 3 and 4 of this book give a detailed treatment of the basic quantum effects with emphasis on their historical origin.

PROBLEMS

3-1   The excess charge on a conductor is measured and found to be $(-1.01 \pm 0.01) \times 10^{-6}$ coulomb. (a) What is the uncertainty in the number of electrons? (b) What is the increase in the mass of the conductor because of the addition of the excess charge, -1.01×10^{-6} coulomb?

3-2 *   Light of intensity 1.0×10^{-10} watt/m² falls normal to a copper surface, where there are two free electrons per atom. The atoms are approximately 1.0 Å apart. Treat the incident radiation classically (as waves) with the energy uniformly distributed over the surface, and assume all the light to be absorbed by the surface electrons. (a) How much energy does each electron gain per second? (b) The binding energy of an electron to a copper surface is 4.50 ev. How long must one wait, after the beam is switched on, until any one electron gains enough energy to overcome its binding energy and be released as a photoelectron? Compare this with the experimental results.

3-3   The threshold wavelength for the emission of electrons from a sodium surface is 5400 Å. (a) Calculate the binding energy, or the work function, ϕ (ev) of an electron to the surface of sodium. (b) What is the maximum kinetic energy (ev) of a photoelectron emitted from the surface when light of 2000 Å strikes this surface?

3-4   (a) What is the maximum velocity of photoelectrons emitted from a potassium ($\phi = 2.20$ ev) surface if ultraviolet light of 2000 Å is used? (b) Does this justify our assumption that $K_{max} = \frac{1}{2} mv_{max}^2$?

3-5   (a) For what wavelength of electromagnetic radiation will the photoelectrons ejected from a tungsten ($\phi = 4.53$ ev) surface have a maximum velocity of one-tenth the speed of light? (b) What photon wavelength would give photoelectrons a maximum velocity of 0.98 c?

3-6 *   A 1000 Å photon strikes a copper target normally and releases a photoelectron from the surface, the photoelectron moving in the opposite direction to that of the incident photon. Assume that essentially all the photon's energy is given to the electron. (a) Calculate the maximum velocity of the released electron. (b) Using the conservation-of-momentum law, find the momentum imparted to the target. (c) If the target has a mass of 100 gm, calculate the fraction of the photon's energy given to the target. This justifies our original assumption that in the photoelectric effect practically all the energy of the photon is transferred to the electron.

3-7 *   A 1.00 Å photon is incident upon a hydrogen atom initially at rest. The photon gives essentially all of its energy to the bound electron, thus releasing it from the atom (binding energy = 13.6 ev). The released electron moves in the same direction as that of the incident photon. (a) Find the kinetic energy and velocity of the photoelectron. (b) What is the momentum and energy of the positive recoil ion?

3-8   What is the maximum wavelength of electromagnetic radiation for which it is possible to separate a carbon monoxide molecule, the binding energy (dissociation energy) between the carbon atom and the oxygen atom being 11 ev?

3-9   In the upper atmosphere of the Earth molecular oxygen, O_2, is dissociated into two oxygen atoms by photons from the Sun. The maximum photon wavelength which permits this process to occur is 1750 Å. What is the binding energy, in ev, between two oxygen atoms forming a molecule? (This phenomenon has been proposed as a possible source of energy for high-altitude flying. The solar radiation splits the molecular oxygen into atomic oxygen, and when the atoms recombine to form molecules, energy is released.)

3-10 *   A photon enters a lead radiator (which "radiates" electrons) and interacts with an inner electron which is bound to a lead atom with a binding energy of 89.1 Kev. The released photoelectron then enters a uniform magnetic field and (Br) is found to be 25.2×10^{-4} weber/m, where r is the radius of curvature of the electron in the magnetic field B. (a) Compute the momentum of the photoelectron. (b) Compute the relativistic kinetic energy of the photoelectron. (c) What is the energy of the incident photon?

3-11   A photon of frequency ν collides head-on with a *free* electron in space. Solve for the photoelectron's speed by using the conservation of momentum and energy and show that: (a) it is impossible for the photoelectric effect to take place if the electron is treated classically $(K = \frac{1}{2} m_0 v^2)$; (b) it is impossible for the photoelectric effect to take place if the electron is treated relativistically $(K = E - E_0)$. Therefore, the photoelectric effect can only occur if a photon collides with a bound electron; the material to which the electron is bound is given a substantial fraction of the photon's momentum, but a trivial fraction of the photon's energy because of the large mass of the material.

3-12   (a) Show that a 300 Mev photon possesses roughly the same momentum as a 50 Mev proton. (b) Calculate the wavelength of this photon and state in what region of the electromagnetic spectrum it lies (see Figure 3-7).

3-13   (a) Compute the energy of a photon which has the same momentum as a 50 Mev electron. (b) Where does this radiation lie in the electromagnetic spectrum?

3-14 A uniform beam of blue light (4000 Å wavelength) has an intensity of 2.5×10^{-9} watt/m². (a) Find the energy (ev) of one photon with this wavelength. (b) How many photons cross a 1 cm² surface, normal to the beam, in one second? (c) How many wavecrests pass this surface in one second?

3-15 Radiation intensities as low as 1.0×10^{-10} watt/m² have been observed experimentally. (This is approximately the minimum light intensity that can be perceived by the eye.) (a) For this intensity, find the number of photons with wavelength 4000 Å crossing an area of 1.0 mm² per second. (b) How many radio photons of frequency 1.0 megacycle/sec would cross 1.0 mm² per second for this same intensity? (c) How many γ-ray photons of wavelength 10^{-4} Å?

3-16 What is the ratio of the number of photons crossing a unit area per second in two monochromatic beams of wavelengths λ_1 and λ_2, respectively, if the intensities of the two beams are equal?

3-17 In an x-ray tube electrons are "evaporated" from the cathode by heating, accelerated to the anode target by a potential difference V, and then decelerated to produce x-rays. (a) What properties of the x-rays change when the temperature of the cathode is increased? (b) What properties of the x-rays change when the potential difference V is increased?

3-18 An electron, initially at rest, is accelerated through a potential difference of 400 kilovolt. (a) What is the final kinetic energy (ev) of the electron? (b) If this electron is suddenly decelerated to rest with the production of one photon, find the photon's wavelength (Å).

3-19 * A mass m whose initial velocity is v_0 ($v_0 \ll c$) collides head-on with a mass M initially at rest. Assuming the collision to be perfectly elastic, show that the energy ΔE transferred to M is given by $\Delta E = (4MmE_0)/(M + m)^2$ where $E_0 = \frac{1}{2} mv_0^2$.

3-20 * (a) Show that the energy ΔE transferred in the collision of Problem 3-19 is a maximum when $m = M$. (b) Plot ΔE as a function of m.

3-21 Photons of wavelength 0.015 Å are scattered by free electrons. What is the wavelength of those scattered photons whose angle of scattering is (a) 45°? (b) 135°? (c) What is the energy transferred to the free electron in parts (a) and (b) above?

3-22 (a) A beam of monochromatic photons ($\lambda = 0.10$ Å) strikes a metal target. If one observes the scattered radiation at an angle of 90° relative to the incident beam, what *two* predominant wavelengths (in Ångstroms), will be detected? (b) What would be the results if the angle were 60°?

3-23 * A monochromatic photon beam is allowed to strike a copper radiator in which Compton collisions occur. It is found that the recoil electrons having a maximum kinetic energy move in a circle of 3.5 cm radius when passing perpendicularly through a magnetic flux density of 0.20 weber/m². What is the energy of the photons?

3-24 A 300 Kev electron strikes a copper target, is brought to rest, and produces a single x-ray photon. This photon then enters a carbon target and is scattered by a free electron. What is the maximum kinetic energy of the recoil Compton electron?

3-25 * Radiation of uniform intensity 2.5×10^{-6} watt/m² falls normally on a surface whose area is 1.0 cm². (a) If the surface is a perfect absorber of this radiation, calculate the average radiation force on the surface. (b) If the surface is a perfect reflector, what is the average force on the surface?

3-26 (a) A flashlight develops 4.0 watts of electrical power when turned on. Assuming that 5.0 per cent of this power is converted into a parallel beam of visible light with an average wavelength of 6000 Å, calculate the recoil force on the flashlight. (b) Prove that, in general, the recoil force is the radiated power divided by c.

3-27 Using Equation 3-20, $I = N(h\nu)$, show that the radiation pressure of an electromagnetic wave is equal to (I/c), the radiation intensity divided by the speed of light. This proves that the radiation pressure is *independent* of the frequency.

3-28 * The intensity of solar radiation at the Earth's surface is 8.0 joule/cm²-min. (a) What is the pressure of the above radiation at the Earth's surface? (b) Compute the total force on the Earth due to this radiation from the Sun (Earth's radius, 4.0×10^3 mi).

3-29 A 0.012 Å photon passing near a gold nucleus (atomic weight = 197) is annihilated and an electron-positron pair is created. (a) Calculate the energy of the photon (Mev) and compare this with the total rest energy of the created pair. (b) If the electron and positron are at rest after being created (the photon having just the threshold frequency), a small part of the photon's energy must be transferred to the gold nucleus because the nucleus must carry away the original photon momentum. Calculate the momentum given to the gold nucleus. (c) Find the energy given to the nucleus and compare this with the original photon energy.

3-30 An x-ray tube accelerates electrons which strike the x-ray target and thus produce radiation. Some of these x-ray photons then enter a lead plate and create electron-positron pairs. (a) Describe the type of energy conversion in each of the processes: electron acceleration, x-ray production, and pair production. (b) What is the minimum potential difference of the x-ray tube necessary to produce photons which can, in turn, produce electron-positron pairs?

3-31 A 130 Mev photon is incident upon a material. What is the largest number of positrons that can be created by this photon?

3-32 * If the radii of curvature of both the electron and positron created when a photon interacts with a heavy nucleus are 2.0 cm when the particles are bent in a uniform magnetic field of 0.25 weber/m², what is the wavelength of the incident photon?

3-33 * An incident photon is scattered by a free electron which is initially at rest. The *scattered* photon may later produce an electron-positron pair. Show that if the angle between the scattered photon and the incident photon is greater than 60°, the scattered photon cannot create an electron-positron pair, no matter how large the energy of the incident photon.

3-34 * A 2.0 Mev photon produces an electron-positron pair, both created particles traveling in the forward direction with equal energies. (a) Calculate the speed of each created particle. (b) Compare the total linear momentum of these two particles with that of the photon.

3-35 An electron-positron pair is created at a distance of 0.010 Å from a heavy nucleus. If the created particles have equal kinetic energies when created, what will be the difference between their kinetic energies $(K^+ - K^-)$ after they are an infinite distance from the heavy nucleus? Assume that the average positive charge of the nucleus "seen" by the electron and positron is 40 electronic charges.

3-36 * An electron and positron are moving together as a system (positronium) at a velocity of $(c/2)$. If these two particles annihilate and two photons are created: (a) What is the energy and momentum of each photon? (b) What is the angle between the direction of motion of these two photons?

3-37 A 5.00 Mev photon produces an electron-positron pair. Assume that the electron and positron move in opposite directions with equal speeds. (a) What is the kinetic energy of the electron (or positron)? (b) What is the speed of the electron? (c) What is the speed of the positron with respect to an observer moving with the electron?

3-38 Gamma-rays with energies of 0.10, 1.0, and 10 Mev, but with equal intensities, are incident on a lead absorber. The absorption coefficients, in lead, for these three energies are 59.9 cm⁻¹, 0.77 cm⁻¹, and 0.61 cm⁻¹, respectively. (a) Calculate the thickness of lead necessary to reduce the intensity of each monoenergetic beam to $\frac{1}{10}$ its original intensity. (b) What is the ratio of the total intensity (of all three photon energies) at any depth x to the total incident intensity?

3-39   The incident intensity of a 0.10 Mev beam of photons on a water surface is I_0. After passing through 10 cm of water, the beam intensity is $I_0/5$. Calculate the absorption coefficient of water for this photon energy.

3-40   A shield of absorbing material is to be designed for polychromatic electromagnetic radiation. Explain which frequencies, low or high, determine the choice of the absorber thickness assuming that there are no photons in the radiation having an energy (a) greater than 4.0 Mev, (b) less than 4.0 Mev.

3-41   A lead plate and an aluminum plate are placed flush against one another. A beam of photons of intensity I_0 is incident on these parallel plates. (a) What is the ratio of the intensity of the emerging beam when the lead plate is first to that when the aluminum plate is first? (b) If a monoenergetic beam of 0.10 Mev photons is incident on this parallel combination, what is the ratio of the thickness of lead ($\mu = 58$ cm^{-1}) to aluminum ($\mu = 0.44$ cm^{-1}) which is necessary so that the intensity is reduced equally in both plates?

3-42   Show that the thickness of absorbing material necessary to reduce the intensity of a beam of radiation to one-half its original intensity is $(ln_e 2)/\mu$ where $ln_e 2$ is the logarithm of 2 to the base e.

F O U R

QUANTUM EFFECTS: THE WAVE ASPECTS
OF MATERIAL PARTICLES

4-1   De Broglie waves   We have seen that electromagnetic radiation has
two aspects: a wave aspect and a particle aspect. Experiments which show
the interference and diffraction of electromagnetic radiation can be explained only if it is assumed to consist of waves. The distinctively quantum effects of electromagnetic radiation, such as the photoelectric and
Compton effects, can be explained only if light is assumed to consist of
particle-like photons, each photon with an energy and momentum which
is precisely specified by the frequency (or wavelength) of the radiation as
follows:

$$\nu = E/h \qquad\qquad\qquad [4\text{-}1]$$

and
$$\lambda = h/p \qquad\qquad\qquad [4\text{-}2]$$

It is interesting to notice that the two quantities, frequency ν and wavelength λ, which have clear meanings only if one is describing a wave, appear on the left-hand sides of Equations 4-1 and 4-2; while on the right-
hand sides of these equations appear the energy E and momentum p,

two quantities which can be localized in a small region of space and thus have the attributes of a particle. Thus, the wave-particle duality of electromagnetic radiation is implied in these fundamental relations, and it is the fundamental constant of the quantum theory, Planck's constant h, which relates the wave characteristics to the particle characteristics. We can say that electromagnetic waves will, under some circumstances, behave as particles, or that photons—zero-rest-mass particles—will, under some circumstances, behave as waves.

It is natural to wonder whether Equations 4-1 and 4-2, which ascribe a wave and a particle nature to electromagnetic radiation, have an even greater generality, and to ask whether these relations apply to *all* particles, that is, to *finite-*, as well as to zero-rest-mass particles. This question was first posed by Louis de Broglie in 1924. De Broglie conjectured that, because of the symmetry of nature, a material particle might well exhibit wave properties. It was further assumed that Equations 4-1 and 4-2, which give the particle characteristics of electromagnetic waves, would hold equally well for a material particle, such as an electron, and would give the wave characteristics, that is, the wavelength λ and the frequency ν, of a material particle. Experiments have emphatically confirmed the correctness of de Broglie's hypothesis, and the wave character of material particles is well established. Because the wavelength can be measured from interference or diffraction effects, we shall concentrate our attention on the second relation, Equation 4-2. The *de Broglie wavelength* λ of a material particle having a momentum $p = mv$, where m is the relativistic mass of the particle and v is its velocity, is given by the *de Broglie relation*

$$\lambda = h/p = h/mv \qquad [4\text{-}3]$$

where h is Planck's constant.

One might well ask, "If an electron is, at least under some circumstances, to be regarded as a wave, what is it that is waving?" In this connection it is useful to recall that the same sort of question was raised concerning the fundamental nature of light. It was not until the electromagnetic theory of Maxwell and the experiments of Hertz that physicists could assert that the wave properties of light correspond to oscillations of the electric and magnetic fields. But this ignorance of light's electromagnetic nature did not prevent physicists from discovering, long before Maxwell and Hertz, the wave-like properties of light, and interpreting certain experiments on this basis. Therefore, to establish whether a material particle has a wave nature, it is *not* necessary to know first what the nature of the wave phenomenon is. To test the de Broglie hypothesis is to determine, on the basis of experiment, whether material particles show interference and diffraction effects. Of course, the question of the physical nature of the wave aspect of a material particle is a crucial one; we will, however, postpone this question until after we have discussed the experiments which confirm that material particles follow the de Broglie relation, $\lambda = h/mv$.

The electron was "discovered" in 1897 by J. J. Thomson. He showed that electrons follow well-defined paths and have a well-defined charge-to-mass ratio (for $v \ll c$)—in short, that electrons show the attributes of a particle. The wave nature of electrons was not discovered until 1927, when the electron-diffraction experiments of C. Davisson and L. H. Germer confirmed the de Broglie relation. Why was the wave nature of electrons discovered many years after their particle nature had been ascertained? We might suspect that the origin of the difficulties in observing the wave properties of electrons is similar to that which arises in observing the wave nature of light; namely, the very small wavelength. We suspect then that the wavelength of an ordinary material particle is very small compared to the dimensions of ordinary objects, so that interference and diffraction effects cannot be easily observed. Therefore, if the wavelength is to be large enough to produce observable wave effects, we see from Equation 4-3 that the mass and the velocity must be small. (Clearly, for a large wavelength, a non-relativistic treatment can be applied, inasmuch as the speed must be small.)

Consider first the wavelength of an ordinary-size object, such as a marble, of mass 0.10 kg, moving with a speed of 1.00 m/sec. We have then from Equation 4-3,

$$\lambda = \frac{h}{mv} = \frac{6.62 \times 10^{-34} \text{ joule-sec}}{(0.10 \text{ kg})(1.00 \text{ m/sec})} = 6.62 \times 10^{-33} \text{ m}$$

$$\lambda = 6.62 \times 10^{-23} \text{ Å}$$

It is hopeless to observe the wave effects of a particle having $\lambda = 10^{-22}$ Å because there exist no slits or diffraction gratings of such a small size.

The diffraction grating having the smallest distance between lines is a crystal, a solid in which the atoms have a regular geometrical arrangement and a regular spacing. We shall see that a typical distance between adjacent atoms is of the order of 10^{-10} m, or 1 Å. The most favorable conditions for observing the diffraction of particles are those in which the particle has a wavelength of approximately 1 Å (see Section 1-7). Inasmuch as the wavelength varies inversely with the particle's mass and velocity, to have the longest wavelength, we must choose a particle having the smallest possible mass; namely, the electron.

Let us compute the kinetic energy that an electron must have so that its wavelength λ is 1.00 Å. We assume that an electron with electric charge e is accelerated from rest by an electrostatic potential difference V, acquiring a final kinetic energy $\frac{1}{2} mv^2$. Therefore,

$$eV = \tfrac{1}{2} mv^2 = \frac{1}{2m} p^2 = \frac{1}{2m} \left(\frac{h}{\lambda}\right)^2$$

$$V = \frac{h^2}{2me\lambda^2} = \frac{(6.62 \times 10^{-34} \text{ joule-sec})^2}{2(9.11 \times 10^{-31} \text{ kg})(1.60 \times 10^{-19} \text{ coulomb})(1.00 \times 10^{-10} \text{ m})^2}$$

$$V = 150 \text{ volts}$$

An electron which is accelerated from rest by a potential difference of 150 volts has a wavelength of 1 Ångstrom. Thus a 150 ev electron has a wavelength which is equal to that of a typical x-ray photon, and we expect that both electrons and x-rays will show similar diffraction effects when passing through a crystal.

4-2 The Bragg law Max von Laue first suggested in 1912 that crystalline solids, in which the arrangement of atoms follows a regular geometrical pattern and in which the distance between atoms is approximately one Ångstrom, might be used as diffraction gratings for measuring x-ray wavelengths.

Consider a crystal of NaCl, which has a particularly simple structure and which is used as a standard material for x-ray diffraction. Examination of the external geometrical features of a rock-salt crystal suggests that the sodium and chlorine atoms (strictly, Na+ and Cl− ions) are arranged in a simple cubic lattice, as shown in Figure 4-1. The sodium and

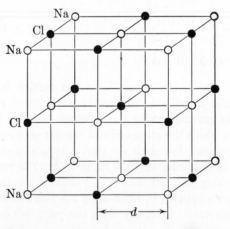

Figure 4-1. Crystal structure of rock-salt (NaCl).

chlorine atoms are located at alternate corners of identical elementary cubes, each with a distance d along an edge.

Let us see how it is possible to compute the *lattice spacing* d from the density of the NaCl crystal and the atomic weights of Na and Cl. If d is the distance in centimeters from a Na atom to a nearest Cl atom, then there are $1/d$ atoms (half Na, half Cl) along an edge of a cube 1 cm long. Further, in a cube of NaCl crystal 1 cm along an edge, there are altogether $1/d^3$ atoms. Therefore, the total number of atoms per unit volume is $1/d^3$. The atomic weight of Na is 23.00, the atomic weight of Cl is 35.46, and hence the molecular weight of NaCl is 58.46. Because Avogadro's number, 6.025×10^{23}, gives the number of atoms in one gm-mole, it follows that there are 6.025×10^{23} Na atoms in 23.00 gm of Na, 6.025×10^{23} Cl atoms in 35.46 gm of Cl, and therefore, $2(6.025 \times 10^{23})$ atoms (half Na, half Cl) in 58.46 gm of NaCl. The measured density of NaCl in the crystalline form of rock-salt is 2.163 gm/cm³. Therefore, we can write

$$\frac{\text{number of atoms}}{\text{volume}} = \frac{2(6.025 \times 10^{23}) \text{ atoms/mole} \times 2.163 \text{ gm/cm}^3}{58.46 \text{ gm/mole}}$$

and

$$\frac{\text{number of atoms}}{\text{volume}} = \frac{1}{d^3}$$

Solving for d gives

$$d = 2.820 \text{ Å}$$

Thus, the lattice spacing of Na and Cl atoms in a crystal of rock-salt is 2.820 Å. This distance, which is typical of the interatomic spacing of atoms in any solid, is, of course, comparable to the wavelength of x-rays or of 150-ev electrons. We shall see how it is possible to determine x-ray

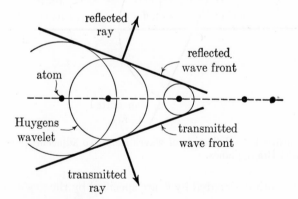

Figure 4-2. Reflection and transmission of a plane wave by a row of atoms.

and electron wavelengths from a knowledge of the lattice spacing and the geometrical character of the atomic arrangement.

We wish first to show that a plane containing atoms behaves as a partially silvered mirror to waves incident on it. Figure 4-2 shows a wave front that has just passed several atoms arranged in a row. Each atom acts as center of scattering, and by Huygens' principle, generates spherical waves. Two envelopes can be drawn for the Huygens' wavelets, one which corresponds to the transmitted wave front proceeding undeviated past the row of atoms, and the second which corresponds to a wave which is reflected from the row of atoms. Thus, a collection of atoms, or scatterers, arranged in a row (for one dimension) or in a plane (for two dimensions) will partially reflect an incident wave as does a half-silvered mirror for light. The angle of reflection equals the angle of incidence. A plane containing atoms is known as a *Bragg plane*, and the reflection of a wave from such a plane is known as a *Bragg reflection*. These terms are named after W. H. Bragg, who, with his son, W. L. Bragg, developed the fundamental theory of x-ray diffraction by crystals in 1913.

Consider now the reflection of waves from two adjacent and parallel Bragg planes as shown in Figure 4-3. The directions of the incident and re-

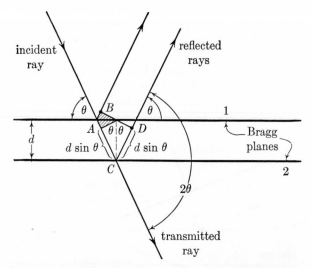

Figure 4-3. Reflection of waves from two adjacent, parallel Bragg planes.

flected waves, both designated by θ, are specified by the angle between the direction of propagation of waves and the Bragg plane (not the normal to the reflecting planes). At each plane the incident wave is partially transmitted in an undeviated direction and partially reflected.

The incident ray is partially reflected at the first Bragg plane, the reflected ray AB making an angle θ with plane 1. That part, AC, of the incident ray which is transmitted through the first plane is partially reflected from the second plane, also in the direction θ. We wish to consider a wave front BD which is perpendicular to the two reflected rays. The reflected rays will constructively interfere at some distant point only if the reflected rays have the same phase at the points B and D. The points B and D on the two reflected rays will be in phase if the path difference, $ACD - AB = 2d \sin \theta$, is an integral multiple of the wavelength λ. The condition, then, for constructive interference of waves reflected from adjacent parallel Bragg planes is

$$n\lambda = 2d \sin \theta \qquad [4\text{-}4]$$

where n, the *order of the reflection*, can have the values 0, 1, 2, ... Equation 4-4†, known as *Bragg's law*, is the basis for all coherent x-ray and electron diffraction effects in crystals. For any angles except those satisfying Equation 4-4, the reflected rays destructively interfere and the incident beam is completely transmitted. The Bragg-law equation forms the basis for measuring wavelengths comparable to interatomic distances. It is obvious that if n, d and θ in Equation 4-4 are known, λ can be determined. We note from Figure 4-3, that the angle between the transmitted and the reflected rays is 2θ, and that the Bragg planes bisect this angle.

4-3 X-ray and electron diffraction The essential elements of an x-*ray spectrometer*, a device to measure x-ray wavelengths, are shown in Figure

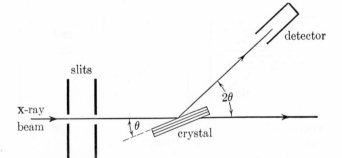

Figure 4-4. Schematic diagram of an x-ray crystal spectrometer.

4-4. A source of monochromatic x-rays shines on a crystal whose structure and interatomic dimensions are known. A detector, such as an ionization chamber which is sensitive to x-ray ionization effects, measures the intensity of x-rays entering it. Both the crystal and the detector are rotatable, but the detector is always set at an angle 2θ from the forward

---

† Equation 4-4 for Bragg's law bears a resemblance to the equation which applies to an ordinary ruled diffraction grating. The two equations, however, are *not* the same.

beam, where θ is the angle between the incident rays and the Bragg planes from which reflection is to be observed. The x-ray intensity in the detector will indicate a pronounced maximum when the conditions of Equation 4-4 for Bragg reflection are satisfied. Thus, since d is known from the crystalline structure and θ is measured, the wavelength λ can be computed. For a given set of Bragg planes, grating space d, wavelength λ, and order n there is a *single direction* 2θ from that of the incident beam for which the reflected beam is strong.

Consider a thin metallic foil through which a monochromatic x-ray beam is sent. The foil consists of a very large number of simple perfect crystals randomly oriented with respect to one another within the foil. Only those particular microcrystals which are so oriented that the Bragg condition is fulfilled will produce a strongly diffracted beam; the other microcrystals do not deviate the incident beam coherently. Therefore, the emerging beam will consist of two parts: an intense, central, undeviated beam, and a scattered beam concentrated in a conical shell, making an angle 2θ with respect to the incident beam as shown in Figure 4-5. The angle θ is uniquely determined, for a given order, by the Bragg relation. When the scattered beam strikes the photographic plate, the intensity

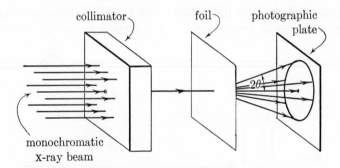

Figure 4-5. Scattering of monochromatic x-rays by a thin metallic foil from one set of Bragg planes.

pattern observed is that of a strong central spot surrounded by a circle. The radius of the circle is easily measured, and by knowing the distance from the scattering foil to the photographic plate, the angle 2θ can be found; finally, the wavelength of the x-rays can be computed from the Bragg relation.

Thus far, we have implicitly assumed that a given crystal has only one set of parallel Bragg planes. In actuality, there are many sets of planes in any single crystal. To see what effect this has on x-ray and electron diffraction, let us consider again the arrangement of atoms in the NaCl crystal. A Bragg plane is any plane that contains atoms, and there are a variety of such planes that can be drawn for a cubic crystal, as shown in Figure 4-6. It is clear that the various planes, of which only a very few are shown in Figure 4-6, will differ in the value of the grating spacing d.

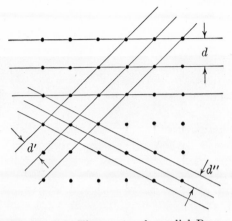

Figure 4-6. Three sets of parallel Bragg planes having different grating spacing.

Consequently, there will be a number of Bragg angles θ, each satisfying the Bragg relation for the particular set of Bragg planes. The x-ray diffraction pattern will, therefore, be somewhat more complicated than that indicated in Figure 4-5. The observed diffraction pattern will commonly show not a single circle, but rather a number of concentric circles, each circle corresponding to diffraction from a particular set of Bragg planes. The intensity of the reflected beam will, however, not be the same for all planes, inasmuch as the number of atoms per unit area in the plane, which determines the intensity of the reflected beam, differs according to the particular plane chosen. Figure 4-7 shows an x-ray diffraction pattern observed for a sample of polycrystalline aluminum. Such a pattern is referred to as a *powder pattern* and the rings are called *Debye-Scherrer rings*.

Our discussion of the diffraction effects from crystals has been in terms of the effects observed for x-rays. Electrons with a kinetic energy of 150 ev have the same wavelength as 1 Å x-rays, and a mono-energetic (therefore, monochromatic) electron beam should and does show essentially the same diffraction effects as do x-rays. Figure 4-8 is an illustration of electron-diffraction from a metallic foil. The observed diffraction pattern is in complete accord with the Bragg and de Broglie relations. In short, *the electron-diffraction experiments confirm the de Broglie relation.*

Electron and x-ray diffraction effects can be used to measure the wavelengths of x-rays or electrons when the crystalline structure is known;

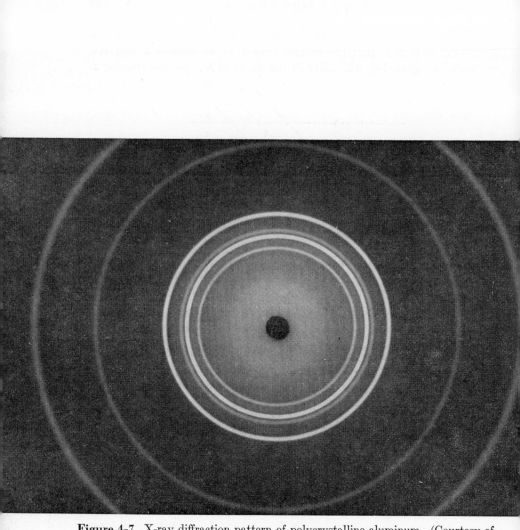

Figure 4-7. X-ray diffraction pattern of polycrystalline aluminum. (Courtesy of Mrs. M. H. Read, Bell Telephone Laboratories, Murray Hill, New Jersey.)

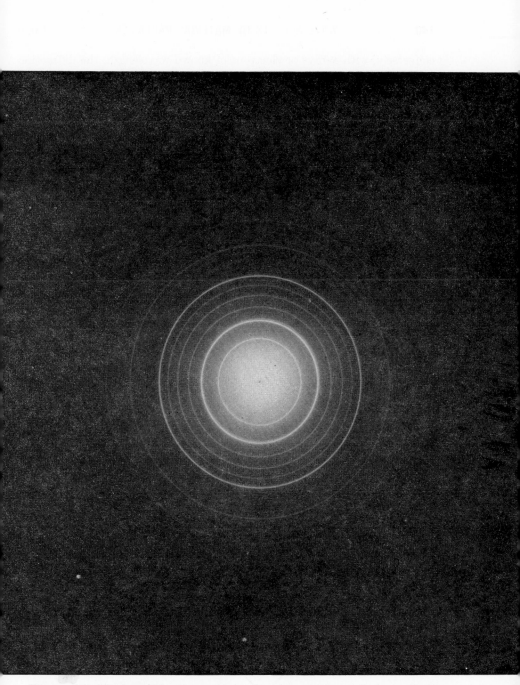

Figure 4-8. Electron-diffraction pattern of polycrystalline tellurium chloride. (Courtesy of RCA Laboratories, Princeton, New Jersey.)

conversely, with x-rays or electrons of known wavelength, the diffraction patterns can be used to deduce the geometry of the crystalline structure and the interatomic spacings of solids. We mention briefly two applications of the principles of x-ray diffraction.

(1) The verification of the Compton effect, in which scattered photons appear with an unmodified and a modified wavelength, can be made by using an x-ray spectrometer to measure the wavelengths of the radiation scattered by a target.

(2) When a beam of x-rays having a continuous range of wavelengths is incident on a crystal, there will be constructive interference leading to a beam deviated by an angle 2θ only if the Bragg law is satisfied. For a given angle θ, order n, and interatomic spacing d, the value of the wavelength is uniquely specified by the Bragg law, Equation 4-4, and only a single wavelength will be strongly reflected at the angle 2θ. A crystal in this arrangement acts as a *monochromater*, in that it selects from the continuous range of wavelengths incident on the crystal a single monochromatic beam emerging at the angle 2θ.

Let us now consider the experiment of Davisson and Germer in which the wave properties of electrons were first confirmed. A beam of 54-ev electrons is incident on a single crystal of nickel. Electrons leave the nickel surface with a kinetic energy of several electron volts for two reasons:

(1) Secondary emission, in which the incident electrons impart their kinetic energy to electrons in the metal, which are then released.

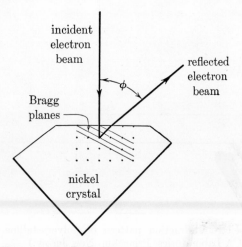

Figure 4-9. Reflection of electron waves by one set of Bragg planes in a crystal of nickel.

(2) Electron diffraction, in which the incident electrons are diffracted within the nickel crystal by reflection from the Bragg planes shown in Figure 4-9.

The number of electrons leaving in the direction ϕ is shown in Figure 4·10. Davisson and Germer found that, in addition to the smooth variation

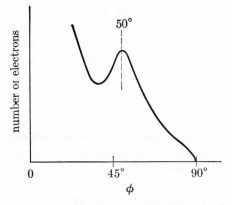

Figure 4-10. Measured distribution of electrons reflected from a crystal of nickel as a function of the angle ϕ (see Fig. 4-9).

in the electron intensity, arising from secondary emission, there was **a** pronounced peak at $\phi = 50°$, which could be attributed to electron diffraction. The computed direction for strong reflection of 54-ev electrons in nickel is 50°. Thus the de Broglie relation was confirmed. The analysis of this electron-diffraction experiment is complicated by the fact that the wavelength of the electrons within the nickel crystal is *not* the same as the wavelength in free space. This difference in de Broglie wavelength arises from a change in the speed of electrons as they enter or leave the nickel surface. The electrons increase their speeds as they pass into the interior of the material because of work done on them at the surface; this work is equal to the work function of the particular material. Because of the change in speed, the electrons are refracted at the surface.

Shortly after Davisson and Germer's confirmation of the de Broglie relation for electrons, G. P. Thomson observed the diffraction rings due to the transmission of electrons through a thin metallic foil, similar to Figure 4-8. It is interesting to note that G. P. Thomson, whose experiments showed the wave properties of electrons (1927), was the son of J. J. Thomson, whose cathode-ray experiments showed the particle properties of electrons (1897).

The fact that a material particle has associated with it a wavelength given by h/mv has been established not only for electrons, but also for atoms, molecules, and the uncharged nuclear particle, the neutron. When a neutron passes through a material it makes collisions with the atoms of the material. At first, the neutron loses kinetic energy at each collision, the struck atoms gaining the energy lost by the neutron; this continues until the neutron's energy is comparable to the thermal energy of the atoms in the material. Then, when the neutron collides with an atom, the neutron can gain energy as well as lose it. When the neutron reaches thermal equilibrium with the material, that is when a typical neutron has an equal probability of gaining or losing kinetic energy in a collision, its behavior is like that of a molecule in a gas, and a temperature can be attributed to neutrons. The relationship between the average kinetic energy and the absolute temperature is given by Equation 1-25

$$K = \tfrac{1}{2} mv^2 = \tfrac{3}{2} kT$$

When the temperature T of the material is room temperature, 300°K, a neutron with this average kinetic energy is said to be a *thermal neutron*. We can compute the de Broglie wavelength of a thermal neutron, whose mass is 1.67×10^{-27} kg.

$$\tfrac{1}{2} mv^2 = \frac{p^2}{2m} = \frac{1}{2m} \left(\frac{h}{\lambda}\right)^2 = \tfrac{3}{2} kT$$

$$\lambda = \frac{h}{\sqrt{3mkT}} = 1.80 \text{ Å} \qquad [4\text{-}5]$$

Therefore, the de Broglie wavelength of a thermal neutron, 1.8 Å, is comparable to the distance between atoms in a crystalline solid, and *neutron diffraction* can be and is observed with thermal neutrons.

It is significant to note that the de Broglie relation attributes a wavelength to *any* material particle which has momentum. Since neutrons can be diffracted, the wave properties are not dependent on the particle's having an electric charge. Furthermore, atoms or molecules, which have an internal structure, show diffraction effects; therefore, the de Broglie relation can be applied even to systems of particles.

4-4   The principle of complementarity   We have seen that it is necessary to attribute both wave characteristics and particle characteristics to electromagnetic radiation and to material particles. Of course, this wave-particle duality makes us uneasy at first sight. We will examine critically the origin of our uneasiness; what is implied when it is stated that electromagnetic radiation or a material particle behaves both as a wave and as a particle; and finally, we will see how this dilemma is resolved by the principle of complementarity.

Let us first recall that the concepts of particle and wave are so basic in physics because they represent the only two possible modes of energy transport (Section 1-8). When energy is transported we can always describe its propagation by waves or by particles. In describing any ordinary large-scale phenomenon of energy transport in classical physics, we are always successful in applying *either* the particle description *or* the wave description. For example, a disturbance which travels on the surface of a pond of water is certainly a wave phenomenon in which energy is transported from one point to another by the wave motion, and a thrown baseball illustrates the transport of energy by a (large) particle. There is never any doubt as to which description we should apply in these two instances, in both of which we can see directly the moving disturbances.

Now let us turn to somewhat less direct illustrations of wave and particle behavior. The propagation of sound through a medium can be understood as a wave disturbance. Here we do not see the waves, as was the case for the water waves. Nevertheless, we apply the wave description to the propagation of sound with confidence because the phenomenon is altogether similar, insofar as the interpretation of diffraction and interference is concerned, to that which we saw directly for water waves. Therefore, when we say that the propagation of sound shows wave aspects, what we are really implying is that sound behaves *as if* it were a wave; the propagation of sound can be explained by a *wave model* because a wave model agrees with *all* experimental observations on sound. Next let us consider particle behavior as it appears in the kinetic theory of gases. We never see the molecules of a gas directly, but we are quite sure that their behavior is like that of very small hard spheres because a variety of experiments can be understood on the basis of assuming that the gas consists of particles, or molecules. Again, what we are saying is that a gas behaves *as if* it consisted of particles and that a *particle model* is the only appropriate means of describing this behavior of the gas. Thus, when we describe phenomena that are somewhat remote from our ordinary experience, whose details we do not "see" directly, we still apply one of the two modes of description, wave or particle, because one of these two descriptions is always successful in accounting for the experimental facts. Moreover, in our ordinary experience with large-size objects we find that any given phenomenon can *always* be described by a *single* model; for example, a baseball is always described as a particle, never as a wave, and a disturbance that travels on a water surface is always described in terms of waves, never in terms of particles.

Furthermore, the wave and particle descriptions are mutually incompatible and contradictory. Let us recall (Section 1-8) that if a wave is to have its frequency or its wavelength given with infinite precision, then the wave must have an infinite extension in space. Conversely, if a wave dis-

turbance is confined to some limited region of space, so that the energy of
the disturbance is localized at any one time to a small region, then the disturbance *resembles* a particle by virtue of its localizability. But such a
wave disturbance, confined to a limited region of space, cannot be characterized by a single frequency and wavelength; instead, a large number of
ideal sinusoidal waves, each with a specific frequency and wavelength,
must be superimposed to give the confined wave disturbance (see Appendix
II). Therefore, an ideal wave, one whose frequency and wavelength are
known with certainty, is altogether incompatible with an ideal particle,
which has a zero extension in space and for which such terms as frequency
and wavelength are meaningless.

Any energy-transport phenomenon, however remote from our direct observation or experience, must be described in terms of waves or of particles.
If the phenomenon is to be *visualized* at all—if we are to have a mental
picture of what goes on in interactions that are inaccessible to direct and
immediate observation—it must be in terms of a wave behavior or a particle behavior; there is no alternative. But, by the mutually contradictory
characteristics of a particle and of a wave, we *cannot simultaneously* apply
a particle description and a wave description; we can and must use one or
the other, never both at the same time.

What strikes us as so disturbing about the wave-particle duality of
electromagnetic radiation and of material particles is the fact that we
apply *both* the wave and particle descriptions. But if we review our interpretation of the experiments discussed thus far, we find that we have *never*
applied both the wave description and the particle description *simultaneously*, which is, as we have seen, logically impossible.

Consider first electromagnetic radiation. The wave model is used to
describe the experiments in interference and diffraction. To interpret these
experiments we say that light consists of waves in that we are confronted
with alternate light and dark bands (that is, alternating regions of large
and small light intensity) which are predicted and accounted for by wave
theory. We never apply the particle description to light propagation when
accounting for the interference and diffraction phenomena, because we
surely cannot apply the principle of superposition and Huygens' principle,
the foundations of wave theory, to particles. Of course, our confidence in
the wave model for the *propagation* of light is strengthened by the fact
that the classical theory of electromagnetic waves of Maxwell predicts all
of the *wave* phenomena observed for light. But it would be rash, in view
of the open-ended, tentative, and incomplete nature of all physical theory,
to conclude that Maxwell's equations are the final equations, or the last
word, on electromagnetic theory. In fact, we have seen that the classical
electromagnetic theory is incomplete to the extent that it cannot account
for quantum effects. In summary, we use the wave model to describe the

propagation of light; we do not, need not, and cannot apply the particle model to interference and diffraction effects.

Now let us turn to the experiments that call for a particle model for electromagnetic radiation. These experiments are the photoelectric effect, the Compton effect, pair production, and pair annihilation. All of these effects show electromagnetic radiation in *interaction* with material particles. We assume that the radiation consists of photons and ascribe to each photon a specific energy and momentum. To ascribe energy and momentum to an electromagnetic *wave* is perfectly possible and indeed necessary (see Section 1-3), but when we ascribe energy and momentum to a photon, we also imply that the energy $h\nu$ and the momentum h/λ are localized at a particular point in space; namely, at the position of the particle-like photon. The quantum effects require the particle description inasmuch as each photon-electron interaction is, in effect, a type of collision; we get agreement with experiment only when electromagnetic radiation is assumed to consist of particles. In short, if we are to visualize the photon-electron interactions by means of a model, it must be the *particle model*. We do not, need not, and cannot apply the wave model of electromagnetic radiation to the photon-electron interactions.

Electromagnetic radiation shows both wave and particle aspects but *not* simultaneously and *not* in the same experiments. Both the wave and the particle aspects are essential features of electromagnetic radiation, and we have no basis for accepting one and rejecting the other. According to the *principle of complementarity*, which was enunciated by Niels Bohr in 1928, *the wave and particle aspects* of electromagnetic radiation *are complementary*. To interpret the behavior of electromagnetic radiation in any one experiment in terms of meaningful visual pictures, we must choose either the particle or the wave description. An experiment showing interference or diffraction requires a wave interpretation, and it is impossible to apply simultaneously a particle interpretation; an experiment showing distinctively quantum effects and the existence of photons requires a particle interpretation, and it is impossible to apply simultaneously a wave interpretation. The wave and particle aspects are *complementary* in that our knowledge of the properties of electromagnetic radiation is partial and incomplete unless both the wave and particle aspects are known; but the choice of one description, which is imposed by the nature of the experiment, precludes the simultaneous application of the second description. We are confronted with a true dilemma in which we must make one of two possible choices. Electromagnetic radiation is a more complicated entity than can be comprehended in the simple and extreme notions of wave and particle, notions which are borrowed from our direct, ordinary experience with large-scale phenomena. Just as the theory of relativity reveals that the common-sense ideas of space, time, and mass are inapplicable in explaining high-speed phenomena, so too the quantum theory, through the wave-particle duality, shows that simple common-sense concepts are inadequate to describe submicroscopic phenomena.

We have seen how Bohr's principle of complementarity elucidates the dual wave-particle aspects of electromagnetic radiation. Let us now examine the wave-particle duality of material particles, such as electrons, to see how the complementarity principle applies. There are many experiments that illustrate the particle-like nature of electrons, but we shall concentrate on J. J. Thomson's cathode-ray experiments, which first showed electrons to be particles. The electrons in a cathode-ray tube follow well-defined paths and indicate their collisions with a fluorescent screen by very small bright flashes. The electrons also are deflected by electric and magnetic fields. It is inferred that electrons are particles (or more properly, that a particle model can be used to describe the behavior of electrons observed in cathode-ray experiments), because all of these effects observed in the cathode-ray experiments make sense if the energy, momentum, and electric charge of the electron are assigned at any one time to a small region of space. When they interact with other objects, electrons behave *as if* they were particles. By the principle of complementarity, we see the particle nature of electrons is revealed in the cathode-ray experiments and, therefore, the wave nature of electrons *must* be suppressed.

The wave nature of electrons appears in the experiments showing electron diffraction, where it is assumed that electrons are propagated as waves having a precisely defined wavelength, h/mv; the de Broglie waves have an indefinite extension in space, and it is, of course, impossible to specify the location of the electron and to follow its motion. In short, the electron-diffraction experiments show the wave nature of electrons; and by the principle of complementarity, the particle nature is suppressed in these experiments. The wave and particle aspects of electrons complement each other; to understand fully the properties of the electron we must accept both descriptions. Again, the electron, or any other material particle, is a more complicated entity than can be fully comprehended in the simple terms, particle or wave; we must use one or the other of these two descriptions to visualize the results of any particular experiment, but we cannot apply both simultaneously.

4-5 The probability interpretation of de Broglie waves The wave nature of electromagnetic radiation is found in the existence of oscillating electric and magnetic fields in space; therefore, the *de Broglie wave associated with a photon is the electromagnetic field*. We wish to inquire more closely into the nature of the de Broglie wave associated with a material particle, and to answer the question, "What is it that is waving when we say that an electron, or any other material particle, shows wave properties?"

Let us consider first a screen which is illuminated uniformly by a mono-chromatic beam of electromagnetic radiation falling perpendicularly on the surface. When the intensity of the light is fairly large, the screen is uniformly illuminated and appears to the eye to have a constant illumination over the entire area; similarly, a photographic plate placed at the screen will show a uniform darkening over the entire area upon being exposed and then developed. The intensity I of the electromagnetic radiation, the energy per unit area per unit time, is given by Equation 1-16:

$$I = \epsilon_0 \mathcal{E}^2 c$$

where ϵ_0 is the electric permittivity of free space and $\mathcal{E}$ is the strength of the electric field at any point on the screen (Section 1-4).†

Now let us suppose that the intensity of the beam is made very weak indeed. If the screen is now examined, we do *not* see a uniformly illuminated area; instead, what is seen is a collection of distinct, bright flashes randomly arranged over the screen. Each bright flash corresponds to the arrival of a single photon. (Strictly, of course, one cannot *see* such a weakly illuminated screen and detect with the eye the arrival of single photons; but one can infer the arrival of photons, one by one, by using more sophisticated detection techniques.) Neither the position nor the time at which a single photon will strike the screen can be predicted in advance; the arrangement of photons on the screen is completely random. One can, however, predict the average number of photons arriving per unit area per unit time; this is the photon flux N. The intensity of a monochromatic beam can be given in terms of the photon flux as follows:

[3-20] $$I = (h\nu)N$$

where $h\nu$ is the energy per photon.

When the intensity of the light beam is large, then the number of photons arriving at the screen is so great that the essentially granular and discrete nature of the electromagnetic radiation is obscured by the great number of photons, and the distinct and randomly arranged bright flashes merge into a seemingly continuous and constant illumination. Instead of increasing the intensity to see a uniformly illuminated screen, one could use a very weak beam, record on the screen the position of each flash, and then find that, after a long time has elapsed, the screen is again uniformly covered.

The situation we have been discussing is somewhat similar to that encountered in the kinetic theory of gases, where one attributes the apparently continuous pressure of a gas on the walls of the container to the combined effects of individual molecular collisions with the walls. The arrival of the molecules is essentially random and discrete; but, because

---

† We could, of course, express the intensity equally well in terms of the magnetic field, rather than the electric field.

of their enormous number under ordinary conditions, the net effect of the molecular collisions appears to be continuous.

Consider a beam of monochromatic light having an intensity of 1.00×10^{-13} watt/m^2 and consisting of visible-light photons each with an energy of 5.00 ev $= 8.00 \times 10^{-19}$ joules. Then the photon flux is

$$N = \frac{I}{h\nu} = \frac{1.00 \times 10^{-13} \text{ watt/m}^2}{8.00 \times 10^{-19} \text{ joule/photon}}$$

$$= 1.25 \times 10^5 \text{ photon/m}^2\text{-sec}$$

or
$$N = 12.5 \text{ photon/cm}^2\text{-sec}$$

This means that, *on the average*, 12.5 bright flashes will be observed over an area of 1 cm^2 during a period of 1 sec. It is, of course, impossible to observe a fraction of a photon, and we will never see 12.5 photons on the 1 cm^2 area in any 1 second interval; only an integral number of photons can be observed. Thus, in one interval, 11 flashes might be seen, then 13, etc.; so that if an average is taken over many intervals, the average number of flashes observed will be 12.5. Furthermore, the spatial distribution of photons over the 1 cm^2 area will *not* be the same for all the 1 second intervals. The flashes will be randomly located and only approach a uniform distribution over a long period of time. Therefore, *the photon flux does not give precise information on the location of any one photon, but gives only the probability for finding a photon.* We can write

$$N \propto \text{probability of finding a photon}$$

We can define the intensity of monochromatic electromagnetic radiation using either the wave description ($I = \epsilon_0 \mathcal{E}^2 c$) or the particle description ($I = h\nu N$). This is highly significant, for we have in the intensity a quantity that has a precise meaning both for the wave and particle descriptions. The intensity bridges the two disparate models.

Let us see what new meaning can be assigned to the square of the electric field strength, $\mathcal{E}^2$, in view of the photon description of light. Our discussion of the weakly illuminated screen has shown us that it is not possible to specify the position or time at which any one photon will strike the screen, but to give only the probability of finding a photon within a given area on the screen within a certain time; that is,

$$N \propto \text{probability of finding a photon}$$

Equating the two expressions for the intensity we have

$$I = h\nu N = \epsilon_0 \mathcal{E}^2 c$$

Therefore,
$$N \propto \mathcal{E}^2$$

Hence,
$$\boxed{\text{probability of finding a photon} \propto \mathcal{E}^2}$$

The probability of finding a photon is proportional to the square of the electric field strength. Thus, the electric field is, from the point of view of the quantum nature of electromagnetic radiation, not merely a quantity which gives the electric force per unit electric charge; it is in addition that quantity, or function, whose square gives the probability of finding a photon at any given place. The classical electromagnetic theory is capable of giving the probability of finding photons, although it is impotent to yield the strictly quantum features of electromagnetic radiation.

We are able to give meaning to the wave nature of a material particle, such as an electron, in the following way: we assume that an exactly analogous relationship exists between the probability for finding a material particle and the square of the amplitude of the de Broglie wave associated with it, as exists between the probability of finding a zero-rest-mass photon and the square of the amplitude of the de Broglie wave (electric field strength) associated with the photon. The displacement of the de Broglie wave associated with a material particle is represented by ψ, which is called simply *the wave function. The wave function ψ is that quantity whose square, ψ^2, is proportional to the probability of finding a material particle;* thus $\psi^2 \, dx$ is the probability of finding the particle between x and $x + dx$.

> probability of finding material particle in the interval $dx \propto \psi^2 \, dx$

The wave function then plays the role for a material particle that is played by the electric field $\mathcal{E}$ for a photon; just as $\mathcal{E}$ will, in general, be a function of both position and time, so too the wave function is in general dependent upon both position and time.

It is impossible to specify with complete certainty the location of a photon in a beam of electromagnetic radiation, but it is possible to specify by $\mathcal{E}^2$ the probability of finding a photon at a particular position at a particular time. Similarly, it is impossible to specify with complete certainty the location of a single material particle, but it is possible to specify by ψ^2 the probability of finding the particle at a particular place at a particular time. Thus, the de Broglie wave function gives, in essence, a *probability interpretation.*

The interpretation of the wave nature of material particles in terms of probabilities was first given in 1926 by Max Born. That branch of quantum physics which deals with the problem of finding the values of ψ is known as *wave mechanics,* or *quantum mechanics.* The two principal originators of the wave mechanics of particles were Erwin Schrödinger (1926) and Werner Heisenberg (1925), who independently formulated quantum mechanics in different but equivalent mathematical forms.

Just as the electromagnetic theory of Maxwell is summarized in the Maxwell equations, which are the basis for computing the value of $\mathcal{E}$, the wave mechanics of matter is governed by the *Schrödinger equation,* which is the basis for computing values of ψ in any problem in quantum physics. Here the parallel stops, however. Whereas the electric field strength, which has its origin in electric charges, gives not only the probability of locating a photon, but also the electric force on a unit positive electric charge, the wave function ψ of the Schrödinger equation has a physical meaning *only* in terms of the probability interpretation. It does *not* indicate any sort of force. The wave function ψ is not directly measureable or observable; it does, however, give the most information one can extract concerning any system of objects, and *all* measurable quantities, such as the energy and momentum, as well as the probability of location, can be found from a knowledge of ψ.

The Schrödinger equation has been extraordinarily successful wherever it has been applied, such as to problems in atomic structure. A simple derivation of the Schrödinger equation is given in Appendix III.

4-6   The uncertainty principle   The principle of complementarity shows the impossibility of applying simultaneously the wave and particle descriptions to a material particle or to a photon. If we choose one description, we preclude the other. If radiation is described in the language of particles, we can locate the photon's position at any instant of time with complete precision; that is, the uncertainties in position Δx and time Δt are both zero. The wave attributes, wavelength and frequency, must be completely uncertain. Therefore, when $\Delta x = 0$ and $\Delta t = 0$, both the uncertainties $\Delta \nu$ and $\Delta \lambda$ must be infinite.

Now let us consider a somewhat less extreme situation, one in which we are content to know the position and time of a photon, not with complete certainty, but with finite uncertainties Δx and Δt.

Our earlier analysis in Section 1-8 showed the limitations with which the frequency or wavelength of a wave can be measured. We saw that if the frequency ν of a wave is measured for a finite period of time Δt, then the frequency will be uncertain by an amount $\Delta \nu$, which is given by Equation 1-28

$$\Delta \nu \, \Delta t \geq 1 \qquad [4\text{-}6]$$

Inasmuch as the energy E of a photon is related to its frequency ν by $\nu = E/h$, then any uncertainty in the frequency will result in an uncertainty in the energy ΔE given by

$$\Delta E = h \, \Delta \nu$$

Using this equation in Equation 4-6, we have

$$\boxed{\Delta E \, \Delta t \geq h} \qquad [4\text{-}7]$$

Thus the product of the uncertainties in the energy and in the time is always at least as large as Planck's constant h. The meaning of Equation

4-7 in words is this: if a photon is known to exist in the energy-state E over a period of time Δt, then this energy is uncertain by, at least, an amount $h/\Delta t$. The energy of a photon can, therefore, be given with infinite precision ($\Delta E = 0$) only if the photon exists for an infinite period of time ($\Delta t = \infty$). Equation 4-7 is one important form of the celebrated *uncertainty principle*, or *principle of indeterminacy*, first introduced by Werner Heisenberg in 1927. We shall explore its fuller meaning after a second formulation of the uncertainty principle.

In Section 1-8 we also saw that if the wavelength λ of a wave is measured over a limited distance Δx, then the wavelength is uncertain by an amount $\Delta\lambda$, which is given by Equation 1-31

$$\Delta\lambda\,\Delta x \geq \lambda^2 \qquad\qquad [4\text{-}8]$$

When we consider the wave aspect of a photon or a material particle, with the wavelength related to the momentum p_x of the photon or material particle by $\lambda = h/p_x$, then any uncertainty in the wavelength will result in an uncertainty in the magnitude of the momentum Δp_x.

$$\Delta\lambda = \frac{h}{p_x^2}\,\Delta p_x$$

Using this equation in Equation 4-8 gives

$$\Delta\lambda\,\Delta x = (h\,\Delta p_x/p_x^2)(\Delta x) \geq \lambda^2$$

$$\Delta p_x\,\Delta x \geq (\lambda p)^2/h$$

$$\boxed{\Delta p_x\,\Delta x \geq h} \qquad\qquad [4\text{-}9]$$

The wavelength was assumed to be measured in the finite distance Δx, along the direction of travel of the wave; from the point of view of the wave-particle duality, the quantity Δx can be interpreted as the uncertainty in the position of the particle. Thus Equation 4-9 shows that the product of the uncertainties in the position and the momentum is equal to or greater than Planck's constant. It is to be noted that the momentum and position referred to in Equation 4-9 are both to be measured along the *same direction*. The uncertainty relation shows that it is impossible to specify simultaneously and with infinite precision the linear momentum and the *corresponding* position of a particle or a photon. Although $\Delta p_x\,\Delta x \geq h$, the product $\Delta p_y\,\Delta x$ can be equal to zero; that is, the uncertainty principle places no restriction on the precise simultaneous measurement of momentum and position which are mutually perpendicular to each other.

The fundamental limitation on the certainty of measurements of energy and time, or of position and momentum, is in harmony with the principle of complementarity. For if the particle nature of, say, an electron is to be perfectly displayed, then both Δx and Δt must be zero. Therefore, when the particle aspect is chosen, the wave aspect is necessarily suppressed.

All of the quantities ν, E, λ, and p are completely uncertain, which follows either from the uncertainty principle or from the principle of complementarity. On the other hand, if the wave characteristics of a material particle or of electromagnetic radiation are to be defined perfectly, that is if $\Delta\nu = 0$ and $\Delta\lambda = 0$ (also $\Delta E = 0$ and $\Delta p = 0$), then by the principle of complementarity or by the uncertainty principle we are precluded from giving simultaneously the distinctively particle characteristics precise location in space and in time. Hence, x and t are completely uncertain.

To derive the uncertainty principle in a different way, we will consider the diffraction of waves by a single parallel slit. A monochromatic plane wave is incident on a slit of width d, and the diffraction pattern is formed

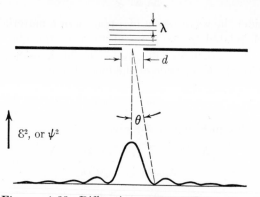

Figure 4-11. Diffraction pattern of a monochromatic plane wave incident on a slit of width d.

on a distant screen as shown in Figure 4-11. The location of the points of zero intensity is given by the equation (see any elementary textbook)

$$\sin \theta = n\lambda/d \qquad [4\text{-}10]$$

where λ is the wavelength and n is 1, 2, 3, ... The total illumination of the central band is much greater than that of any of the other bands, inasmuch as the area under it far exceeds that under any of the others. In fact, the area under the central hump is approximately three times the total area under the remaining ones; therefore, roughly three-fourths of the energy passing through the slit falls within this central region. The limits of this central region are given by Equation 4-10 for $n = 1$.

$$\sin \theta = \pm\lambda/d \qquad [4\text{-}11]$$

We have not yet specified what sort of wave is diffracted by the single slit. If the incident wave consists of monochromatic electromagnetic radiation, then the intensity of the resultant diffraction pattern is proportional to $\mathcal{E}^2$, where $\mathcal{E}$ is the resultant electric field strength at the screen.

If, on the other hand, the wave incident on the slit consists of a mono-energetic, and therefore monochromatic, beam of electrons, the intensity of the diffraction pattern is proportional to ψ^2, where ψ^2, the square of the wave function at the screen, gives the probability of finding an electron at any point along the screen. Whether the waves are electromagnetic waves or the de Broglie waves of a material particle, the diffraction effects will be pronounced only when the wavelength is comparable to the slit width d (see Figure 1-18). If the wavelength is much less than the slit width, the intensity pattern on the screen corresponds to the geometrical "shadow" of the slit.

Suppose now that the intensity of the electromagnetic radiation, or the number of electrons, is reduced so that the number of particles passing through the slit is drastically decreased. Then, if the screen is observed, we no longer see smooth variations in the intensity of the light or in the number of electrons along the screen; instead, we see the photons or electrons arriving one by one. The intensity of the diffraction pattern is given by $\mathcal{E}^2$ for photons and by ψ^2 for electrons; thus, in Figure 4-11, the intensity represents the *probability* for finding a photon or an electron on the screen. There is a 75 per cent probability of finding a photon within the central hump in the diffraction pattern, a smaller probability of finding the photon in any other hump, and *zero* probability of finding the photon at the zeros in the intensity pattern. Under very low illumination of the slit by photons or electrons, bright flashes appear over a large area of the screen. As time passes, more and more particles accumulate on the screen, and the distinct bright flashes merge to form the smoothly varying intensity pattern predicted by wave theory.

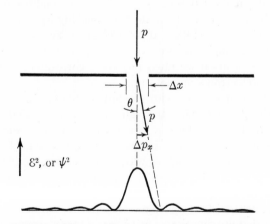

Figure 4-12. An illustration of the uncertainty of momentum for a particle which has passed through a single slit.

There is *no* way of predicting in advance where any one electron or photon will fall on the screen. All that wave mechanics permits us to know is the probability for finding the particle at any one point. Before the particles pass through the diffraction slit, their momentum is known with complete precision both in magnitude (monochromatic waves) and in direction (vertically downward). When they pass through the slit, their position along the X direction—which was completely uncertain before reaching the slit—is now known with an uncertainty $\Delta x = d$, the slit width. However, we do not know precisely where any one particle will strike the screen. Any particle has approximately a three-to-one chance of falling anywhere within the central region, whose boundaries are given by Equation 4-11. Therefore, there will be an uncertainty in the X component of the momentum, p_x, which is *at least* as big as $p \sin \theta$, as is easily seen in Figure 4-12. Therefore,

$$\Delta p_x \geq p \sin \theta \qquad [4\text{-}12]$$

Using Equation 4-11, we have

$$\Delta p_x \geq p\lambda/\Delta x$$

and since $p = h/\lambda$, we have

$$\Delta p_x \, \Delta x \geq h$$

the *Heisenberg uncertainty relation.*

Now suppose that Δx in our example is very large; that is, that the slit is very wide. Then the uncertainty in the position is increased, and we cannot be certain as to where an electron is located. But the uncertainty in the momentum is reduced correspondingly, which is shown by the fact that the diffraction effect becomes less pronounced and essentially all electrons fall within the geometrical shadow (see Figure 1-18). Conversely, as the slit is reduced in width and Δx becomes very small, the diffraction pattern is expanded along the screen, and for the increase in our certainty of the electron's position we must pay by a correspondingly greater uncertainty in the electron's momentum.

We see that when the slit width is much greater than the de Broglie wavelength, the particles pass through the slit undeviated to fall within the geometrical shadow. This is in agreement with classical mechanics, where the wave aspect of material particles is ignored. Thus there is a close parallel in the relationship of wave optics to ray optics, and of wave mechanics to classical mechanics. Ray optics is a good approximation of wave optics whenever the wavelength is much less than the dimensions of obstacles or apertures that the light encounters; similarly, classical mechanics is a good approximation of wave mechanics whenever the de Broglie wavelength is much less than the dimensions of obstacles or apertures encountered by material particles. Symbolically, we can write

| Limit (wave optics) = (ray optics) |
| $\lambda/d \to 0$ |
| Limit (wave mechanics) = (classical mechanics) |
| $\lambda/d \to 0$ |

It is to be noticed that no ingenious subtlety in the design of the diffraction experiment will remove the basic uncertainty. Here we do *not* have, as in the large-scale phenomena encountered in classical physics, a situation in which the disturbances on the measured object can be made indefinitely small by ingenuity and care. The limitation here is rooted in the fundamental quantum nature of electrons and photons; it is intrinsic in their complementary wave and particle aspects.

As an illustration of the uncertainty principle, we will compute the uncertainty in the momentum of a 1000 ev electron whose position is uncertain by no more than 1 Å $= 1.0 \times 10^{-10}$ m, the approximate size of atoms. From $\Delta p_x \geq h/\Delta x$ it follows that $\Delta p_x = 6.6 \times 10^{-24}$ kg-m/sec. Now let us compare this uncertainty in the momentum with the momentum itself, $p_x = (2mK)^{1/2} = 17 \times 10^{-24}$ kg-m/sec. Therefore, the fractional uncertainty in the momentum is $\Delta p_x/p_x = 6.6/17 = 39$ per cent! Thus, the uncertainty principle makes it impossible to specify with even moderate accuracy the momentum of an electron confined to atomic dimensions.

Consider now the uncertainty that arises when we have a 10.0 gram body moving at a speed of 10.0 cm/sec. This is an ordinary-size object moving at an ordinary speed. Let us further assume that the position of the object is uncertain by no more than 1.0×10^{-3} mm. We wish to find the uncertainty in the momentum, and more especially, the fractional uncertainty in the momentum. We find $\Delta p_x = 6.6 \times 10^{-28}$ kg-m/sec, and $p_x = 1.0 \times 10^{-3}$ kg-m/sec. Therefore, $\Delta p_x/p_x = 6.6 \times 10^{-25}$! The fractional uncertainty in the momentum arising in this example of a large-scale, macroscopic body is so extraordinarily small as to be negligible compared with all possible experimental limitations. The uncertainty principle imposes a fundamental limitation on the certainty of measurements only in the microscopic domain, where the wave-particle duality is important. In the macroscopic domain, the uncertainties are, in effect, trivial (see Figure 1-1).

Figure 4-13 shows the momentum of the electron in our example above plotted against position. The uncertainty principle requires that the shaded area in this figure, which gives the product of the uncertainties in the momentum and position, be equal in magnitude to Planck's constant h. If the position is known with high precision, the momentum is rendered highly uncertain; if the momentum is specified with high certainty, the position must necessarily be highly indefinite. Thus it is impossible to predict and follow in detail the path of an electron confined to essentially atomic dimensions. Newton's laws of motion, which are known to be com-

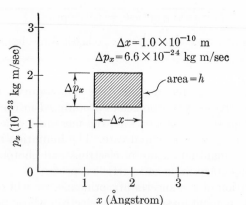

Figure 4-13. Representation of the uncertainties in the simultaneous measurement of position and momentum of an electron.

pletely satisfactory for giving the paths of large-scale particles, are inapplicable here; in order to predict the future course of any particle it is necessary to know not only the forces that act on the particle but also its initial position and momentum. Because *both* position and momentum cannot be given simultaneously without uncertainty, it is not possible to predict the future path of the particle in detail. Instead, *wave mechanics* must be used to find the probability of locating the particle at any future time.

Figure 4-14 shows the momentum and position of the 10.0 gram body moving at 10.0 cm/sec (our earlier example). The area h, representing the

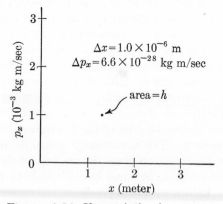

Figure 4-14. Uncertainties in momentum and position of a 10 gm body. (On the scale of this drawing the uncertainty (area h) has been exaggerated by a factor of 10^{26}!)

product of the uncertainties in momentum and position, is so tiny for these macroscopic circumstances that it appears as an infinitesimal point on the figure. Here, the classical laws of mechanics can be applied with no appreciable uncertainty.

We see here another example of the correspondence principle as it applies to relations between classical physics and quantum physics. It is the finite size of Planck's constant that is responsible for quantum effects. These quantum effects are subtle because Planck's constant is very small —but not zero. (Recall that the relativity effects are subtle because the speed of light is very large—but not infinite.) If Planck's constant were zero, the quantum effects would disappear; thus classical physics is the correspondence limit of quantum physics as h is imagined to approach zero. Symbolically,

$$\underset{h \to 0}{\text{Limit}} \text{ (quantum physics)} = \text{(classical physics)}$$

4-7   The quantum description of a confined particle   A particle which is completely free of any external influence will, by Newton's first law, move in a straight line with a constant momentum. In the language of wave mechanics, such a free particle, having a constant, well-defined momentum, must be represented by a monochromatic sinusoidal wave with a well-defined de Broglie wavelength. If the wave associated with the material particle is to have a precisely defined wavelength, the wave must have an infinite extension in space. Therefore, in accordance with the uncertainty principle, when the wavelength, and therefore the momentum of the particle is specified precisely, the position of the particle is altogether uncertain and indeterminate.

In Section 3-1, we saw an example from classical physics of waves with perfectly defined wavelengths, yet confined to a limited region of space. This example was that of resonant standing waves on a string fixed at both ends. The wave on the string is repeatedly reflected from the boundaries, or fixed ends, and constructively interferes with itself. Resonance is achieved only when the length of the string is some integral multiple of half wavelengths (see Figure 3-2c).

We now wish to study an elementary wave-mechanical problem which is analogous to that of standing waves on a string. Let us confine the de Broglie wave associated with a particle in the same way that a transverse wave on a string is confined within the boundaries. We shall assume that the particle moves freely back and forth along the X axis, but that it encounters an infinitely hard wall at $x = 0$ and $x = L$. The particle is then confined between these boundaries. The infinitely hard walls correspond to an infinite potential for all values of x less than zero and greater than L. Because the particle is free between zero and L, its potential energy in this region is constant, chosen to be zero. The situation we have described is that of a *particle in a one-dimensional box*. Because the walls are infinitely hard, the particle imparts none of its kinetic energy to the walls, the total energy of the particle remains constant, and it continues to bounce back and forth between the walls indefinitely.

From the point of view of wave mechanics, we can say that if the particle is to be confined within the limits stated, then the probability for finding the particle at any points outside these limits is zero. Therefore, the wave function ψ, whose square represents the probability of finding the particle, must be zero for $x \leq 0$ and $x \geq L$.

We can summarize mathematically the conditions of our problem as follows:

$$V = \infty \quad \text{for} \quad x < 0, \quad x > L$$

$$V = 0 \quad \text{for} \quad 0 < x < L$$

$$\psi = 0 \quad \text{for} \quad x \leq 0, \quad x \geq L$$

Only those wave functions that satisfy these conditions are allowed. Inasmuch as the particle is free in the entire region between the walls, we know that its de Broglie wave is a sinusoidal wave. But to satisfy the conditions at the boundaries, only those wavelengths are allowed which will permit an integral number of half wavelengths to be fitted between $x = 0$ and $x = L$. The condition for the existence of *stationary*, or standing, de Broglie waves is then:

$$L = n(\lambda/2) \qquad [4\text{-}13]$$

where λ is the de Broglie wavelength and n is the *quantum number*, having the possible values, $n = 1, 2, 3, \ldots$.

Figure 4-15a shows the wave function ψ plotted against x for the first three possible *stationary states* of the particle in a box. The square of the wavefunction gives the probability of finding the corresponding particle, and Figure 4-15b shows ψ^2 plotted against x for each of the first three states. Note that, whereas ψ can be negative as well as positive, ψ^2 is always positive.

The probability distribution is such that ψ^2 is always zero at the boundaries. For the first state, $n = 1$, we see that the probability for finding the particle is greatest at a point midway between the two walls. For the state in which $n = 2$, the probability for finding the particle at the point $x = L/2$ is zero, and it is impossible for the particle to be found at this point!

The imposition of the boundary conditions on ψ, that is the fitting of the de Broglie waves between the walls, has restricted the wavelength of the particle to those values given by Equation 4-13. But if only certain wavelengths are permitted, the magnitude of the momentum is also re-

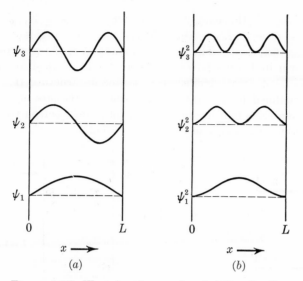

Figure 4-15. Wave functions and probability distributions for the first three stationary states of a particle in a one-dimensional box.

stricted to certain allowed values, since $p = h/\lambda$. Therefore, the permitted momenta are given by

$$p = h/\lambda = hn/2L \qquad [4\text{-}14]$$

Finally, the kinetic energy K (and therefore, the total energy E of the particle, since the potential energy is zero), is given by

$$K = E = \tfrac{1}{2} mv^2 = p^2/2m = (hn/2L)^2/2m$$

$$\boxed{E_n = n^2 \frac{h^2}{8mL^2}} \qquad [4\text{-}15]$$

where m is the mass of the particle. (Equation 4-15 holds only for nonrelativistic speeds.) The energy is written with the subscript n inasmuch as Equation 4-15 shows that the possible values of the energy depend only on the quantum number n for fixed values of m and L. By Equation 4-15 we see that the *energy* of the particle in the one-dimensional box is *quantized*. The particle cannot assume any energy or speed, but is restricted to those particular energies and speeds which satisfy the boundary conditions on the wave function. The quantization of the energy is analogous to the classical quantization of the frequencies of waves on a string fixed at both ends.

Let us compute the possible values of the energy, assuming that an electron with $m = 9.1 \times 10^{-31}$ kg is confined to move back and forth within a distance of $L = 4$ Å $= 4 \times 10^{-10}$ m. Substituting these values in Equation 4-15 gives for the energy of the first state $(n = 1)$, $E_1 = 2.3$ ev. Because $E_n = n^2(h^2/8mL^2) = n^2E_1$, the next possible energies of the particle are $4E_1$, $9E_1$, $16E_1$, ... The permitted energies of the electron in a 4 Å box are shown in Figure 4-16. Such a diagram is called an *energy-level diagram*. It is of great significance for atomic structure that when an electron is confined to a distance approximately the size of the atom, it has possible energies that are in the range of a few electron volts, comparable to the binding energy of electrons in atoms.

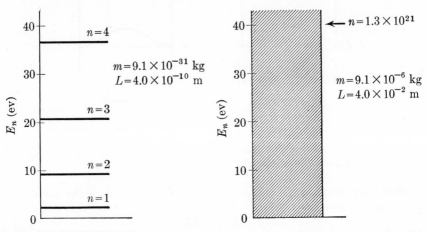

Figure 4-16. Allowed energies of an electron confined to a one-dimensional box of atomic dimensions.

Fig. 4-17. Allowed energies of a 9.1 mg particle confined to a one-dimensional box of 4 cm.

Now consider the allowed energies of a relatively large-size object confined in a relatively large box; assume that $m = 9.1$ milligram $= 9.1 \times 10^{-6}$ kg, and $L = 4$ cm $= 4 \times 10^{-2}$ m. Equation 4-15 shows that for these values of m and L, $E_1 = 2.3 \times 10^{-41}$ ev, a fantastically small energy! Figure 4-17 shows the energy-level diagram for these circumstances, where the energy is plotted to the *same* scale as in Figure 4-16. Of course, the spacing between adjacent possible energies in the energy-level diagram is so very small that the energy is effectively continuous for such macroscopic conditions. This is why we never see any obvious manifestation of the quantization of the energy of a macroscopic particle—the quantization is there, but it is too fine to be discerned. This result is, of course, in agreement with the correspondence principle, which requires that the discrete energies of a bound system must appear continuous for large-scale phenomena.

We note that the lowest possible energy of the particle in the box is *not* zero, but E_1. This is in accord with the uncertainty principle. If the energy were zero, the particle would be at rest somewhere within the box ($\Delta x = L$), and both p and Δp would equal zero. But, $\Delta p \geq h/\Delta x = h/L$, not zero. For an electron confined to an atomic dimension, E_1, the energy of the *ground state* is a few electron volts. We find that the electron is never at rest, but bounces back and forth between the confining walls. This is, of course, true for any particle. Let us compute the speed of the 9.1 milligram particle confined to 4 cm; since $E_1 = 2.3 \times 10^{-41}$ ev $= \frac{1}{2} mv^2$, we find that $v = 9.0 \times 10^{-28}$ m/sec $\simeq 10^{-7}$ Ångstrom/millennium. Of course, it is experimentally impossible to measure so small a speed, and the particle is observed to be essentially at rest.

The problem of the particle in the box is a somewhat artificial one. There is no such thing as an infinitely large potential energy, and a particle cannot easily be compelled to move in a straight line, free of all external influences; nevertheless, the particle in the box is an important problem because it reveals the quantization of the energy. Energy quantization occurs because only certain discrete values of the de Broglie wavelength can be fitted between the boundaries (see Appendix III for the solution, using the Schrödinger equation, of the problem of the particle in the box). As we have mentioned earlier, it is the task of the rather difficult field of quantum mechanics, or wave mechanics, to determine the wave function of a particle or system of particles under more realistic conditions than discussed above.

4-8 Summary Every particle (whether of finite or zero rest mass) has associated with it a de Broglie wave whose wavelength and frequency are given by

$$\lambda = h/p \quad \text{and} \quad \nu = E/h$$

The de Broglie wavelength of a 150-ev electron, the de Broglie wavelength of a thermal ($\frac{1}{25}$ ev) neutron, the interatomic spacing of a solid, and a typical x-ray wavelength are *all* of the order of *one Ångstrom*.

When a wave is reflected from a single Bragg plane, the angles of incidence and reflection are equal. When a wave is reflected from a set of parallel Bragg planes separated by a distance d, constructive interference occurs when

$$n\lambda = 2d \sin \theta$$

X-rays and electrons can be diffracted by crystalline solids. X-ray and electron diffraction can be used to: measure x-ray or electron wavelengths, measure interatomic spacings in crystals, produce monochromatic beams.

The wave and particle aspects of electromagnetic radiation and of material particles are complementary according to Bohr's principle of complementarity.

The probability for finding a photon is proportional to $\mathcal{E}^2$, the square of the electric field strength; the probability for finding a particle is proportional to ψ^2, the square of the wave function.

Heisenberg's uncertainty principle requires that

$$\Delta p_x \, \Delta x \geq h$$

$$\Delta E \, \Delta t \geq h$$

A free particle can exist in a one-dimensional box only if an integral multiple of half-deBroglie-wavelengths can be fitted between the boundaries. The particle can assume only those energies given by

$$E_n = n^2 h^2 / 8mL^2$$

where n is the quantum number, 1, 2, 3, ...

By the correspondence principle,

$$\underset{h \rightarrow 0}{\text{Limit}} \text{ (quantum physics)} = \text{(classical physics)}$$

REFERENCES

Albert Einstein: Philosopher-Scientist, ed. P.A. Schilpp. Evanston, Illinois: The Library of Living Philosophers, Inc., 1949. Chapter 7, "Discussion with Einstein on Epistemological Problems in Atomic Physics," written by Niels Bohr, contains a penetrating discussion of the implications of the uncertainty principle and some hypothetical machines devised in an attempt to violate this principle. It emerges that all these attempts fail.

Born, M., *Atomic Physics*. New York: Hafner Publishing Co., Inc., 1957. Born's probability interpretation of the wave function and a discussion of complementarity are found in Chapter 4.

The Crystalline State, Vol. 1, ed. W. H. Bragg and W. L. Bragg. New York: The Macmillan Co., 1948. The fundamentals of crystalline structure and x-ray diffraction are discussed here.

Heisenberg, W., *The Physical Principles of the Quantum Theory*. Chicago, Illinois: The University of Chicago Press, 1930. The first four chapters concentrate on the uncertainty principle, its meaning and consequences.

Heitler, W., *Elementary Wave Mechanics*. Oxford: The Clarendon Press, 1946. The first hundred pages give a mathematically simple development of the wave mechanics and the Schrödinger equation.

Semat, H., *Introduction to Atomic and Nuclear Physics*. New York: Rinehart & Company, Inc., 1954. Chapter 6 gives a detailed analysis of the Davisson-Germer experiment, and also discusses the diffraction of atoms, molecules, and neutrons.

PROBLEMS

4-1   (a) What is the speed of a proton whose de Broglie wavelength is 0.12 Å? (b) Through what potential difference must a proton be accelerated from rest to acquire this speed?

4-2   What is the kinetic energy (ev) of an electron whose de Broglie wavelength is equal to that of light of wavelength 5000 Å?

4-3   Calculate the wavelength of a monochromatic beam of hydrogen atoms, each with a kinetic energy of 1.50×10^{-3} ev.

4-4   What is the wavelength of an electron which has been accelerated from rest through a potential difference of 15 kilovolts?

4-5   (a) Solve for the de Broglie wavelengths of a particle in terms of its rest mass m_0 and its velocity v by eliminating the relativistic mass. (b) Plot λ versus v on a diagram and show that the slope $d\lambda/dv$ approaches minus infinity both at $v = 0$ and $v = c$.

4-6   Show that the ratio of the de Broglie wavelength to the Compton wavelength (h/m_0c) of a particle can be written as $\sqrt{(c/v)^2 - 1}$.

4-7   Show that the speed of the de Broglie wave associated with a material particle is given by: speed $= E/p = c^2/v$.

4-8 *   In an electron microscope a beam of electrons replaces a light beam, and electric and magnetic focusing fields replace the refracting lenses. The smallest distance (resolving power) that can be resolved by any microscope is approximately equal to the wavelength used in the microscope. A typical electron microscope might use 50 Kev electrons. (a) Calculate the ultimate resolving power for this microscope. (b) By what factor is the actual resolving power of 20 Å (presently attained in well-designed microscopes) different from the ultimate resolving power, which is limited by the wave properties of electrons?

4-9 *   The ultimate resolving power of *any* microscope depends solely on the wavelength. If we wish to study an object having dimensions of 0.10 Å, what is the minimum energy and momentum required if we use: (a) electrons? (b) photons? Comparing the required photon energy with that of the electron, we can see why these high resolving powers can be achieved in an electron microscope, but not in a gamma-ray microscope.

4-10   Calculate the lattice spacing of a K Cl (sylvite) crystal (simple cubic structure). The atomic weights of potassium and chlorine are 39.1 and 35.5 gm/mole, respectively, and the density of K Cl is 1.98 gm/cm³.

4-11   Monochromatic x-rays are incident on a Na Cl crystal, whose lattice spacing is 2.82 Å. First-order Bragg reflection is observed at an angle of 80° with respect to the incident direction. What is the wavelength of the x-rays?

4-12   In the study of x-ray diffraction we have assumed that the scattered x-rays have the same wavelength as the incident x-rays. (a) Justify this by considering Compton scattering of x-ray photons by the crystal a*toms*. (b) What would be the maximum fractional change in wavelength of 0.10 Å x-rays due to the scattering from Na and Cl atoms?

4-13   A narrow beam of thermal neutrons produced by a nuclear reactor falls upon a crystal whose lattice spacing is 1.80 Å. (a) For first-order diffraction, at what angle must the Bragg planes of the crystal be placed with respect to the incident neutron beam so that neutrons of energy 0.0300 ev will be strongly diffracted? (b) What is the angle between this diffracted monochromatic beam and the incident neutron beam?

4-14   A narrow beam of x-rays of wavelength 0.500 Å is incident on K Cl powder, which consists of randomly oriented K Cl microcrystals. The lattice spacing of K Cl is 3.14 Å. A flat photographic plate is placed 10.0 cm behind the powder target and is perpendicular to the incident beam. (a) What is the radius of the circle on the photographic plate due to the first-order reflections from the sets of Bragg planes which are 3.14 Å apart? (b) What is the radius of the circle due to second order reflections from these same planes? (See Figure 4-5.)

4-15   Refer to Problem 4-14. (a) Calculate the radius on the photographic film of the first-order Bragg reflection from the set of 45° Bragg planes (see Figure 4-6), and compare this with the results of Problem 4-14.

4-16 * A monochromatic beam of 0.85 Å x-rays passes through two narrow slits and strikes a target of randomly-oriented crystals whose lattice spacing is 3.0 Å. The target is at the center of a cylindrical sheet of photographic film, as shown in Figure 4-18. (a) At what

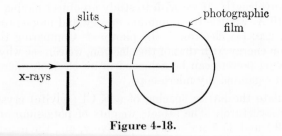

Figure 4-18.

angle will there be first-order reflection from this set of planes? (b) If there are Bragg reflections from several sets of Bragg planes, what is the general appearance of the diffraction pattern on the photographic film when it is laid flat?

4-17 *   What is the longest wavelength radiation which can be observed using the x-ray spectrometer of Problem 4-16? Assume first-order reflection from Bragg planes having a lattice spacing of 3.0 Å.

4-18 *   A narrow polychromatic beam of electrons passes between two closely spaced horizontal parallel plates of electric field strength $\mathcal{E} = 5.5 \times 10^5$ volt/m. A uniform magnetic field of 0.10 weber/m², directed into the paper, is superimposed on the electric field. Those electrons which emerge from the parallel plates next pass through a thin gold foil (composed of microscopic metallic crystals) and strike a fluorescent screen. (a) What is the angle (with respect to the incident electron beam) of the first-order ring due to Bragg reflections from those Bragg planes which are 4.07 Å apart? (b) What would this angle become if the magnetic flux density B were doubled? (c) What is the ratio of the electron wavelengths of parts (a) and (b) above?

4-19   Two parallel wire-mesh screens are separated by 0.20 m. What is the frequency of the radio waves which will show strong first-order Bragg reflection with an angle of 30° between the transmitted and reflected beams?

4-20   A monochromatic beam of electromagnetic radiation has an intensity of 1.00 watt/m². What is the average number of photons per cubic centimeter for (a) 1.00 kilocycle/sec radio waves and (b) 10.0 Mev gamma rays?

4-21   Show that the photon density, the average number of photons per unit volume, is given by N/c, where N is the photon flux.

4-22   In Section 4-5 we saw that $Nh\nu = \epsilon_0 \mathcal{E}^2 c$. Using this with the results of Problem 4-21 above, show that the density of monochromatic photons is just the energy density of an electromagnetic wave (Equation 1-15) divided by the energy per photon.

4-23   What is the photon flux in photons/mm²-sec at a distance of 100 m from a light source emitting 20.0 watt of visible light (6000 Å)?

4-24   A screen of 10.0 cm² is oriented with the normal to the plane making an angle of 85° to the direction of a 6000 Å wavelength light beam having an intensity of 2.0×10^{-4} watt/m². How many photons will strike the screen in an interval of 1.00 microsecond?

4-25 *   A 35-mm camera takes a photograph of a 50-watt light source 100 m away, with a setting of $f/2$ and 1/100 sec. If the focal length of the lens is 50 mm, what is the number of photons (assume all 6000 Å) striking the film during the exposure? (The f/number gives the focal length divided by the effective lens diameter.)

4-26   We wish to measure simultaneously the position and velocity of a one-gram mass. If the velocity in the X-direction is measured to an accuracy of 10^{-5} cm/sec, what is the limit of accuracy with

which we can simultaneously locate the particle (a) along the
X-axis? (b) along the Y-axis?

4-27   If in Problem 4-26 we replace the one-gram mass by an electron,
what will be the limit of accuracy with which the X- and Y-components of position can be determined for the same desired accuracy in the velocity?

4-28   The wavelength of a photon is measured to an accuracy of one
part in a million ($\Delta\lambda/\lambda = 10^{-6}$). What is the uncertainty Δx in
the simultaneous measurement of the location of the photon for
(a) a light photon of wavelength 5000 Å? (b) An x-ray photon of
wavelength 1.00 Å? (c) A gamma-ray photon of wavelength
1.00×10^{-5} Å?

4-29   Consider a particle with energy $E = \frac{1}{2}mv^2$ traveling along the
X-direction. Its x-coordinate is uncertain by Δx. Show that if
$\Delta p_x \Delta x \geq h$, then $\Delta E \Delta t \geq h$, where $\Delta t = \Delta x/v$.

4-30   The size of an atom is around 10^{-10} meter. If we wish to locate
an electron within the atom, we must be able to measure distances
smaller than this. Therefore, to locate an electron to within a distance of, say, 10^{-11} meter, we must use electromagnetic radiation
whose wavelength is at least this small. (a) What would be the
energy of a photon with this wavelength? (b) How large would
the uncertainty in the electron's momentum be if $\Delta x = 10^{-11}$
meter? This is one of Heisenberg's hypothetical experiments illustrating the uncertainty principle and is called the *gamma-ray
microscope* experiment.

4-31   The diameter of the nucleus of an atom, composed of protons and
neutrons, is of the order of 10^{-14} meter. (a) What will be the uncertainty in the momentum of a proton (or a neutron) confined to
this small region of space? (b) The momentum of the particle
must be at least as large as the uncertainty in the momentum.
Assuming the momentum to be equal to the uncertainty in momentum, calculate the kinetic energy and velocity of a proton
bouncing about within the nucleus.

4-32 *   (a) What would be the minimum kinetic energy and velocity of
an electron confined to a nuclear dimension of 10^{-14} m? (b) What
would be the minimum energy of a photon confined to this region
of space?

4-33   What is the lowest possible energy of a neutron confined to a "one-dimensional box" of 1.0×10^{-14} m (the approximate size of the
nucleus)?

4-34   At what temperature is the average kinetic energy of a gas of
hydrogen atoms equal to the energy of an electron confined to a
1.00 Å one-dimensional box?

4-35 * For a particle in a box show that the probability of finding the particle at any interval between x and $x + dx$ is proportional to $\sin^2 (\pi nx/L) \, dx$, where n is the quantum number and L is the length of the box.

4-36 * Show that, for very large quantum numbers, the probability for finding a particle in any small interval in a one-dimensional box is independent of the position within the box. This is in agreement with the classical expectation that the probability of finding a particle which moves with a constant speed is the same for all points.

4-37 Consider a particle in a one-dimensional box. Show that the following two facts are consistent with the uncertainty principle: the energy in any of the permitted energy states is perfectly sharp, and the particle must remain in this energy state for an infinite time.

4-38 Confining an electron to a one-dimensional box of length L introduces an uncertainty in the momentum, and therefore the energy, of the electron. Show that the uncertainty in the energy is of the same order of magnitude as the lowest allowed energy E_1 of the electron (Equation 4-15).

4-39 * (a) Show that the fractional difference in energy between two adjacent energy levels for a particle in a box is given by $(\Delta E/E_n) = (2n + 1)/n^2$. (b) What does this reduce to for large quantum numbers n (i.e., $n \gg 1$)? This is another example of the correspondence principle, $\lim_{n \to \infty} (\Delta E/E) = 0$, inasmuch as a continuous distribution of energies occurs for $n \gg 1$.

4-40 Consider the example in the text of the large-size object of mass 9.1 milligram confined to a region of 4 cm. For simplicity, assume the mass and length to be known exactly. If the particle is known to be moving with a velocity of $(1.00 \pm .01)$ cm/sec: (a) Calculate the average quantum number n. (b) Calculate the uncertainty Δn in the quantum number, and therefore the uncertainty as to which discrete energy the particle has.

4-41 * (a) Show that the wave functions for a particle in a box are given by $\psi_n = A_n \sin (\pi nx/L)$. (b) Imposing the requirement that the total probability for finding the particle between 0 and L is 100 per cent, $\int_0^L \psi_n^2 \, dx = 1$, show that $A_n = \sqrt{\dfrac{2}{L}}$.

FIVE

THE STRUCTURE OF THE HYDROGEN ATOM

5-1 **Alpha-particle scattering** Our modern concept of the structure of an atom posits these essential features: a *nucleus* occupying a very small region of space, in which all of the positive charge and practically all of the atom's mass are concentrated; and negatively charged electrons, which surround this nucleus. Let us examine the evidence for this concept of atomic structure, first proposed by Ernest Rutherford in 1911.

At the end of the nineteenth century it was known that the negative electric charge of the atom is carried by electrons, whose mass is but a small fraction of the total mass of the atom. Because atoms as a whole are ordinarily electrically neutral, it follows that if we were to remove all of an atom's electrons, then what would remain would contain all of the positive electric charge and essentially all of the mass. The question is then, "How is the mass and positive charge distributed throughout the volume of the atom?" Atoms are known, on the basis of a variety of experiments, to have a "size" of the order of one Ångstrom unit; and because the positive charge and mass are confined to at least this small a region, it is impossible by any direct measurement to see and observe any details of the atomic structure. Therefore, an indirect measurement must be resorted to. One of the most powerful methods of studying the distribution of matter or of electric charge is the method of *scattering*, and it was by the α-particle scattering experiments, suggested by Rutherford, that the existence of atomic nuclei was established.

We can best grasp the strategy of the scattering method by considering first a simple example which illustrates the principal features of a scattering experiment. Suppose that we are confronted with a large, black box, the mass of whose contents is known. We are not allowed to look inside to examine its internal structure, yet we are asked to determine how the mass is distributed throughout the interior of the box. The box might, for instance, be filled completely with some material of relatively low density, such as wood; or it might be only partially filled with a high-density material. How can we find out whether one of these two possibilities represents the actual distribution of material within the box? We can use a very simple expedient: shoot bullets into the box and see what happens to them. If we find all of the bullets emerging in the forward direction, possibly with reduced speeds, then it might be inferred that the box is filled with some such material as wood, which deviates the bullets only slightly as they pass through. On the other hand, if we find a few bullets strongly deflected by ricochets from their original direction, it is fair to assume that these bullets have been scattered from the forward direction by virtue of collisions with small, hard, and very dense objects, widely dispersed throughout the box. It is possible, then, by studying the distribution of the scattered bullets to learn much concerning the arrangement of material within the black box. Note that it is *not* necessary to aim the bullets; instead, the shots are fired randomly over the front surface of the box. This is the essence of scattering experiments as applied to atomic and nuclear structure.

Rutherford suggested that the mass and positive charge of an atom are confined to essentially a point-charge and mass-point, called the *nucleus*, and located at the center of the atom. He further suggested that this nuclear hypothesis could be tested by shooting high-speed, positively charged particles (the bullets) through a thin metallic foil (the black box), and then examining the distribution of the scattered particles. At the time of Rutherford's suggestion, the only available and suitable charged particles were α-particles emitted with energies of several Mev from radioactive materials. Rutherford had shown earlier that α-particles consist of doubly-ionized helium atoms; therefore, the α-particle has a positive electric charge twice the magnitude of the electron charge and a mass several thousand times greater than the mass of an electron but considerably smaller than the mass of such heavy atoms as gold. To confirm Rutherford's nuclear hypothesis, H. Geiger and E. Marsden in 1913 scattered α-particles from thin gold foils.

The essentials of a scattering experiment are shown in Figure 5-1. A collimated beam of incident particles strikes a thin foil of scattering mate-

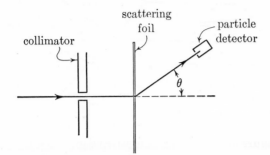

Figure 5-1. Schematic arrangement for a simple scattering experiment showing the collimator, scattering material, and the detector.

rial; a detector counts the number of particles scattered at some scattering angle θ from the incident direction. The experiment consists in measuring the relative number of scattered particles at various scattering angles θ. In one of their experiments, Geiger and Marsden used α-particles having a kinetic energy of 7.68 Mev from a radioactive source (polonium); these α-particles struck a gold foil, whose thickness was 6.00×10^{-5} cm. The rotatable detector consisted of a zinc sulfide screen, viewed by a microscope; α-particles striking the screen produced bright flashes, or scintillations, which could be observed and counted for any angle θ.

Consider the behavior of the α-particles as they traverse the interior of the scattering foil. We can dismiss as inconsequential any encounters the α-particles have with electrons within the material; this is proper because the mass of the α-particle is very much greater than that of an electron, and therefore, the α-particle is essentially undeviated by collisions with electrons and a negligible fraction of its energy is transferred to any one electron. Thus, the α-particles are deflected and scattered only by encounters with the nuclei. These nuclei, in a gold foil, have a mass which is considerably greater (50 times) than that of the α-particle; therefore, the gold nuclei do not recoil appreciably in the collision, and can be assumed to remain at rest. Since the α-particles and the nuclei are both positively charged, the α-particles are repelled by the nuclei. Rutherford assumed not only that the nuclei were point charges, but also that the *only* force acting between the nucleus and an α-particle was the Coulomb electrostatic force. This Coulomb electric force varies inversely with the square of the distance between the charges; therefore, although the Coulomb force is never zero (except for an infinite separation between the charges) the α-particle is acted upon by a *strong* repulsive force only when it comes quite close to a nucleus. In Figure 5-2 are shown a number of paths of

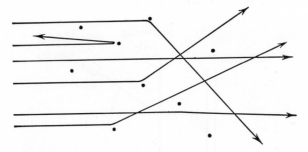

Figure 5-2. Scattering of α-particles by the nuclei of a material. The number of α-particles scattered through sizable angles is greatly exaggerated.

α-particles as they move through the interior of a scattering foil. We see that most of the particles pass through the material with only a slight deviation from their original direction; the chances for a close encounter with a nucleus, or scattering center, are fairly remote. On the other hand, those few α-particles which barely miss a head-on collision are scattered by sizeable angles, and those extremely rare events in which an α-particle makes a head-on collision result in its being scattered through 180°; that is, the α-particle is brought to rest momentarily and then returned along its incident path.

Now let us consider in somewhat greater detail the collision of a positively charged particle (e.g., an α-particle) with a heavy nucleus (e.g., gold). The incident particle is scattered by the angle θ, as shown in Figure 5-3. The particle moves in a nearly straight line until it comes fairly close

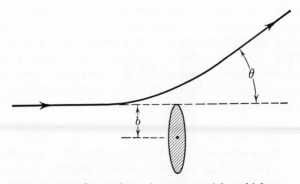

Figure 5-3. Scattering of an α-particle which approaches a heavy nucleus with an impact parameter b.

to the scattering center. In the vicinity of the nucleus, it is deflected and then recedes in a nearly straight line, the path being that of an hyperbola. The incident particle misses making a head-on collision with the nucleus by a distance b; that is, the incident particle would pass the nucleus at a distance b if there were no force to deflect it. This distance is called the *impact parameter*.

We see from Figure 5-3 that all of those incident particles aimed so as to strike at the circumference, or rim, of the circle drawn about the nucleus and having a radius b, will be deflected by an angle θ. Furthermore, any particles so aimed as to strike within the shaded area πb^2 will be deflected by an angle greater than θ. A particle which makes a head-on collision with the nucleus is, of course, scattered by 180°. The area, πb^2, is called the *cross section* and ordinarily designated by σ.

$$\sigma = \pi b^2 \qquad [5\text{-}1]$$

With every scattering center can be associated an area such that an incident particle aimed to strike (but scattered, of course, before reaching the immediate vicinity of the nucleus) within this area will be scattered through an angle θ or greater. We shall soon see that for scattering through an appreciable angle the cross-section σ is extraordinarily small; that is, for large-angle scattering, each nucleus presents a very small target to an incident particle.

Let us now calculate the total target area presented by *all* of the scattering centers, or nuclei, within a foil of area A and thickness t. We assume the foil to be so thin that the cross-sectional area presented by any one nucleus does not overlap, or cover, that of any other nucleus, as shown in Figure 5-4. If the total shaded area of Figure 5-4 is very small compared to the area of the foil A, there is a very low probability that an incident particle will be perceptibly scattered by more than one nucleus. Thus, if the foil is sufficiently thin and the cross section σ is small, *single scattering*, rather than multiple scattering, will occur. The total number of nuclei, or scattering centers, per unit volume is n. The value of n can be computed

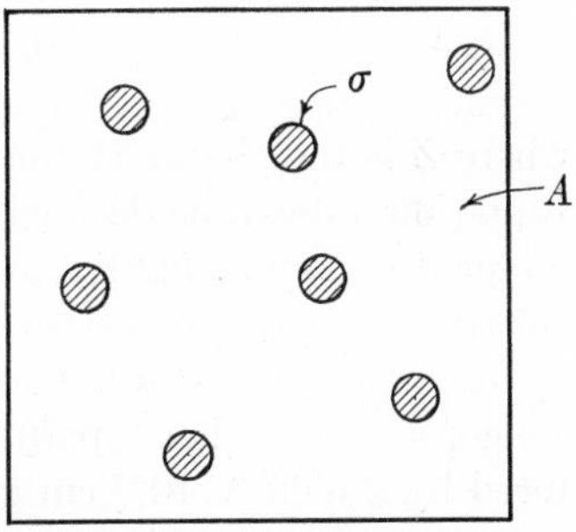

Figure 5-4. Target areas presented by the scattering centers of a thin foil resulting in the scattering of incident particles through an angle θ or greater.

from Avogadro's number N_0, and the density ρ and atomic weight w of the scattering foil by $n = N_0\rho/w$ (see a similar computation on page 133). If there are n nuclei per unit volume, then in a foil having a volume At there are altogether nAt nuclei. The total shaded area of Figure 5-4, representing the target area for scattering by at least an angle θ, is then $\sigma(nAt)$.

We wish to calculate the fraction of the particles incident on the foil that will be scattered by an angle of θ or more. The incident beam cannot be aimed to strike any one nucleus in the foil because it is spread over an area that is very large compared to that of σ. Therefore, the probability that any one incident particle will be scattered through an angle greater than θ is simply the ratio of the shaded target area, σnAt, to the total area A of the foil. For a large number of incident particles, the fraction scattered is then given by $(\sigma nAt)/A$, or

$$\boxed{\frac{N_s}{N_i} = \sigma nt} \qquad [5\text{-}2]$$

where N_i is the number of particles incident on the foil, and N_s is the number of particles scattered by an angle of θ or more.

Equation 5-2 is the basis for all scattering experiments with thin foils. The number of incident and scattered particles, N_i and N_s, can be directly measured by experiment, and their ratio can be compared with that computed from Equation 5-2. But to compute N_s/N_i, one must know the value of the cross section σ. Following Equation 5-1, σ can be computed from a knowledge of the impact parameter b, which in turn can be related to the angle of scattering θ.

The relation between b and θ for the scattering of α-particles, having a charge $+2e$ and kinetic energy K_α, by an infinitely massive point-charge is given by

$$b = \frac{Ze^2}{(4\pi\epsilon_0)K_\alpha} \cot\left(\frac{\theta}{2}\right) \qquad [5\text{-}3]$$

where Z is the charge of the nucleus in multiples of the electron charge e. Equation 5-3 can be derived† from the conservation laws of energy, linear momentum, and angular momentum and by assuming that a Coulomb electrostatic force acts between the particles.

Let us apply the α-particle scattering theory to the conditions of the Geiger and Marsden experiment in which 7.68 Mev α-particles were scattered by a 6.00×10^{-5} cm gold foil. We use the following known values, and choose θ to be $90°$:

$\theta = 90°$ $\qquad\qquad\qquad$ $1/4\pi\epsilon_0 = 8.99 \times 10^9$ newton-m²/coulomb²

$t = 6.00 \times 10^{-7}$ m $\qquad\qquad$ $\rho = 1.93 \times 10^4$ kg/m³

$K_\alpha = 7.68$ Mev $\qquad\qquad\quad$ $w = 197.2$

$Z = 79$ $\qquad\qquad\qquad\quad$ $N_0 = 6.02 \times 10^{26}$ atoms/kg-mole

$e = 1.60 \times 10^{-19}$ coulomb

† For a derivation of Equation 5-3, see the references at the end of this chapter.

Equations 5-3, 5-1, and 5-2 then yield:

$$b = 1.48 \times 10^{-14} \text{ m}$$

$$\sigma = 6.88 \times 10^{-28} \text{ m}^2$$

$$N_s/N_i = 2.43 \times 10^{-5}$$

For the conditions of this experiment, any incident α-particle which is originally moving so as to miss a head-on collision with a nucleus by no more than 1.48×10^{-14} m will be scattered by at least 90°. Notice that this impact parameter is much less than the distance between gold nuclei, approximately 3×10^{-10} m, which is also the size of a gold atom. The cross section, or target area, for the scattering of α-particles through an angle greater than 90° is 6.88×10^{-28} m², a very small area. Finally, we see that N_s/N_i is 2.43×10^{-5}; the theory of α-particle scattering predicts that, for the stated conditions, slightly more than two out of every 10^5 incident α-particles will be scattered by 90° or more. The assumption that the shaded area of Figure 5-4 is much less than the total area of the foil is amply justified, and therefore only single scattering is important for such a large scattering angle. If the computation is repeated for $\theta = 1°$, one finds $N_s/N_i = 0.320$; therefore, 32.0 per cent of the incident α-particles are expected to be scattered by 1° or more.

On the assumption of nuclear point-charges, most of the incident particles are predicted to be scattered only slightly; however, a small but significant fraction of the α-particles are scattered through large angles. If we had chosen an entirely different model, in which the positive charge of the atom is assumed to be uniformly spread over a substantial region of the atom's volume, the predicted probability for scattering through a large angle with a thin foil is zero. This model was proposed by J. J. Thomson, but scattering experiments showed it to be incorrect. The nuclear hypothesis of Rutherford was confirmed by the experiments of Geiger and Marsden in that the measured distribution of the scattered α-particles was in agreement with the distribution predicted by assuming Coulomb-force scattering by point charges. They confirmed the Rutherford theory for a variety of α-particle energies, and foil materials and thicknesses.

Alpha-particle scattering experiments show, however, that the Coulomb-force law between the nucleus and the α-particle does *not* hold when the charged particles are separated by distances somewhat less than 10^{-14} m. Discrepancies between Coulomb-scattering theory and experiment are found when one compares the observed with the predicted number of high-energy α-particles scattered through very large angles. For such collisions, the α-particles may come closer than 10^{-14} m to the nucleus. Thus, for distances of this order or smaller, the force between the nucleus and an α-particle is not given merely by Coulomb's law, and a distinctive, additional *nuclear force* must be assumed to act between the particles. These deviations from the Coulomb force are crucial for our considerations in nuclear physics; at the moment we shall be content to note that the mass and the positive electric charge are located in any atom within a region no larger than 10^{-14} m. In a typical atom, having a size of approximately 10^{-10} m, it is certainly proper to assume that the electrons are subject to a strictly electrostatic Coulomb force of attraction originating from a point charge, the nucleus.

5-2 The classical planetary model The existence of the nucleus as the center of positive charge and mass in the atom, along with the fact that the force between charged particles at atomic dimensions is given by the Coulomb electrostatic force, can be used as the basis of a very simple, classical model of atomic structure. Let us consider the structure of hydrogen, the simplest of all atoms, in terms of this classical model. Ordinary hydrogen consists of a nucleus having a single positive charge (a proton) and one electron. The proton, which is 1836 times more massive than the electron, attracts the electron by a Coulomb electrostatic force which varies inversely as the square of the distance between them. (The gravitational force between these particles is 10^{39} times smaller than the electric force and can, therefore, be neglected.) The situation here is similar to that of our solar system; the Sun's mass greatly exceeds that of a planet, and the planet and Sun attract one another by an inverse-square gravitational force. For the planet to be bound to the Sun, the planet must move in an elliptical or circular path about the Sun. The atomic planetary model assumes that an atom is, in effect, a miniature solar system, where the nucleus replaces the Sun, an electron replaces a planet, and the Coulomb force replaces the gravitational force. The model is strictly classical; no wave aspects are ascribed to the electron, and all quantum effects are excluded. It is assumed, for simplicity, that the electron moves in a circular orbit about the hydrogen nucleus, the nucleus remaining at rest.

We shall compute the energy of the hydrogen atom and the frequency of the orbital motion. The total mechanical energy E of the system (excluding the rest energies of the electron and proton) is the sum of the kinetic energy K and the electrostatic potential energy P. An electron with a mass m is assumed to move at a (non-relativistic) speed v in a circle of radius r; both the electron and proton carry an electric charge of magnitude e. Therefore,

$$E = K + P$$
$$E = \tfrac{1}{2} mv^2 + (-ke^2/r) \qquad [5\text{-}4]$$

where $k = 1/4\pi\epsilon_0 = 8.99 \times 10^9$ newton-m²/coulomb².† The centripetal

† The *mks* system of units is used throughout this book. To change to the *cgs* system it is necessary merely to put $k = 1$, leaving all other quantities in the formulas of atomic structure unchanged.

orce maintaining the electron in its circular orbit is supplied by the electric force due to the nucleus. Thus,

$$F = ma$$
$$ke^2/r^2 = mv^2/r$$

or
$$mv^2 = ke^2/r \qquad [5\text{-}5]$$

We see from Equations 5-5 and 5-4 that, for circular orbits, *the kinetic energy is one-half the magnitude of the potential energy.*
Substituting Equation 5-5 in Equation 5-4 gives

$$E = \tfrac{1}{2}(ke^2/r) - (ke^2/r)$$

$$\boxed{E = -(ke^2/2r)} \qquad [5\text{-}6]$$

Equation (5-6) shows that the total energy of the system is negative. As the radius of the electron orbit is increased, E approaches zero; this means that the electron is most tightly bound to the nucleus when it moves in a small circular orbit and that the electron is free of the binding with the nucleus only when it is separated from it by an infinite distance. Thus, when E is negative, the electron and proton form a bound system. It is known that a hydrogen atom has a diameter of approximately one Ångstrom and that the electron in hydrogen is bound to the nucleus with an energy of 13.6 ev. Putting $E = -13.6$ ev in Equation 5-6 gives $r = 0.53$ Å; therefore, the hydrogen planetary model shows, thus far, agreement with experimental facts.

Let us now consider the electromagnetic radiation from a classical planetary hydrogen atom. Classically, electromagnetic waves are produced by an accelerated electric charge, and the frequency of the waves is precisely the frequency of oscillation of this charge. The electron in a planetary hydrogen atom is continuously accelerated as it moves in a circular path; therefore, the atom radiates continuously. The frequency of the radiation, moreover, is expected to be equal to the frequency f of the electron's orbital motion. The orbital frequency is given by

$$f = \omega/2\pi = v/2\pi r \qquad [5\text{-}7]$$

where ω is the angular velocity of the electron. Equation 5-5 can be written in the form

$$v/r = \sqrt{ke^2/mr^3} \qquad [5\text{-}8]$$

Substituting Equation 5-8 in 5-7 gives

$$f = (1/2\pi)\sqrt{ke^2/mr^3} \qquad [5\text{-}9]$$

If one puts $r = 0.5$ Å in Equation 5-9, the orbital frequency f is found to be 7×10^{15} sec^{-1}. Therefore, for this radius, the radiated frequency is predicted to lie in the ultraviolet region of the electromagnetic spectrum.

But if the atom radiates, the total energy E of the atom must decrease and become even more negative. From Equation 5-6 we see that, if the total energy E decreases, the radius r of the orbit must decrease. By Equation 5-9, f increases as r decreases. In short, when energy is radiated, E decreases, r decreases, the orbital frequency f increases, and hence the radiated frequency continuously increases.

The classical planetary theory then predicts that, starting from some initial orbit, the electron will spiral inward toward the nucleus. The atom radiates a *continuous* spectrum, the frequency increasing as the electron's radius decreases. (See Figure 5-5.) Calculations show that the electron in

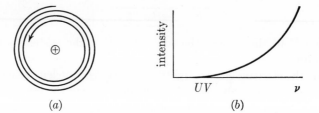

(a) (b)

Figure 5-5. (a) Classical collapse of an atom arising from the continuous radiation by the orbiting electron. (b) The resulting classical intensity distribution of the electromagnetic radiation as a function of frequency.

this classical model reaches the nucleus and combines with it in less than 10^{-8} sec. The atom collapses! The classical planetary atomic model is clearly untenable on two important counts: it predicts that atoms are unstable, and it predicts a continuous radiated spectrum. This is completely in disagreement with the experimental facts: atoms *are* stable, and atoms radiate a *discrete* spectrum of frequencies, as we shall see in the next section.

5-3   The hydrogen spectrum   We have seen that the strictly classical planetary model leads to the expectation that a continuous range of electromagnetic radiation will be emitted from hydrogen atoms and that the atom collapses. Because the observed radiation from hydrogen is not continuous, and because hydrogen atoms are stable, this classical model is fundamentally defective. A correct model must be able to account in detail for the observed radiation, or spectrum, and for the stability of hydrogen atoms. Before we discuss the model which gives an adequate description of the structure of hydrogen atoms, the natural starting point for any theoretical description of atomic structure, let us set down the known facts concerning the hydrogen spectrum.

In order to observe the spectrum radiated by individual hydrogen atoms it is necessary to use gaseous atomic hydrogen, for which the atoms are so far apart that each atom behaves as an isolated system. Molecular hydrogen (H_2) radiates a spectrum which reflects some aspects of two hydrogen atoms bound together, and thus this spectrum is complicated by the molecular structure. An even more complicated spectrum is radiated by liquid or solid hydrogen, in which any one hydrogen atom is always strongly acted upon by forces arising from neighboring atoms; and the spectrum is now characteristic of large groups of interacting atoms. Therefore, whenever we wish to observe radiation characteristic of individual, isolated atoms, the spectrum of atoms in the gaseous state must be studied.

The visible spectrum emitted by hydrogen can be studied with a spectrometer of the type shown schematically in Figure 5-6. The hydrogen gas is excited by an electrical discharge and emits radiation. Some of this

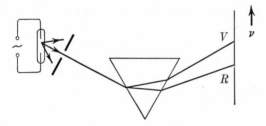

Figure 5-6. Schematic diagram of a prism spectrograph.

radiation passes through a narrow slit S, and is dispersed by the glass prism, with the short wavelengths, or high frequencies, being deviated the most. The radiation, now separated into its various frequency components, falls on a photographic plate, thus making it possible to measure the frequencies and intensities of the *emission spectrum*.

An instrument which disperses and photographs the spectrum, as illustrated in Figure 5-6, is known as a *spectrograph;* if the visible spectrum is viewed directly by the eye, the instrument is called a *spectroscope.* Any instrument which can disperse and measure the various wavelengths of a beam of electromagnetic radiation is called a *spectrometer.* Figure 5-6 shows the principal elements of a prism spectrometer; a diffraction grating can also be used in a spectrometer to disperse light (visible, ultraviolet, etc.). Suitable spectrometers exist for the study of each of the several regions of the electromagnetic spectrum (e.g., radiofrequency, x-ray, gamma-ray spectrometers). That branch of physics which deals with the study of electromagnetic radiation emitted or absorbed by substances is called *spectroscopy.* Spectroscopy is a very powerful method of inquiry into atomic, molecular, and nuclear structure; it is characterized by very high precision (frequencies or wavelengths easily measured to 1 part in 10^7) and very high sensitivity (emission or absorption from samples as small as fractions of a microgram).

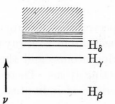

The spectrum from atomic hydrogen, as observed on a photographic plate, consists of lines, these lines being images of the slit. In fact, the spectrum of all chemical elements in mono-atomic gaseous form is composed of a group of lines, each spectrum being characteristic of the particular element; such a spectrum is known as a *line spectrum*. Each line spectrum consists of a number of sharp, discrete bright lines on a black background. Therefore, the *emission spectrum* from atomic hydrogen is a line spectrum characteristic of hydrogen. Inasmuch as each chemical element has its own characteristic line spectrum, spectroscopy serves as a particularly sensitive method for identifying elements.

The line spectrum from atomic hydrogen in the visible region is shown in Figure 5-7. The lines are identified by the labels H_α, H_β, H_γ, etc., in the order of decreasing wavelength or increasing frequency. Ordinarily the H_α line is much more intense than H_β, which is in turn more intense than H_γ, etc. The spacing between adjacent lines decreases as the frequency increases, and the discrete lines approach a *series limit*, above which there appears a weak continuous spectrum. This group of hydrogen lines is known as the *Balmer series* because in 1885 J. J. Balmer arrived at a simple empirical formula from which all of the observed wavelengths in the visible region could be computed. This formula, giving the wavelength λ for all spectral lines in this series, can be written in the form

Figure 5-7. Frequency distribution of radiation from atomic hydrogen in the visible region. This particular group of spectral lines comprises the Balmer series.

$$\frac{1}{\lambda} = R\left(\frac{1}{2^2} - \frac{1}{n^2}\right) \qquad [5\text{-}10]$$

where $R = 1.0967758 \times 10^7$ m$^{-1} \simeq 1.0968 \times 10^{-3}$ Å^{-1} and n is an integer having the values $3, 4, 5, \ldots$. Putting $n = 3$ in Equation 5-10, the Balmer formula, gives $\lambda = 6564.7$ Å, the H_α line; similarly, putting $n = 4$ gives $\lambda = 4862.7$ Å, the H_β line. The wavelength of the series limit is given by Equation 5-10 when $n = \infty$. The constant R is known as the *Rydberg constant*; its value is chosen by trial to give the best fit for the measured wavelengths. In atomic spectroscopy, spectral lines are typically specified by their wavelengths, rather than their frequencies, because it is the wavelength that is measured; we shall find, however, that the frequency is a more fundamental quantity from the theoretical viewpoint.

In addition to the Balmer series in the visible region, hydrogen radiates a series of lines in the ultraviolet and several series of lines in the infrared region. Each of these observed spectral series can be represented by a formula similar to the Balmer equation; and, in fact, one general formula can be written from which *all* of the spectral lines of hydrogen can be computed. This general formula, known as the *Rydberg equation*, is

$$\frac{1}{\lambda} = R \left(\frac{1}{n_l^2} - \frac{1}{n_u^2} \right)$$

[5-11]

where $n_l = 1$ and $n_u = 2, 3, 4, \ldots$, for the *Lyman* series in ultraviolet region;

$n_l = 2$ and $n_u = 3, 4, 5, \ldots$, for the *Balmer* series in the visible region;

$n_l = 3$ and $n_u = 4, 5, 6, \ldots$, for the *Paschen* series in the infrared region;

and so on, to further series lying in the far infrared. The value of R, the Rydberg constant, in Equation 5-11 is *precisely* the same as that in Equation 5-10; in fact, Equation 5-11 becomes Equation 5-10 when n_l is set equal to 2. The choice of u (for upper) and l (for lower) as subscripts for the integers in the Rydberg formula will become obvious in Section 5-4. The several series of hydrogen lines are named after their discoverers. Although Equation 5-11 is remarkably successful in summarizing the wavelengths radiated by atomic hydrogen, it must be recognized that this formula is merely an empirical relation which in itself supplies no information on the fundamental nature of the atomic structure of hydrogen. On the other hand, a truly successful theory of the hydrogen atom must be capable of predicting the hydrogen spectral lines; that is, it must be able to yield the Rydberg formula as a result.

We have been concerned with the spectrum emitted by hydrogen when it is excited by an electrical discharge or by extreme heating. Atomic hydrogen at room temperature does *not*, by itself, emit appreciable electromagnetic radiation; however, unexcited hydrogen, such as hydrogen gas at room temperature, can selectively absorb electromagnetic radiation, giving rise to an *absorption spectrum*. The absorption spectrum of atomic hydrogen is observed when a beam of white light (all frequencies present) is passed through atomic hydrogen gas and the spectrum of the transmitted light is examined in a spectrometer. What is found is a series of dark lines superimposed on the spectrum of white light, and such an absorption spectrum is known as a *dark-line spectrum*. The gas is transparent to all frequencies except those corresponding to the dark lines, for which it is opaque; that is, the atoms of the gas absorb only certain discrete, sharp frequencies from the continuum of frequencies passing through the gas. This absorbed energy is very quickly radiated by the excited atoms, but in *all directions*, not just in the incident direction. The dark lines in the absorption spectrum of hydrogen occur at precisely the same frequencies as do the bright lines in the emission spectrum, as shown schematically in Figure 5-8, where the intensity is plotted as a function of frequency for

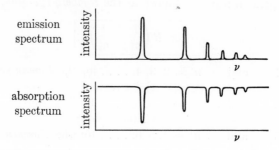

Figure 5-8. Intensity variation of the emission and absorption line spectra of hydrogen as a function of frequency.

the emission and absorption spectra. Hydrogen is a radiator of electromagnetic radiation only at the specific frequencies or wavelengths given by the Rydberg formula; it is an absorber of radiation only at the same characteristic frequencies.

What we have discussed concerning the emission and absorption spectra of atomic hydrogen holds equally well for the line spectra of all elements. A characteristic set of frequencies is emitted when the atoms radiate energy; the same set of frequencies is absorbed by the atoms when a continuous frequency band of electromagnetic radiation is sent through the gas.

5-4 . The Bohr theory of atomic structure   The planetary model of the hydrogen atom developed in Section 5-2 is a classical one in that it treats the electron strictly as a particle and the electromagnetic radiation strictly as continuous waves. The failure of this classical model can be attributed to the fact that quantum effects have not been included; that is, this model neglects the wave properties of the electron and the particle properties of the radiation. Any successful model must take these quantum effects into account.

The first theory of the hydrogen atom incorporating quantum effects was developed in 1913 by Niels Bohr, a student of Rutherford. The photon nature of electromagnetic radiation had been established prior to this time, but the wave aspects of material particles were not to be recognized until 1924. Nevertheless, in our treatment of the Bohr atom we shall take advantage of the now well-established wave nature of particles. We will see shortly that this is equivalent to Bohr's original procedure, which did not explicitly use the electron's wave nature.

The Bohr model of the atom was the first step toward a complete, thoroughgoing, wave-mechanical treatment of atomic structure; it should be realized at the start, however, that the Bohr theory has limited applicability. It retains enough classical features so that the atomic structure can be readily visualized in terms of a particle model, and it introduces enough quantum features to give a fairly accurate description of atomic spectra. Bohr's theory is, therefore, transitional between classical mechanics and the wave mechanics developed during the 1920's. A comprehensive wave-mechanical treatment is mathematically and conceptually difficult and does not lend itself to simple visual pictures; and inasmuch as the unique and fundamental quantum features of atomic structure are revealed by the Bohr theory, it is instructive to study it in detail.

The Bohr theory of the hydrogen atom is based on three basic postulates, containing the essential quantum aspects. Rather than list these postulates now, we shall see that the Bohr theory can be developed in a quite natural way from the fundamental quantum effects discussed in Chapters 3 and 4. The Bohr postulates will then appear as the necessary consequences of the wave nature of particles and photon nature of radiation. At the end of this section, we will list the three postulates explicitly.

As in the strictly classical model of the hydrogen atom, the Bohr theory also assumes that the proton is at rest and the electron moves in a circular orbit about it. The centripetal force maintaining the electron in its orbit is the attractive electrostatic Coulomb force between the electron and proton; thus, Equation 5-5, $mv^2 = ke^2/r$, applies. The total energy of the atom is again given by Equation 5-6, $E = -ke^2/2r$. So far, nothing new.

We can see how the wave aspect of the electron can be incorporated into the theory by first recalling an analogous situation involving transverse waves on a wire. If the wire is fixed at both ends, only certain frequencies or wavelengths will lead to resonant oscillations. The allowed wavelengths will be those in which the length of the wire is an integral multiple of one half-wavelength. In these allowed states, standing, or stationary, waves exist. The wire, when in an allowed state, can in principle oscillate with a constant energy for an indefinite period of time. Now let us imagine that the wire is bent into a closed circular loop. If transverse waves are to be propagated around the loop, then the wave will destructively interfere with itself unless it joins *smoothly* onto itself. Therefore, for a circular loop, standing, or stationary, resonant waves exist only if an *integral* number of whole wavelengths can be fitted around the circumference.

Let us apply similar conditions to the electron orbits in hydrogen. We have seen that a wave aspect must be attributed to a material particle, with the wavelength given by $\lambda = h/mv$, the de Broglie relation. Thus, if an electron's motion in a circular orbit is to be described in terms of the wave motion associated with the electron, we can regard the electron *as a wave* that is propagated around the circular orbit. But stationary waves can exist only for those certain orbits in which the circumference ($2\pi r$) is some integral multiple of the electron wavelength, as shown in Figure 5-9. All other orbits, that is, all other radii, are ones in which this condition

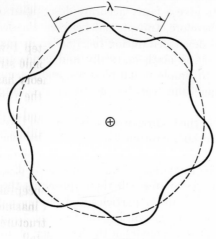

$$2\pi r = 6\lambda$$

Figure 5-9. One possible stationary electron orbit. In this example, the circumference of the orbit is six wavelengths.

cannot be satisfied. In the *stationary states* the energy of the atom is assumed, despite classical radiation theory, to remain constant, and thus *the atom does not radiate electromagnetic waves while in any one of these stationary states.* The stationary orbits, for which there is no radiation, are those satisfying the relation

$$n\lambda = 2\pi r \qquad [5\text{-}12]$$

where $n = 1, 2, 3, \ldots$, and is called the *principal quantum number*. Of course, if the electron radius is restricted to certain permitted values, then the energy E of the atom is also restricted to certain discrete values by Equation 5-6. We note that, by taking into account the wave properties of the electron, we have been able to bring stability to an atomic model!

When we replace the wavelength λ by (h/mv) in Equation 5-12 we have

$$nh/mv = 2\pi r$$

or

$$mvr = n\left(\frac{h}{2\pi}\right) = n\hbar \qquad [5\text{-}13]$$

where the symbol $\hbar$ ("h-bar"), defined as Planck's constant divided by 2π, is introduced for convenience. The left-hand side of Equation 5-13 is just the angular momentum, $(mv)r$, of the electron in its orbital motion about the nucleus. Thus, the condition for the existence of stationary states can be stated in a different way: for those particular states, or orbits, of the hydrogen atom in which the orbital angular momentum of the atom is $\hbar, 2\hbar, 3\hbar, \ldots$, the atom is stable and does not radiate electromagnetic energy.

In these stationary states, that is, in any one of the allowed or permitted orbits, the atom must have a constant, precisely-defined energy. Let us calculate the allowed energies E_n, and the allowed orbital radii r_n, and speeds v_n; these quantities are identified by the subscript n because the respective values will depend on the quantum number n. Equations 5-5 and 5-13 can be solved simultaneously for r and v. Solving Equation 5-13 for the tangential speed of the electron, we have

$$v_n = n\hbar/mr_n \qquad [5\text{-}14]$$

and using Equation 5-14 in Equation 5-5, we have

$$m(n\hbar/mr_n)^2 = ke^2/r_n$$

or,
$$r_n = n^2\hbar^2/kme^2 \qquad [5\text{-}15]$$

where again $n = 1, 2, 3, \ldots$.

The smallest allowed radius, the so-called "radius of the first Bohr orbit," is given by

$$\boxed{\begin{aligned} r_1 &= \hbar^2/kme^2 \\ r_1 &= 0.528 \text{ Å} \end{aligned}} \qquad [5\text{-}16]$$

where the known atomic constants are substituted in the right side of Equation 5-16. The Bohr model predicts, then, that the "size" of the hydrogen atom in its smallest stationary orbit is of the order of 1 Å, in good agreement with experimental determinations. We can express all of the allowed radii in a simpler form, by substituting Equation 5-16 into Equation 5-15:

$$\boxed{r_n = n^2 r_1} \qquad [5\text{-}17]$$

The radii of the stationary orbits are, therefore, $r_1, 4r_1, 9r_1, \ldots$.

The orbital speed of the electron in the stationary states can be found immediately from Equation 5-14, where we now replace r_n by $n^2 r_1$. Thus,

$$v_n = n\hbar/m(n^2 r_1)$$
$$v_n = (1/n)(\hbar/mr_1) = (1/n)(ke^2/\hbar) \qquad [5\text{-}18]$$

For $n = 1$, $v_1 = \hbar/mr_1 = ke^2/\hbar$. Equation 5-18 can be written as

$$\boxed{v_n = v_1/n} \qquad [5\text{-}19]$$

The permitted orbital speeds are $v_1, v_1/2, v_1/3, \ldots$, the electron having a maximum speed v_1 in the first Bohr orbit. The ratio of this speed to the speed of light, v_1/c, is defined by the symbol α. Equation 5-18 yields

$$\alpha = v_1/c = ke^2/\hbar c \qquad [5\text{-}20]$$

Substitution of the known values of the constants on the right-hand side of Equation 5-20 shows that $\alpha = 1/137.0$; thus, the electron in the first Bohr orbit moves at 1/137th the speed of light. The quantity α, which appears frequently in the theory of atomic structure, is known as the *fine-structure constant*.† To treat the Bohr hydrogen atom as a nonrelativistic problem is not unreasonable; but, because of the very high precision of wavelength measurements in spectroscopy, relativistic effects can be observed and would have to be included in a more complete theory.

The allowed values of the total energy (excluding the rest energies of the proton and electron) of the hydrogen atom can now be determined easily from Equations 5-6 and 5-17

$$E_n = -ke^2/2r_n = -(1/n^2)(ke^2/2r_1) \qquad [5\text{-}21]$$

We represent the quantity, $(ke^2/2r_1)$, by E_I; then Equation 5-21 becomes

$$\boxed{E_n = -\frac{E_I}{n^2}} \qquad [5\text{-}22]$$

Thus, the only possible energies of the hydrogen atom, the bound electron-proton system, are $-E_I$, $-E_I/4$, $-E_I/9$, ... The permitted energies are discrete and, therefore, *the energy is quantized*. The lowest energy (that is, the most negative energy) occurs for the state in which the principal quantum number n equals 1. In the lowest, or *ground state*, $E_1 = -E_I$, and the computed value of this ground-state energy, from Equations 5-21 and 5-16, is

$$E_n = -(k^2e^4m)/(2n^2\hbar^2) \qquad [5\text{-}23]$$

$$E_1 = -E_I = -13.58 \text{ ev} \qquad [5\text{-}24]$$

Figure 5-10 shows an *energy-level diagram* for a hydrogen atom. We note from Equation 5-22 that for bound states $E < 0$, and only discrete energies are allowed. As n approaches infinity, the energy difference between adjacent energy levels approaches zero. For $n = \infty$, $E_n = 0$, and the hydrogen atom is dissociated into an electron and proton, separated by an infinite distance and both at rest. In this condition, the atom is said to be ionized, and the energy which must be added to the atom when in its lowest, or ground, state $(n = 1)$ to bring it up to $E_n = 0$, is just E_I, the so-called *ionization energy*. The value predicted by the Bohr theory, $E_I = 13.58$ ev, is in complete agreement with the value obtained by experiment. When

---

† The fine-structure constant plays a crucial role in quantum electrodynamics because it gives a relationship involving the fundamental constants of electromagnetism (e), quantum theory $(\hbar)$, and relativity (c).

the total energy of the "atom" is positive, and the electron and proton are unbound, the electron does not move in a closed orbit. There is, therefore, no restriction on the electron's wavelength imposed by the fitting of an integral number of wavelengths around the path; hence there is no restriction on the total energy. *All* possible positive energies are allowed, and there is thus a continuum of energy levels for $E > 0$.

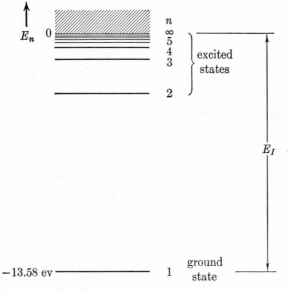

Figure 5-10. Energy-level diagram for a hydrogen atom.

It is interesting to compare Figure 5-10, the energy-level diagram for a two-particle system following the quantum theory, with Figure 3-1c, the energy-level diagram for a two-particle system following the classical theory. Whereas there is a continuum of energies for both the bound and unbound system in the classical theory, the quantum theory requires quantized states (e.g., energy, angular momentum) for a bound system.

Each of the permitted or quantized energies of Figure 5-10 corresponds to a stationary state in which the atom can exist without radiating. All of those energy states, $n = 2, 3, 4, \ldots$, above the ground state are called *excited states* because an atom in one of these states tends to make a transition downward to some lower stationary state. In such a downward transition, the electron can be imagined to jump from some one stationary orbit to a smaller permitted orbit. Let us assume that an atom is in some excited state and has an energy E_n, as shown in Figure 5-10. The amount by which the energy of this excited state exceeds that of the ground state is called the *excitation energy*. The energy that must be added to an atom in any state to free the bound particles, making $E_n = 0$, is called the *binding energy* (see Section 2-9).

Consider an atom, initially in an upper excited state and having an energy E_u (u for upper), which makes a transition to a lower-energy state E_l (l for lower). Then, when the transition occurs, the energy of the atom falls from E_u to E_l, an amount of energy ($E_u - E_l$) having been lost by the atom. Bohr assumed that, in such a transition, a single photon having an energy $h\nu$ is created and emitted by the atom. By the conservation of energy,

$$\boxed{h\nu = E_u - E_l} \qquad [5\text{-}25]$$

We see here that the Bohr theory incorporates the particle nature of electromagnetic radiation by assuming that a single photon is created whenever the atom makes a transition to a lower energy. This theory gives no details of the electron's quantum jump to the lower stationary state, nor of the photon-creation process. The situation here is like that encountered in photon-electron interactions (photoelectric effect, Compton effect, etc.) in that we did not concern ourselves with the details of the interactions but merely applied the conservation laws of energy, momentum, etc., to the collision.

Let us compute the frequencies (and wavelengths) of the photons that can be radiated by a hydrogen atom according to the Bohr model. Using Equations 5-25 and 5-22, we have

$$\nu = \frac{E_u - E_l}{h} = \left(\frac{-E_I}{n_u^2 h}\right) - \left(\frac{-E_I}{n_l^2 h}\right)$$

$$\nu = \frac{E_I}{h}\left(\frac{1}{n_l^2} - \frac{1}{n_u^2}\right) \qquad [5\text{-}26]$$

where n_u and n_l are the quantum numbers for the upper- and lower-energy states, respectively. The wavelengths, $\lambda = c/\nu$, of emitted photons can thus be written in the form

$$\frac{1}{\lambda} = \frac{E_I}{hc}\left(\frac{1}{n_l^2} - \frac{1}{n_u^2}\right) \qquad [5\text{-}27]$$

Equation 5-27 is of precisely the same mathematical form as the empirically derived Rydberg formula, Equation 5-11:

$$\frac{1}{\lambda} = R\left(\frac{1}{n_l^2} - \frac{1}{n_u^2}\right)$$

By comparing these two equations for $(1/\lambda)$ we can evaluate the Rydberg constant R from known atomic constants and compare it with the experimentally determined value, 1.0968×10^{-3} Å^{-1}, for hydrogen.

$$R = \frac{E_I}{hc} = \frac{k^2 e^4 m}{4\pi\hbar^3 c} \qquad [5\text{-}28]$$

To arrive at the last term of Equation 5-28, we have used $E_I = ke^2/2r_1$, Equation 5-21, and $r_1 = \hbar^2/kme^2$, Equation 5-16. Using the known values of the physical constants in Equation 5-28, the value of R is computed to be 1.0974×10^{-3} Å^{-1}, in close agreement with the experimental spectroscopic value. Thus, the Rydberg formula, which summarizes the emission and absorption spectrum of hydrogen, follows as a *necessary consequence* of the Bohr atomic model.

The observed spectral lines of hydrogen can now be understood in terms of the energy-level diagram, Figure 5-11. The vertical lines represent transitions between stationary states; the lengths of these lines are pro-

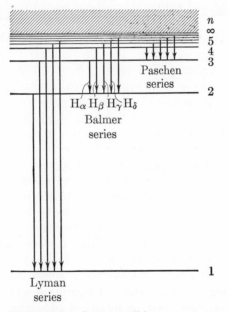

Figure 5-11. Some possible energy transitions in atomic hydrogen.

portional to the respective photon energies and, therefore, to the frequencies. The Lyman-series lines correspond to those photons produced when hydrogen atoms in any of the excited states ($n_u = 2, 3, 4, 5, \ldots$) undergo transitions to the ground state ($n_l = 1$). Transitions from the unbound states ($E > 0$) to the ground state account for the observed continuous spectrum, lying beyond the series limit. We account in a similar way for the Balmer series, which is produced by transitions from the $n_u = 3, 4, 5, \ldots$ excited states to the first excited state ($n_l = 2$). Still further emission series involve downward transitions to $n_l = 3$, $n_l = 4$, etc., these series falling progressively toward longer wavelengths.

We have examined the emission from a single hydrogen atom. Such an atom can exist in only *one* of its quantized energy states at any one time; and when this atom makes a transition from an excited state to a lower-energy state, it emits a *single* photon. When the entire emission spectrum from an excited hydrogen gas, a collection of a very large number of hydrogen atoms, is observed in a spectroscope, we see the simultaneous emission of many photons produced by downward transitions from each of the excited states. Therefore, to observe the entire emission spectrum, we must have a very large number of hydrogen atoms in each of the excited states, making downward transitions to all lower-energy states.

At room temperature, essentially all of the atoms of a hydrogen gas are in the ground state, and noticeable emission cannot occur. Let us see why this is true. The average kinetic energy per atom, $\frac{3}{2}kT$, of a gas at room temperature is 1/25 ev; thus, there are very few atoms which have a translational kinetic energy of 10.2 ev, the minimum energy necessary to raise the atom from the ground state ($n = 1$, $E_1 = -13.6$ ev) to the first excited state ($n = 2$, $E_2 = -3.4$ ev). Thermal excitation of atoms occurs when some of the translational kinetic energy of two colliding atoms is transformed into *internal* excitation energy of one or both of the atoms; *translational* kinetic energy is *not* conserved in such a collision and, thus, the collision is *inelastic*. When the gas temperature is raised to the point where the average translational kinetic energy of the atom $\frac{3}{2}kT$ is approximately equal to some possible excitation energy, appreciable numbers of atoms can absorb enough energy in inelastic collisions to raise them to this higher-energy state. To excite atoms by heating requires very high temperatures. A simpler method involves the use of an electric discharge, by which electrons and ions are accelerated to very high kinetic energies by an external electric field; this is accomplished in practice by applying a potential difference between two electrodes which are placed in a glass chamber containing the gas. Thermal excitation and, more commonly, electrical excitation thus serve as means of producing emission spectra.

We can now see why, in the kinetic theory of gases, gas molecules and atoms can be regarded as inert particles having no internal structure and making perfectly elastic collisions with each other when the gas is at moderate temperatures. Unless the average translational kinetic energy per atom is comparable to the energy difference between the ground state and the first excited state, the internal structure of the atom cannot change, the total translational kinetic energy in a collision is conserved, and the collision is perfectly elastic. If the gas temperature is sufficiently high, inelastic collisions occur. Some atoms are thereby excited, and they can no longer be considered as inert particles incapable of undergoing internal change.

We now have a basis for understanding the characteristics of absorption spectra. As shown in Figure 5-8, an absorption spectrum shows dark lines on a white background with the same wavelengths as the bright lines on the black background of the corresponding emission spectrum. When white light, consisting of photons having all possible frequencies, or energies, passes through a gas, those particular photons having energies equal to the energy difference between stationary states can be removed from the beam. Those photons with the appropriate energies are annihilated, thereby giving their radiant electromagnetic energy to the internal excitation energy of the atoms. The same set of quantized energy levels participate in both emission and absorption; consequently, the frequencies of the emission and absorption lines are identical. (Ordinarily all hydrogen atoms are in the ground state and consequently only the Lyman series is observed in absorption.)

We have used the fundamental postulates of the Bohr theory implicitly in our development of the hydrogen-atom model. It is helpful, however, to isolate these basic assumptions inasmuch as they are retained in their essential forms in more complete wave-mechanical treatments of atomic structure.

(1) *A bound atomic system can exist without radiating only in certain discrete stationary states.*

(2) *The stationary states are those in which the orbital angular momentum, mvr, of the atom is an integral multiple of $\hbar$, Planck's constant divided by 2π. This is a natural consequence of the wave properties of the electron confined to a semi-classical circular orbit.*

(3) *When an atom undergoes a transition from an upper energy state E_u to a lower energy state E_l, a photon of energy $h\nu$ is emitted, the conservation of energy requiring that $h\nu = E_u - E_l$. If a photon is absorbed, the atom will make a transition from the lower- to the higher-energy state, following the same relation.*

5-5　Extensions of the Bohr model

HYDROGENIC ATOMS　　It is natural to attempt extending this model to systems other than hydrogen atoms, and the Bohr theory can successfully be applied to other atomic systems containing only one electron. Such a one-electron system is known as a *hydrogenic atom*. A hydrogenic atom has a nuclear positive electric charge Ze; Z is the *atomic number*, and e is the magnitude of the charge of the electron. Any neutral atom becomes a hydrogenic atom if it is stripped of all its electrons, save one. For example, helium has a double nuclear charge $(Z = 2)$ and contains two electrons when neutral. If one electron is removed, that is, when the atom is singly-ionized to form He^+, we have a hydrogenic atom. Other simple hydrogenic atoms are: doubly-ionized lithium, Li^{++} $(Z = 3)$; triply-ionized beryllium, Be^{+++} $(Z = 4)$; etc. All of these ions differ from the hydrogen atom $(Z = 1)$ by virtue of the size of their nuclear charges, Z. Therefore, the Coulomb force between the nucleus and the electron is enhanced by a factor Z and is given by kZe^2/r^2. We can very simply extend the Bohr theory to hydrogenic atoms by replacing the quantity e^2, wherever it occurs in any equation for the hydrogen atom, by the quantity Ze^2. Therefore, Equations 5-15, 5-18, 5-23, and 5-26 for the radii, speeds, energies, and photon frequencies become respectively

$$\left. \begin{array}{ll} r_n = \dfrac{n^2\hbar^2}{km(Ze^2)} & v_n = \dfrac{k(Ze^2)}{n\hbar} \\[3ex] E_n = \dfrac{-k^2(Z^2e^4)m}{2n^2\hbar^2} & \nu = \dfrac{c}{\lambda} = cZ^2R\left(\dfrac{1}{n_l^2} - \dfrac{1}{n_u^2}\right) \end{array} \right\} \quad [\text{5-29}]$$

The energies of the permitted states are all lowered by a factor Z^2, that is, a factor 4 for He^+, a factor 9 for Li^{++}, etc. By the same token, the energy differences between stationary states and the corresponding photon frequencies are increased by Z^2. Thus the hydrogenic atoms emit spectra whose lines are shifted toward the ultraviolet with respect to the hydrogen lines. Figure 5-12 shows the energy-level diagrams for hydrogen, He^+, and Li^{++}. We note that certain emitted lines are expected to have the same frequency; for example, the transition from the $n = 2$ to the $n = 1$ state in hydrogen $(Z = 1)$ should produce a photon of the same wavelength as the transition from the $n = 4$ to the $n = 2$ state in singly-ionized helium $(Z = 2)$. The observed spectra agree closely, but not completely, with this expectation for reasons that we will now explore.

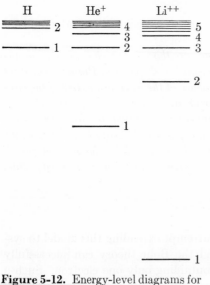

Figure 5-12. Energy-level diagrams for hydrogen, singly-ionized helium, and doubly-ionized lithium.

MASS CORRECTION OF THE RYDBERG CONSTANT   It has been assumed that the nucleus of hydrogen or hydrogenic atoms is so massive compared to the electron that the nuclear mass could be taken as infinite. This approximation is a fairly good one inasmuch as the nucleus of ordinary hydrogen has a mass that is 1836 times that of the electron. But extremely high resolution of measured wavelengths is possible in spectroscopy; for example, the wavelengths of visible spectral lines can be measured to a 1/1000th of an Ångstrom! To reach really good agreement between theory and experiment, it is necessary to take into account the fact that the nuclear mass is non-infinite.

The electron and the nucleus of an atom are both in motion with respect to the center of mass, which lies on a line between the nucleus and the electron. The ratio of the center-of-mass's distances to the nucleus (r_M) and to the electron (r_m) is in the inverse ratio of their masses, M and m respectively. Therefore, $r_M/r_m = m/M$. The electron and nucleus both move in circles about the center of mass, as in Figure 5-13. Now, the motions of any two particles about their center of mass can be reduced into the motion of a single particle having an effective mass called the *reduced mass*. We may retain all of the relations of the Bohr theory if the electron mass m is

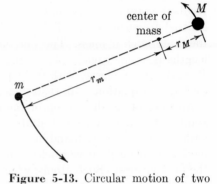

Figure 5-13. Circular motion of two bound particles about their center of mass.

replaced, wherever it occurs, by the *reduced mass* μ of the electron-nucleus system. It can be shown that μ is related to m and M by the formula

$$\frac{1}{\mu} = \frac{1}{m} + \frac{1}{M}$$

Therefore,

$$\mu = \frac{mM}{m + M} = \frac{m}{1 + (m/M)} \qquad [5\text{-}30]$$

Equation 5-30 shows that μ is always less than m. When M is very much greater than m, $\mu \simeq m$.

The Rydberg constant, R, which was derived earlier [Equation 5-28] and which is proportional to the electron mass m, can more properly be labeled R_∞, since it applies only for a nucleus with an infinitely large mass.

$$R_\infty = \frac{k^2 e^4 m}{4\pi \hbar^3 c} \qquad [5\text{-}31]$$

The Rydberg constant for any nuclear mass M is labeled R_M, and from Equations 5-30 and 5-31, R_M is given by

$$R_M = R_\infty(\mu/m) = \frac{R_\infty}{1 + (m/M)} \qquad [5\text{-}32]$$

Therefore, R_M is always less than R_∞ by an amount that becomes increasingly smaller as the nuclear mass increases; for example, with hydrogen and helium,

$$R_H = \frac{1.09737 \times 10^{-3} \text{ Å}^{-1}}{1 + \frac{1}{1836}} = 1.09678 \times 10^{-3} \text{ Å}^{-1}$$

$$R_{He^+} = \frac{1.09737 \times 10^{-3} \text{ Å}^{-1}}{1 + [1/(4 \times 1836)]} = 1.09722 \times 10^{-3} \text{ Å}^{-1}$$

Because the allowed energies and frequencies are proportional to the Rydberg constant, the energy levels are shifted slightly upward, and the emitted frequencies are slightly reduced when a correction is made for the finite nuclear mass. The shift in the energy levels and photon wavelengths is a small one, but one that can clearly be observed. Note that R_H agrees more closely with the experimentally determined value of R for hydrogen, Equation 5-10, than does R_∞.

An interesting illustration of the mass correction to the Rydberg constant is found in a careful study of the spectrum of natural hydrogen. Ordinary hydrogen as found in nature consists of two kinds of atoms, both having a single positive nuclear charge but with two different values of the nuclear mass. Such atoms, which are chemically identical but differ in the value of the nuclear mass, are called *isotopes*. The stable isotopes of hydrogen are: H^1, whose nucleus is a proton with a mass 1.00759 amu; and H^2, *deuterium*, or *heavy hydrogen*, whose nucleus is a proton and a neutron bound together, with a nuclear mass 2.01419 amu. The nucleus of the deuterium atom—called a *deuteron*—therefore, has a mass which is approximately twice that of a proton. The natural abundance of H^2 as compared to H^1 is approximately one part in 6000.

The isotopes, H^1 and H^2, are expected to yield identical spectra apart from the small correction arising from their different nuclear masses. From Equation 5-32, the corresponding Rydberg constants, R_H and R_D are

$$R_H = 1.09678 \times 10^{-3} \text{ Å}^{-1}$$

$$R_D = 1.09707 \times 10^{-3} \text{ Å}^{-1}$$

Because R_D is slightly larger than R_H, the spectral lines of deuterium are shifted slightly to shorter wavelengths. Thus for the more abundant isotope, H^1, the wavelength of the H_α line is 6562.80 Å; for H^2, the H_α line is 6561.01 Å. This small wavelength difference, 1.79 Å, can be easily observed with high-resolution spectrometers (capable of measuring wavelengths to 0.001 Å). In fact, deuterium was first discovered (1932) through the observation of the closely-spaced pairs of hydrogen spectral lines. The deuterium lines are not ordinarily seen in the hydrogen spectrum because of the low abundance of H^2; the intensity of the deuterium lines can be enhanced by increasing the concentration of the heavy isotope.

5-6 The correspondence principle According to the quantum theory, the frequency of the photon emitted or absorbed in a transition is determined solely by the energy difference between the two participating stationary states,

$$\nu = \frac{E_u - E_l}{h}$$

or
$$\nu = cR\left(\frac{1}{n_l^2} - \frac{1}{n_u^2}\right) \qquad [5\text{-}33]$$

We recall from Section 5-2 that in the classical planetary model of the hydrogen atom it was assumed, following classical electromagnetism, that the frequency of the electromagnetic waves generated by the accelerated electric charge is precisely the frequency of motion of the electron about the nucleus (Equation 5-9)

$$f = (1/2\pi)\sqrt{ke^2/mr^3} \qquad [5\text{-}34]$$

Equations 5-33 and 5-34 were derived from different assumptions, and we see clearly that the two predicted frequencies are quite dissimilar. The quantum theory is the correct theory; it not only explains short-wavelength visible radiation from atomic systems, but it also describes correctly long-wavelength radiation, such as radio waves. The classical electromagnetic theory fails, of course, to explain atomic spectra; on the other hand, it agrees with experiments performed with long-wavelength radio waves, where it is found that the frequency of the radiation is equal to the frequency of oscillation of the electric charges.

The correspondence principle (Section 1-6) requires that the quantum theory, the more general theory, *must* yield the *same* results as the more restricted classical theory under those circumstances for which the classical theory suffices. Therefore, for atomic radiation the frequencies ν of the emitted photons must approach the orbital frequencies f when and only when the hydrogen atom can be regarded as approximating the conditions for which the classical theory applies. We wish to show, then, that $\nu = f$ in the correspondence limit.

It is easy to show that the Bohr atom approaches classical conditions as the principal quantum number n becomes a very large integer and small quantum jumps are involved. As n becomes large, the discrete energy levels crowd more closely together, approaching the continuum characteristic of the classical bound system (see Figure 5-10). For a large n, the photons emitted in transitions between adjacent energy levels are then of very long wavelength. Furthermore, as n increases, the radii of the stationary orbits become large, and the hydrogen atom approaches a macroscopic system, for which classical physics is adequate.

Let us compute the frequency of a photon emitted in the transition between the adjacent states, $n_u = n$ and $n_l = n - 1$, where $n \gg 1$. Rewriting Equation 5-33 we have

$$\nu = cR \left(\frac{n_u^2 - n_l^2}{n_u^2 n_l^2} \right)$$

$$\nu = cR \frac{(n_u - n_l)(n_u + n_l)}{n_u^2 n_l^2} \qquad [5\text{-}35]$$

But $n_u - n_l = 1$; $n_u + n_l \simeq 2n$; and $n_u^2 n_l^2 \simeq n^4$. Therefore Equation 5-35 becomes, for large n,

$$\nu = \frac{2cR}{n^3} \qquad [5\text{-}36]$$

We wish to show that this quantum frequency, ν, equals the classical orbital frequency, f. Using Equation 5-29 for the electron radius r in Equation 5-34, there results

$$f = (1/2\pi)\sqrt{ke^2/mr^3} = (1/2\pi)\sqrt{(ke^2/m)(ke^2m/n^2\hbar^2)^3}$$

$$f = (2c/n^3)(k^2e^4m/4\pi\hbar^3c) \qquad [5\text{-}37]$$

The quantity in the parentheses in Equation 5-37 is simply the Rydberg constant R, following Equation 5-28. Therefore, Equation 5-37 reduces to

$$f = \frac{2cR}{n^3} \qquad [5\text{-}38]$$

By comparing Equations 5-36 and 5-38 we see that

$$\nu = f$$

as the correspondence principle requires.

Actually, it was by applying the correspondence principle that Bohr arrived at the quantization of orbital angular momentum, the second postulate of his atomic theory. In his original paper on the quantum theory of the hydrogen atom, Bohr stated:

"... we only assume (1) that the radiation is sent out in quanta $h\nu$, and (2) that the frequency of the radiation emitted during the passing of the system between successive stationary states will coincide with the frequency of revolution of the electron in the region of slow vibrations."†

Bohr showed that the quantization of orbital angular momentum is a necessary consequence of these two postulates. In our treatment of the Bohr atom we have reversed the procedure; that is, we assumed the electron's wave properties, which we found to be equivalent to the quantiza-

† Philosophical Magazine *26*, 1 (1913): "On the Constitution of Atoms and Molecules."

tion of orbital angular momentum, and the identity of the photon- and orbital-frequencies in the correspondence limit was a result.

5-7   The successes and failures of the Bohr theory   The following features of the Bohr atomic theory are general ones, which apply to *any* comprehensive theory of atomic structure: (1) The existence of non-radiating, stationary states. (2) The quantization of the energy of a bound system of particles. (3) The quantization of angular momentum. (4) The emission or absorption of photons in transitions between stationary states.

More specifically, the Bohr theory explains: (1) The stability of atoms. (2) The wavelengths of the emission and absorption spectra of hydrogenic atoms. (3) The measured ionization energies of one-electron atoms. (4) The assumption, in the kinetic theory, that molecules are inert particles, making only elastic collisions for moderate temperatures.

The Bohr atomic theory has, however, certain serious shortcomings: (1) It is non-relativistic. (2) It gives no method of calculating the intensities of the spectral lines. (3) It is incapable of explaining the spectra of atoms having more than one electron. (4) It does not explain the binding of atoms to form molecules, liquid, and solids. (5) It fails to account for the fine details of the hydrogen spectrum. (High-resolution spectrographs show that each "line" predicted by the Bohr theory consists of two or more very closely-spaced lines, or fine-structure.) (6) The orbital angular momentum, although quantized, follows more complicated rules than that given by the Bohr theory.

All of these defects are corrected in a relativistic wave-mechanical treatment, which, because of its mathematical sophistication, lies beyond the intent of this book. A fundamental reason why the Bohr theory is defective is that it still overemphasizes the classical particle nature of the electron; the electron is considered to move in a well-defined circular path, having precisely defined radii, speeds, and orbital frequencies. In the correct treatment, the electron *wave* must be allowed to extend and move throughout the whole region of space surrounding the nucleus; the problem is solved in the more general quantum theory by fitting three-dimensional electron waves to produce quantized stationary states.

5-8   Summary   The α-particle scattering experiments of Rutherford showed that all of the positive charge and essentially all of the mass of an atom are confined to a very small region of space (no greater than 10^{-14} m), called the nucleus.

The fraction of incident particles, N_s/N_i, scattered by a foil of thickness t containing n scatterers per unit volume is

$$N_s/N_i = \sigma n t$$

where σ is the cross section associated with each scattering center.

The classical planetary model predicts that atoms are unstable and that they emit a continuous spectrum.

An excited hydrogen gas is observed to emit and absorb spectral lines given by

$$\frac{1}{\lambda} = R\left(\frac{1}{n_l^2} - \frac{1}{n_u^2}\right)$$

where R is the Rydberg constant.

The Bohr atomic theory assumes: (1) Stationary states. (2) Orbital angular momentum is $n\hbar$. (3) In a transition, $h\nu = E_u - E_l$.

The radii, speeds, energies, and frequencies predicted by the Bohr theory are given by

$$r_n = \frac{n^2\hbar^2}{k\mu(Ze^2)} \qquad\qquad v_n = \frac{k(Ze^2)}{n\hbar}$$

$$E_n = \frac{-k^2(Z^2e^4)\mu}{2n^2\hbar^2} \qquad\qquad \nu = cZ^2R_M\left(\frac{1}{n_l^2} - \frac{1}{n_u^2}\right)$$

where
$$R_M = \frac{k^2e^4\mu}{4\pi\hbar^3c}$$

and
$$\mu = \frac{m}{1 + (m/M)}$$

REFERENCES

Blanchard, C. H., C. R. Burnett, R. G. Stoner, and R. L. Weber, *Introduction to Modern Physics*. Englewood Cliffs, New Jersey: Prentice-Hall, Inc., 1958. An elegant derivation of the relation between the impact parameter and the scattering angle for Rutherford scattering is given in Chapter 6. This derivation does not depend upon a detailed knowledge of the equation of the path.

Heitler, W., *Elementary Wave Mechanics*. Oxford: The Clarendon Press, 1946. A simple wave-mechanical treatment of the hydrogen atom is to be found in Chapters 2 and 3 of this small book.

Richtmeyer, F. K., E. H. Kennard, and T. Lauritsen, *Introduction to Modern Physics*. New York: McGraw-Hill Book Company, Inc., 1955. Chapter 5 contains a fairly detailed treatment of the origin of spectral lines and their theoretical interpretation.

White, H. E., *Introduction to Atomic Spectra*. New York: McGraw-Hill Book Company, Inc., 1934. A discussion of the early discoveries in spectroscopy is given in Chapter 1.

PROBLEMS

5-1   The density and atomic weight of gold are 19.3 gm/cm³ and 197.2, respectively. Compute the number of gold nuclei per cubic centimeter.

5-2   A black cubical box, one meter along an edge, contains 10,000 small hard spheres, each 0.0010 m in diameter. The spheres are dispersed randomly throughout the interior of the box. (a) If 1.00×10^8 bullets (very small compared to the spheres) are shot randomly into a broad face of the box, how many bullets can be expected to be scattered by collisions with the spheres? (b) How many bullets will be scattered if the diameter of the bullets is equal to that of the spheres?

5-3   An α-particle makes a head-on elastic collision with a gold nucleus at rest. Calculate the fractional energy transferred to the gold nucleus.

5-4   What is the speed and the de Broglie wavelength of a 7.0 Mev α-particle?

5-5   (a) What is the minimum distance that an 8.0 Mev α-particle can approach a gold nucleus (charge, $+79e$) assuming that the gold nucleus remains fixed? (b) What is the potential energy of this α-particle when momentarily at rest at this distance of closest approach? (c) What is the minimum distance that an 8.0 Mev *proton* can approach a gold nucleus?

5-6 * A 6.0 Mev α-particle makes a head-on elastic collision with an electron. (a) What are the kinetic energies of the electron and α-particle after the collision? (b) How many such collisions would be required to bring the α-particle to rest?

5-7   A 5.0 Mev α-particle is scattered 10° by a gold nucleus. (a) What is the corresponding impact parameter? (b) If the gold foil is 4.0×10^{-5} cm thick, what fraction of the incident α-particles are expected to be scattered by more than 10°?

5-8   Suppose that a 10.0 Mev α-particle collides with a number of atoms in a gas, and the α-particle loses, on the average, 5.0 ev per collision. How many collisions would be necessary to bring the α-particle to rest?

5-9   Prove that a positively charged particle scattered by a heavy nucleus approaches the nucleus more closely for the impact parameter b equal to zero (head-on collision) than for any other b. (*Hint:* use conservation of energy.)

5-10   (a) What thickness of gold foil must be used with 6.5 Mev α-particles so that one out of every 10^3 incident particles is scattered by an angle of 90° or more? (b) Is it safe to assume that only *single* scattering occurs for 6.5 Mev α-particles scattered by more than 1°?

5-11 * A 5.0 Mev α-particle approaches an iron nucleus ($Z = 26$, atomic weight $= 56$). (a) Calculate the kinetic energy (Mev) and potential energy (Mev) of the α-particle when it is a distance of

3.2×10^{-14} m from the iron nucleus. (b) If this is the smallest separation between the α-particle and the iron nucleus during this non-head-on collision, show that the impact parameter b must be 2.3×10^{-14} m. (Hint: conserve angular momentum.)

5-12 * (a) Calculate the electric current for an electron moving in the first Bohr orbit. (b) Calculate the magnetic field at the nucleus of a hydrogen atom arising from the orbital motion of the electron in the first Bohr orbit. (c) Is the magnetic field aligned with or opposite to the orbital angular momentum vector of the electron with respect to the nucleus?

5-13 Verify that the radius of the first Bohr orbit of hydrogen is $r_1 = 0.529$ Å.

5-14 Calculate the radius and speed of the electron in the first Bohr orbit of doubly-ionized lithium, and compare these values with those of hydrogen.

5-15 Compute the numerical value of the fine-structure constant $\alpha = v_1/c = ke^2/\hbar c$.

5-16 (a) Prove that the product of the first Bohr orbit and the fine-structure constant is equal to the Compton wavelength of the electron divided by 2π. (b) Prove that the radius of the first Bohr orbit is equal to $(\alpha/4\pi R_\infty)$.

5-17 (a) What is the energy, momentum, and wavelength of the photon that is emitted when a hydrogen atom undergoes a transition from the state $n = 10$ to $n = 1$? (b) With what speed does the hydrogen atom recoil when the photon is emitted?

5-18 The lifetime (the average time before a photon is created) of an excited state of a typical atom is 10^{-8} sec. On the basis of the Bohr theory, how many revolutions about the nucleus are made, on the average, by an electron in the first excited state of hydrogen before it jumps to the ground state?

5-19 (a) Show that the de Broglie wavelength of an electron in the nth Bohr orbit of hydrogen is given by $\lambda_n = n\lambda_1$, where λ_1 is the wavelength in the ground state. (b) Determine the ratio of λ_n to the circumference $(2\pi r_n)$ of the orbit, and show that the limit of this ratio approaches zero as n approaches infinity.

5-20 A hydrogen atom is in an excited state having a *binding energy* of 0.85 ev in this state. The atom makes a transition to a state whose *excitation energy* is 10.2 ev. Calculate the wavelength of the photon emitted in this transition.

5-21 Evaluate numerically the short *and* long wavelength limits of the radiation emitted by hydrogen for (a) the Lyman series, (b) the Balmer series.

5-22   The visible spectrum extends from about 3800 Å to 7700 Å. Express these limits in terms of the energy (ev) of the corresponding photons.

5-23   The most intense line emitted by atomic hydrogen has a wavelength of 1216 Å. What transition of the hydrogen atom is responsible for this radiation?

5-24 * Assume that a gas composed of 3000 hydrogen atoms initially has all the atoms in the $n = 4$ energy state. The atoms then proceed to make transitions to lower energy states. (a) How many spectral lines will be observed from this particular gas? (b) What is the *total* number of photons emitted when all the atoms have finally arrived at the ground state, assuming that, from any given excited state, all possible downward transitions are equally probable?

5-25 * A hydrogen atom (in its ground state) moving with a kinetic energy of 12 ev collides *head-on* with another hydrogen atom at rest and in the ground state. (a) By using the principles of energy and momentum conservation show that an inelastic collision cannot take place and, therefore, an elastic collision must occur. (b) Show that the minimum initial energy necessary in part (a) above to raise one of the atoms to the first excited state is $\frac{3}{2}$ times the ionization energy!

5-26 * (a) What is the minimum potential difference necessary to accelerate electrons from rest, so that, upon colliding with hydrogen atoms, some of the atoms are just excited out of the ground state? Assume that all the electron's kinetic energy is transferred to internal energy of the hydrogen atom in such collisions. (b) What minimum potential difference would be necessary to *ionize* the hydrogen atoms?

5-27   Verify the magnitude (to 3 significant figures) and units of the Rydberg constant by substituting the values of the atomic constants in Equation 5-27.

5-28   Assume that an appreciable number of atoms in a gas are ionized, because of thermal collisions, when the average kinetic energy per atom, $\frac{3}{2}kT$, is equal to (or exceeds) the ionization energy. (a) To what temperature must a hydrogen gas be heated to fulfill this condition? (b) If the gas were at atmospheric pressure at room temperature, what would be the pressure at the ionization temperature, the volume being held constant?

5-29   Show that the mass of a proton and electron exceeds that of a hydrogen atom in the ground state by an amount $m\alpha^2/2$, where m is the electron's mass and α, the fine-structure constant, is the speed of the electron in the first Bohr orbit in units of c, the speed of light.

5-30   (a) Should the visible spectrum (3800 Å to 7700 Å) from Li^{++} show more or fewer lines than the visible spectrum of hydrogen, assuming that all possible transitions occur? (b) How many complete series lie in the ultraviolet region for Li^{++}?

5-31   (a) What transition in the singly-ionized helium spectrum corresponds to the first Lyman line ($n = 2$ to $n = 1$) of hydrogen, assuming, for simplicity, both spectra have the same Rydberg constant, R_∞? (b) Using the proper values for the Rydberg constants, R_H and R_{He}, find the wavelength difference (Ångstroms) between the two lines above, and state whether the ionized helium line has a longer or shorter wavelength than the hydrogen line.

5-32   Suppose that a relative heavy atom such as tungsten (atomic number 74) has all but one of its electrons removed. (a) How much energy (ev) is required to remove completely this last electron if it is in the ground state? (b) What is the wavelength of the photon emitted if this remaining electron were to make a transition from $n = 2$ to $n = 1$? (c) In what portion of the electromagnetic spectrum is such a photon found?

5-33   Calculate the (a) excitation energy and (b) binding energy of a singly-ionized helium atom when it is in the $n = 5$ state.

5-34   Compute the wavelength difference between the H_α lines of hydrogen and deuterium arising from the difference in their mass.

5-35   Using the same energy scale, sketch side by side, the Lyman and Balmer series for hydrogen, deuterium, and doubly-ionized lithium.

5-36   Show that failing to take into account the non-infinite nuclear mass results in the theoretical prediction of spectral lines having a frequency which is too large, and a wavelength which is too small, both by a fraction m/M, where m is the electron mass and M is the nuclear mass.

5-37   An atom of "positronium" is formed when a positron and an electron combine to form a bound system. Compute (a) the Rydberg constant, (b) the wavelength of the first line in the Balmer series, and (c) the ionization energy for positronium.

5-38   A μ^- meson is an elementary particle with charge $(-e)$ and mass $207\,m$, where m is the rest mass of an electron. There is evidence that occasionally a μ meson is captured by a proton (also by other nuclei) forming a "μ-mesic" atom. (a) Calculate the radius of the first Bohr orbit for this atom. (b) What is the ionization energy? (c) Compare the speed of the μ meson in the first orbit with that of the electron in the first orbit.

5-39   (a) Compute the frequency of the photon emitted from a hydrogen atom undergoing a transition from $n = 100$ to $n = 99$. (b) Compute the orbital frequency of the electron in the $n = 100$ state and the fractional difference between the orbital and photon frequencies.

5-40 * The energy difference between two stationary states is strictly equal to the sum of the photon energy and the recoil kinetic energy of the atom. (a) Show that when this recoil is taken into account, the frequency of the emitted photon is reduced by a fraction, $h\nu/2Mc^2$, where M is the mass of the atom. (b) What is the fractional correction to the frequency of the H_α line when this effect is taken into account?

5-41 The Doppler effect for light occurs when the frequency ν' measured by an observer differs from that of the "true" frequency ν emitted by the source. For the special case where the observer and source move away or toward one another along the same line, with a relative velocity v ($v \ll c$), the *apparent* frequency ν' is given by $\nu' = \nu[1 \pm (v/c)]$, where ν is the frequency of the source, the minus sign is for recession, and the positive sign for approach. Thus, the fractional change (magnitude only) in frequency or wavelength is $(\Delta\nu/\nu) = (\Delta\lambda/\lambda) = (v/c)$. The Doppler effect is one reason why spectral lines are not infinitely sharp. The various radiating atoms have different velocity components along the direction of propagation of the light that enters the spectrometer. Assuming that an excited hydrogen gas is at a temperature of 6000°K, and taking the maximum velocities of recession and approach to be given approximately by $\frac{1}{2}mv^2 = \frac{3}{2}kT$, calculate the width (in Ångstroms) of the H_α line due to Doppler broadening.

5-42 See the above paragraph in Problem 5-41. A star is receding from the Earth with a speed of 1000 km/sec. Calculate the wavelength of the H_α line emitted by hydrogen atoms in the star when observed on Earth.

5-43 The Sun, like the Earth, is approximately a sphere rotating about an axis. However, it does not rotate as a rigid body: the nearer the equator, the faster is the rotation. When one compares the absorption spectrum of the H_α radiation coming from opposite sides of the Sun's equator, there is a wavelength difference of 0.0914 Å. Assuming this difference to be due to the Doppler effect (see Problem 5-41), show that the period of rotation at the equator is 25 days. The diameter of the Sun is 865,000 miles.

5-44 * Show that the fraction of the incident α-particles scattered between the angle θ and $\theta + d\theta$ is given by

$$\frac{\pi Z^2 e^4 nt \cos(\theta/2)\, d\theta}{(4\pi\epsilon_0)^2\, K_\alpha^2 \sin^3(\theta/2)}$$

S I X

MANY-ELECTRON ATOMS

Although the Bohr atomic theory is incapable of describing completely
and in detail the structure and spectra of atoms, some of its essential quantum features are found to hold for many-electron atomic systems. These
unchanged features include the existence of stationary states, the quantization of energy, and the quantization of angular momentum. A strictly
correct treatment of the many-electron atom can be made only by using
a rigorous wave-mechanical theory; such a treatment is mathematically
difficult and does not lend itself to a simple visualization of the atomic
structure. In this wave-mechanical treatment, the electron of an atom
must be regarded as a three-dimensional wave surrounding the nucleus;
therefore, it is incorrect and impossible to assign a well-defined path to
the electron's motion. Instead, wave mechanics yields only the probability
of finding an electron at a particular location. Nevertheless, with these
limitations in mind, we can gain certain insights into the results of wave
mechanics by examining a completely classical particle model. We shall
first discuss the classical problem of a particle moving under an inverse-
square attractive force. Next, we will give, without proof, a few important

results of wave mechanics. It will then be possible to interpret these results of wave mechanics, by analogy with the corresponding classical model.

6-1   The constants of the motion for a classical planetary system   We will show in this section that the *constants of the motion*—those quantities which remain unchanged throughout the motion—of a particle bound under the influence of an inverse-square attractive force, are *the total energy, the magnitude of the orbital angular momentum, the component of the orbital angular momentum vector along any direction in space,* and *the rotational (spin) angular momentum vector.* The reader is reminded that in this section we shall be discussing a *strictly* classical model, altogether devoid of any quantum connotations.

Any particle of mass m which is bound by an inverse-square force of attraction to an infinitely heavy mass moves in an elliptical orbit about the fixed center of force located at one focus of the ellipse. A familiar example is that of a planet moving under the influence of the inverse-square gravitational attractive force of the Sun. The simplest elliptical path is a circle, orbit C in Figure 6-1, for which the eccentricity† is zero, and the center of force is at the center c of the circle. The diameter of the circle is defined as $2a$. Orbit D is an ellipse which has a major axis also equal to

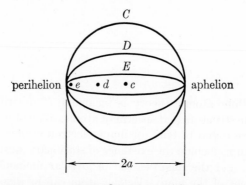

Figure 6-1. Elliptical orbits with the same major axis $(2a)$; c, d, and e are the respective centers of force for a particle moving in the orbits C, D, and E.

$2a$, the center of force now being at d. Orbit E is highly eccentric (eccentricity slightly less than one) with the center of force e very close to the perihelion position and essentially a distance $2a$ from the aphelion position.

Let us now discuss the classical constants of the motion for a bound

---

† The eccentricity is defined as $[1 - (b/a)]^{1/2}$, where a and b are the semi-major and semi-minor axes respectively.

electron-proton system, which is *classically* equivalent to the planet-sun system.

ENERGY   For the circular orbit C, the particle moves with a constant speed and at a constant distance a from the center of force; therefore, both the kinetic and the potential energies are separately constant. The total energy E is also a constant and is given, for the electron-proton system, by $E = -ke^2/2a$ (Section 5-2). For any elliptical orbit neither the speed nor the distance from the particle to the center of force is constant. The kinetic and potential energies are not, therefore, separately constant throughout the particle's motion. Nevertheless, from Newton's laws it can be shown that the total energy of the planetary system, the sum of the kinetic and potential energies, is always constant. Another general result of Newtonian mechanics is that *the total energy of a planetary system is independent of the eccentricity and depends only on the major axis,* 2a. Therefore, the total energies of all three orbits, C, D, and E, of Figure 6-1, are identical. Let us prove this for a special case.

We assume that orbit E has so large an eccentricity (approximately one) that the particle can be regarded to move along an essentially straight line. When the electron is at the aphelion position, reversing the direction of its motion, it is momentarily at rest, with zero kinetic energy. The total energy at this point is just the potential energy, $E = P = -ke^2/r$, where r is now equal to 2a. Thus, $E = -ke^2/2a$. This is exactly the total energy of the system for the circular orbit, C. More detailed analysis shows that *all* elliptical orbits having the same major axis have the same total energy. If we take the major axis as a measure of the size of the orbit, then the total energy of the planetary system depends solely on the size of the orbit, but not on its shape (eccentricity). Classically, all sizes and, hence, all energies, are possible.

MAGNITUDE OF THE ORBITAL ANGULAR MOMENTUM   A second important constant of the motion is the orbital angular momentum, $\boldsymbol{P}_\theta$, of the particle about the center of force. The magnitude of $\boldsymbol{P}_\theta$ is mvr', where v is the speed of the particle and r' is the *perpendicular* distance between the direction of the velocity and the center of the force. The direction of the angular momentum is perpendicular

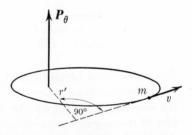

Figure 6-2. Angular momentum, $P_\theta = mvr'$, of a particle in an elliptical orbit.

to the plane of the orbit and points in the direction of advance of a right-hand screw turning in the sense of the particle's orbital motion (see Figure 6-2).

We wish to prove that the angular momentum deriving from the orbital motion of the particle about the fixed center of force is constant. Consider the torque acting on the moving particle. Inasmuch as the force is a *central force*, acting along the line connecting the particle and the center of force, the torque L about the focal point due to this force is obviously zero (see Figure 1-3). By Newton's second law, $L = dP_\theta/dt$; therefore, since the torque L is here equal to zero at all times, the orbital angular momentum P_θ must remain constant, both in magnitude and in direction.

The orbital angular momentum for any elliptical orbit is a constant. In contrast to the fact that three different orbits C, D, and E, of Figure 6-1 have the same energy, the angular momenta of these orbits differ. The angular momentum of the highly eccentric orbit E is essentially zero. This follows from the fact that $P_\theta = mvr'$ and $r' \simeq 0$. The angular momentum is a maximum for a circular orbit C; all other orbits have an intermediate value of the angular momentum, depending on the eccentricity, or shape, of the orbit. For a given energy, or size, of the orbit, there is a continuum of possible angular momenta, ranging from zero (eccentricity = 1) to a maximum value (eccentricity = 0), as shown in Figure 6-3.

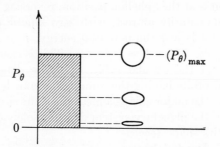

Figure 6-3. Classical allowed values of the angular momentum for elliptical orbits of the same major axis but different eccentricity.

THE COMPONENT OF THE ANGULAR MOMENTUM ALONG A DIRECTION IN SPACE We have seen that a planetary system has an angular momentum which is a constant of the motion, both in magnitude and in direction. The direction of the angular momentum vector, P_θ, is perpendicular to the plane of the elliptical orbit. Therefore, the component P_z of P_θ along any direction in space, Z, is also a constant of the motion. From Figure 6-4, $P_z = P_\theta \cos \theta$. There is no restriction on the choice of the direction Z; hence, the possible values of the component P_z range continuously from $-P_\theta$ to $+P_\theta$, as shown in Figure 6-5. Stated differently, the angle θ can take on any value from $-180°$ to $+180°$; or, once a direction Z is chosen, there is no restriction on the direction of the angular momentum vector P_θ. This seemingly trivial consideration of the constancy of P_z is of the utmost significance in the quantum-mechanical analog.

We have seen that the size, shape, and orientation of the orbit for an inverse-square force, corresponding respectively to the energy, the magnitude of the orbital angular momentum, and the component of the angular momentum along any direction in space, are all constants of the motion.

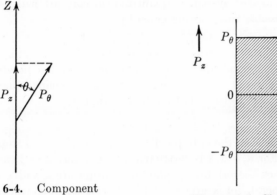

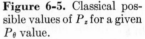

Figure 6-4. Component of the angular momentum along some arbitrary Z direction.

Figure 6-5. Classical possible values of P_z for a given P_θ value.

SPIN ANGULAR MOMENTUM   The angular momentum that we have discussed above is the orbital angular momentum of the planetary particle. In the planetary system of the Sun and the Earth there is still another contribution to the *total* angular momentum of the entire system. It is the angular momentum originating in the rotating, or spinning, of the Earth and of the Sun about their respective axes of rotation. The spin, or rotational, angular momentum of an extended body, such as the Earth, is the product of the angular velocity and the moment of inertia about the axis of rotation; the direction of the spin angular momentum is along the direction of this axis (North, for the Earth). To obtain the total angular momentum of a planetary system, one must add vectorially the orbital and spin angular momenta. We shall see that an angular momentum analogous to the spin angular momentum of the Earth must be assigned to such particles as electrons.

6-2   The quantization of orbital angular momentum   The Bohr theory for a one-electron atom introduces the principal quantum number n, whose integral value determines the total energy of the atom following $E_n = -E_I/n^2$, where E_I is the ionization energy. This theory also shows that if the electron is imagined to be moving in a circular orbit, the value of n specifies, as well, the magnitude of the angular momentum P_θ arising from the orbital motion of the electron, according to $P_\theta = n\hbar$. It is, however, *not* proper from the point of view of wave mechanics, to visualize the electron as moving in such a clear-cut path, and the Bohr rule for the quantization of the magnitude of the orbital angular momentum is *not* correct.

Wave mechanics shows that the magnitude of orbital angular momentum P_θ of an atomic system is quantized, in contrast with the classical theory, the possible values being given by

$$P_\theta = \sqrt{l(l + 1)}\hbar \qquad\qquad [6\text{-}1]$$

where l is an integer called the *orbital angular momentum quantum number*. The possible values of l, for a given value of the principal quantum number n, go from zero to $(n - 1)$ by integers; that is,

$$l = 0, 1, 2, 3, \ldots (n - 1)$$

Thus, when $n = 1$, the only possible value for l is 0; and from Equation 6-1, $P_\theta = 0$. For $n = 2$, l is restricted to the values 0 or 1, and the permitted values of the orbital angular momentum are 0 and $\sqrt{2}\hbar$, respectively. In general, for a given n, there are n possible values of l, and therefore n possible values of the angular momentum. The integral values of the quantum number l are often represented by letter symbols (for reasons that are of historical origin) as follows:

$$l = 0, 1, 2, 3, 4, 5, \ldots$$
$$\text{symbol} = S, P, D, F, G, H, \ldots$$

Whereas in the Bohr theory the state of an atom is specified by giving the quantum number n (hence, the radius of the circular orbit, or the total energy), in wave mechanics the state of an atomic system is completely specified by giving the values of *all* the appropriate quantum numbers. To each state there corresponds a distinctive wave function ψ. Those states for which, say, $n = 3$ and $l = 0$, 1, or 2 are designated as $3S$, $3P$, and $3D$ states, respectively. The corresponding values of the orbital angular momentum in these states are 0, $\sqrt{2}\hbar$, and $\sqrt{6}\hbar$. (The Bohr theory would specify the angular momentum as $3\hbar$, a value that is slightly higher than that of the $3D$ state.) The possible magnitudes of P_θ for $n = 3$ are shown in

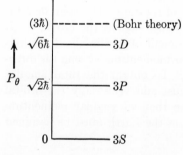

Figure 6-6. Quantum allowed values of the magnitude of the orbital angular momentum for $n = 3$.

Figure 6-6. The $3S$, $3P$, and $3D$ states have a common value for the principal quantum number n; for a single electron under the influence of a Coulomb force from a nucleus assumed to be a point charge, all three states have identical energies, but differ in angular momentum. Such states which have identical values of the total energy, but differ in some other respect are said to be *degenerate*.

It is useful to recall that in classical planetary motion the total energy of the bound system depends only on the value of the major axis of the ellipse, but not on the eccentricity of the orbits, or on the orbital angular momentum. A similar situation obtains in the quantum theory: for a given value of n, which specifies the energy of the atom, there are n possible values of l, each l specifying a possible value of the orbital angular momentum. There are also important differences: the classical theory places no restriction on the possible values of orbital angular momentum; the quantum theory limits the orbital angular momentum to discrete, quantized values.

In the classical theory the orbits corresponding to the small values of angular momentum are those of high eccentricity, the circular orbit having the largest angular momentum for a given major axis, or energy. One can paraphrase this by saying that the orbits of small angular momentum are those in which the orbiting particle spends an appreciable fraction of time in each cycle close to the center of force, whereas the orbit of large angular momentum (the circle) is that in which the circulating particle is always at a maximum distance from the center of force for a given major axis, or energy. The corresponding wave mechanical situation is analogous: one can say that for a $1S$ state ($l = 0$) with $P_\theta = 0$, the electron in the hydrogen atom spends a large fraction of the time near the nucleus and a small

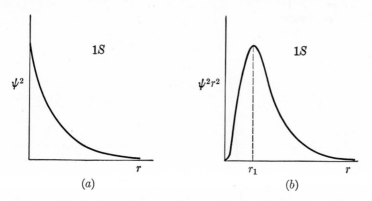

(a)   (b)

Figure 6-7. (a) Probability of finding the electron in the volume element dV as a function of r for the $1S$ state of hydrogen. (b) Probability of finding the electron at a distance r from the nucleus for the $1S$ state of hydrogen.

fraction of the time at large distances from it; conversely, for the state in which the orbital angular momentum is a maximum, with $l = n - 1$, the electron spends a smaller fraction of its time near the nucleus.

Wave mechanics yields solutions to the quantum wave equation, giving the wave function ψ as a function of position in space (see Section 4-5). The quantity $\psi^2 dV$ gives the probability of finding the electron in a small volume element dV. Figure 6-7a shows $\psi^2 dV$ as a function of r, the distance from the nucleus to the electron in the hydrogen atom, for the $1S$ state. It is possible to plot the probability as a function of the one coordinate r alone, inasmuch as the wave functions for all S states are always spherically symmetrical. Clearly, the probability *per unit volume element* of finding the electron at or near the nucleus is greater than that for any more-distant points. One might ask what the probability is of finding the electron at a

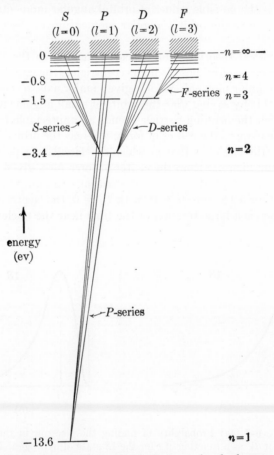

Figure 6-8. Energy-level diagram for hydrogen, showing the S, P, D, and F series.

distance r from the nucleus; that is, between r and $r + dr$. This is equivalent to asking what is the probability of finding the electron in a spherical shell of area $4\pi r^2$ and thickness dr, or in a volume element $dV = 4\pi r^2 dr$. Therefore, the probability for finding the electron *at* a distance r is equal to $\psi^2 4\pi r^2 dr$, or proportional to $\psi^2 r^2$. Figure 6-7b shows $\psi^2 r^2$ as a function of r for the $1S$ state. The maximum probability occurs precisely at a distance equal to the radius of the first Bohr orbit, $r_1 = \hbar^2/kme^2$.

Let us consider again an energy-level diagram of hydrogen, where the quantization of angular momentum, as well as the quantization of energy, is displayed. Figure 6-8 shows the states of the hydrogen atom, identified according to energy and angular momentum; that is, according to the n and the l quantum numbers. The n-fold degeneracy for each energy is shown. The diagonal lines connecting states represent the possible transitions between stationary states leading to the emission of photons. Only those transitions connecting states between adjacent columns, in which the orbital angular momentum quantum number l changes by one unit, are shown. The wave-mechanical theory selects from all of the possible combinations of stationary states those particular combinations for which appreciable radiation can take place; a transition between other pairs of states will not lead to the emission (or absorption) of a photon. Transitions for which l changes by 1, that is, for which the change in the orbital angular momentum quantum number, Δl, is $+1$ or -1 are said to be *allowed transitions*. The *selection rule* for allowed transitions is then

$$\Delta l = \pm 1 \qquad\qquad [6\text{-}2]$$

All other combinations of states correspond to *forbidden transitions*. (Such forbidden transitions are not absolutely prohibited, but have a probability of occurrence which is at least 10^6 times smaller than the allowed transitions.) There is *no* selection rule restricting the possible changes in the quantum number n.

The selection rule requires, in effect, that the orbital angular momentum of the atom must change when a photon is emitted or absorbed. By the conservation-of-angular-momentum law, the angular momentum of an atom in an excited state must equal the angular momentum of the atom-plus-photon after photon emission. Because the angular momentum of the atom alone changes in photon emission or absorption, *the photon must itself carry angular momentum*. Thus, a photon carries energy, linear momentum, and angular momentum. There is a classical analog to the angular momentum of the photon in that angular momentum must be ascribed to the electromagnetic field, or to (circularly polarized) electromagnetic waves (see Section 1-4).

The frequencies, or wavelengths, of the photons in the hydrogen spectrum, as indicated in Figure 6-8, are precisely those which the Bohr theory predicts. The Bohr theory is completely in accord with wave mechanics in giving the allowed energies as $E_n = -E_I/n^2$. The transitions are, however, arranged in groups, or series, which are labeled according to the l value of the *originating* state in a downward transition. Thus, the P series shown consists of the transitions, $2P \rightarrow 1S$, $3P \rightarrow 1S$, $4P \rightarrow 1S$, etc. (Other P series, such as $3P \rightarrow 2S$, $4P \rightarrow 2S$, $5P \rightarrow 2S$, etc., occur, but ordinarily with less intensity than the indicated series.) We note that for hydrogen, or for one-electron hydrogenic atoms, many of the indicated allowed transitions give rise to photons having the same energy, or wavelength: the H_a line arises from the transitions, $3S \rightarrow 2P$, $3P \rightarrow 2S$, and $3D \rightarrow 2P$. We shall see that the distinction between these transitions, although formal for hydrogenic atoms, is significant for many-electron atoms.

6-3 Hydrogen-like atoms There are a number of many-electron atoms which resemble a hydrogen atom. One such group consists of the elements listed in the first column of the periodic table of chemical elements, called the *alkali metals*; a second group consists of the singly-ionized elements in the second column of the periodic table, called the *alkaline earths*. Table 6-1 shows these elements, together with the rare gases, which precede the respective alkali metals in the periodic table. The *atomic number*, which gives the number of electrons in a neutral atom, and therefore, also the positive charge of the nucleus in multiples of the electron charge, is given as a presubscript to the chemical symbol.

Table 6-1

Rare gases	Alkali metals	Singly-ionized alkaline earths
	$_1$H	$_2$He$^+$
$_2$He	$_3$Li	$_4$Be$^+$
$_{10}$Ne	$_{11}$Na	$_{12}$Mg$^+$
$_{18}$Ar	$_{19}$K	$_{20}$Ca$^+$
$_{36}$Kr	$_{37}$Rb	$_{38}$Sr$^+$
$_{54}$Xe	$_{55}$Cs	$_{56}$Ba$^+$

Let us recall some properties of these elements. All of the elements in each group (or column), are chemically similar. The rare gases are chemically inert, and can be ionized only by energies which considerably exceed those for other elements. The alkali metals show a valence of $+1$ and are extremely active chemically; the alkaline earths typically show a valence of $+2$. It is reasonable to suppose that, in a rare-gas element, such as neon, the 10 electrons are somehow arranged so as to form a relatively inert configuration around the nucleus. When the atom is in its lowest energy state, it is only with great expenditure of energy that any electron can be extracted from this stable arrangement. The electrons in a rare gas can be imagined to form a tightly-bound shell of negative charge around the nucleus.

An atom of sodium has 11 electrons when neutral; because its valence is +1 and the atom is very active chemically, the sodium atom can be regarded as a neon atom whose nuclear charge has been increased by 1 and whose number of electrons has also been increased by 1. This last electron is very loosely bound to the atom, inasmuch as the atom readily loses it to form a positively charged ion. It is then tempting to imagine that the first 10 electrons of sodium form a relatively inert closed shell about which the eleventh electron moves.

Finally, the neutral atom of magnesium with atomic number 12, valence +2, can be thought of as consisting of a shell of 10 inactive electrons surrounded by two chemically active electrons. When magnesium is singly ionized, only one valence electron remains outside the closed shell. The structure of sodium and singly-ionized magnesium thus bears a strong resemblance to that of hydrogen in that the chemical properties result from a single electron held to an inert core.

Consider an atom of the alkali-metals group, having an inner core composed of a nucleus of positive charge Ze, immediately surrounded by $(Z - 1)$ relatively tightly-bound electrons, and, located outside this inert inner core, a single valence electron (see Figure 6-9). The valence electron experiences an electric force due to the combined effects of the positive nucleus and negative-core electrons. When the valence electron is truly outside the core it will feel the effects of the positive nuclear charge Ze and negative electron charges $(Z - 1)e$; thus, the valence electron, in effect, is subject to an equivalent charge of $Ze - (Z - 1)e$, or $+1e$. This is exactly the charge of the nucleus of the hydrogen atom, and therefore, the states and energies of the atom for the various possible configurations of the valence electron should resemble closely those of hydrogen.

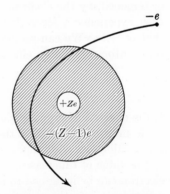

Figure 6-9. Representation of the penetration of the valence electron of a hydrogen-like atom into the inner electron core. Charge of the nucleus, $+Ze$; charge of the inner electron core, $-(Z - 1)e$.

There will, however, be significant differences. The valence electron, even when outside the inert inner core, will not be subject to the force of a single *point* charge originating at the nucleus, inasmuch as the valence electron will attract the nucleus but repel the electrons of the inert shell, thereby displacing the center of the electron shell from the nucleus. An even more significant difference is that the valence electron cannot properly be regarded as a particle which is at all times outside the inert core.

Suppose that the hydrogen-like atom is in the state for which $n = 3$; there are then three possible values for the orbital angular momentum quantum number, $l = 0, 1,$ or 2. The relative fractions of time the electron spends in the vicinity of the nucleus are different for the $3S$, $3P$, and $3D$ states. For the $3D$ state, which corresponds to the classical circular orbit, the valence electron spends only a small fraction of the time in the immediate vicinity of the nucleus; therefore, for the $3D$ state, the assumption that the valence electron is outside the inner core is a fairly good approximation. In the $3S$ state, on the other hand, the valence electron will have a high probability of being found in the vicinity of the nucleus and it will penetrate the inner core a large fraction of the time. The assumption that the valence electron is completely outside the inert electron core is, for this $3S$ state, not warranted.

How will the energy of the atom be changed from that which is predicted for the hydrogen atom, for which the penetration of an electron core is, of course, impossible? When the valence electron is found inside the closed shell, the nuclear electric charge will be less completely shielded, or screened, by the electrons in the closed shell. The valence electron will then experience a force that arises from an effective charge *greater* than that of $+1e$. We can see how the energies are affected by recalling that, for an atom having an effective nuclear charge Ze, the energy is given by

$$E_n = -\frac{Z^2 E_I}{n^2}$$

If the effective charge Ze is greater than $+1e$, the corresponding energy is made more negative, and the energy level is displaced downward. For a given value of n, this effect is greatest for the S state, and less pronounced as the value of l increases. Furthermore, when n is increased, the valence electron can be imagined to move in orbits of increased size; therefore, the penetration effects are correspondingly reduced and the downward displacement of the energy levels arising from penetration is also reduced.

We can see these effects by examining the energy-level diagram for the hydrogen-like atom, $_{11}$Na, as shown in Figure 6-10. The energy levels for hydrogen are shown for comparison by dotted lines. We first note that energy levels corresponding to $n = 2$, or 1, are *not* found; the reasons why the valence electron in $_{11}$Na is excluded from the $n = 2$, or 1, states will be explored in Section 6-8. All of the $n = 3$ states, $3S$, $3P$, and $3D$, for sodium have a lower energy than the corresponding states in hydrogen. Furthermore, the $3S$ state is lower than $3P$, which is, in turn, lower than $3D$. These three states in sodium all have different energies and are, therefore, *not* degenerate, despite the fact that they have the same quantum

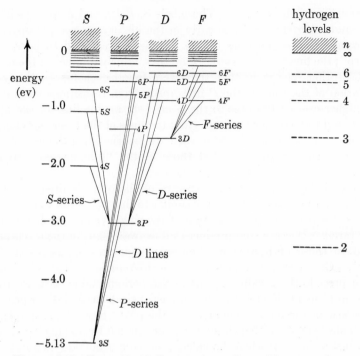

Figure 6-10. Energy-level diagram for sodium. For comparison, the hydrogen energy levels are shown on the right.

number n. The energy of the atom for $_{11}$Na is *not* independent of the value of the orbital angular momentum, but is, instead, lowest for the smallest orbital angular momentum. For higher values of n, corresponding to larger orbits of the valence electron, the energy levels more closely approach those of hydrogen.

The allowed transitions, following the selection rule, $\Delta l = \pm 1$, are classified into S, P, D, and F series, as in Figure 6-8 for hydrogen; but the corresponding emitted photons are of different frequencies, and do *not* coincide. For example, the $4S \rightarrow 3P$ and $4P \rightarrow 3S$ transitions in sodium yield two *different* spectral lines. The most prominent line in the spectrum of Na is the so-called sodium D line,† which arises from the $3P \rightarrow 3S$ transition,

† The label D for the strong yellow lines of sodium has no connection with the symbol D designating $l = 2$. Fraunhofer discovered in 1809 that the spectrum from the Sun contains a number of dark absorption lines (*Fraunhofer lines*) which arise from the absorption of radiation from the interior of the Sun by elements in the Sun's atmosphere. These lines were labeled A, B, C, D, etc. The D Fraunhofer line corresponds to absorption by sodium vapor in the $3P \rightarrow 3S$ transition. Under close observation it is seen that this transition consists of two closely spaced yellow lines having wavelengths of 5890 Å and 5896 Å; other lines in the sodium spectrum show a similar fine structure, whose origin will be treated in Section 6-6. Also appearing in the Sun's absorption spectrum were lines identified with the element helium, and named for the Sun (*helios*); helium was later isolated and identified on Earth.

the first line of the P series. The labels S, P, D, and F were assigned to these and other similar series early in the history of spectroscopy for the following reasons: the lines of the S series were found to be relatively "*sharp*"; the lines of the P series were the "*principal*" lines in the emission or absorption spectra in that they were found even for relatively small excitation of the source (the P lines result from transitions from the *first* and higher excited P states to the *ground* state); the lines of the D series were found to be rather "*diffuse*"; and the lines of the F series, lying in the infrared, were so named because their frequencies were the lowest of any series and closely approximated those of the "*f*undamental" hydrogen atom.

The emission and absorption spectra of the other hydrogen-like alkali metals, $_3$Li, $_{19}$K, $_{37}$Rb, and $_{55}$Cs, are similar to that of $_{11}$Na in that the energy levels corresponding to penetration of the single valence electron into the inert rare-gas electron core produces a downward shift of the levels as compared to those of hydrogen. The singly-ionized alkaline earths, $_4$Be$^+$, $_{12}$Mg$^+$, $_{20}$Ca$^+$, $_{38}$Sr$^+$, and $_{56}$Ba$^+$, are *iso-electronic* with the alkali metals, which precede the alkaline earths in the periodic table, because the two adjacent elements have the same number of electrons and differ principally in the size of the nuclear charge. For the $_{12}$Mg$^+$ atom, for example, there are 10 electrons in a $_{10}$Ne closed shell surrounded by a single valence electron, just as in $_{11}$Na; when the valence electron is outside the inert core, it sees a net positive charge of $+2e$. Thus, the non-penetrating energy states of $_{12}$Mg$^+$ will correspond closely with those for the one-electron atom, $_2$He$^+$, for which the nuclear electric charge is also $+2e$. The states having small values of n, and especially those for which the orbital angular momentum quantum number is small, will, however, be displaced downward with respect to those of $_2$He$^+$.

We have seen that it is possible to understand qualitatively the energy levels and spectra from hydrogen-like atoms by assuming that the excited states arise from the last, valence electron of the atom, moving around a tightly-bound, inert electron shell. We shall see that the stability and inertness of the closed shells arise in a natural way from fundamental principles. The problems involved in the study of the energy levels and spectra of atoms containing more than one active valence electron are much more complex.

6-4 Space quantization In the classical planetary model, the total energy, the magnitude of the orbital angular momentum, and the component of the orbital angular momentum along any direction in space are constants of the motion. According to the quantum theory, the energy of a one-electron atom is quantized and identified by the principal quantum number n; and the orbital angular momentum is quantized, its possible values depending on the value of the orbital angular momentum quantum number, l. We will see here that the third classical constant of the motion, the component of the orbital angular momentum along a fixed direction in space, is also quantized and, therefore, specified by a quantum number. This quantization, which can be derived formally by wave mechanics, is closely related to magnetic effects in atoms.

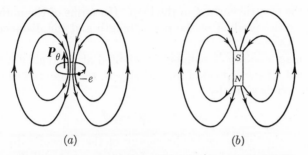

(a)   (b)

Figure 6-11. (a) Magnetic field of magnetic dipole comprised of a circulating negative electric charge. (b) Magnetic field of a magnetic dipole comprised of a small permanent magnet.

Let us examine the magnetic effects associated with the orbital motion of an electron. The orbital angular momentum P_θ of a charged particle moving in an elliptical orbit is represented by a vector oriented at right angles to the plane of the orbit. A circulating negative electric charge comprises an electric current loop, which, by the Oersted effect, produces a magnetic field in the vicinity of the equivalent current loop. The magnetic field at any point is proportional to the magnitude of the current, and the field configuration is shown in Figure 6-11a. This magnetic-field configuration is altogether equivalent to that arising from a small permanent magnet (Figure 6-11b), or magnetic dipole (N- and S-pole); and it is useful to define a *magnetic dipole moment*, μ, which is a measure of the magnetic effects of the current loop. The direction of the magnetic-dipole-moment vector is taken to be that of the equivalent permanent magnet, pointing from the S- to the N-pole, with the arrow at the N-pole. Therefore, for a *negative* electric charge, the orbital angular momentum vector P_θ and the magnetic-dipole vector μ are both perpendicular to the plane of the orbit

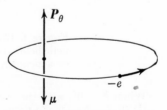

Figure 6-12. Orbital angular momentum P_θ and magnetic moment μ of an orbiting electron.

and point in *opposite* directions, as seen in Figure 6-12. The right-hand screw rule relates the direction of P_θ to the sense of rotation of the moving particle; similarly, the direction of μ is given by the right-hand screw rule, where the sense of rotation of the *conventional* (positive charge) electric current I is used. The magnitude of the magnetic moment of a current loop enclosing a plane of area A is given by

$$\mu = IA \qquad [6\text{-}2]$$

where I is the electric current in the loop. (In terms of the notion of pole strength, μ is the product of the pole strength and the distance between magnetic poles.)

When a magnetic dipole moment is oriented at an angle θ with respect to an external magnetic field of flux density B, the potential energy of the magnetic dipole is given by

$$\Delta E_m = -\mu B \cos\theta \qquad [6\text{-}3]$$

where ΔE_m is the difference in the energy of the magnetic dipole, or the associated current loop, between that for any angle θ and that for which the dipole is perpendicular to the magnetic lines (see Figure 6-13a). (Equations 6-2 and 6-3 are derived in elementary texts.)

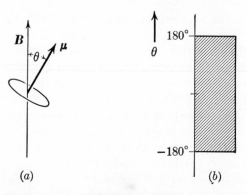

(a)   (b)

Figure 6-13. (a) Magnetic dipole in a magnetic field. (b) Classical permitted orientations of a magnetic dipole in a magnetic field.

When $\theta = 0°$, and the North pole of the equivalent magnet is aligned with the magnetic field lines, $\Delta E_m = -\mu B$, a minimum. When $\theta = 180°$, and the magnetic dipole is anti-aligned with the external field, $\Delta E_m = +\mu B$, a maximum. Classically, all orientations of the dipole between 0° and 180° are allowed, of course, as shown in Figure 6-13b.

The relation between the magnetic dipole moment arising from a circulating electric charge and the orbital angular momentum is of particular significance. Let us suppose, for simplicity, that an electron moves in a circle of radius r with a frequency f. Then the current is given by $I = ef$, and using Equation 6-2 there results

$$\mu = IA = (ef)(\pi r^2) \qquad [6\text{-}4]$$

The orbital angular momentum is simply the product of the moment of inertia of the electron, mr^2, and the angular velocity, $2\pi f$; therefore,

$$P_\theta = (mr^2)(2\pi f) \qquad [6\text{-}5]$$

Combining Equations 6-4 and 6-5 we have

$$\boxed{\mu = \left(-\frac{e}{2m}\right) P_\theta} \qquad [6\text{-}6]$$

The minus sign is introduced into Equation 6-6 because the magnetic moment and angular momentum vectors point in opposite directions. We see that μ is directly proportional to P_θ, the proportionality constant, customarily called the *gyromagnetic ratio*, depending only on the mass and charge of the moving particle. Precisely the same relation, Equation 6-6, is found when the calculation is extended to non-circular loops.

Calculations using the wave mechanics show that the gyromagnetic ratio of an electron in an atom with an orbital angular momentum $P_\theta = \sqrt{l(l + 1)}\hbar$, is given by precisely the same relation, Equation 6-6, found for the classical case, despite the fact that it is impossible to visualize the connection between the magnetic and the angular-momentum effects of an electron simply in terms of a well-defined orbital motion. Therefore, since P_θ depends on the orbital angular momentum quantum number l, so must the magnetic moment μ. Using Equation 6-1 in Equation 6-6 we can write

$$\mu_l = \frac{e}{2m} \sqrt{l(l + 1)}\hbar \qquad [6\text{-}7]$$

where the subscript denotes the magnetic moment associated with l.

Consider now the change in the energy ΔE_m of an atom which results when the atom is placed in an external magnetic field of flux density $\boldsymbol{B}$. Combining Equations 6-3 and 6-7, and taking θ to be the angle between $\boldsymbol{P_\theta}$ and $\boldsymbol{B}$, rather than the angle between μ and $\boldsymbol{B}$, we have

$$\Delta E_m = \frac{e\hbar}{2m} \sqrt{l(l + 1)}B \cos \theta \qquad [6\text{-}8]$$

Thus, when the atom is subject to an external magnetic field, the energy of the atom is changed depending on the value of the angle θ between the orbital angular momentum vector and the external field. If there were no restriction on the angle θ, the component of the orbital angular momentum along the direction of the magnetic field could assume any value between $P_\theta = \sqrt{l(l + 1)}\hbar$ and $-P_\theta = -\sqrt{l(l + 1)}\hbar$. Similarly, with no restriction on the angle θ, ΔE_m could assume any value between $+(e\hbar/2m)B\sqrt{l(l+1)}$ and $-(e\hbar/2m)B\sqrt{l(l+1)}$, following Equation 6-8. In short, if there were no rule restricting, or quantizing, the values of P_θ along the direction of the magnetic field, there would exist a *continuum* of possible energies, quite unlike the situation that has heretofore been found to hold for bound systems. Furthermore, the emission lines from atoms having magnetic moments located in a magnetic field would be continuously broadened, and not split into discrete component lines.

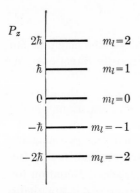

The emission lines from atoms placed in a strong external magnetic field were first studied in 1896 by P. Zeeman, who found that the spectral lines *appeared* to be broadened. Under higher resolution it was found that a single line in the absence of a magnetic field, is split, by the application of an external field, into two or more closely-spaced, sharp component lines. This splitting of the spectral lines into discrete components by a magnetic field is known as the *Zeeman effect*. Its proper understanding lies in the wave-mechanical phenomenon, called *space quantization*.

Figure 6-14. Quantum permitted values of the component of the orbital angular momentum along the direction of a magnetic field for a D state.

According to wave mechanics, an orbital angular momentum vector $\boldsymbol{P_\theta}$ *cannot* assume *any* direction in an external magnetic field; rather, it is restricted to those particular orientations for which the *component of $\boldsymbol{P_\theta}$ along the direction of the magnetic field is an integral multiple of $\hbar$*. The direction of the external magnetic field is chosen, by convention, to be the Z-direction. The possible values of the Z-component of $\boldsymbol{P_\theta}$ are given by (see Figure 6-14)

$$\boxed{P_z = m_l \hbar}$$ [6-9]

where m_l, the *orbital magnetic quantum number*, can assume, for a given value of l, the integral values

$$m_l = l, l-1, l-2, \ldots 0, \ldots -l$$ [6-10]

For example, in a D state with $l = 2$, the possible values of m_l are $+2$, $+1$, 0, -1, or -2. In this state then, P_z can have the possible values $2\hbar$, $1\hbar$, 0, $-1\hbar$, $-2\hbar$, whereas $P_\theta = \sqrt{6}\hbar$. Figure 6-15 shows the possible orientations of the orbital angular momentum vector with respect to an external magnetic field (compare with Figure 6-5). Because the angular momentum vector is restricted to certain discrete orientations in space

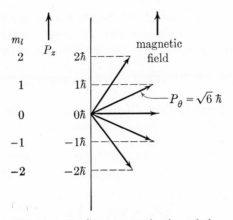

Figure 6-15. Space quantization of the orbital angular momentum vector for a D state.

the angular momentum is said to be *space quantized.* Inasmuch as P_z is $P_\theta \cos \theta$, the rule governing the orientation of the P_θ vector, that is the rule for space quantization, is

$$\cos \theta = \frac{m_l}{\sqrt{l(l + 1)}} \qquad\qquad [6\text{-}11]$$

It is interesting to note that the maximum component of P_θ along the direction of space quantization, $P_z = l\hbar$, is always less than the magnitude of $P_\theta = \sqrt{l(l + 1)}\hbar$; therefore, the orbital angular momentum vector is never completely aligned along the direction of an external magnetic field. The wave mechanics permits the magnitude of P_θ and its Z-component along a direction in space to be precisely specified, but paradoxically, precludes a knowledge of the X- and Y-components of P_θ. It is customary to regard the P_θ vector to *precess* around the Z-direction at a constant angle θ, thereby tracing out a cone about the Z-axis for any particular allowed value of m_l; in this way, the magnitude and Z-component of P_θ are known, but the X- and Y-components are indefinite.†

Consider now the situation for a very large value of the orbital angular momentum. If P_θ is very large, so is the quantum number l and, therefore, the number of orientations, $(2l + 1)$. The orientations θ of the P_θ vector are still quantized, following Equation 6-11; but the differences between adjacent values of θ are so small that the classical condition, for which *all* possible orientations are allowed, is approached. Furthermore, for a very large value of l, the maximum component of P_θ along the Z-direction differs

† The indefiniteness of the X- and Y-components can be shown to be a necessary consequence of the uncertainty principle.

from P_θ by only a small amount, and the angular momentum and magnetic moment vectors are permitted to assume (almost) a continuum of orientations (Figure 3-1d). We have here another example of the correspondence principle, which requires that the quantum theory reduce to the classical theory in the limit of large orbital angular momenta.

6-5   The normal Zeeman effect   If the orientation of the angular momentum vector is quantized, the possible orientations of the magnetic dipole moment vector μ_l associated with P_θ are similarly quantized, and so too is the magnetic potential energy ΔE_m of the state. Using Equation 6-11 in Equation 6-8 for the change in the energy of a state with quantum numbers l and m_l, we have

$$\Delta E_m = m_l \left(\frac{e\hbar}{2m} \right) B \qquad [6\text{-}12]$$

Figure 6-16 shows the change in the energies of S, P, D, and F states, having respectively 1, 3, 5, and 7 *magnetic sublevels*. In general, the number of Zeeman components for a given l is equal to $(2l + 1)$. The energy separation between adjacent magnetic sublevels, equal to $(e\hbar/2m)B$ and independent of the value of l, is shown in Figure 6-16. The quantity

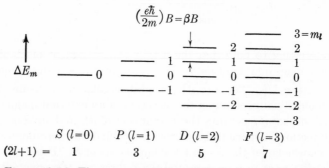

Figure 6-16. The energy splitting of S, P, D, and F states of an atom in a magnetic field.

$(e\hbar/2m)$ has the units of magnetic moment and is known as the *Bohr magneton*, β, because β is the magnetic moment of the electron in the first circular orbit of the hydrogen atom according to the Bohr theory.

Bohr magneton† $= \beta = e\hbar/2m = .9273 \times 10^{-23}$ joule/(weber/m²)

We will now analyze the spectrum of lines emitted from excited atoms by transitions between a D state and a P state in the presence of a magnetic field. When $B = 0$, the energy of the D state is E_D (for all five m_l values), the energy of the P state is E_P (for all three m_l values), and photons hav-

† In emu units, the gyromagnetic ratio for orbital angular momentum of an electron is $(e/2mc)$, and the Bohr magneton is $(e\hbar/2mc) = 9.273 \times 10^{-21}$ erg/gauss.

ing the single frequency ν_0 are emitted following $h\nu_0 = E_D - E_P$, as shown in Figure 6-17a. When the field is turned on, the D state splits into five

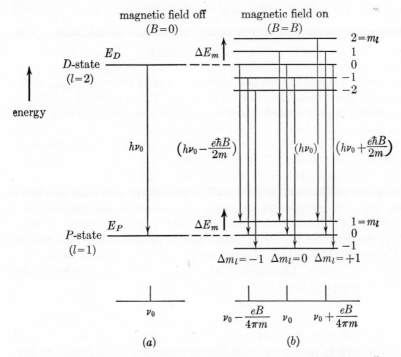

Figure 6-17. The energy levels and spectra for the normal Zeeman effect in a $D \rightarrow P$ transition. (a) Zero magnetic field; (b) non-zero field.

equally-spaced magnetic sublevels and the P state splits into three equally-spaced magnetic sublevels, the energy difference between any two adjacent magnetic sublevels being $(e\hbar/2m)B$. The transitions between the D $(l = 2)$ state and the P $(l = 1)$ state satisfy the selection rule, $\Delta l = \pm 1$. The selection rule for transitions between magnetic sublevels is

$$\boxed{\Delta m_l = 0, \quad \text{or} \quad \pm 1} \qquad [6\text{-}13]$$

That is, only those transitions are allowed for which the magnetic quantum number m_l is unchanged, or for which m_l changes by one unit. The permitted transitions and the spectrum of the emitted lines are shown in Figure 6-17b. It can be seen that the energy differences $h\nu$ for allowed transitions have one of three possible values.

$$h\nu = h\nu_0 - (e\hbar/2m)B$$

or

$$h\nu = h\nu_0$$

or

$$h\nu = h\nu_0 + (e\hbar/2m)B$$

$$\left. \right\} \qquad [6\text{-}14]$$

Dividing both sides of Equation 6-14 by h gives the frequencies of the emitted radiation:

$$\nu = \nu_0 - (e/4\pi m)B$$

or $\qquad\qquad\qquad \nu = \nu_0$ $\qquad\qquad\qquad\qquad\qquad$ [6-15]

or $\qquad\qquad\qquad \nu = \nu_0 + (e/4\pi m)B$

Thus, the single line in the spectrum is split by an external magnetic field into three equally-spaced components: the original line of frequency ν_0, and two equally-spaced satellite lines, whose separation, $(e/4\pi m)B$, from ν_0 is proportional to the magnetic flux density B. For a relatively strong magnetic field, $B = 1.0$ weber/m^2 (10,000 gauss), the energy difference between adjacent Zeeman levels is only 9.3×10^{-23} joule, or 5.8×10^{-4} ev, using Equation 6-14. Since the typical energy difference between levels giving rise to emission in the visible region of the spectrum is a few electron volts, the energy, frequency, or wavelength is changed by less than 1 part in 10^3 when a strong magnetic field is applied. Thus, observation of the Zeeman effect requires spectrometers of moderately high resolution.

Figure 6-17b shows the energy levels and allowed transitions between a D and a P state. The value of ΔE_m and the selection rules governing m_l are both independent of l. *Therefore, all transitions for which $\Delta l = \pm 1$ will give rise to an identical Zeeman effect, with three equally-spaced Zeeman component lines.* This is called the *normal Zeeman effect*, and the observed splittings of some lines for *some* elements, such as calcium, or mercury, are in complete agreement with the spectrum shown in Figure 6-17b. On the other hand, most elements do *not* show a normal Zeeman effect in that the magnitude of the splittings and the number of Zeeman components is *not* in accord with the theory presented here. Such Zeeman spectra are said to be *anomalous*, inasmuch as the emitted radiation cannot be accounted for simply in terms of the space quantization of the *orbital* angular momentum vector and the associated magnetic effects.

The space quantization rule, which limits the value of the component of the orbital angular momentum along any direction in space to integral multiples of $\hbar$, holds whether or not a magnetic field is applied. When the magnetic field is turned on, its direction specifies the direction for space quantization, and the energies of the several states differ according to the value of m_l. When the magnetic field is turned off, the space quantization persists; now, however, the energies of the states corresponding to the several possible values of m_l are all identical. Therefore, in the absence of a magnetic field any state with orbital angular momentum quantum number l has $(2l + 1)$ substates, which are identical in energy $(\Delta E_m = 0)$ *and* in the magnitude of the orbital angular momentum $(\sqrt{l(l + 1)}\hbar)$, but differ in the component of the angular momentum vector $(m_l\hbar)$ along a direction in space. Thus, in the absence of a magnetic field, there is a $(2l + 1)$-fold degeneracy in the energy of the states.

6-6   Electron spin   We have seen that three of the classical constants of the motion for a particle subject to an inverse-square force of attraction—the energy, the magnitude of the orbital angular momentum, and the component of the orbital angular momentum along a fixed direction in space—are quantized in the quantum theory. Classically speaking, the energy of the particle in an elliptical orbit is determined by the *size* of the orbit; that is, by the major axis of the ellipse. The magnitude of the orbital angular momentum is determined, for a given major axis, by the *shape* of the elliptical orbit; that is, by the eccentricity of the elliptical path. The component of the orbital angular momentum along a direction in space is determined by the *orientation* of this elliptical orbit. To these constants of the motion there correspond, quantum mechanically, the quantum numbers n, l, and m_l. In this section we shall introduce the fourth and final quantum number, which is associated with the concept of electron spin.

We have remarked that the strongest emission from sodium comes from the $3P \rightarrow 3S$ transition. When this radiation is examined with a spectrometer of moderately high resolution, it is seen that there are two closely-spaced yellow lines (5890 Å and 5896 Å), called the sodium D lines. In fact, each of the spectral lines of sodium exhibits a *fine structure* in that, for each transition in Figure 6-10, there are actually two or three distinct lines, separated from one another by no more than a very few Ångstroms in wavelength. Indeed, the occurrence of fine structure in emission and absorption spectra is a common feature of all atomic line spectra. Furthermore, the Zeeman splitting of the components from elements exhibiting this fine structure is *anomalous* in that it cannot be accounted for on the basis of the *normal* Zeeman effect (discussed in Section 6-5). Apparently, a distinctive, additional feature of atomic structure is manifest in the occurrence of fine structure, which cannot be accounted for in terms of the quantum numbers n, l, and m_l.

The fine structure bears a resemblance to the Zeeman effect in that both result in a splitting of otherwise single lines into multiple lines. Unlike the Zeeman effect, however, the fine-structure does *not* require the presence of an *external* magnetic field for its observation, and it is suggestive to attribute the fine-structure splitting to an *internal* Zeeman effect within the atom. Such an effect would require the presence of an internal magnetic field and a new source of magnetic moment and angular momentum within the atom. The orbital angular momentum of the atom has already been taken into account. What other contribution to the angular momentum can be imagined?

It was first suggested in 1925 by S. A. Goudsmit and G. E. Uhlenbeck that an intrinsic angular momentum might be associated with an electron which was quite apart from its orbital motion. Such an intrinsic angular momentum is named *electron spin*, for it can be visualized as analogous to the intrinsic angular momentum that a planet has by virtue of rotation, or spin, about its own axis of rotation. Now it is, of course, not proper, on the basis of wave mechanics, to regard an electron as a simple sphere of electric charge; but for the sake of identifying the electron-spin angular momentum with some sort of model that can be visualized, it is useful to imagine the electron as having a definite spherical extension in space, and continuously spinning around an axis of rotation. Then, the electron spin is the angular momentum arising from the rotation of the charge cloud about an axis of rotation fixed with the electron. Furthermore, because negative electric charge is rotating, a magnetic field will be produced by the spinning electron, and a magnetic moment, μ_s, opposite in direction to that of the spin angular momentum P_s, can be attributed to the electron spin, as shown in Figure 6-18a.

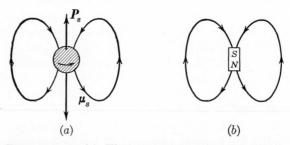

$$(a) \qquad\qquad (b)$$

Figure 6-18. (a) Electron-spin angular momentum (P_s) and magnetic moment (μ_s), with the associated magnetic field. (b) Equivalent permanent magnet.

We can visualize the spin magnetic moment as a permanent magnet, as shown in Figure 6-18b. If the electron, with its spin magnetic moment, finds itself in a magnetic field produced by internal effects within the atom, one would expect to find the electron spin to be space-quantized by this internal field. The spin axis would be restricted to certain quantized orientations, and the energy of the atom would differ according to the particular orientation.

A simple model showing the source of the internal magnetic field which can act on the spin magnetic moment of the electron is shown in Figure 6-19a. The electron is visualized as a small, spinning sphere carrying a spin angular momentum P_s and a spin magnetic moment μ_s. At the same time, the electron is imagined to move in an elliptical orbit about a fixed positively-charged nucleus, and thus has an orbital angular momentum P_θ and an orbital magnetic moment μ_l. In Figure 6-19b the orbital- and spin-

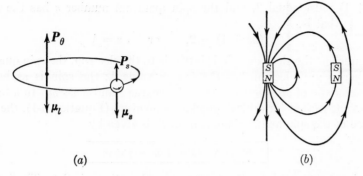

Figure 6-19. Schematic representation of spin-orbit interaction. (a) Orbital and electron-spin angular momenta and magnetic moments. (b) Magnetic interaction between the orbital and spin magnetic moments.

magnetic moments are represented by small magnets. The magnetic field of the orbital magnetic moment μ_l is shown. The spinning electron is immersed in the magnetic field produced by μ_l, and the energy of the system will differ slightly, depending on the orientation of the spin magnetic moment with respect to this magnetic field. The magnetic interaction between the spin and the orbital motion is aptly called *spin-orbit interaction*. The strength of the magnetic field at the site of the electron spin obviously depends on the orbital magnetic moment μ_l and thus on the orbital quantum number l (see Equation 6-7). For all orbital states except the S-state ($l = 0$), there will be a magnetic field acting on, and thereby space-quantizing, the electron-spin magnetic moment.

We now turn to spectroscopic evidence to find the allowed values of the spin angular momentum P_s and the spin magnetic moment μ_s. A study of the spectral lines from a single-valence-electron atom, such as sodium in the *absence* of an *external* magnetic field, indicates that each of the orbital energy levels (except the S state) is split into two components (a doublet), the S state remaining unsplit (a singlet). It is for this reason that the $3P \rightarrow 3S$ transition in sodium consists of the two closely-spaced D lines: the $3S$ state is a singlet; the $3P$ state is a doublet. How can the doubling of all states (except the S states) be interpreted in terms of an internal magnetic field space-quantizing the electron-spin angular momentum? In the normal Zeeman effect any state having an orbital quantum number l is split, under the influence of an external magnetic field, into $(2l + 1)$ sublevels. Similarly, we assume that a state having a *spin angular momentum quantum number*, s, is split into $(2s + 1)$ components under the influence of an internal magnetic field. Because the multiplicity of all those fine-structure states with a non-zero orbital momentum is always 2,

$(2s + 1)$ must equal 2, and the spin quantum number s has the *single value* $\frac{1}{2}$; that is,

$$(2s + 1) = 2, \quad \text{or} \quad s = \frac{1}{2}$$

Inasmuch as the electron spin is intrinsic to the electron, the spin quantum number of any electron has the *unique* value $\frac{1}{2}$; the spin is as basic a characteristic of an electron as are the charge e and mass m. In a fashion analogous to that for orbital angular momentum (Equation 6-1), the magnitude of the spin angular momentum $\boldsymbol{P}_s$ is given by

$$\boxed{P_s = \sqrt{s(s + 1)}\hbar = \frac{1}{2}\sqrt{3}\hbar} \qquad [6\text{-}16]$$

It is easy to see why the S state of sodium is a singlet. The internal magnetic field due to orbital motion is zero $(l = 0)$; hence, the two spin states are unsplit, and therefore *degenerate*. This degeneracy can, however, be removed by an external magnetic field.

In the presence of a magnetic field, the electron spin is space-quantized such that the component $P_{s,z}$ of the spin angular momentum along the direction of the magnetic field is

$$\boxed{P_{s,z} = m_s\hbar} \qquad [6\text{-}17]$$

where the *spin magnetic quantum number*, m_s, has two possible values, $m_s = +\frac{1}{2}$ or $m_s = -\frac{1}{2}$. The space-quantization of the electron-spin angular momentum by a magnetic field, as shown in Figure 6-20, restricts the orientation of the electron-spin vector to those two possible states in which

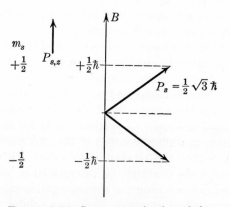

Figure 6-20. Space quantization of electron-spin angular momentum.

the component of the angular momentum along Z is $+\frac{1}{2}\hbar$ or $-\frac{1}{2}\hbar$. For the $m_s = +\frac{1}{2}$ state, the angular momentum vector points more nearly in the direction of the magnetic field than against it, and the magnetic moment is more nearly anti-aligned with the field than aligned with it. Thus, the magnetic potential energy arising from the orientation of the electron-spin magnetic moment is higher for the $m_s = +\frac{1}{2}$ state than for the $m_s = -\frac{1}{2}$ state. Roughly speaking, for the higher-energy $m_s = +\frac{1}{2}$ state, the electron-spin magnetic moment $\boldsymbol{\mu}_s$ is anti-aligned with the field; for the lower-energy $m_s = -\frac{1}{2}$ state, the electron-spin magnetic moment $\boldsymbol{\mu}_s$ is aligned with the magnetic field.

As in the case of electron orbital motion, the electron-spin magnetic moment $\boldsymbol{\mu}_s$ lies along the same line, but oppositely directed, to that of the spin angular momentum vector. A detailed study of the Zeeman effect for atoms having fine structure shows that the gyromagnetic ratio associated with the electron spin is given by

$$\frac{\mu_s}{P_s} = 2(e/2m) \qquad [6\text{-}18]$$

where e and m are the electron charge and mass, respectively. The gyro-magnetic ratio of the electron spin is closely two times that of electron orbital motion; that is, electron spin is twice as effective in magnetic effects as orbital motion. The magnetic potential energy ΔE_s of an electron-spin magnetic moment in a magnetic field having a flux density B is given by

$$\boxed{\Delta E_s = m_s[2(e\hbar/2m)]B} \qquad [6\text{-}19]$$

which is closely analogous to Equation 6-12.

It is interesting to calculate the strength of the magnetic flux density produced by the orbital motion which results in the split-ting of the P state in sodium. Figure 6-21 shows the $3S$ and $3P$ states of sodium. The two transitions giving rise to the sodium D lines differ by $\Delta\lambda = 6$ Å; therefore, the energy difference ΔE between the two P states is given by

$$\Delta E = \frac{hc\,\Delta\lambda}{\lambda^2} = 2\mu_{s,z}B$$

or $B = 20$ weber/m² $= 200{,}000$ gauss

—a very strong field.

The fine-structure splitting of energy lev-els, arising from spin-orbit interaction, is in complete agreement with the experimental observations on fine structure in spectral lines. For quantitative details the more sophisticated methods of quantum mechanics are required. The internal magnetic fields are extraordinarily

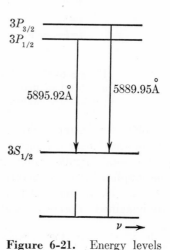
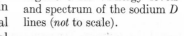

Figure 6-21. Energy levels and spectrum of the sodium D lines (*not* to scale).

strong; and the spin-orbit interaction, or fine-structure splitting, of typically 10^{-3} ev, or a few Ångstroms in wavelength for visible lines, is in accord with the theoretical predictions.

In an atom there are, then, two contributions to the *total angular momentum* of the atom: the orbital angular momentum and the electron-spin angular momentum.† The quantum theory correctly predicts that the total angular momentum P_j, of a single-valence-electron atom is characterized by the quantum number j and has a magnitude given by

$$P_j = \sqrt{j(j+1)}\hbar \qquad [6\text{-}20]$$

This total quantum number j can assume the two values $j = l + s$, or $j = l - s$, where l and s are the orbital- and spin-quantum numbers. Figure 6-22 shows the vector relations between the orbital, spin, and total

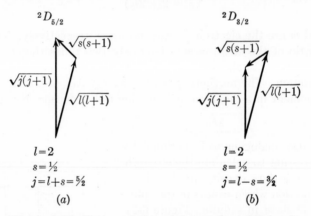

Figure 6-22. Vector relations between the orbital, spin, and total angular momentum vectors for (a) the $^2D_{5/2}$ and (b) the $^2D_{3/2}$ states.

angular momentum vectors. It can be seen that the orbital and spin vectors are never completely aligned or anti-aligned along the direction of the total angular momentum vector. A useful method of illustrating the coupling between the orbital and spin angular momenta is the so-called *vector model*. In this model, the spin and orbital angular momentum vectors are imagined to precess about the P_j vector at a constant rate.

† The nucleus of an atom may also have intrinsic angular momentum, or *nuclear spin*, which is added to the angular momentum arising from that of the electrons. *Hyperfine structure (hfs)*, consisting of very closely-spaced spectral lines (typically less than 10^{-3} Å) has its origin in the interaction of the nuclear spin and nuclear magnetic moment with those of the electrons. The magnetic moments associated with nuclei are always smaller than the Bohr magneton by a factor, approximately 10^3, and hence the hyperfine energy splitting is correspondingly less than that of the fine-structure splitting. On the planetary model, the nuclear spin is analogous to the spin angular momentum of the Sun.

For a single-valence-electron atom, all electrons within a closed shell are so arranged, as we shall see in Section 6-8, that the total angular momentum and magnetic moment of the closed shell are zero. As an illustration of the combination of the orbital and spin quantum numbers, consider the $3P$ states of sodium. We have $l = 1$ and $s = \frac{1}{2}$; therefore, the total angular momentum quantum number j has the two possible values, $j = 1 + \frac{1}{2} = \frac{3}{2}$ and $j = 1 - \frac{1}{2} = \frac{1}{2}$. These two states are represented in the conventional spectroscopic notation as $3^2P_{3/2}$ and $3^2P_{1/2}$ (read as "three doublet P three-halves" and "three doublet P one-half"). The pre-superscript gives the value of $(2s + 1)$; the post-subscript gives the value of j. In a similar

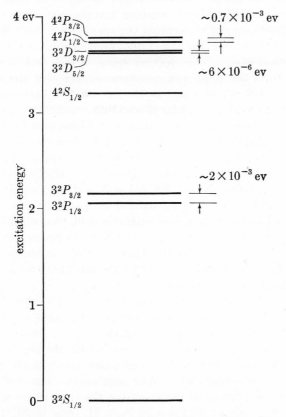

Figure 6-23. Fine-structure splitting arising from spin-orbit interaction for several states of sodium. (The splittings are grossly exaggerated.)

fashion, we have $3^2D_{5/2}$ and $3^2D_{3/2}$ states for sodium. Figure 6-23 shows the splitting from spin-orbit interaction for several of the states of sodium. In an external field (less than 10^5 gauss) the total angular momentum

P_j, having a magnitude $\sqrt{j(j + 1)}\hbar$, now is space-quantized along the field direction, which is again chosen to be the Z axis. The permitted $(2j + 1)$ components of P_j are

$$P_{j,z} = m_j\hbar$$

where $m_j = j, j - 1, \ldots, -j$. This is the basis of the *anomalous Zeeman effect*, which arises whenever spin-orbit interaction occurs. The computation of the anomalous Zeeman splitting requires a knowledge of the magnetic moment μ_j associated with the *total* angular momentum P_j; the computation of μ_j is somewhat involved, and we shall not pursue it further.

The analysis of the structure of atoms having more than one valence electron becomes increasingly complex because one must combine the spin- and orbital-angular momentum vectors of two or more electrons. We wish to note here merely how the normal Zeeman effect can arise in a two-valence-electron atom. The electron spins, $s_1 = \frac{1}{2}$ and $s_2 = \frac{1}{2}$, usually combine to form the total spin quantum number S of the atom, where $S = s_1 + s_2 = 1$ (parallel spins) or $S = s_1 - s_2 = 0$ (anti-parallel spins). Similarly, the two orbital angular momentum quantum numbers, l_1 and l_2, form a total orbital angular momentum quantum number L. The resultant spin and orbital quantum numbers, S and L, then combine to form the total angular momentum quantum number J of the atom. Consider a state for which $S = 0$; such a state is known as a "singlet" state inasmuch as $2S + 1 = 1$. (When $S = 1$, $2S + 1 = 3$, and the state is described as a "triplet.") For $S = 0$, $J = L$, and the total angular momentum is due only to orbital motion; thus, in a magnetic field, the transitions between *singlet* states exhibit the *normal* Zeeman effect. In general, the occurrence of the *normal* Zeeman effect requires that the *total spin* angular momentum of the atom be *zero*, which is to say, that the atom have an *even* number of electrons grouped in pairs of anti-aligned spins.

In a non-relativistic wave-mechanical analysis of atomic structure, the three quantum numbers, n, l, and m_l emerge in a natural way by the fitting of three-dimensional waves representing the electrons into the region surrounding the nucleus. The electron spin, which has no classical analog (except for the fictitious model of a spinning sphere of charge), is *not* a consequence of non-relativistic wave mechanics. The first relativistic wave-mechanical treatment, incorporating the three space and one time coordinates, was successfully made by P. A. M. Dirac in 1928; in this relativistic quantum theory the electron spin, having an angular momentum of $\sqrt{s(s + 1)}\hbar$ and a gyromagnetic ratio $2(e/2m)$, emerges in a natural way, together with the three other quantum numbers. Another consequence of the Dirac wave mechanics was the first prediction of the existence of the electron's anti-particle, the positron. The positron has an electron spin s of $\frac{1}{2}$ and a gyromagnetic ratio of $2(e/2m)$, just as the electron. The spin angular momentum and the magnetic dipole moment vectors of the positron point in the same direction, however, by virtue of the positive electric charge.

6-7 The Stern-Gerlach experiment The space-quantization phenomenon, which limits the possible orientation of the angular momentum and magnetic moment vectors in a magnetic field, is demonstrated *directly* in the experiment of O. Stern and W. Gerlach, performed first in 1921. Using atoms of silver, for which the only contribution to the angular momentum of the atom is that arising from the electron-spin of a single electron, Stern and Gerlach showed that the silver atoms are space-quantized in a magnetic field.

To see the basis of the direct experimental confirmation of space-quantization, let us consider the behavior of a magnetic dipole in an inhomogeneous magnetic field. A magnetic dipole moment in a uniform magnetic field—one for which the magnetic lines of force are uniformly spaced and parallel—is subject to a torque tending to orient the magnet along the lines of force; in a uniform magnetic field, the dipole is *not*, however, subject to a *net* force tending to displace the magnet as a whole. Consider now the behavior of a magnetic dipole in an *inhomogeneous* magnetic field, which shows a divergence of the magnetic lines of force. We see from Figure 6-24 that when a magnet is located in an inhomogeneous magnetic

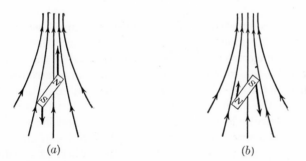

(a)   (b)

Figure 6-24. Forces on a magnetic dipole in an inhomogeneous magnetic field for two orientations.

field, the force on the N-pole of the magnet is *not* equal to the force on the S-pole; therefore, there is a *net* force acting on the magnet which tends to pull the magnet into the region of strong magnetic field when the magnet is more nearly aligned with the field. A net force tends to push the magnet out of the region of strong field when the magnet is more nearly anti-aligned with the magnetic lines of force.

Let us compute the force on an atom having an electron-spin magnetic moment when it is in an inhomogeneous magnetic field. The magnetic potential energy of the electron spin is given by Equation 6-19

$$\Delta E_s = m_s[2(e\hbar/2m)]B$$

Inasmuch as the force is just the negative space derivative of this potential energy, the magnetic force acting on the dipole is given by

$$F = -(d/dz)(\Delta E_s) = -m_s(e\hbar/m)(dB/dz) \qquad [6\text{-}21]$$

where the Z-direction is the direction of symmetry of the inhomogeneous magnetic field, as well as the direction for space-quantization of the electron spin.

Consider now the circumstances of the original Stern-Gerlach experiment, as shown in Figure 6-25. Silver atoms leave an oven at relatively

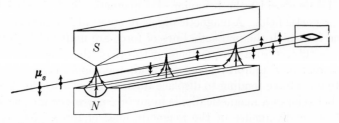

Figure 6-25. Schematic representation of the Stern-Gerlach experiment.

high speeds, are collimated by slits, pass through an inhomogeneous magnetic field, and fall upon a photographic plate, where their final location is recorded. The electron spins are space-quantized in the magnetic field into orientations for which the component of the electron-spin angular momentum is $+\frac{1}{2}\hbar$ or $-\frac{1}{2}\hbar$, for $m_s = +\frac{1}{2}$ or $-\frac{1}{2}$, respectively. For the $m_s = +\frac{1}{2}$ state, the silver atoms are deflected downward; for the $m_s = -\frac{1}{2}$ state, the atoms are deflected upward. On the photographic plate there will *not* be a continuous spread in the position of the arriving atoms, as would be expected if space-quantization did not occur; rather, two distinct lines, corresponding to the silver atoms in the two allowed spin orientations are observed. In general, if the atom has a total angular momentum quantum number J, the number of lines appearing on the plate is $(2J + 1)$; thus, the method of atomic beams can be used to evaluate angular momentum quantum numbers.

6-8   The Pauli exclusion principle and the periodic table   To specify the complete state of an electron in an atom is, in the quantum theory, to specify the values for each of the four quantum numbers: n, l, m_l, and m_s. (The electron spin, $s = \frac{1}{2}$, need not be indicated, for there is no other possible value.) By the procedures of the quantum mechanics, it is possible to compute such properties of an atom as its energy in the absence or presence of an external magnetic field, the angular momentum and magnetic moment of the atom, as well as other measurable characteristics. Indeed, it is possible, *in principle*, to predict theoretically from the quantum theory *all* properties of the chemical elements; in practice, such a program cannot be easily carried out, inasmuch as formidable mathematical difficulties arise when dealing with systems having many component particles. In fact, only the problem of the simplest atom, hydrogen, has been solved completely using the relativistic-quantum theory. For this atom, there is essentially perfect agreement with experiment. Even though solutions for the other atomic elements are not known exactly, the quantum theory does provide a wealth of information on the chemical and physical properties of atoms. One of its greatest achievements is the fundamental basis for understanding the ordering of the chemical elements as they appear in the periodic table. It should be recalled that the periodic table was first constructed merely by listing elements in the order of their atomic weights, and it was found that remarkable periodicities in the properties of the elements were thereby revealed.

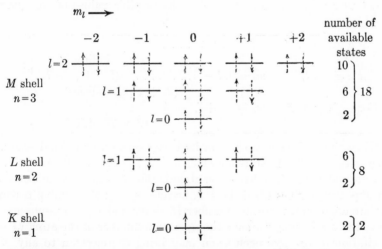

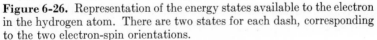

Figure 6-26. Representation of the energy states available to the electron in the hydrogen atom. There are two states for each dash, corresponding to the two electron-spin orientations.

In this section we first discuss the method by which this ordering can be understood in terms of the quantum theory. We shall see that the key to understanding the periodic table is a principle proposed by W. Pauli in 1924, the *Pauli exclusion principle*. This principle, together with the quantum theory, can be used to predict many of the known chemical and physical properties of atoms.

Consider again the energy levels available to the single electron in the hydrogen atom. These energy levels are shown schematically (but *not* to scale) in Figure 6-26, where each horizontal dash corresponds to a particular possible set of values for the quantum numbers n, l, and m_l. For each dash there are two possible values of the electron-spin quantum number, $m_s = \pm\frac{1}{2}$. Hereafter, the occupancy of an available state by an electron will be indicated on a diagram by an arrow whose direction (↑ or ↓) will also show the electron spin orientation ($m_s = +\frac{1}{2}$, ↑ ; $m_s = -\frac{1}{2}$, ↓). For brevity, only the energy levels for the principal quantum numbers, $n = 1$, 2, and 3, are shown. For a given value of n, the S states are lowest, then the P states, etc. For a given value of the orbital angular momentum quantum number l, the possible values of the magnetic orbital angular momentum quantum number m_l are horizontally displaced from one another. Every one of the states (two for each dash) in Figure 6-26 is available to the electron in the hydrogen atom. Some of the states are degenerate, having the same total energy; such states are, nevertheless, distinguishable when a strong magnetic field or other external influence is applied to the atom.

Let us review the rules governing the possible values of the quantum numbers.

$$
\begin{array}{lll}
\text{For a given } n, & l = 0, 1, 2, \ldots n - 1 & \\
 & \qquad (n \text{ possibilities}) & \\
\text{For a given } l, & m_l = l, l - 1, \ldots 0, \ldots -(l - 1), \quad -l & \quad [6\text{-}22] \\
 & \qquad (2l + 1 \text{ possibilities}) & \\
\text{For a given } m_l, & m_s = \quad +\frac{1}{2}, \quad -\frac{1}{2} & \\
 & \qquad (2 \text{ possibilities}) &
\end{array}
$$

When the hydrogen atom is in its lowest-energy, or ground, state, the single electron is in the state for which $n = 1$, $l = 0$, $m_l = 0$, and $m_s = -\frac{1}{2}$. The ground state of the hydrogen atom is thus $1^2S_{1/2}$, where, as before, the pre-superscript gives $(2s + 1) = 2$, because $s = \frac{1}{2}$; the symbol S denotes the atom's orbital quantum number ($L = 0$); and the post-subscript denotes the total angular momentum quantum number of the atom ($J = \frac{1}{2}$). Excitation of the hydrogen atom may bring the electron to any of the higher-lying available states, from which the atom will then decay to the ground state by the emission of one or more photons. Figure 6-27a is an energy-level diagram of a hydrogen atom in its ground state, with the electron in the lowest state. A hydrogen atom in an excited, $3D$ state is shown in Figure 6-27b.

Consider next the element helium, $_2$He, which has two electrons to be arranged in the energy levels shown in Figure 6-26. The separation between the energy levels is, however, *not* precisely the same as for hydrogen because the nuclear charge $(+2e)$ is different, and also because there are three interacting particles rather than two. Nevertheless, the order of the states is the same for helium as for hydrogen. When the helium atom is in its ground, or normal, state we expect both of the electrons to be in the $1S$ state with $n = 1, l = 0$, and $m_l = 0$. For each electron two possible values of m_s exist, $\frac{1}{2}$ and $-\frac{1}{2}$.

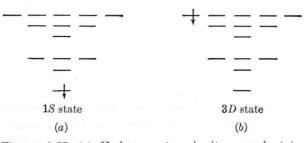

1S state

(a)

3D state

(b)

Figure 6-27. (a) Hydrogen atom in its ground state. (b) Hydrogen atom in a $3D$ state.

The key to the possible values of m_s for each of the two electrons in helium is found in the study of its spectrum. When this spectrum is examined and from it the various states of the helium atom are inferred, it is found that there is *no* 1^3S_1 state, although there does exist a 1^1S_0 state. A $^3S_1$ state would represent the situation in which the two electron spins are coupled together in alignment to form a total spin $S = 1$ (for which $2S + 1 = 3$, a triplet); the post-subscript, giving the total angular momentum quantum number J is 1, inasmuch as the orbital angular momentum is zero. Because only the 1^1S_0 state is found for helium, a state for which the total spin $S = 0$, it follows that the two electron spins must be anti-aligned; that is, have different m_s values. Therefore, the two electrons, which have identical values for n, l, and m_l, must have magnetic spin quantum numbers, $m_s = \frac{1}{2}$ and $m_s = -\frac{1}{2}$, respectively. We can interpret the non-occurrence of the 1^3S_1 state in helium in the following way— two electrons in a helium atom cannot have the same set of four quantum numbers; that is, the two electrons cannot exist in the same state.

Spectroscopic evidence from all elements shows that atoms simply never occur in nature with two electrons occupying the same state. The *Pauli exclusion principle formalizes this experimental fact: no two electrons in an atom can have the same set of quantum numbers* $(n, l, m_l,$ *and* $m_s)$; or, *no two electrons in an atom can exist in the same state.* No exceptions to the exclusion principle, which applies also to other systems than atoms and which holds for some particles besides electrons, have ever been found. The Pauli principle is analogous, but not equivalent, to the classical assertion that no two particles can be in the same place at the same time (regarding the particles as impenetrable).

Thus, the two electrons in helium in the normal state occupy the two lowest available states of Figure 6-26. No further electrons can be added to the $n = 1$, or K, shell; in helium the K shell is filled, or closed. With the electron spins oppositely aligned in the 1^1S_0 state, the helium atom has no magnetic moment or angular momentum. Furthermore, the two electrons are tightly bound to the nucleus, and a considerable energy is required to excite one of them to a higher-lying energy state. It is primarily for these reasons that helium is chemically inactive.

When the values of the quantum numbers for each and every electron in an atom are given, the *electron configuration* of the atom is known. There is a simple convention for specifying an electron configuration, which we illustrate for helium. When this atom is in its ground state, the two electrons each have $n = 1$ and $l = 0$, and their configuration is represented by $1s^2$, where the leading number specifies the n value; the lower case letter, s, p, d, f, . . ., designates the orbital quantum number l of *individual* electrons, and the post-superscript gives the number of electrons having the particular values of n and l. An energy-level diagram for neutral helium is shown in Figure 6-28. The energy levels are segregated according

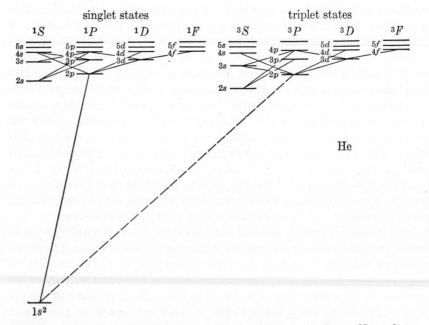

Figure 6-28. Energy-level diagram and transitions for helium. Note that the 1 $^3S_1$ state does *not* exist.

to whether they are singlets or triplets; that is, whether the two electron spins are anti-aligned or aligned, respectively. Note that the 1^3S state does *not* exist. When the atom is in an excited state, one of the electrons remains as a $1s$ electron in the K shell, and the second electron occupies any of the higher excited levels. For example, the electron configuration for the first excited state in helium is $1s^12s^1$. Typically, transitions occur only between singlet states or only between triplet states.

The element with the next higher atomic number ($Z = 3$) is lithium, $_3$Li. Of the three electrons in this atom, two occupy the two available $n = 1$ states; therefore, the exclusion principle requires that, when lithium is in its ground state, the third electron must go to the lowest of the remaining available levels. The next lowest available level, following the K shell, is $n = 2$ and $l = 0$; and the configuration for the electrons in $_3$Li is $1s^22s^1$, indicating two electrons in the closed K shell and one electron in the incomplete $l = 0$ subshell of the L shell (see Figure 6-29).

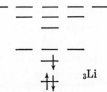

Figure 6-29. Electron configuration of lithium in the ground state.

Proceeding in this fashion, adding one electron as the nuclear charge or atomic number is increased by one unit, but always with the restriction that no two electrons within the atom can have the same set of quantum numbers, we can find the electron configurations for further atoms. We see from Figure 6-26 that 2 electrons can be accommodated in the s subshell of the L shell, 6 electrons will fill the p subshell, after which the L shell is completely occupied, holding its full quota of 8 electrons. Table 6-2 gives the electron configuration for the elements $_4$Be through $_{11}$Na.

We shall shortly discuss the chemical properties of a number of the elements in Table 6-2 in terms of electron configurations and the occurrence

Table 6-2

ELEMENT	ELECTRON CONFIGURATION FOR THE GROUND STATE			
$_4$Be	$1s^2$	$2s^2$		
$_5$B	$1s^2$	$2s^2$	$2p^1$	
$_6$C	$1s^2$	$2s^2$	$2p^2$	
$_7$N	$1s^2$	$2s^2$	$2p^3$	
$_8$O	$1s^2$	$2s^2$	$2p^4$	
$_9$F	$1s^2$	$2s^2$	$2p^5$	
$_{10}$Ne	$1s^2$	$2s^2$	$2p^6$	
$_{11}$Na	$1s^2$	$2s^2$	$2p^6$	$3s^1$

of closed shells and subshells. But before proceeding further with the periodic ordering of the elements, let us note some of the properties of $_{11}$Na which can be directly related to its electron configuration $1s^22s^22p^63s^1$. There is a single electron outside the closed L shell, the lowest state available to this valence electron being the $3s$ state. The total angular momentum and magnetic moment arising from the inner closed subshells is zero, inasmuch as the orbital angular momenta as well as the spin angular momenta of the electrons in the closed subshell are paired off. These closed shells, $1s^22s^22p^6$, are chemically inert and correspond to the electron configuration of the inert gas, $_{10}$Ne. The reason that sodium behaves approximately as a hydrogen atom is clear: a single valence electron moves about inner, inert electron shells. The optical spectrum of sodium originates from the change in the state of the valence electron, the 10 electrons in the inner closed shells remaining in their same states.

There is a continuous filling of the sublevels in the expected order $1s$, $2s$, $2p$, $3s$, and $3p$, the last element being argon, $_{18}$Ar, whose complete electron configuration in the ground state is $1s^22s^22p^63s^23p^6$, or $3p^6$ for short. After the completion of the $3p$ subshell with $_{18}$Ar, we might expect that the succeeding elements would fill, in sequence, the 10 available states of the $3d$ sublevel. Spectroscopic and chemical evidence indicates, however, that the $4s$ subshell is filled first, because the 2 electrons in the $4s$ subshell are more tightly bound to the atom than $3d$ electrons. We can understand this apparent anomaly in the following way. Classically, an electron in the $4s$ subshell moves in a highly eccentric orbit, penetrates deeply into the inner core, and is thereby strongly bound. On the other hand, even though a $3d$ electron moves in a "circular" orbit of smaller size ($n = 3$, rather than 4), it remains more nearly outside the closed shells. The increase in the binding of a $4s$-electron, due to penetration, is greater than the increased binding energy of a $3d$ electron, due to its smaller orbit. For example, the electron configuration of the element potassium, $_{19}$K, is $1s^22s^22p^63s^23p^64s^1$; the outer electron is a $4s$-electron, not a $3d$-electron. In *hydrogen*, the $3d$ state lies *below* the $4s$ state; but for potassium, with 19 electrons, the reverse is true. From experimental evidence, the general order in which the electron subshells are filled is as follows:

$$1s, 2s, 2p, 3s, 3p, 4s, 3d, 4p, 5s, 4d, 5p, 6s, 4f, 5d, 6p, 7s, 6d$$

The periodic table of chemical elements is shown in Figure 6-30. The electron configuration of the outer electrons in the atom is given with the atomic number and chemical symbol for each element. The arrangement is such that any column contains elements that have a common orbital state and also have the same number of electrons in this orbital state. For example, carbon, $_6$C, which has an outer electron configuration of $2p^2$, is found above silicon, $_{14}$Si, which has an outer electron configuration, $3p^2$;

Figure 6-30. Periodic table of the chemical elements.

																	2p
1s	1s²																
2s	3 Li 2s¹	4 Be 2s²							5 B 2p¹	6 C 2p²	7 N 2p³	8 O 2p⁴	9 F 2p⁵	10 Ne 2p⁶			3p
3s	11 Na 3s¹	12 Mg 3s²							13 Al 3p¹	14 Si 3p²	15 P 3p³	16 S 3p⁴	17 Cl 3p⁵	18 Ar 3p⁶			4p
4s	19 K 4s¹	20 Ca 4s²	21–30 see (a)						31 Ga 4p¹	32 Ge 4p²	33 As 4p³	34 Se 4p⁴	35 Br 4p⁵	36 Kr 4p⁶			5p
5s	37 Rb 5s¹	38 Sr 5s²	39–48 see (b)						49 In 5p¹	50 Sn 5p²	51 Sb 5p³	52 Te 5p⁴	53 I 5p⁵	54 Xe 5p⁶			6p
6s	55 Cs 6s¹	56 Ba 6s²	57–80 see (c)						81 Tl 6p¹	82 Pb 6p²	83 Bi 6p³	84 Po 6p⁴	85 At 6p⁵	86 Rn 6p⁶			
7s	87 Fr 7s¹	88 Ra 7s²	89–102 see (d)														

(a) Transition elements: the ten 3d states follow the 4s states.

3d	21 Sc 3d¹	22 Ti 3d²	23 V 3d³	24 Cr 4s¹3d⁵	25 Mn 3d⁵	26 Fe 3d⁶	27 Co 3d⁷	28 Ni 3d⁸	29 Cu 4s¹3d¹⁰	30 Zn 3d¹⁰

(b) 4d elements: the ten 4d states follow the 5s states.

4d	39 Y 4d¹	40 Zr 4d²	41 Nb 5s¹4d⁴	42 Mo 5s¹4d⁵	43 Tc 4d⁵	44 Ru 5s¹4d⁷	45 Rh 5s¹4d⁸	46 Pd 5s⁰4d¹⁰	47 Ag 5s¹4d¹⁰	48 Cd 4d¹⁰

(c) Rare earths and 5d elements: the ten 5d states and fourteen 4f states follow the 6s states.

5d	57 La 5d¹	72 Hf 5d²	73 Ta 5d³	74 W 5d⁴	75 Re 5d⁵	76 Os 5d⁶	77 Ir 5d⁷	78 Pt 6s¹5d⁹	79 Au 6s¹5d¹⁰	80 Hg 5d¹⁰

4f	58 Ce 5d¹4f¹	59 Pr 5d⁰4f³	60 Nd 5d⁰4f⁴	61 Pm 5d⁰4f⁵	62 Sm 5d⁰4f⁶	63 Eu 5d⁰4f⁷	64 Gd 5d¹4f⁷	65 Tb 5d⁰4f⁸	66 Dy 5d⁰4f¹⁰	67 Ho 5d⁰4f¹¹	68 Er 5d⁰4f¹²	69 Tm 5d⁰4f¹³	70 Yb 5d⁰4f¹⁴	71 Lu 5d¹4f¹⁴

(d) Actinides and transuranic elements: The 6d and 5f states follow the 7s states.

6d	89 Ac 6d¹	90 Th 6d²	91 Pa	92 U	93 Np	94 Pu	95 Am	96 Cm	97 Bk	98 Cf	99 Es	100 Fm	101 Md	102 No

5f	91 Pa	92 U	93 Np	94 Pu	95 Am	96 Cm	97 Bk	98 Cf	99 Es	100 Fm	101 Md	102 No

Figure 6-30. Periodic table of the chemical elements. The ground-state configuration for the outermost electron shell is given (except when an inner shell is incomplete, in which case both shell configurations are given). For example, the element, $_{42}$Mo, has the complete electron configuration of $1s^2\ 2s^2\ 2p^6\ 3s^2\ 3p^6\ 3d^{10}\ 4s^2\ 4p^6\ 5s^1\ 4d^5$.

both have two electrons in an incomplete p subshell. The groups of elements, corresponding to the filling of the $3d$, $4d$, $4f$, and $5d$ subshells are listed separately. It is clear why the main body of the periodic table has periodicity of eight: the total number of electrons completing an s-subshell (2) plus the p-subshell (6) is 8.

We now see that the fundamental basis for the chemical properties of elements is the electron configurations of the atoms. Atoms that have similar electron configurations show remarkably similar chemical behavior; the periodicity of chemical properties contained in the periodic table reflects the periodicity of the electron configurations. Let us consider some of these properties in more detail.

THE RARE GASES   The inert, or noble, rare gases are $_2$He, $_{10}$Ne, $_{18}$Ar, $_{36}$Kr, $_{54}$Xe, and $_{86}$Rn. We see from Figure 6-30 that all of these elements have configurations in which the outermost electrons complete a p subshell (except $_2$He). All rare gases have a ground state of $^1S_0$. The total angular momentum of the atom, from orbital motion and from electron spin, is zero; therefore, the total magnetic moment of the atom is zero. The atoms are chemically inert because there is no excess of electrons beyond a closed subshell and no deficiency of electrons in a subshell. The "last" electron within the p subshell is very strongly bound; thus, the ionization energy for elements in this group is particularly high. These atoms fail to form chemical compounds, or non-monatomic molecules; furthermore, the rare gases have very low electrical conductivities and liquefaction (boiling) points.

THE ALKALI METALS   The elements in this group (the first column of the periodic table) are $_3$Li, $_{11}$Na, $_{19}$K, $_{37}$Rb, $_{55}$Cs, and $_{87}$Fr. For every alkali metal there is a single electron outside a closed rare-gas subshell; in the ground state, this electron is in an s subshell. The chemical activity of these elements can be attributed to this single electron and the resulting valence of $+1$. The binding energy of the valence electron is relatively low, and the electron can be removed easily from the neutral atom to form a singly-charged positive ion. Clearly, the alkali metals can be regarded as hydrogen-like, with spectra which resemble that of hydrogen.

THE ALKALINE EARTHS   The elements in this group (the second column of the periodic table) are $_4$Be, $_{12}$Mg, $_{20}$Ca, $_{38}$Sr, $_{56}$Ba, and $_{88}$Ra. All of these elements have two s-electrons outside a closed p subshell when in the normal state. The two electrons have a relatively small binding energy and are responsible for the valence of $+2$. When these elements are singly ionized, they become hydrogen-like in that their spectra are similar to those of the alkali metals.

THE HALOGENS   The halogen group consists of the elements in the seventh column of the periodic table: $_9$F, $_{17}$Cl, $_{35}$Br, $_{53}$I, and $_{85}$At. The atoms of elements in this group lack one electron of completing a closed p subshell; therefore, the atoms have a valence of -1. The halogen elements are highly active chemically and form stable compounds when combined with elements of the alkali-metal group; for example, the compound NaCl. When two respective atoms of these groups are in close proximity, the halogen atom, lacking one electron of a closed subshell is, so to speak, "desirous" of acquiring an additional electron to complete its p subshell; the alkali-metal atom, having one electron outside a closed subshell, is particularly ready to relinquish its last valence electron. When halogen and alkali-metal elements unite to form compounds, each atom increases the stability of its electron configuration. The formation of the molecule, in this case, by the combination of ions, is called *ionic binding* (Section 11-1).

THE TRANSITION GROUP   This group, consisting of elements in which the $3d$-subshell is being filled, includes $_{21}$Sc, $_{22}$Ti, $_{23}$V, $_{24}$Cr, $_{25}$Mn, $_{26}$Fe, $_{27}$Co, $_{28}$Ni, $_{29}$Cu, and $_{30}$Zn. The electrons in the $3d$-subshell are responsible for some important properties of these elements. Many of these substances are either paramagnetic (weakly magnetic) or ferromagnetic (strongly magnetic) as elements or when in compounds. Their magnetism has its origin in the incomplete $3d$-subshell, whose total magnetic moment is *not* zero. On the other hand, the chemical activity of the transition elements is primarily a result of the outer $4s$ electrons.

THE RARE EARTHS   The atoms of the fourteen elements in this group, $_{58}$Ce through $_{71}$Hf, have incomplete $4f$-subshells. The inner $4f$ electrons are well shielded from the outer $6s$-valence electrons. Thus, the chemical properties of the rare-earths result primarily from the $6s$ electrons; and, for this reason, the rare-earth elements are chemically nearly indistinguishable.

We have discussed just a few chemical properties which are related to the electron configurations of atoms. Many more chemical and physical properties of materials can be accounted for using the procedures of quantum mechanics and the exclusion principle.

6-9   Summary   For a particle subject to an inverse-square force of attraction, the constants of the motion (classically having a continuum of possible values) are quantized in the quantum theory of the atom.

The possible values of the quantum numbers:

$$n = 1, \ 2, \ 3, \ 4, \ \ldots$$
$$K, \ L, \ M, \ N, \ \ldots$$

For a given n:

$$l = 0, 1, 2, 3, 4, \ldots n - 1 \qquad (n \text{ possible values})$$
$$s, p, d, f, g, \ldots$$

For a given l:

$$m_l = l, l - 1, \ldots -(l - 1), -l \qquad (2l + 1 \text{ possible values})$$

For a given m_l:

$$m_s = +\tfrac{1}{2}, -\tfrac{1}{2} \qquad (2 \text{ possible values})$$

When the valence electron of a hydrogen-like atom is in a state for which l is small, the electron penetrates the inner core, and the energy of the atom is thereby lowered with respect to the corresponding hydrogen energy-level. The selection rule, giving the allowed transitions, is: $\Delta l = \pm 1$.

The ratio of the magnetic moment to the angular momentum, called the gyromagnetic ratio, is:

$$\mu_l/P_\theta = (e/2m) \qquad \text{for orbital motion}$$
$$\mu_s/P_s = 2(e/2m) \qquad \text{for electron spin}$$

Table 6-3

CONSTANT OF THE MOTION	QUANTUM NUMBER	ALLOWED VALUES
Energy (size of elliptical orbit)	n (principal quantum number)	$E_n = -Z^2 E_I/n^2$ (for a single electron, ignoring spin-orbit interaction)
Magnitude of the orbital angular momentum (shape of orbit)	l (orbital angular momentum quantum number)	$P_\theta = \sqrt{l(l + 1)}\hbar$
Component of the orbital angular momentum along Z (orientation of orbit)	m_l (orbital magnetic quantum number)	$P_z = m_l\hbar$ Space quantization: $\cos\theta = m_l/\sqrt{l(l + 1)}$
Magnitude of the spin angular momentum	s (electron-spin quantum number)	$P_s = \sqrt{s(s + 1)}\hbar = \tfrac{1}{2}\sqrt{3}\hbar$
Component of the spin angular momentum along Z	m_s (spin magnetic quantum number) $s = \tfrac{1}{2}$	$P_{s.z} = m_s\hbar$ Space quantization: $\cos\theta = m_s/\sqrt{s(s + 1)}$

The change in the magnetic energy of a state for a magnetic moment in a field of flux density B is:

$\Delta E_m = m_l (e\hbar/2m)B = m_l \beta B$ for an orbital magnetic moment

$\Delta E_s = m_s 2(e\hbar/2m)B = m_s (2\beta) B$ for an electron-spin magnetic moment

β (the Bohr magneton) $= (e\hbar/2m) = .9273 \times 10^{-23}$ joule/(weber/m^2)

The normal Zeeman effect arises from the interaction of the *orbital* magnetic moment with a magnetic field. The magnetic field causes each line in the spectrum to split into three equally spaced lines. The selection rule for allowed transitions is $\Delta m_l = 0, \pm 1$.

The Stern-Gerlach experiment, in which atoms with a net electron spin pass through an inhomogeneous magnetic field, shows directly the phenomenon of space-quantization.

The fine structure of spectral lines has its origin in the internal magnetic interaction between P_θ and P_s, spin-orbit interaction.

The spectroscopic nomenclature for atoms having one or more electrons is:

total orbital angular momentum quantum number	$L = 0, 1, 2, 3, 4, \ldots$ $S, P, D, F, G, \ldots$

total spin angular momentum quantum number	$S = \frac{1}{2}$ (for one electron) $S = 0,$ or 1 (for two electrons) $J = j = l + s,$ or $l - s$

The Pauli exclusion principle, the basis for understanding the periodic table of chemical elements, specifies that no two electrons in the same atom can have the same set of four quantum numbers (n, l, m_l, m_s).

REFERENCES

French, A. P., *Principles of Modern Physics*. New York: John Wiley & Sons, Inc., 1958. See Section 8.2 of this text for a simple derivation of the magnitude of the spin-orbit interaction.

Lindsay, R. B., *Physical Mechanics*. Princeton, New Jersey: D. Van Nostrand Company, Inc., 1950. Chapter 3 gives a formal analysis of the motion of a particle under an inverse-square attractive force.

Richtmeyer, F. K., E. H. Kennard, and T. Lauritsen, *Introduction to Modern Physics*. New York: McGraw-Hill Book Company, Inc., 1955. A fairly detailed discussion of atomic spectra and atomic structure. The classical (normal) Zeeman effect is treated on page 80.

Slater, J. C., *Modern Physics*. New York: McGraw-Hill Book Company, Inc., 1955. This text emphasizes historical aspects in the development of spectroscopy and the quantum theory of atomic structure.

Sproull, R. L., *Modern Physics, A Textbook for Engineers*. New York: John Wiley & Sons, Inc., 1956. A descriptive discussion of atomic structures, using wave mechanics, is found in Chapter 6.

White, H. E., *Introduction to Atomic Spectra*. New York: McGraw-Hill Book Company, Inc., 1936. A standard reference on the experimental and theoretical aspects of atomic spectra. See page 71 for photographic patterns representing the wave functions for a number of states of the hydrogen atom.

PROBLEMS

6-1 * Prove that, for a group of elliptical paths having the same major axis but different eccentricities, the orbital angular momentum is proportional to the area of the ellipse.

6-2 * Show that the period (or frequency) of a particle moving in an elliptical orbit under the influence of an inverse-square central force depends on the major axis (the size), but not on the eccentricity (the shape), of the ellipse.

6-3 In the study of spectral lines from a *given* element, Rydberg found that the sharp and diffuse series had a common short-wavelength series limit. Explain this result by reference to an energy-level diagram.

6-4 Calculate the energy necessary to ionize the $2s$, or valence, electron of a $_3$Li atom, assuming the atom to be hydrogen-like (the valence electron orbiting about an inert core). Compare this with the experimental value of 5.39 ev, and explain the difference qualitatively.

6-5 For the potassium atom, which transition has the shorter wavelength: $5S \rightarrow 4P$ or $5D \rightarrow 4P$? Explain why.

6-6 Calculate the approximate wavelength of the radiation from triply-ionized titanium $_{22}$Ti^{+++} for the transition $5S \rightarrow 4P$.

6-7 What is the effective positive charge, Z_{eff}, seen by the valence electron of $_{11}$Na when the atom is in the ground state? The ionization energy to remove this electron is 5.14 ev.

6-8 List all possible downward transitions which can occur for sodium atoms in the $5P$ state (see Figure 6-10).

6-9 In general an atom in an excited state spends such a short time there (around 10^{-8} sec) that the probability of its absorbing a photon and jumping to a still higher state is very small. On this basis, explain why only the P series (*principal* series) is ordinarily observed in the *absorption* spectrum of sodium.

6-10 The magnetic pole strength m of a permanent magnet is defined so that the force on a N-pole is mB and the force on the S-pole is $-mB$. Show that the magnetic potential energy of the magnetic dipole, oriented at an angle θ to the magnetic field (see Figure 6-13), is given by $\Delta E_m = -\mu B \cos \theta$ where $\mu = ml$ and l is the distance between the poles.

6-11 * Show that, for an electron moving in an *elliptical* orbit, the magnetic moment $(\mu = I\ A)$ is related to the orbital angular momentum by: $\boldsymbol{\mu} = (-e/2m)\ \boldsymbol{P}_\theta$.

6-12 Prove that an electron moving about a proton in a circular orbit of radius r has a magnetic moment whose magnitude is $\mu = (ke^4r/4m)^{1/2}$.

6-13 (a) Show that the Bohr magneton is just the magnetic moment of the hydrogen atom in the first Bohr orbit. (b) What is the magnetic moment of a hydrogen atom for the $n = 3$ circular orbit in units of the Bohr magneton?

6-14 Verify that the Bohr magneton, $(e\hbar/2m)$, has the units (joules)/(weber/m²).

6-15 Show that the frequency difference between adjacent magnetic sublevels in the normal Zeeman splitting can be written as $(\beta B/h)$.

6-16 * Assume that an electron moves in a circular orbit with an angular velocity ω_0 under the influence of a centripetal force $m\omega_0^2 r$. A magnetic field of flux density B is applied at right angles to the plane of the orbit. (a) Show that, if the radius r remains the same,

$$\omega = \sqrt{\omega_0^2 + (eB/2m)^2} \pm (eB/2m)$$

(b) In all cases of interest, $(eB/2m) \ll \omega_0$. Using this approximation, show

$$\nu = \nu_0 \pm eB/4\pi m$$

(Comparing this result with Equation 6-15, we see that the frequencies for the *normal* Zeeman effect can be accounted for by a classical calculation.)

6-17 (a) How large must the orbital angular momentum quantum number l be so that the difference between the magnitude of the orbital angular momentum and its maximum component is no more than 1 part in 10^6? (b) Compare this angular momentum with that of a mosquito (1.0×10^{-3} gm) flying in a circle of 1.0 cm radius at a speed of 10 cm/sec.

6-18 The separation between adjacent Zeeman (normal) components is measured to be 0.0283 Å for emitted radiation at 4500 Å when in a magnetic field of 0.300 weber/m². What is the value of (e/m) deduced from this experiment? (The fact that the value of (e/m) deduced from Zeeman-effect experiments agreed with the value obtained from cathode-ray experiments indicated early in the development of atomic theory that electrons in atoms played an important role in the radiation from atoms.)

6-19   Draw an energy-level diagram and compute the energy separation (ev) between the adjacent (normal) Zeeman components for a transition between the $5F$ and $4D$ states. The magnetic field is 0.500 weber/m².

6-20   Suppose a certain spectrometer can resolve spectral lines in the visible region (say, 6000 Å) separated by 0.1 Å. What magnetic flux density B is necessary to permit the resolution of the normal Zeeman effect?

6-21   Show that the maximum component of the spin magnetic moment of the electron along the direction of the magnetic field is equal to the Bohr magneton.

6-22 *   Assume the electron to be a spherical shell with all its mass and charge uniformly distributed over the surface of the sphere of radius 2.8×10^{-15} m. (a) What must the angular velocity of the spinning electron be so that the spin angular momentum is $\frac{1}{2}\sqrt{3}\hbar$. (b) Compute the spin magnetic moment of the electron for this (fictitious) model, and compare this with the correct value, Equation 6-18.

6-23   A beam of free electrons passes perpendicularly into a uniform magnetic field B whose magnitude is 6.0×10^{-1} weber/m². Calculate the difference in energy between electrons aligned and antialigned with the magnetic field.

6-24   As mentioned in the footnote on page 232, the spin angular momentum and spin magnetic moment of the *nucleus* of the atom can give rise to a hyperfine splitting of the emitted spectral lines. Using a typical nuclear spin magnetic moment of $10^{-3}\ \beta$, where β is the Bohr magneton of the electron, (a) show that the energy separation due to nuclear spin is of the order of 10^{-5} ev; (b) show that the corresponding wavelength separation for 5000 Å light is $\Delta\lambda \simeq 10^{-2}$ Å.

6-25 *   There is spin-orbit interaction for an electron in which $l = 20$ and s is, of course, $\frac{1}{2}$. (a) Show that the spin and orbital angular momentum vectors are nearly at right angles to one another for both of the two possible j values. (b) Show that the total magnetic moment vector lies essentially along the line of the total angular momentum vector.

6-26   (a) Show that the energy difference between the two electron-spin orientations in a magnetic field of flux density B is given by $2\beta B$. (b) What is the frequency of radiation which can cause transitions (spin flips) between these two states when $B = 0.30$ weber/m²?

This effect, in which electrons absorb radiation that flips the electron spins, and for which $h\nu = 2\beta B$, is called *electron-spin-resonance*.

6-27 Under certain circumstances, transitions (magnetic-dipole) can be produced between the fine-structure components arising from spin-orbit interaction by the absorption of radiation. Because these energy differences are very small, the corresponding frequencies are low; and the *direct* study of fine structure and hyperfine structure by transitions between the multiplet components lies in the area of radio-frequency or microwave spectroscopy. What is the frequency of radiation that can induce transitions between the $3^2P_{1/2}$ and $3^2P_{3/2}$ states of sodium? The wavelengths of the sodium D lines are 5889.95 Å and 5895.92 Å.

6-28 * An "anti-atom," is one in which *all* of the elementary particles are replaced by their corresponding anti-particles. Show that the lower energy state of a spin-orbit doublet is still $j = l - s$. (An "anti-universe" cannot be distinguished from the universe simply by examining the electromagnetic radiation from it and inferring the energy levels of atoms.)

6-29 Find the angle between the total and the orbital angular momentum vectors for the $^2D_{3/2}$ state.

6-30 Show with a vector diagram that the vector sum of the orbital magnetic moment μ_l and spin magnetic moment μ_s does *not* lie along the same line as the sum of the orbital angular momentum P_θ and spin angular momentum P_s.

6-31 * Using the law of cosines, show that

$$\cos (l,j) = \frac{j(j + 1) + l(l + 1) - s(s + 1)}{2\sqrt{j(j + 1)}\sqrt{l(l + 1)}}$$

where $\cos (l,j)$ is the cosine of the angle between the orbital and total angular momenta.

6-32 * (a) Knowing that $\mu_l = (e\hbar/2m)\sqrt{l(l + 1)}$ and $\mu_s = 2(e\hbar/2m)\sqrt{s(s + 1)}$, show that the component of the total magnetic moment, μ_j, along the direction of the total angular momentum P_j is given by

$$\mu_j = \sqrt{j(j + 1)}(e\hbar/2m)\left[1 + \frac{j(j + 1) + s(s + 1) - l(l + 1)}{2j(j + 1)} \right]$$

The quantity in the square brackets is called the Landé g-factor. (b) Show that the Landé g-factor gives the magnetic moment in units of the Bohr magneton divided by the total angular momentum in units of $\hbar$.

6-33   In the Stern-Gerlach experiment, silver atoms are heated to 1000°K. The atoms pass through an inhomogeneous magnetic field whose gradient is 0.50 (weber/m^2)/cm for a distance of 8.0 cm, and continue through field-free space for 10.0 cm before being deposited on a collector plate. Show that the two lines on the plate have a maximum separation of 0.25 mm. Assume that all atoms have the same speed, given by $\frac{1}{2}\,mv^2 = \frac{3}{2}\,kT$.

6-34   Compute the total angular momentum and total magnetic moment of $_{13}$Al and $_{19}$K in their ground states.

6-35   A trivalent atom in its lowest electronic configuration is in the $3P$ state. What is the element?

6-36   (a) For a neutral lithium ($_3$Li) atom in its ground state, write down the four quantum numbers for each of the three electrons. (b) What are the quantum numbers of the third electron for the two lowest excited states of this atom?

6-37   Show that the total angular momentum and total magnetic moment of an element with closed subshells is zero.

6-38   (a) Show on an energy-level diagram the occupied states of the oxygen atom when it is in the lowest energy configuration. (b) Give the electron configuration for this atom.

6-39   Which of the following elements can show a normal Zeeman effect: $_{11}$Na, $_{12}$Mg, $_{13}$Al?

6-40   In the absence of an external magnetic field, the 2^1P_1 and 3^1D_2 states of cadmium are singlet states; and in the $3^1D_2 \rightarrow 2^1P_1$ transition, light of 6439 Å wavelength is emitted. If a magnetic flux density of 1.000 weber/m^2 is then applied, what wavelengths will be emitted?

6-41   Show that Equation 6-21, giving the force on an electron spin in the Stern-Gerlach experiment, is consistent with Figure 6-24 insofar as the magnetic force is in the direction of the region of strong magnetic field.

SEVEN

X-RAY SPECTRA

7-1 Excitation and ionization of atoms by photon and electron collisions
An atom free of external influences is ordinarily in its *normal state*, that state for which the electron configuration is such as to give the atom its lowest possible energy. For an atom of relatively large atomic number, 2 electrons will fill the innermost shell ($n = 1$, or K shell), 8 electrons will fill the next shell ($n = 2$, or L shell), 18 electrons will fill a still larger shell ($n = 3$, or M shell), and so on until the last and outermost electrons possibly find themselves in unfilled shells. The innermost electrons, exposed to the strong electric field of the nucleus, have the largest binding energy (typically, over 10^3 ev); the last electron, shielded from the nucleus by the inner electrons, has the least binding energy (typically, a few ev). Therefore, when the atom gains relatively small amounts of energy, all electrons in inner closed shells remain in their original shells, while the last, or valence, electron jumps to higher unoccupied energy levels. It is the transitions of the outermost electrons of atoms that gives rise to the *optical spectra*, the spectra falling in or near the visible region of the electromagnetic spectrum.

Consider the energy-level diagram of Figure 7-1, which represents, schematically and without regard for such details as fine structure, the energy states available to the most-loosely-bound outer electron of a relatively heavy atom. Figure 7-1 represents as well the possible energies of the atom when all the other electrons remain in their respective initial states. The lowest, or ground, state g has an energy E_g, which is negative, inasmuch as the last electron is *bound* to the atom; the higher lying excited states, $h, i, j, \ldots$, have energies (all negative) of $E_h, E_i, E_j, \ldots$, respectively. When the atom exists in the excited state i, its excitation energy E_e is $(E_i - E_g)$, and the binding energy E_b of the last electron to the remainder of the atom in this state is $-E_i$. The ionization energy of the atom, the energy that must be added to remove this least-tightly-bound electron from its ground state is E_I; therefore, $E_I = -E_g$. For any particular state, the sum of the excitation and binding energies equals the ionization energy. For the state i

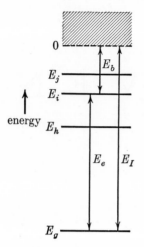

Figure 7-1. Generalized energy-level diagram giving the energy states available to the least-tightly bound electron of an atom.

$$E_e = E_i - E_g, \qquad E_b = -E_i, \qquad E_I = E_e + E_b$$

$$[7-1]$$

as shown in Figure 7-1. For example, the energy of a mercury atom in its ground state is $E_g = -10.42$ ev; in the first excited state the energy is $E_h = -5.54$ ev. Therefore, the ionization energy for removing the least-tightly-bound electron is 10.42 ev. The excitation energy of this outer electron, E_e, in the first excited state is $(-5.54) - (-10.42)$ ev $= 4.88$ ev; and the binding energy of the electron in this state is 5.54 ev.

An atom can be excited from its ground state to any one of its optical excited states by absorbing a photon whose energy equals the energy of excitation of the particular excited state. To excite an atom to the first excited state requires that a photon of energy $h\nu_{gh}$ be absorbed, where

$$h\nu_{gh} = E_h - E_g \qquad [7-2]$$

After absorbing a photon of this energy, the excited atom then decays to the ground state by emitting a photon of precisely the same frequency; excitation of the atom to the first excited state can produce radiation of no other frequency. The process in which the absorption of a photon by an atom is followed by the emission of a photon of the *same* energy is known as *resonance radiation.*

A gas of atoms in the ground state is completely transparent to photons whose energies are less than the first excitation energy of the atoms, as the quantum condition, Equation 7-2, requires. On the other hand, if atoms in the gas are excited to still higher-lying states $(E_i, E_j, \ldots)$ by absorbing photons of the appropriate energy, more downward transitions are possible and the emitted spectrum contains more than one spectral line. This process, in which a single photon is absorbed by the atom and two or more photons are emitted when the atom decays quickly to its ground state, is called *fluorescence*. The fluorescent radiation must, of course, consist of longer wavelengths (smaller photon energies) than the radiation incident upon the absorbing atoms. Finally, if photons having an energy $h\nu_I$, equal to the ionization energy,

$$h\nu_I = E_I \qquad [7\text{-}3]$$

are absorbed, *all* transitions corresponding to the optical spectrum can be emitted as the ionized atom recaptures an electron, and returns by quantum jumps to the ground state. The ionization energies of all neutral atoms, regardless of significant differences in their energy levels and spectra, are all of the order of a few electron volts; therefore, excitation of outer electrons and subsequent emission typically involves the absorption of visible or ultraviolet photons.

An atom can absorb a photon which, if it has sufficient energy, not only ionizes the atom but also imparts kinetic energy to the freed electron. This process is simply the photoelectric effect, and the conservation of energy requires

$$h\nu = E_b + \tfrac{1}{2}\, mv^2 \qquad [7\text{-}4]$$

where $h\nu$ is the energy of the absorbed photon, E_b is the binding energy of the electron to the atom, and $\tfrac{1}{2}\, mv^2$ is the kinetic energy of the photoelectron. When an atom is in the ground state, E_b is just the ionization energy E_I.

Excitation and ionization of atoms can, however, take place in still another way; namely, by the collision of electrons, or other particles, with the atoms. Suppose that an electron having a relatively small kinetic energy, $\tfrac{1}{2}\, mv^2$, makes a collision with an atom in the normal, or ground, state. The atom will remain in the ground state, and the collision will be perfectly elastic, if the electron's energy is *less* than the excitation energy of the first excited state; for, the atom can absorb energy in only quantized, discrete amounts corresponding to the energy differences between stationary states. When the kinetic energy of the electron is just equal to the energy which the atom must gain to raise it to the first excited state, the collision of the electron with an atom initially in the ground state can result in a completely inelastic collision. In such a collision, the electron is brought essentially to rest and the atom gains *internal* energy—the excitation energy—equal to the kinetic energy lost by the electron. Furthermore, when the electron's kinetic energy exceeds the excitation energy of the first excited state, an inelastic collision can occur in which the atom is raised to an excited state and the electron leaves the site of the collision with some remaining energy. In short, the excitation of the atom to a state i requires that the kinetic energy of the electron equal, or exceed, the excitation energy:

$$\frac{1}{2}\,mv^2 \geq (E_i - E_g) = E_e \qquad \text{[7-5]}$$

As in the case of excitation by photon absorption, the excitation by electron collision is signaled by the emission of radiation as the atom returns to the ground state.

The experiment of J. Franck and G. Hertz in 1914 first demonstrated that the excitation of atoms by electron bombardment was governed by the quantization of energy. In the Franck-Hertz experiment, electrons were made to collide with atoms in a mercury vapor. The wavelength of radiation corresponding to the transition from the first excited to the ground state in mercury is 2536 Å; the equivalent photon energy, which therefore equals the excitation energy, is 4.88 ev. Franck and Hertz observed that electrons of at least 4.88 ev energy were required to produce excitation of mercury atoms. This was inferred from the fact that the collisions between the electrons and the mercury atoms were *completely elastic* when the electrons had an energy of *less* than 4.88 ev, but some *inelastic* collisions occurred when the electron's energy *exceeded* 4.88 ev. It was also found that mercury atoms emit radiation of 2536 Å if, and only if, electrons having at least the excitation energy, 4.88 ev, collide with mercury atoms.

Excitation potential is that potential difference through which an electron must be accelerated from rest so that its final kinetic energy is equal to the excitation energy; therefore, the first excitation potential for mercury is 4.88 volt. A variety of experiments, similar to that of Franck and Hertz, have demonstrated that the energy of the atom is quantized. The minimum electron energies for excitation in all instances are found to equal the excitation energies of the respective atoms.

Ionization, as well as excitation, can be produced by electron collisions. When an electron having an energy equal to or exceeding the ionization energy collides with an atom, the least-tightly-bound, outer electron can be removed in the inelastic collision. These ionized atoms then collect electrons and are de-excited, with the emission of *all* lines in the emission spectrum as the outermost electrons return by downward quantum jumps to the ground state. The condition for ionization by electron collisions is thus

$$\frac{1}{2}\,mv^2 \geq E_I \qquad \text{[7-6]}$$

This relation, Equation 7-6, is experimentally confirmed by noting that, when electrons of energies equal to or exceeding the ionization energy strike an atom, the entire emission spectrum is radiated and the electrical conductivity of the gas drastically increases with the onset of ionization.

7-2 The production of the characteristic x-ray spectra When atoms are bombarded by electrons having only a few electron volts kinetic energy, the resulting excitation or ionization involves a change in the state of one or more weakly-bound outer electrons. The optical emission spectrum is induced by such collisions, while the more-tightly-bound electrons in the inner closed shells remain in their initial states. But an inner electron can also be excited or removed, provided sufficiently great energy is added to the atom. In this section we will discuss the spectrum emitted when an inner electron is displaced; that is, we shall be concerned with the *characteristic x-ray spectrum.*

Consider the schematic diagram, Figure 7-2, which indicates the elec-

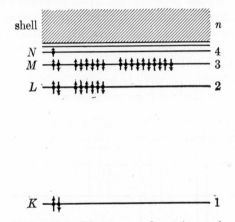

Figure 7-2. Electron configuration and energy levels of $_{29}$Cu in the ground state.

tron configuration and energy levels of copper, $_{29}$Cu. Again, the fine-structure splittings are ignored. The K, L, and M, shells are completely filled with their respective quotas of electrons (2, 8, and 18), as specified by the exclusion principle. There is a single electron, the valence electron, in the N shell. Near the top of the diagram are the unfilled shells, or optical levels; transitions of the outer electron among these states give rise to the optical spectrum. The energy separation between any of the unfilled levels and the onset of the continuum, at the energy $E = 0$, is very small compared to the energy difference between that of the K shell and $E = 0$.

Suppose now that a very energetic electron strikes the atom, thereby removing one of the two electrons in the K shell to an available higher-lying energy level. The electron cannot be accepted in the filled L and M shells; it must go, then, to one of the unfilled high-lying energy states close to the ionization limit, or it may be removed completely, thereby ionizing the atom. In any event, the minimum energy necessary to excite a K electron is close to the ionization energy *for the K shell.* The removal of the single K electron has increased the energy of the atom drastically (well over 10^3 ev), and a vacancy exists in the K shell. This "hole" thus created in the K shell can be filled by a transition of one of the eight electrons from the L shell to the K shell, where the electron will then be more tightly bound. Such a transition, which reduces the energy of the atom by more than 1 Kev, gives rise to the emission of the so-called K_α x-ray line. Because the energy difference is typically of the order of thousands of electron volts, the photon created in such a transition is an x-ray photon with a wavelength between 0.1 and 10 Å.

Still other transitions can occur. The vacancy in the K shell can be filled by a somewhat less probable transition in which an electron in the M shell jumps inward to the K shell; the corresponding emitted photon is labeled K_β. The transitions from successively higher energy states, all *terminating* at the K shell, are identified in the x-ray notation by K_α, K_β, K_γ, K_δ, . . . , as shown in Figure 7-3. This group of K x-ray lines comprise the K *series.* The energy required to produce ionization from the K shell, that is to remove a K electron from the atom with both the freed electron

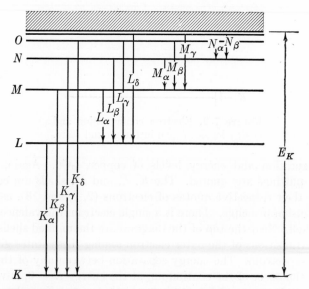

Figure 7-3. Inner-electron transitions giving rise to characteristic x-ray lines.

and the atom at rest, is represented by E_K. Note that E_K is only *slightly* greater than the energy required to bring a K electron to *any* of the unoccupied optical levels.

Other series of x-ray lines, having smaller energies, or longer wavelengths, also occur. When an L electron jumps to the K shell to fill the vacancy created by the removal of a K electron, a vacancy or hole is created in the L shell. This hole can be filled by electron transitions from still higher-lying states, giving rise to the L series, L_α, L_β, L_γ, L_δ, . . . , where all such transitions terminate in the L shell and originate at progressively higher-energy states.

The energy required to produce x-ray ionization from the L shell is designated E_L. Because E_K is ordinarily much greater than E_L, the x-ray lines in the K series have appreciably shorter wavelengths than the lines in the L series. (For Z greater than 30, lines in the K series have wavelengths less than 1 Å, and L series lines have wavelengths less than 10 Å.) Ordinarily much weaker and longer-wavelength lines originate in still higher transitions in the M, N, . . . , series.

The x-ray ionization energies, E_K, E_L, . . . , are related in a simple way to the frequencies of the x-ray emission lines. Consider the K_α line. If an energy E_K is required to remove a K electron, the atom then has its energy increased by the amount E_K; similarly, if an electron is removed from the L shell, the atom's energy is increased by E_L. These two conditions correspond respectively to the situations before and after the emission of the K_α line. The energy of an emitted K_α photon, $h\nu$ is simply $E_K - E_L$. Therefore,

$$K \text{ series} \quad \left\{ \begin{array}{l} h\nu_{K\alpha} = E_K - E_L \\ h\nu_{K\beta} = E_K - E_M \\ h\nu_{K\gamma} = E_K - E_N \\ \text{etc.} \end{array} \right\} \qquad [7\text{-}7]$$

$$L \text{ series} \quad \left\{ \begin{array}{l} h\nu_{L\alpha} = E_L - E_M \\ h\nu_{L\beta} = E_L - E_N \\ h\nu_{L\gamma} = E_L - E_O \\ \text{etc.} \end{array} \right\} \qquad [7\text{-}8]$$

We can now interpret the characteristic x-ray lines that appear when the target in an x-ray tube is struck by electrons having energies of the order of thousands of electron volts. Figure 7-4 shows the measured intensity of emitted x-radiation from a molybdenum target. There is a smooth variation in the continuous x-ray spectrum, caused by *Bremsstrahlung* collisions of electrons with the target; the continuous spectrum has a well-defined maximum which is determined solely by the accelerating potential of the x-ray tube (see Section 3-3). Superimposed on the x-ray continuum are sharp peaks in the intensity, comprising the characteristic x-ray spectrum of molybdenum. Those peaks can be identified as the K_α and K_β lines; the lines of the L series, having much longer wavelengths and lower intensity, do not appear.

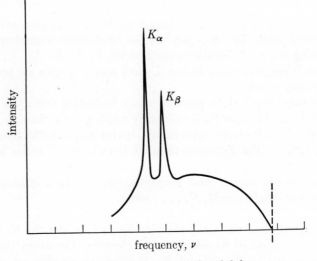

Figure 7-4. X-ray spectrum of molybdenum.

The analysis of an x-ray spectrum into its component wavelengths is typically made with an x-ray crystal spectrometer. A single crystal having a known crystalline structure can be used to separate the various wavelengths, and the wavelength can be computed from the Bragg law (see Section 4-2). The intensity of the x-radiation can be measured with a chamber which is sensitive to the ionization effects produced by x-rays passing through a gas (Section 8-5). X-ray spectroscopy, by which the wavelengths of characteristic lines in x-ray spectra are measured, is the basis of our knowledge of the energies of electrons in the innermost shells of atoms.

Clearly, the lines of the K series cannot be emitted unless the electrons striking the target of an x-ray tube have enough energy to raise a K electron to one of the high-lying, available energy states. Thus, there is a critical electron kinetic energy, $\frac{1}{2} mv^2$, that must be reached, or a *critical potential*, V_c, of the x-ray tube which must be exceeded, before x-ray emission of the K series can occur, where $\frac{1}{2} mv^2 = eV_c$. When the potential applied to the x-ray tube equals or exceeds the critical potential, V_c, electrons will be raised to available excited states; a hole is then produced in the K shell, and the K-series lines, as well as *all* other lines (K, L, M, . . . series) in the x-ray spectra, will be emitted. The value of eV_c is just slightly less (usually no more than 1 part in 10^3) than E_K, the energy necessary to free completely the K electron (see Figure 7-2). Therefore, we can write approximately,

$$\boxed{eV_c \simeq E_K} \tag{7-9}$$

The critical potential for excitation of the L series and higher lying series is, of course, less than that for the K series. When the potential of the x-ray tube is raised to the value of the critical potential for removing an L electron, then the L, M, and higher-lying series will be emitted, but the K series will *not*.

The frequency, or wavelength, of the K_α line can be calculated on the basis of a rather simple theoretical analysis, involving only the Bohr atomic theory. The wavelength λ of lines emitted by one-electron, or hydrogenic, atoms, is given by the Rydberg formula, Equation 5-29:

$$\frac{1}{\lambda} = RZ^2 \left(\frac{1}{n_l^2} - \frac{1}{n_u^2}\right) \tag{7-10}$$

where R is the Rydberg constant, $(k^2e^4m/4\pi\hbar^3c)$, n_u and n_l are the principal quantum numbers of the upper and lower states of the transition respectively, and Z is the atomic number of the one-electron atom. When a K $(n = 1)$ electron has been removed from the atom of a heavy element, an electron in the L $(n = 2)$ shell will "see" the nuclear electric charge (Ze) shielded by the charge $(-e)$ of the one remaining K electron. Electrons in the M, N, and higher-lying states do not penetrate appreciably into the region between the K and L shells. Therefore, an L electron will approximate closely the single electron in a hydrogenic atom, moving in the electric field of the nucleus-plus-K-electron, which has an *effective* atomic number $(Z - 1)$. Equation 7-10 then yields for the frequency $\nu_{K\alpha}$ of the K_α line emitted when an L electron jumps to the hole in the K shell:

$$\nu_{K\alpha} = \frac{c}{\lambda} = cR(Z - 1)^2 \left(\frac{1}{1^2} - \frac{1}{2^2}\right)$$

or

$$\boxed{\nu_{K\alpha} = (3cR/4)(Z - 1)^2} \tag{7-11}$$

A plot of $(\nu_{K\alpha})^{1/2}$ against the atomic number Z of the emitting x-ray elements should, therefore, yield a straight line. The first comprehensive study of the characteristic x-ray frequencies was made by H. G. J. Moseley in 1913. Moseley found that Equation 7-11 represented the data for the K lines very well. In fact, it was his measurements that first established clearly the values of the atomic numbers of the elements. If the chemical elements of the periodic table are listed in the order of their atomic *weights*, the resulting list is (with a few notable exceptions) identical with a listing by atomic number. One such exception is found in the ordering of the elements, cobalt and nickel. The work of Moseley established that, although $_{27}$Co has a greater atomic *weight* than $_{28}$Ni (58.94 as against 58.69), the atomic *numbers* are in the reverse order.

Our discussion of atomic structure has heretofore been restricted to free atoms of a gas, and not to atoms strongly interacting with one another, as in molecules, liquids, or solids. Why is it proper, in the theory of x-ray emission, to consider the atoms as essentially free when actually x-ray target materials are in the form of solids? The answer is, of course, that x-ray transitions involve the innermost, tightly-bound electrons, not the outer electrons. The outer electrons have their configurations and energies changed when atoms are brought close together in a solid, but the inner, tightly-bound electrons are hardly influenced by the state of the material, whether solid, liquid, or gas.

We must again remind the reader that in our simplified treatment of x-ray emission spectra we have ignored such details as the fine structure which occurs in x-ray lines, as well as in optical lines. This fine structure results from the fact that electrons in any one shell may be bound with slightly different energies. For this reason, a given atom will have three values, $E_{L\ I}$, $E_{L\ II}$, and $E_{L\ III}$, rather than a single value E_L, for the energies required to ionize the L-electrons. We can see why there are three ionization energies for removing L electrons in the following way. In a filled L shell, with an electron configuration, $2s^2 2p^6$, one can remove either a $2s$ or a $2p$ electron. The energy to ionize a $2s$ electron is not the same as the energy to ionize a $2p$ electron because they are in different subshells. Furthermore, the energy to remove a $2p$ electron will depend upon whether the electron spin is aligned or anti-aligned with the orbital angular momentum in the p state; of course, in any $2s$ subshell ($l = 0$) both electrons are bound with the same energy because there is no spin-orbit interaction. Thus, there are three distinct ionization energies for L electrons designated $E_{L\ I}$, $E_{L\ II}$, and $E_{L\ III}$, where $E_{L\ I} > E_{L\ II} > E_{L\ III}$. In a similar manner, it follows that there is a single value E_K for K-shell ionization, and that for the M shell there are altogether five possible values.

7-3 X-ray absorption When high-speed electrons strike the target of an x-ray tube the inner electrons of the target atoms can be knocked out. These incident electrons relinquish at least a portion of their kinetic energy to the atom in producing x-ray excitation or ionization. The excitation or ionization can, of course, also be produced by the absorption of photons of the appropriate energies. All the lines of the x-ray emission spectrum will be emitted if a material absorbs photons having enough energy to bring a K electron to one of the higher-lying available energy levels. Such a process, whereby the absorption of a relatively high-energy photon by a material results in the emission of a number of lower-energy x-ray photons, is called x-*ray fluorescence*. A K electron will be ionized if the energy of the incoming photon is at least as large as the K ionization energy, E_K. Clearly, a photon which is capable of causing excitation or ionization of atoms by removing an electron from the K shell must be an x-ray photon; that is, a photon having a wavelength in the general vicinity of 1 Å.

Let us note that x-ray ionization, the removal of an inner, bound electron from an atom by the absorption of an x-ray photon, is simply the photoelectric effect (Section 3-2). The equation of the photoelectric effect is

$$h\nu = E_b + \tfrac{1}{2}\,mv^2 \qquad\qquad [7\text{-}12]$$

where $h\nu$ is the energy of the absorbed photon, E_b is the binding energy of the electron to the material, and $\tfrac{1}{2}\,mv^2$ is the kinetic energy of the released photoelectron. The threshold for photoelectric emission will be reached or exceeded if $h\nu \geq E_b$. (The relativistic form for the electron kinetic energy, $(m - m_0)c^2$, must be used if very-high-energy photons impart a kinetic energy to the electron which becomes comparable to the electron rest energy.)

In the x-ray photoelectric effect, the binding energy E_b is simply the energy, E_K, E_L, E_M, . . . , required to ionize an electron from the K, L, M, . . . shells, respectively. Equation 7-12 can thus serve as a means of measuring the binding energies of inner electrons if the energy of the incoming x-ray photon is known and the kinetic energy of the photoelectrons is measured. The electron kinetic energy is most simply determined by measuring the momentum mv of the released photoelectrons as they pass through a magnetic field of flux density B in a radius of curvature r, using the relation, Equation 2-40,

$$mv = eBr \qquad\qquad [7\text{-}13]$$

Inasmuch as $E_K > E_L > E_M$ . . . , if a photon is energetic enough to remove electrons from the K shell, then it will also be energetic enough to remove less tightly-bound electrons from the L, M, N, . . . shells. Therefore, monochromatic high-energy photon beams with $h\nu > E_K$ will produce the photoemission of K, L, M, . . . , photoelectrons, whose kinetic energies will increase in that order. Furthermore, the photoelectric effect will be accompanied by the emission of the entire characteristic x-ray spectrum.

A simple experimental arrangement for studying the x-ray photoelectric effect is shown in Figure 7-5. Monochromatic x-rays strike an absorber A; several groups of the photoelectrons travel in circular arcs through a vacuum chamber and at right angles to an externally applied magnetic field. The least energetic electrons, arising from the photoelectric emission of a K electron, travel in the smallest radius of curvature; photoelectrons from higher-lying shells, having greater energies, move in larger radii of curvature. From measurements of the incident x-ray photon energy, or wavelength, and a knowledge of the energies of the photoelectrons, the binding energies of electrons in the several inner shells can be determined.

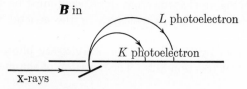

Figure 7-5. Experimental arrangement for measuring the energy of photoelectrons produced by x-ray absorption.

The photoelectric effect is just one of three processes by which photons can be removed from a beam passing through a material (see Section 3-7). For photon energies above approximately 1 Mev, the pair-production process can occur; for somewhat lower energies, the Compton effect is important in photon absorption; and in the x-ray region, roughly 1 to 100 Kev, the photoelectric effect is usually the dominant process for photon absorption (see Figure 3-23).

The absorption of photons in materials is governed by the relation (Section 3-7)

$$I = I_0 e^{-\mu x} \qquad [7\text{-}14]$$

where: I_0 is the intensity of the photon beam incident normally on an absorber of thickness x; I is the intensity of the beam after passing through the absorber; and μ, the absorption coefficient, gives a measure of the effectiveness of the absorber in removing photons from the incident beam. The absorption coefficient is determined experimentally by measuring the intensities, I_0 and I, of the x-ray beam reaching a detecting device without and with an absorber of thickness x interposed. In the x-ray region the absorption coefficient for any given material shows a decrease as the energy of the x-rays increases, as indicated in Figure 3-23. We wish to examine in more detail certain abrupt changes that occur in μ at x-ray photon energies.

The x-ray photon absorption process, which has its origin in the x-ray photoelectric effect, can be treated most simply by considering separately photon absorption by the K and by the L electrons. Although it is impossible to produce photo-emission from the K shell without, at the same time, producing photo-emission from the $L, M, \ldots$ shells, we shall consider first the absorption by K-shell ionization alone. Suppose that the energy of the incident monochromatic photon beam passing through an absorber is varied, and the absorption coefficient is measured. The results of such an hypothetical experiment would be as shown in Figure 7-6a. At low energies the photons will not be capable of dislodging K electrons, the material will be transparent (for K electron ionization), and the absorption

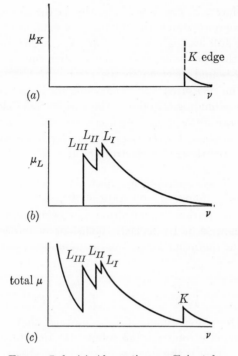

Figure 7-6. (a) Absorption coefficient for
photoelectric emission from the K shell as
a function of frequency. (b) Absorption
coefficient for photoelectric emission from
the L shell. (c) Total absorption coefficient
as a function of frequency, showing the
$K, L_{\mathrm{I}}, L_{\mathrm{II}},$ and L_{III} absorption edges.

coefficient will be zero. When the photon energy reaches the point for
which $h\nu = E_K$, the threshold for photoelectric emission, the photoelectric
effect can occur, and the measured μ will show an abrupt rise. This discontinuity in the absorption coefficient is called the K-*edge*. Absorption by
K-electron ionization persists to still higher photon energies, but falls off
as the photon energy increases, finally reaching zero.

Now we turn to photoelectric absorption by the removal of L electrons.
Again, there is no absorption by L electrons until the photon energy is
equal to the threshold energy, $E_{L\ \mathrm{III}}$, which is also the binding energy of
the least-tightly-bound electrons in the L shell, as shown in Figure 7-6b.
After the L_{III} edge is exceeded, the absorption coefficient decreases until
the threshold energies, $E_{L\ \mathrm{II}}$ and $E_{L\ \mathrm{I}}$, for removal of L_{II} and L_{I} electrons
are reached. The total absorption coefficient for photon removal by $K, L,$
and still other bound electrons, is simply the sum of the separate absorption

coefficients, as shown in Fig. 7-6c. The rise in the absorption coefficient towards lower energies shows the influence of photon absorption by the ionization of the still less-tightly-bound M, N, . . . electrons. Ultimately, of course, when the photon energy is smaller than the excitation energy of the least-tightly-bound electron in the atom, the absorption coefficient falls to zero. The measured wavelengths of the x-ray absorption edges give the respective binding energies of the various inner electrons in heavy atoms. Thus x-ray absorption measurements, as well as x-ray emission spectra, provide a means for studying the energy levels and atomic structure of the inner, tightly-bound electrons of atoms.

7-4 Summary The excitation or ionization of an atom at low energies (few ev) involves a change in the state of the least-tightly-bound, outer electrons. Such excitation or ionization can be produced either by the absorption of photons or by inelastic collisions of particles with atoms, both following the quantum conditions, as follows:

EXCITATION   IONIZATION

$h\nu = E_e$ $h\nu \geq E_I$

$\tfrac{1}{2}\,mv^2 \geq E_e$ $\tfrac{1}{2}\,mv^2 \geq E_I$

The Franck-Hertz experiment first confirmed that the excitation of atoms by inelastic electron collisions follows the quantum condition.

Resonance radiation is the process in which the absorption of a photon by an atom is followed immediately by re-emission of a photon of the same energy; fluorescent radiation is that process wherein the absorption of a high-energy photon is followed immediately by the re-emission of two or more lower-energy photons.

The characteristic x-ray lines are emitted when a tightly-bound, inner electron of an atom is removed by electron collision or photon absorption, and other electrons within the atom move inward to fill the electron vacancy. X-ray lines are arranged in series and classified according to the convention: the letter symbol (K, L, M, . . .) specifies the shell ($n = 1, 2, 3, . . .$) in which the transition terminates, and the subscript (α, β, γ, . . .) denotes the originating shell. Thus K_α denotes the x-ray line resulting from the electron transition $n = 2$ to $n = 1$.

The critical potential, V_c, applied to an x-ray tube for exciting emission lines in the K- and longer-wavelength series is given by

$$eV_c \simeq E_K$$

where E_K is the K-electron ionization energy.

The frequencies of K_α lines, $\nu_{K\alpha}$, are given closely by the Moseley relation

$$\nu_{K\alpha} = (3cR/4)(Z - 1)^2$$

where Z is the atomic number of the atom.

X-ray absorption occurs principally through the photoelectric emission of inner electrons. The K-edge, the discontinuity in the absorption coefficient arising from the onset of K-electron photoelectric ionization, has a threshold which is given by

$$h\nu = E_K$$

REFERENCES

Clark, G. L., *Applied X-rays*. New York: McGraw-Hill Book Company, Inc., 1955. See Chapter 6 for a discussion of x-ray spectra and atomic structure.

Compton, A. H., and S. K. Allison, *X-Rays in Theory and Experiment*. Princeton, New Jersey: D. Van Nostrand Company, Inc., 1935. This is a standard text in the theory and practice of x-rays.

Encyclopedia of Physics (Handbuch der Physik), ed. S. Flügge, Vol. XXX, *X-rays*. Berlin: Springer-Verlag, 1957. The article by A. E. Sandström includes comprehensive tables on x-ray excitation potentials, emission-line wavelengths, and the wavelengths of absorption edges.

Richtmeyer, F. K., E. H. Kennard, and T. Lauritsen, *Introduction to Modern Physics*. New York: McGraw-Hill Book Company, Inc., 1955. A fairly comprehensive treatment of x-ray spectra and structure is given in Chapter 8.

PROBLEMS

7-1 Spectroscopic observation shows that the sodium D lines (mean wavelength 5893 Å) are emitted when a sodium vapor is bombarded by electrons which has been accelerated by a potential difference of 2.11 volt. Compute the value of h/e.

7-2 It is found that when sodium vapor is illuminated with radiation of 3303 Å, the wavelength of the *second* line in the principal series of sodium, emission of the D lines ($3P \rightarrow 3S$) occurs. What additional lines should be simultaneously emitted? (See Figure 6-10.)

7-3 The longest wavelength that can be used to produce resonance radiation in mercury is 2536 Å. What is the first excitation potential of mercury?

7-4 The average wavelength of the sodium D lines is 5893 Å. (a) What is the first excitation potential for sodium? (b) If emission of the D lines is produced by heating a sodium vapor, what approximate temperature must be reached to produce strong emission?

7-5 * (a) Show that the maximum wavelength of radiation required to produce fluorescent radiation (optical) of the entire spectrum from potassium is equal to the wavelength of the P series limit of potassium. (b) Show that it is *not* possible to produce fluorescent radiation of *all* S-series lines in potassium when potassium vapor is irradiated with monochromatic photons whose wavelength is equal to that of the S-series limit.

7-6 (a) What is the minimum kinetic energy of electrons that can produce emission of the H_α line when they strike hydrogen atoms in the ground state? (b) What is the minimum kinetic energy of the electrons for causing emission of *all* lines in the hydrogen spectrum?

7-7 * The phenomena of the *aurora borealis* (Northern Lights) and the *aurora australis* (Southern Lights), the luminous displays in the sky near the Earth's poles, are produced when charged particles (protons, etc.) thrown out by the Sun collide with oxygen and nitrogen 100 km or more above the Earth's atmosphere, thereby exciting and ionizing these atoms. (a) The charged particles are known to travel the 93 million miles from the Sun to the Earth in about 24 hours. Assuming that the exciting particles are protons, what is their average kinetic energy? (b) Why do these charged particles produce appreciable ionization of oxygen and nitrogen only in the vicinity of the Earth's poles? (*Hint:* consider the effect of the Earth's magnetic field on the path of charged particles approaching the Earth.)

7-8 Show that the complete x-ray fluorescence of a material can be emitted only if the material is irradiated with characteristic x-rays produced by a target of *higher* atomic number.

7-9 Show that the frequency of the K_α line from a heavy element of atomic number Z is given approximately by $\nu = 0.246 \times 10^{16}\, Z^2$ sec^{-1}.

7-10 (a) Find the energy (Kev) of a K_α x-ray photon emitted from uranium. (b) What is the fractional decrease in mass of a uranium atom arising from this photon emission?

7-11 The critical voltage required to produce the emission of the K series from $_{74}$W is found to be 69.5 kilovolt; a measurement of the wavelength of the K-absorption edge in tungsten yields 0.178 Å. Compute the ratio, h/e.

7-12 Calculate the wavelength of the K_α line in silver, $_{47}$Ag.

7-13 * Is the slope of the line obtained by plotting the square-root of the frequencies of K_α lines (as ordinates) against the corresponding atomic numbers greater than or less than the slope of the line for the K_β frequencies?

7-14 (a) With what speed would a free copper ($_{29}$Cu) atom recoil upon emitting a K_α photon? (b) What is the ratio of the kinetic energy of the recoiling atom to the energy of the emitted K_α photon?

7-15 * The K_α "line" is found to consist of two very-closely-spaced lines, called the $K_{\alpha 1}$ and $K_{\alpha 2}$ lines. Why are there two lines in the K_α fine structure, rather than three?

7-16 * (a) How many M absorption edges would one expect to find in an element of moderately large atomic number? (b) Show that sodium has only one M absorption edge.

7-17 Show that the removal of a K electron from an atom leaves the K-shell with an angular momentum of $\sqrt{\frac{1}{2}(\frac{1}{2} + 1)}\hbar$ and a magnetic moment of $\sqrt{3}$ Bohr magnetons.

7-18 The critical potential V_c for excitation of all lines in the x-ray spectrum never exceeds by a large amount the energy of the K_α photon divided by the electron charge. Show that this is a good rule-of-thumb.

7-19 The K absorption edge in copper is found to have a wavelength of 1.377 Å. What is the minimum potential difference that can be applied to an x-ray tube with a copper target to excite the emission of the characteristic lines of the K series?

7-20 Prove that the wavelength corresponding to the K absorption edge in a given element is always slightly smaller than the wavelength of any x-ray emission line in the K series from the same element.

7-21 The K_α emission line of uranium has a wavelength of 0.0126 Å, and the K absorption edge is 0.0107 Å. What is the approximate wavelength for the L absorption edges in uranium?

7-22 * Monochromatic x-rays of wavelength 1.00 Å strike a $_{29}$Cu target producing x-ray photoelectric emission from copper. A narrow beam of these photoelectrons is projected at right angles into a uniform magnetic field of flux density 1.00×10^{-2} weber/m^2 and strikes a photographic plate, as shown in Figure 7-5. The K and L absorption edge wavelengths in copper are approximately 1.38 Å and 13.3 Å, respectively. Calculate the distance of separation on the photographic plate resulting from the photoelectron emission from the K and L shells.

7-23 When lead is irradiated with x-rays, the K edge in the absorption coefficient occurs at 0.140 Å. (a) If this material were irradiated with monochromatic 0.100 Å x-rays, what would be the maximum kinetic energy of photoelectrons emitted from the K shell? (b) Are these photoelectrons relativistic?

7-24 The L_I absorption edge of lead has a wavelength of 0.780 Å. (a) Calculate the maximum kinetic energy of photoelectrons emitted from the L_I shell when lead is bombarded with 0.085 Å x-rays. (b) Calculate the speed of these emitted photoelectrons.

7-25　*　The wavelength of the K absorption edge in element A lies between the wavelengths of the K_α and K_β lines in element B. If the characteristic x-radiation from a target of B is passed through an absorber of element A, which of the K lines will be more strongly attenuated?

7-26　The L_I absorption edge of tungsten occurs at 1.02 Å. What is the minimum photon energy that can produce fluorescent radiation of the L and M series in tungsten?

7-27　The K_α line from tungsten has a wavelength of 0.213 Å; the K absorption edge of copper is 1.378 Å. If K_α radiation from tungsten is absorbed in copper by the photoelectric effect in the K shell, and the emitted photoelectrons pass at right angles through a magnetic field of flux density 0.020 weber/m², what is the radius of their path?

7-28　The critical voltage which must be applied to an x-ray tube before the $_{47}$Ag target of the tube emits the characteristic K lines is 25.8 kv. (a) Calculate the wavelength of the K absorption edge in silver. (b) Why is the wavelength of this absorption limit slightly shorter than that of the K_α line of silver?

E I G H T

INSTRUMENTS AND ACCELERATING MACHINES USED IN NUCLEAR PHYSICS

8-1 Introduction The progress of atomic and nuclear physics has depended on the development of devices for studying submicroscopic phenomena. Our understanding of the structure of the atom, derived in large measure from the study of emitted spectral lines, was as much dependent on experimental physics and spectroscopic observations as it was on theoretical physics and the development of the quantum theory. Nuclear physics had its origin in the discovery by H. Becquerel in 1896 that uranium salts can produce a darkening of photographic plates. But the development of a detailed, if still somewhat incomplete, knowledge of the nucleus—an object so minute that it is observed only indirectly—has come from a variety of experiments using instruments which are, in some instances, remarkably simple, and in others, extraordinarily complex, subtle, ingenious, and costly. In this chapter we shall set forth the physical principles on which a number of important devices in nuclear physics operate: detectors of nuclear radiation; devices for measuring the mass, velocity, and momentum of charged particles; and high-energy particle accelerators.

We shall confine our discussion to the fundamental physical laws underlying the operation of these devices; technical matters, however important they may be in the actual construction of these instruments, are beyond the scope of this book.

The particles with whose detection, control, and acceleration we shall be primarily concerned in this chapter are the electron, proton, deuteron, α-particle, and photon. The *deuteron* is the nucleus of the deuterium, or heavy-hydrogen, atom; its charge is equal to that of the proton, but its mass is approximately twice that of the proton. The α-particle has already been described as one of the particles emitted, typically with energies of a few Mev, from radioactive materials; it is the nucleus of the helium atom with a double positive charge and a mass roughly four times that of the proton. The nuclei of radioactive materials may also emit β-rays and γ-rays. The β-particles are high-speed electrons (or positrons) with energies up to several Mev; the γ-rays are photons emitted from nuclei with energies measured in Kev or Mev. The properties of radioactive nuclei and the characteristics of their nuclear radiation are treated in some detail in Chapter 9.

We shall not discuss in this chapter the detection of the *neutron,* a fundamental constituent of atomic nuclei. The neutron has a zero electric charge and a mass closely equal to that of the proton. Because the neutron carries no charge and is incapable of producing ionization effects, or of being deflected by electric or magnetic fields, it cannot be studied with the same directness as charged particles. For this reason we postpone discussion of neutron detection (Section 10-6) until the nuclear reactions in which it participates have been explored.

The term *nuclear radiation* is used to describe *all* types of particles, including the photons of electromagnetic radiation, which are emitted from the nuclei of atoms. Experiments in nuclear physics must deal with the measurement of such properties as the identity, mass, energy, and momentum of particles of nuclear radiation. These particles generally have energies up to several Mev, and since such energies are considerably higher than those encountered in atomic physics (never higher than a few hundred Kev), the detection and measurement of nuclear radiation require special experimental techniques.

8-2   Ionization and absorption of nuclear radiation   It is worth realizing at the outset that the detection and control of nuclear particles pose difficult experimental problems by virtue of the very minuteness of the particles encountered. For example, the mass of the electron is a mere 9.11×10^{-31} kg and its electric charge is only 1.60×10^{-19} coulomb; therefore, the electron, as well as other particles, is never "seen" directly, and moreover, its influence on instruments is usually an extremely subtle one. The chief problem of experimental atomic and nuclear physics is to infer the properties and structure of essentially unobservable submicroscopic particles, and to control such particles, all with macroscopic apparatus.

If a particle is identified and known to be at a certain place at a certain time, we can say that we have detected the particle. The problem of detection is, then, one of being able to establish that a particle is present or absent in some detecting device. What is required is that the detecting instrument be of such sensitivity that a minute change produced by the particle's presence can influence a large-scale, easily observed characteristic of the instrument. We have an example of such a delicate measurement in the Millikan oil-drop experiment, by which the electron charge was first directly determined. A tiny, uncharged oil-drop, observed through a microscope, falls slowly under the influence of gravity (and the retarding effects of the air). If the drop acquires a single electron charge and a known electric field is applied to produce an upward electric force on the oil drop, the downward gravitational force can be balanced out, the drop can be brought to rest, and the electric charge on the drop can be computed by equating the two forces.

The most simple and frequently used detectors of electrons, protons, and α-particles depend in their operation on the fact that these particles have an electric charge. The photon has, of course, no electric charge, but it does interact strongly with charged particles, thereby producing ionization. Thus, an instrument sensitive to the effects of electric charges can detect the presence of a photon if the instrument responds to electrons produced by photons in the photoelectric effect, the Compton effect, or in pair production (see Section 3-7).

Consider the ionization effects produced by a charged particle, such as a proton, deuteron, or α-particle, whose mass is large compared to that of the electron. The probability that such a particle will make a close encounter or collision with a nucleus in the material is very slight (see Section 5-1), and therefore, its energy is influenced almost entirely by the interactions with the electrons of the atom. The positively charged particle collides with and transfers some energy to the electrons it encounters. Thus, an electron initially bound to an atom can be excited to a higher-lying energy state; or, if the energy gain is sufficiently great, it can even be removed from its parent atom, thereby producing an ion. Inasmuch as the heavy particle does work on the electrons in exciting or removing them, the energy of the heavy particle is decreased and it is slowed down. The direction of travel is, however, not appreciably changed by such a collision with an electron because the mass of the ionizing particle is much greater than the electron mass. As the particle continues through the material in a nearly straight line, it leaves a track of ionized atoms and freed electrons along its path. The heavy particle is finally brought to rest when all of its energy has been transferred to atoms of the material. The number of ion pairs (an ion plus the detached electron comprise one ion-pair) produced by the heavy particle increases greatly near the end of the path. This occurs because the heavy particle is then moving more slowly, and the time it spends in the vicinity of any one atom increases correspondingly.

When heavy particles of nuclear radiation are stopped in a material through the process just described, we can say that the particles have been absorbed in the material. It is found that, for mono-energetic heavy charged particles, there is a well-defined *range* in a given absorbing material. There is a close relationship between the energy of the charged particle and its range in any particular absorbing material; this serves as a simple means of determining the initial energy of the heavy particle. As might be expected from the difference in density of the absorbing materials, the range of a heavy particle with a given energy is considerably greater in a gas than in a solid. An α-particle with an energy of several Mev is stopped by a few centimeters of air, but by a fraction of a millimeter in an absorber of relatively high density, such as aluminum. The difference in the range results directly from the fact that the ionizing particle encounters more atoms in a solid than in a gas.

It is a remarkable fact that *all* charged particles lose close to 32 ev energy in creating a single ion pair in *any* gas. This does not mean that the ionization potential for gas atoms and molecules is exactly 32 ev, but rather that the ionizing particle loses, on the average, 32 ev in all collisions (excitation, as well as ionization) in order to produce a single ion pair. Because the number of ion pairs produced in a gas is directly proportional to the energy lost by an ionizing particle, a measurement of the ionization in gases can lead to a simple means of detecting charged particles of nuclear radiation and also of determining their energies.

The absorption of β-rays, or electrons, in an absorbing material is more complicated than the absorption of heavy, charged particles because the ionizing particle now has a mass which equals the mass of the atomic electrons to which the energy is imparted in collisions. The path of a high-speed electron through an absorbing material is not strictly a straight line, the electron being appreciably deflected in collisions. Therefore, an electron does not have a well-defined path. Nevertheless, the range of a collection of mono-energetic electrons can be defined as the thickness of the absorber required to stop all electrons. For electrons, the range is nearly inversely proportional to the density of the absorber. For a given energy and absorber, the range of an electron is appreciably greater than that of a heavy, charged particle.

The processes by which photons are absorbed—the photoelectric effect, the Compton effect, and pair-production—were discussed in Section 3-7.

It should be recalled here that it is impossible to assign a precise range to a photon of a particular energy. Instead, the absorption of photons is characterized by the absorption coefficient, μ, whose reciprocal gives the thickness of absorber for which the number of photons in a beam is reduced to $(1/e)^{\text{th}}$ (36 per cent) of the initial number. The intensity of a photon beam is reduced by exponential attenuation, following $I = I_0 e^{-\mu x}$. Therefore, the intensity of the beam is truly zero, *all* photons having undergone collisions, only after the beam has passed through an infinite absorber thickness. It is worth noting the fundamental difference between the absorption of a photon beam and the absorption of charged particles. Photons lose all of their energy in a single collision;† they always move at the speed c; and their absorption can *not* be characterized by a range. On the other hand, charged particles lose their energy little by little in many collisions; their speed gradually decreases; and their absorption *is* characterized by a definite range. Photon absorption is accompanied by ionization effects, which can be detected electrically. The electrons (and positrons, in pair-production), which are given energy in photon collisions, can themselves produce ions in the absorbing material. Photons are far more penetrating than electrons or heavy, charged particles of the same energy.

Table 8-1 shows the ranges for the absorption of α-particles, protons, and electrons in air and in aluminum for the energies, 1, 5, and 10 Mev. For comparison, the table also gives that thickness of aluminum absorber, $(1/\mu)$, required to reduce the intensity of a photon beam to $(1/e)^{\text{th}}$ of its initial intensity. It is seen that: (1) all types of nuclear radiation are more readily absorbed in a solid than in a gas; (2) γ-rays are more penetrating than β-rays, which are, in turn, more penetrating than protons or α-rays; (3) the penetration of charged particles increases with increasing energy.

Table 8-1

Ranges, in centimeters, of α-particles, protons, and electrons in air and in aluminum. The values listed for γ-ray absorption in aluminum are the reciprocals of the absorption coefficients, $(1/\mu)$.

Ranges in Air (STP)　　　　　　*Ranges in Aluminum*

ENERGY	α-PARTICLE	PROTON	ELECTRON	ENERGY	α-PARTICLE	PROTON	ELECTRON	γ-RAY
1 Mev	0.5	2.3	314	1 Mev	0.0003	0.0014	0.15	6.1
5 Mev	3.5	34	2000	5 Mev	0.0025	0.019	0.96	13.1
10 Mev	10.7	117	4100	10 Mev	0.0064	0.063	1.96	16.0

† In a Compton collision, the incident photon is annihilated and the scattered photon is created.

8-3 Cloud and bubble chambers The cloud chamber, invented by C. T. R. Wilson in 1907, is one of the earliest and most useful detecting instruments in nuclear physics. With a cloud chamber it is possible to see and photograph the trail of ions produced by a charged particle moving through a gas.

The operation of the cloud chamber depends on the behavior of a supersaturated vapor. When a vapor is in equilibrium with a liquid, the pressure of the vapor is the saturated vapor pressure. Typically, the saturated vapor pressure increases as the temperature of the liquid is raised. If this saturated vapor is suddenly expanded adiabatically, so that no heat leaves or enters the vapor, the temperature of the vapor falls. But, the pressure of the expanded vapor is now too high for the reduced temperature, and if dust particles or ions are present, the vapor will condense around these condensation centers at such a rate that the pressure of the vapor is always equal to the saturated vapor pressure. If, however, the vapor is dust- and ion-free, liquid droplets cannot form and the adiabatic expansion will produce a *super*saturated vapor. An ionizing particle passing through such a supersaturated vapor leaves a trail of ions along its wake about which condensation can take place, and the tracks of these liquid droplets (radius, 10^{-3} cm) may easily be seen, or, better still, photographed.

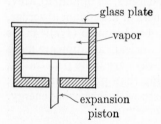

Figure 8-1. Simple elements of a cloud chamber.

The simple elements of a cloud chamber are shown in Figure 8-1. The chamber, whose volume may be as large as one cubic foot, operates with such mixtures as air and water vapor, or argon gas with alcohol. To produce a supersaturated vapor the piston is suddenly retracted and the vapor is thereby cooled by adiabatic expansion to a supersaturated condition. If a charged particle, such as an α-particle, passes through the chamber, ions are formed along its path upon which droplets are condensed. At this instant the chamber is illuminated with a light source, and a camera photographs the track. (If two cameras are used, tracks can be analyzed in three dimensions.) The tracks can then be studied at leisure to determine such characteristics as the length of the track, from which the energy of the ionizing particle can be determined. Because the ionization increases as a heavy particle is brought to rest, its track is more dense near the end of its path. Electrons are easily deflected, and their tracks are usually irregular. Cloud-chamber photographs are particularly useful in studying collisions between particles. The photograph will record the path of the incident particle, as well as the paths of any emerging charged particles. The angles between the directions of the particles can be measured directly,

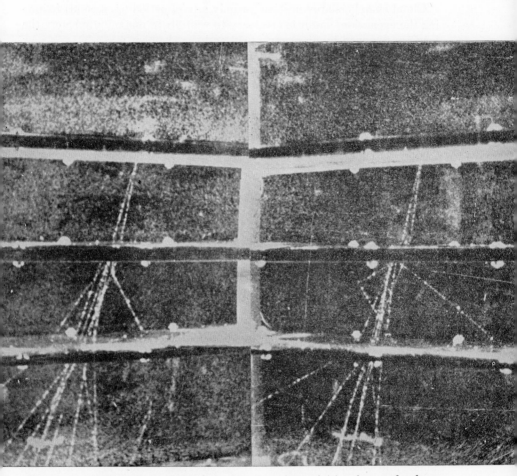

Figure 8-2. Cloud-chamber photographs of a photon-initiated cascade shower taken simultaneously with two cameras to permit three-dimensional analysis of the tracks. A 700 Mev photon (producing no track) enters from the top, and a positron-electron pair is created at the uppermost thin lead plate. Photons are created at the lower plates by *Bremsstrahlung* collisions, and these photons create more pairs, leading to a shower of electrons, positrons and photons. (Courtesy of Professor J. C. Street, Harvard University.)

and the collisions studied in detail. Figure 8-2 is an illustration of a cloud-chamber photograph.

Cloud chambers are sensitive for only a brief period ($\frac{1}{10}$ second) following the expansion, and it is necessary to wait about one minute before the ions are removed by an electric field and the chamber is ready for a new expansion. In order that photographs are taken only when interesting events take place, Geiger counters, or other detectors, can be placed above the cloud chamber. Pulses from the Geiger counters, signaling the passage of a charged particle headed for the cloud chamber, trigger the expansion piston, the light source, and the camera. In addition, an external magnetic field may be applied so that the sign of the charge and the momentum of a charged particle can be deduced from the curvature of the track, following $mv = qBr$.

A second type of cloud chamber, the *diffusion cloud chamber*, has come into common use since 1950. In such a chamber the bottom is maintained at a considerably lower temperature than the top; a heavy gas fills the chamber and a light vapor is introduced into the top. As the vapor diffuses downward, it is cooled and becomes supersaturated. The chamber is then continuously sensitive in the region of the supersaturated vapor.

An important disadvantage of a cloud chamber is that the density of a gas is so low that the chance for a very-high-energy particle to be stopped within the chamber is remote; furthermore, the probability that an incoming particle will make a collision with another nuclear particle within the gas is slight. The *bubble chamber*, invented by D. A. Glaser in 1952, overcomes these disadvantages by using a *superheated liquid*, rather than a supersaturated vapor. This increases the density by a factor of approximately 10^3, and thereby increases the probability of nuclear collisions and interesting events. In a typical bubble chamber, liquid hydrogen is maintained at 27°K by a pressure of about 5 atmospheres; hydrogen boils at 20°K at atmospheric pressure. If the pressure is suddenly reduced, the liquid becomes *superheated*, its temperature being greater than the boiling point, and bubbles can be formed by ions lying in the wake of a passing ionizing particle. The bubble tracks can be photographed and studied as in the case of the cloud chamber. It is interesting to note that a bubble chamber is, so to speak, a cloud chamber turned "inside out," with vapor bubbles formed in a liquid, rather than liquid droplets formed in a vapor. A bubble-chamber photograph is shown in Figure 8-3. Bubble chambers have been constructed having a volume up to 10 cubic feet, and the recycling time between successive operations can be less than one second.

8-4   Nuclear emulsions   Apart from the human eye, the most common detector of photons having wavelengths in the visible portion of the electromagnetic spectrum is the photographic plate or film. A photographic

Figure 8-3. Bubble-chamber photograph. The incoming beam consists of protons and π-mesons; heavy tracks stopping in the chamber are protons, and light tracks passing through the chamber are π-mesons. A number of nuclear collisions can be discerned in this picture. (Courtesy of Brookhaven National Laboratory.)

emulsion contains grains of silver bromide, which are changed, upon development, into grains of pure silver. Some particles dissipate energy slowly upon traversing such an emulsion; they produce track images as they lose energy and finally come to rest. Photons, however, die instantly upon striking the emulsion, producing only point images rather than tracks. Photographic emulsions are sensitive to higher-energy photons—ultraviolet, x-ray, and γ-ray photons. Electrons, protons, deuterons, and α-particles can also be detected photographically. An emulsion prepared for use in recording the paths of charged particles is made thicker and more sensitive than those used in ordinary photography. Emulsions which can record the tracks of nuclear particles are called *nuclear emulsions*.

When a charged particle passes through a nuclear emulsion it produces a latent image along its path, which appears as a track when the emulsion is developed. The range of a particle is dependent on its energy, and therefore the energy can be found by measuring the length of the track with a microscope. In a typical emulsion, a 10 Mev proton produces a track approximately 0.5 mm long, and a 20 Mev proton produces a 2.0 mm track. Heavy particles, such as protons, or α-particles, produce much more dense tracks (because of their greater effectiveness in exciting AgBr crystals) than do electrons. This allows the masses of charged particles to be evaluated by counting the number of grains in the track. Furthermore, the density of the tracks increases near the end of the path, so that it is possible to determine in what direction the particle has traveled. Nuclear emulsions are particularly useful in studying collisions between nuclear particles inasmuch as the energies, masses, and directions of the participating particles can all be measured quite directly. Like cloud and bubble chambers, nuclear emulsions record particle tracks; nuclear emulsions are, however, continuously sensitive and because of their small size, weight, and simplicity, they are commonly used in high-altitude balloons to study cosmic radiation. A photograph of a nuclear emulsion is shown in Figure 8-4.

8-5   Gas-filled detectors   The simplest detector sensitive to the ionization effects of nuclear radiation through a gas is an electroscope. When an electroscope is charged, a gold leaf or some other such light conductor is displaced by the mutual electrical repulsion from the fixed conductor to which it is attached, the amount of the displacement being a measure of the charge on the electroscope. Nuclear radiation passing through an initially charged electroscope ionizes the air, and the electroscope is discharged as ions are collected on and neutralize the charged conductors. Because electroscopes are relatively insensitive, they are not used as commonly as the three gas-filled detectors to which we now turn. The three

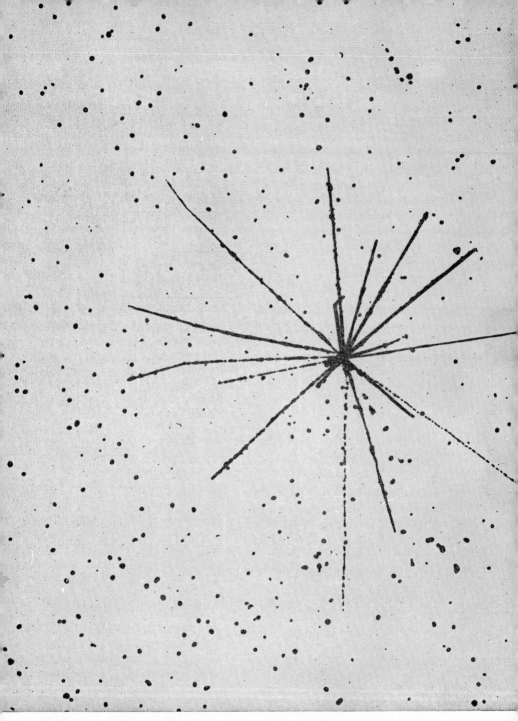

Figure 8-4. A nuclear emulsion showing a star event. A 2 Bev neutron (producing no track) has hit and exploded a nucleus into 17 different track-forming particles. The heavy lines are made by relatively slow particles, such as protons; the lighter lines are made by lighter particles, such as mesons. Magnification, 200X. (Courtesy of Brookhaven National Laboratory.)

most important types of gas-filled detectors are the *ionization chamber,* the *proportional counter,* and the *Geiger counter.*

Consider the device shown in Figure 8-5. The gas-filled chamber has

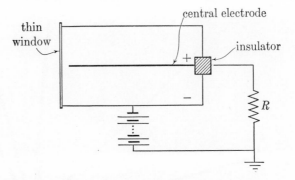

Figure 8-5. Simple elements of a gas-filled detector.

two electrodes, consisting of an outer cylinder and a thin wire along the cylinder axis. The wire is maintained at a high positive voltage with respect to the cylinder. The wall of the chamber, whether of glass, thin metal, or mica, is sufficiently thin to permit the entry of charged particles or photons from the outside. Various gases may be used in the chamber, and the pressure may range from a fraction up to several atmospheres. The electric field between the two electrodes is highly inhomogeneous, being very large near the central wire. All gas-filled detectors operate on the following principle: (a) nuclear radiation ionizes some of the gas molecules within the chamber; (b) the electric field pulls the charged particles to the electrodes, producing a current in the circuit; and (c) the resulting current, or current pulses, through the resistor R in Figure 8-5 are measured by electrical instruments.

A plot showing the number of ions collected as a function of the applied voltage for a typical gas-filled detector is shown in Figure 8-6. The number of ions collected at any particular applied voltage will depend on the volume of the detector, and the details in the curve may differ according to the gas used. At the lower voltages two curves are shown, one for the ionization effects from α-particles and one for the ionization effects of high-speed electrons, or β-particles. The various types of gas-filled detectors can best be understood by considering separately each of the four regions, A, B, C, and D, of Figure 8-6.

In region A, where the applied voltage is relatively small, ions formed within the detector are subject to a relatively weak electric field. Therefore, some of the ions created by the passage of charged particles or photons recombine into neutral atoms or molecules before the ions reach an

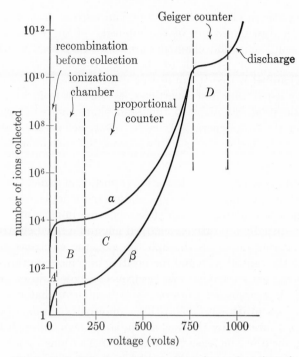

Figure 8-6. Number of ions collected as a function of
the applied voltage for a typical gas-filled detector of
α-particles and β-particles.

electrode. The current due to ion collection is so small that a gas-filled
detector cannot operate efficiently in region A. The number of ion pairs
produced by α-particles greatly exceeds the number of ions produced by
β-particles because an α particle is more effective in ionizing the gas. (See
Table 8-1.)

THE IONIZATION CHAMBER   In region B, where the applied voltage has
been increased, practically all ions formed by the ionizing radiation are
collected on the electrodes before recombination can take place. A gas-
filled detector operated in this region is called an *ionization chamber*. It is
worth recalling that the energy required to produce one ion pair by *any*
type of ionizing radiation is close to 32 ev.

Ordinarily the ionization produced by the nuclear radiation is so small
that the formation of a single ion pair and its collection on the electrodes
can not be detected as a single, abrupt change in the current through the
circuit. Instead, the combined ionization effects of many nuclear particles
produce a nearly constant current that can be read directly on a sensitive
galvanometer, or measured as a voltage drop across the high resistance R

by amplification in electronic circuits. The current from the ionization chamber is a direct measure of the intensity of the ionizing radiation; for this reason, ionization chambers are frequently used to measure the intensity of x-ray or γ-ray photon beams. Another characteristic of an ionization chamber is that it can distinguish between the ionization effects produced by α-particles and β-particles; we can see from Figure 8-6 that an α-particle produces about a thousand times more ion pairs than a β-particle of the same energy. It is, however, *not* possible to determine the *energy* of charged particles by measuring the current in an ionization chamber.

THE PROPORTIONAL COUNTER   In the lower portion of region C of Figure 8-6, the applied voltage, and therefore also the electric field, is increased further. In this region, ion multiplication effects become important. Thus, when an ion pair—usually a positively charged ion and a freed electron—is produced by nuclear radiation through the gas, the electrons are strongly attracted to the central wire and the positive ions move towards the outer negatively-charged electrode. An electron can now gain enough kinetic energy as it is accelerated towards the central wire that it can produce still further ions in its collisions with gas molecules. In fact, a single ion pair, produced directly by the nuclear radiation, can, through the process of gas-multiplication, increase the total number of ions by a factor of 10^5 to 10^6. This increase in ion pairs produces a markedly enhanced current, or voltage, pulse; and it is possible under these conditions to register incoming nuclear particles one-by-one; that is, to count the pulses directly. The lower portion of region C is the basis of operation of the *proportional counter*. The counter is so named because the size of the pulse is directly proportional to the energy of the ionizing particle. The pulses from α-particles are considerably greater than those from β-particles, and it is possible to arrange the circuitry of the proportional counter so that only those large pulses resulting from α-particles are registered while the much smaller pulses resulting from β-particles are discriminated against.

THE GEIGER COUNTER   The behavior of the gas-filled detector for relatively high voltages (in the vicinity of 1000 volts) changes fundamentally in region D, the Geiger-Müller region. The applied voltage and the electric fields within the detector are now so high that *any* nuclear particle producing a single ion pair within the gas can initiate an avalanche of electrons by gas-multiplication. The pulses of current through the detector are now *independent* of the energy of the initiating particle. It is not possible to distinguish between various types of nuclear radiation with a Geiger counter (sometimes also called a G-M tube, or Geiger-Müller counter) inasmuch as an α-particle, β-particle, or γ-ray all produce similar electron avalanches (see Figure 8-6). Geiger counters are used to count β-, x-, and

γ-rays, but usually not α-rays; β-ray counters must have relatively thin windows to permit the easily stopped electrons to enter the tube. A typical Geiger counter is filled with an inert gas, such as argon, to a pressure of about 10 cm Hg; a small amount (0.1 per cent) of halogen gas, such as bromine, is added to quench the discharge in the tube quickly after it has been registered. The duration of a pulse is around 10^{-6} sec.

We see from Figure 8-6 that when voltages greater than those in region D, the Geiger-Müller region, are applied to a gas-filled chamber, the gas is continuously conducting. This occurs because spontaneous electrical discharge can take place between the electrodes without any ionizing radiation entering the chamber.

8-6 Scintillation detectors The operation of a scintillation detector depends on the fact that certain materials, called *phosphors*, emit visible light when they are struck by particles or irradiated by ultraviolet light or x-rays. When a particle collides inelastically with a phosphor, it excites an electron to a higher-lying energy state. A phosphor has the distinctive property of remaining in an excited state for a much longer period of time (10^{-7} sec or greater) than an ordinary material (10^{-8} sec); the de-excitation of the phosphor by the emission of visible-light photons is delayed because the downward transitions do not follow the usual selection rules, and the transition is, to some degree, "forbidden."

A familiar example of scintillation, or phosphorescence, is found in the cathode-ray oscilloscope or television picture tube, in which high-speed electrons strike a phosphor and thereby excite the emission of visible radiation. One of the earliest means of detecting the presence of α-particles was through their scintillation effects on the phosphor, zinc sulphide. The Rutherford scattering experiments employed a zinc-sulfide screen viewed by a microscope as an α-particle detector (see Section 5-1). The tedium and relative insensitivity of this early scintillation detecting method, in which the scintillations are counted by the direct visual observation of bright light flashes, are eliminated in modern scintillation detectors which employ a remarkable electronic tube, the *photomultiplier*.

Consider the schematic diagram of a scintillation detector and photomultiplier tube shown in Figure 8-7. The phosphor may be such a material as sodium iodide (with small traces of thallium), or an organic substance, such as anthracene; these materials are transparent to visible light. The phosphors produce light flashes after their atoms are excited by collisions with particles or photons. The phosphor is sealed in a light-tight envelope, and photons reach the photo-cathode of the photomultiplier tube, possibly after internal reflections within the phosphor.

A photon striking the photomultiplier cathode undergoes a photoelectric collision, one electron being dislodged from the cathode surface. This

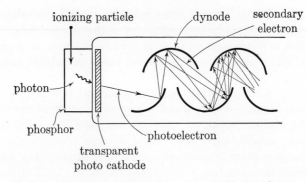

Figure 8-7. Schematic diagram of a scintillation detector and photomultiplier tube.

photoelectron is accelerated by a potential difference of about 100 volts to the first *dynode* of the photomultiplier tube. When the single electron, now with kinetic energy of at least 100 ev, collides with the surface of the dynode, secondary electron emission occurs, with two or more electrons released from the surface by the kinetic energy they gain from the initial electron. These secondary electrons are then accelerated to the second dynode, through another 100-volt potential difference. At this surface, electron multiplication by secondary emission again occurs. A typical photomultiplier tube has ten dynodes, or ten stages of electron amplification. The original single photoelectron can produce, at the final dynode, a pulse of current due to the arrival of up to 10^6 electrons; this pulse can be readily measured.

A particularly important feature of the scintillation detector is that the voltage pulse emerging from the photomultiplier tube is closely proportional to the energy of the particle or photon which initiates the original scintillation in the phosphor. Therefore, particles can not only be detected, but also their energies can be measured, which is not possible in G-M counters. Scintillation detectors are superior to Geiger tubes in other respects. The duration of a pulse may be as short as 10^{-9} sec; thus, the scintillation detector is capable of handling very high counting rates. Furthermore, the efficiency for counting γ-rays is much higher than that of Geiger counters (20 versus 2 per cent).

8-7   Radiation dose   A common unit for measuring the ionization effects of x-rays and, by extension, other radiation is the *roentgen unit*, r. An x-ray beam has a *radiation dose* of 1 r if it creates 2.08×10^9 ion pairs in 1 cm^3 of air under standard conditions (0°C, 1 atmosphere). Because *any* ionizing radiation loses close to 32 ev in producing a single ion pair in *any* gas, 1 roentgen corresponds to an energy absorption of $32 \times 2 \times 10^9$ ev, or 64 Bev, by the 1 cm^3 of ionized gas.

The roentgen is a useful radiation unit because it measures only the energy absorbed in a material, not the energy that passes through; therefore, the roentgen is a measure of disruptive ionization events, but *not* of the intensity of the ionizing radiation. For this reason, the roentgen is an appropriate unit for measuring radiation dose; the damage to biological systems by nuclear radiation arises from ionization and disruption of molecules and cells. (Radiation passing unabsorbed through living matter produces no ill effects.)

The presently accepted tolerance rate, or radiation dose, for humans is 0.3 r, or 300 milliroentgen (mr), per week. The radiation dose is readily measured with ionization chambers which register the total number of ion pairs produced in a known volume of gas.

8-8   Velocity-, momentum-, and mass-measuring devices   It is possible to measure such characteristics of a particle as its identity and energy from its absorption in materials or from its ionization effects in detecting instruments. All such measurements are, however, of rather limited precision. In this section we shall discuss the physical principles of instruments with which the velocity, momentum, mass, and energy of charged particles can be measured with very high precision.

A charged particle can be appreciably influenced in its motion through a vacuum only by electric and magnetic forces arising from external electric and magnetic fields. All of the velocity-, momentum-, and mass-measuring devices will involve merely the use of electric and magnetic fields, singly or in combination, to determine the path of a charged particle. Each instrument will consist of three parts: a source of charged particles, a region in which electric or magnetic fields act on the particles, and a detector for registering their arrival. What is done in every instrument is to set up, so to speak, an obstacle course for the charged particles in such a way that, if the particles succeed in moving from the source to the detector, one can infer, from a knowledge of the electric or magnetic fields acting on the particle, some quantity of interest, such as the velocity of the particle.

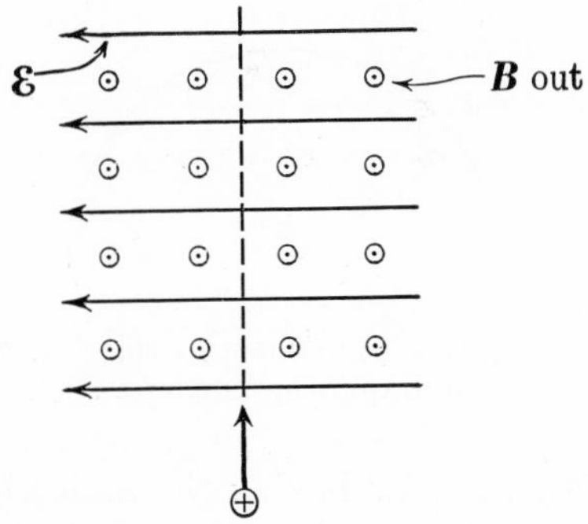

Figure 8-8.   A velocity selector, consisting of crossed electric and magnetic fields.

THE VELOCITY SELECTOR   Let us consider first a velocity selector. A narrow beam of charged particles is projected into a region of space where a uniform electric field of intensity ε (see Figure 8-8) acts to the left, and simultaneously a uniform magnetic field of flux density B is applied in the direction out of the paper. The incident beam is composed of particles having a variety of masses, charges, and velocities. A particle of mass m and charge $+q$ which enters the region of the *crossed* electric and magnetic fields at right angles to both $\mathcal{E}$ and $\boldsymbol{B}$, will be acted on by an electric force $q\mathcal{E}$ to the left and a magnetic force qvB to the right, where v is the speed of the particle. If the particle is to travel through the selector undeflected, the net force on it must be zero; that is,

$$q\mathcal{E} = qvB$$

$$\boxed{v = \mathcal{E}/B} \qquad [8\text{-}1]$$

Thus, only those particles moving with velocities equal to the ratio $(\mathcal{E}/B)$ will emerge from the selector without having been deflected to the left or right. When $\mathcal{E}$ and $\boldsymbol{B}$ are known, so too are the velocities of particles in the emerging beam; this device, which employs crossed electric and magnetic fields, is known as a *velocity selector*. Notice that *all* particles having the same velocity, despite differences in their masses or (positive) charges, pass through undeviated. This method of velocity measurement was used by J. J. Thomson in his cathode-ray experiments (1897) to measure the speed of electrons.

THE MOMENTUM SELECTOR   We consider now a device for measuring the momentum of a charged particle. For this instrument, only a magnetic field is required. This field is directed into the paper in Figure 8-9, and (negatively) charged particles move at right angles to the magnetic lines of force. The magnetic force acts at right angles to the velocity, deflecting the particle into a circular path of radius r, where

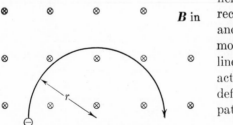

B in

$$qvB = mv^2/r$$

$$\boxed{mv = qBr} \qquad [8\text{-}2]$$

Figure 8-9. A momentum selector, consisting of a uniform magnetic field.

All particles having the same charge q and momentum mv will move in paths having the same radius of curvature. It should be recalled that the mass m appearing in Equation 8-2 is the relativistic mass and, therefore, mv is the relativistic momentum (see Section 2-8). The momentum selector does not distinguish mass or velocity separately; it merely selects from the initial beam, those particles whose momentum follows Equation 8-2. The quantity Br, to which the relativistic momentum is proportional for particles of a given charge q, is sometimes called the magnetic rigidity.

THE BETA-RAY SPECTROMETER   A particularly useful form of the momentum selector is the *beta-ray spectrometer*. This instrument is used to measure the momentum of electrons emitted from nuclei with energies up to a few Mev; or, it can be used to measure the momentum of electrons released by the photoelectric or Compton collisions of x- and γ-rays. The value of e/m_0, the ratio of the electric charge to rest mass, as well as the rest mass m_0, is well established for electrons. Therefore, if a momentum selector gives the value of the momentum, $mv = eBr$, the total relativistic energy E, and the relativistic kinetic energy $K = E - m_0c^2 = E - E_0$, can be directly computed using

$$E^2 = (K + E_0)^2 = (pc)^2 + E_0^2 = (mv)^2c^2 + E_0^2 = (eBr)^2c^2 + E_0^2 \qquad [8\text{-}3]$$

following Equation 2-45 in Section 2-8 (also see Problem 2-31).

It is worthwhile considering why the kinetic energy of a high-energy electron is most easily determined by measuring the momentum of the electron. In principle, one could measure the energy of 1 Mev electrons by finding that the electrons were brought to rest by a retarding potential of one million volts. This is, of course, practically impossible, because of insuperable difficulties encountered in executing such an experiment; and one must resort, instead, to the indirect determination of the energy of a very-high-speed particle by a measurement of its momentum.

A simple arrangement of a beta-ray spectrometer is shown in Figure 8-10. A beam of poly-energetic electrons from a source passes through

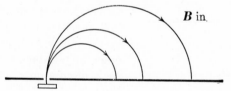

Figure 8-10. Simple form of β-ray spectrometer with a fixed magnetic field.

collimating slits to enter a uniform magnetic field acting into the paper. Electrons will fall, after being turned through 180°, on a photographic plate; the distribution of the momenta, and therefore the energies, is related to the positions of electrons along the plate. In the β-ray spectrometer shown in Figure 8-10, the magnetic flux density is maintained constant, and electrons having different momenta move in paths of different radii.

Another common design of a β-ray spectrometer is that in which the radius is fixed and the magnetic flux density is adjusted to bring electrons of any particular momentum to the detector, as shown in Figure 8-11. An exit slit, through which electrons reach a detector (such as a Geiger

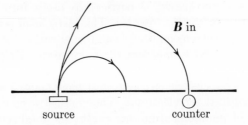

Figure 8-11. Simple form of a β-ray spectrometer with a variable magnetic field.

counter) is at a fixed distance, $2r$, from the entrance slit. We see from Equation 8-2 that the momentum is directly proportional to the flux density B for a fixed radius; and, if the magnetic field is supplied by an electric current through a coil or solenoid, the momentum of the electron is proportional to the current. In this fashion it is possible to measure the distribution in energy, or energy spectrum (hence β-ray *spectrometer*), of the β-particles, the number of electrons passing through the exit slit being proportional to the counting-rate of the detector.

THE MASS SPECTROMETER   A device with which one can measure the masses of ionized atoms is the *mass spectrometer*. We shall see that in nuclear physics it is of great importance to know the masses of atoms to an accuracy of a few parts in 10^5; such high precision can be achieved in mass spectrometry. Although mass spectrometers can take on a variety of forms, we shall discuss only one of the simpler types.

Inasmuch as the linear momentum p is equal to mv, it is obvious that a mass selector, or spectrometer, can be constructed by combining a velocity selector with a momentum selector. Consider the device shown schematically in Figure 8-12. Ions from a source pass through a slit S_1 and are accelerated through a potential difference V. After passing through slit S_2,

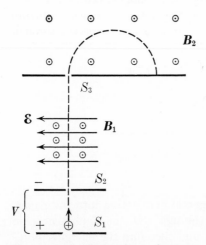

Figure 8-12. Simple form of a mass spectrometer, consisting of a velocity selector followed by a momentum selector.

the ions enter a velocity selector. Only those ions moving with a velocity $(\mathcal{E}/B_1)$ emerge through slit S_3, where $\mathcal{E}$ is the intensity of the uniform electric field between the vertical plates, and B_1 is the magnetic flux density of the uniform magnetic field (out of the paper), which is confined to the region of the velocity selector. The surviving ions leaving S_3 enter a second uniform magnetic field of flux density B_2 directed out of the paper, and are deflected to move in a circle of radius r. We have, from Equations 8-1 and 8-2,

$$\frac{m}{q} = \frac{B_2 r}{v} \qquad [8\text{-}4]$$

where $v = \mathcal{E}/B_1$. The mass-to-charge ratio, m/q, can be computed directly from Equation 8-4. If the charge of the ion is known (for a singly-ionized atom, $q = e$), the mass itself can be evaluated. The mass m is directly proportional to the radius r. The mass of the ions is directly measured; but, if one corrects for the deficiency of electrons, the mass of the neutral atom can be determined. When ions of various masses fall on a photographic plate (*mass spectrograph*), the mass spectrum of the ions is recorded. Alternatively, if the ions are collected in a detector located behind a slit at a fixed distance, $2r$, from the entrance slit S_3, a plot of collector current against the variable flux density B_2 yields the mass spectrum.

There are many different types of mass spectrometers. Although differing in design features, they all employ an electric field and a magnetic field, either simultaneously or in succession. The first mass spectrometer was developed by J. J. Thomson in 1912; in this spectrometer the electric and magnetic fields acted along the same direction in the region through which the ions were sent. Thomson found, by using this mass spectrometer, that any given chemical element may consist of atoms having several discrete values of the atomic mass. Such atoms, which have the same atomic number Z and are, therefore, chemically indistinguishable, but which have different values of the atomic mass, are known as *isotopes*.

As an example, consider the element chlorine, $_{17}Cl$, whose chemical atomic weight, as found in nature, is 35.457. Mass spectrometry shows that chlorine has two isotopes with the atomic masses, 34.980 and 36.978 amu. By definition, the mass of one neutral atom of the oxygen isotope O^{16} is 16.0000 amu; we recall from Section 2-10 that 1 amu = 1.660×10^{-27} kg. Natural chlorine, a mixture of the two isotopes, has a mass of 35.457, not

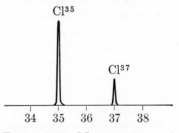

Figure 8-13. Mass spectrum of natural atomic chlorine.

at all close to an integer, whereas the two separate isotopes, Cl^{35} and Cl^{37}, with relative abundances of 75.53 and 24.47 per cent respectively, have atomic masses which are very close to the integers, 35 and 37, the so-called *mass numbers*. The mass spectrum of chlorine, as measured by a mass spectrometer, is shown in Figure 8-13. Indeed, any chemical element is found to consist of one or more isotopes, whose masses in amu are very close, but not equal, to integers. These slight departures of the atomic masses from integral values yield valuable information on the structure of nuclei.

8-9   Why high-energy accelerators?   Our understanding of the structure of matter has always advanced hand-in-hand with the development of machines capable of accelerating electrons and ions to increasingly higher energies. We have already seen that the bombardment of atoms by electrons accelerated to several electron-volts can excite or ionize the outer, weakly-bound electrons of the atom and induce the optical photon emission. From such experiments the excitation and ionization energies of atoms can be determined, and the outer-electron structure of the atom deduced. When atoms are bombarded with electrons accelerated to energies of 10^3 or 10^4 ev, the innermost, tightly-bound electrons of atoms can be dislodged and x-ray photon emission can be induced; from these experiments the inner-electron structure of the atom can be determined. In all such experiments the nucleus of the atom behaves as a positively-charged, but otherwise inert, mass point having no internal structure. This does not mean, of course, that the atomic nucleus is truly a simple point charge and mass point, but rather that the bombardment of the atom by particles having energies no greater than several Kev produces no perceptible changes in any internal structure the nucleus may possess.

We shall see in Chapter 9 that the constituent particles of the nucleus are bound together with energies of several million electron volts. Therefore, if the internal structure and arrangement of the nuclear constituents is to be altered, so that the structure of the nucleus may be studied, the nucleus must be given energies of the order of Mev, *millions* of electron volts. The most direct means of altering the nuclear structure is to bombard targets containing atoms (and therefore also nuclei) with particles which have been accelerated to very high energies. Progress in nuclear physics has, therefore, depended upon the invention and design of machines which accelerate charged particles to energies measured in Mev, or even Bev (billion ev).

An ideal accelerating machine produces a beam of charged particles with a well-defined high energy and with a high beam intensity, or large number of particles. The beam energy must be high, because only then can the nuclear structure be appreciably changed when the accelerated particles collide with target nuclei; the beam intensity should, ideally, be high because the probability for a collision between an incoming particle and a target nucleus is very low by virtue of the extremely small target area.

All charged-particle accelerators are based on the fact that a charged particle has its energy changed when it is acted on by an *electric* field. A *constant* magnetic field does *no* work on a moving charged particle and cannot change its energy; but a constant magnetic field can cause a particle to move in a circular orbit at a constant speed, and thereby influence its path. On the other hand, a *changing* magnetic field produces an electric field, which can, in turn, be used to accelerate a charged particle. Therefore, all high-energy accelerators change the energy of charged particles by subjecting them to an *electric field* or a *changing magnetic field*.

Before we turn to descriptions of the several types of accelerators, we point out two formidable technical problems that must be solved in the design of any high-energy accelerator. These are the maintenance of a very high *vacuum* in the interior of the machine, and the *focusing* of the beam of accelerated particles by electric or magnetic fields. A high vacuum reduces the probability of collisions with gas molecules and the consequent loss from the useful beam intensity. Focusing insures that those accelerated particles which deviate slightly from the desired path (which can be many miles from the ion source to the target) will be returned to this path, and thus kept in the useful beam. Although high-vacuum and focusing problems are crucial in the design of all accelerators, we shall concern ourselves only with the basic principles underlying the acceleration of particles.

There are two general classes of charged-particle accelerators: *linear accelerators*, in which the charged particles move along a straight line; and *cyclic accelerators*, in which the charged particles move in curved paths and are recycled.

8-10   Linear accelerators   A number of relatively low-energy (no greater than 1 Mev) accelerating machines utilize such conventional devices as step-up transformers, or capacitors charged in parallel and then connected in series, to achieve high voltages. All such machines are, however, ultimately limited—by electrical discharge—to approximately one million volts.

THE VAN DE GRAAFF GENERATOR   The most successful machine for accelerating charged particles along a straight line by applying a *single*, large potential difference is the *Van de Graaff electrostatic generator*. This machine, invented by R. J. Van de Graaff in 1931, was one of the first to accelerate charged particles to energies greater than 1 Mev. Many Van de Graaff accelerators are in current operation, and although these machines are restricted to relatively low energies (10 Mev or less), their chief virtues are a large beam intensity (a few milliamperes) and a precisely controlled energy (0.1 per cent).

The physical principle on which this machine is based is that electric charge placed within a metal conductor must always move to the surface, irrespective of the quantity of charge already residing on this surface. The essential parts of a Van de Graaff accelerator are shown in Figure 8-14. An insulated fabric belt is stretched over two rollers, B and T, at the bottom and top respectively; roller B is motor-driven and roller T is contained

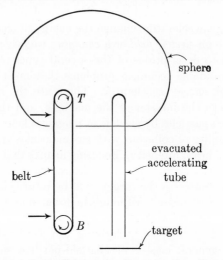

Figure 8-14. Schematic diagram of a Van de Graaff accelerator.

within a hollow metal sphere S, supported on an insulated column. A large electric field in the vicinity of the sharp needle points, close to the belt at the bottom, produces ionization, and positive ions are collected on the belt. These charged ions are carried upward and into the hollow metal sphere, where they are readily removed by a second set of needle points connected to the metal surface by a conductor. Any charge collected on the upper needle points moves immediately to the sphere's outer surface.

Positive charge is transported from the base to the sphere, and the sphere's potential increases until finally the loss of charge from the sphere, because of insulation leakage and corona discharge through the air, equals the gain in charge from the swiftly moving belt. Losses from the upper conductor are minimized by the spherical shape and also by enclosing the generator in a shell containing a gas under very high pressure. The ions to be accelerated are produced by an ion source within the sphere and at the potential of the sphere; these ions are accelerated downward through an evacuated tube to strike a target at ground potential.

If a proton, deuteron, or electron (each particle having a single unit of electric charge e) is accelerated from rest by a potential difference of, say, 5×10^6 volt, its final energy is 5 Mev, although the velocities will differ because of the difference in mass. An α-particle accelerated by 5×10^6

volt will, however, gain an energy of 10 Mev because it carries a double electron charge.

The Van de Graaff accelerator can produce a continuous, high-intensity beam of positive ions, or of electrons, with energies as high as 10 Mev. It can serve as an x-ray generator when high-energy electrons are brought to rest at the target. The voltage difference in this accelerator can be precisely controlled by adjusting the leakage of charge from the sphere. The output energy of the Van de Graaff accelerator is limited finally by the unavoidable charge leakage.

The Van de Graaff generator is the most commonly used D.C. machine, in which particles are accelerated with a *single* application of a potential difference. All accelerators operating at energies above 10 Mev must accelerate and energize charged particles by *multiple* applications of an electric field.

THE DRIFT-TUBE LINEAR ACCELERATOR   There are two general types of linear accelerators in which charged particles are multiply accelerated along a straight line: the *drift-tube accelerator* (R. Wideröe, 1929) and the *waveguide accelerator* (D. W. Fry, 1947). In the drift-tube type, the particles are subjected to an electric field between insulated conductors; and in the waveguide type, the particles are accelerated by the electric field of very-high-frequency electromagnetic waves guided through a hollow conductor. We will discuss only the first type.

Charged ions (α-particles, protons, deuterons) or electrons from a relatively low-energy (1 to 5 Mev) accelerator enter a long, straight, evacuated tube within which are a number of hollow conducting cylinders of increasing length (Figure 8-15). These cylinders are connected alternately to the

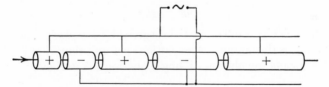

Figure 8-15. Simple form of a drift-tube linear accelerator.

opposite terminals of a radio-frequency generator, and thus a sinusoidal alternating electric field exists in the region between any two adjacent cylinders. Positive ions drift through the first cylinder at a constant velocity. If these ions pass into the gap between the first two cylinders during that part of the cycle for which the second cylinder is negative with respect to the first cylinder, the ions will be accelerated to the right by the electric field in the region between the cylinders. The ions then enter and pass through the second cylinder at a constant, but higher velocity. The length of the second cylinder is so chosen that, when the ions emerge from it, the polarity of the cylinders will have reversed, that is, the time of travel through the cylinder is exactly one-half cycle; again, the electric field accelerates the particles to the right.

Of course, whenever a charged particle is in the interior of a cylinder, it is shielded from the electric field. Because the frequency of the alternating voltage applied to the drift tube is constant, it is necessary that the ions spend the same period of time drifting through each tube, so that the particles may arrive at the spaces between the tubes at just the right time to be still further accelerated. To achieve this, the drift tubes are made progressively longer.

The final energy of the particles when they strike the target is dependent, of course, on the over-all length of the accelerator and on the energy gained at each gap. The beam striking the target consists of pulses of particles; the number of such pulses arriving at the target during each second is equal to the frequency of the alternating voltage applied to the drift tubes. The highest particle energies, 68 Mev (total length, 120 ft) for protons and 700 Mev (total length, 260 ft) for electrons, have been attained with linear accelerators operating on the waveguide principle.

8-11 Cyclic accelerators A whole class of particle accelerators, sometimes known as cyclic accelerators, apply *multiple* accelerations to charged particles confined to move in circular orbits by a magnetic field. A few general remarks can be made which apply to all cyclic accelerators. Consider a charged particle of mass m, charge q, moving at right angles to the magnetic lines of force of flux density B, with a speed v, and in a circle of radius r. From Equation 8-2,

$$mv = qBr \qquad [8\text{-}5]$$

The angular velocity ω of the particle is given by $\omega = 2\pi f = 2\pi/T$, where f, the frequency, is the number of revolutions per unit time, and T, the period, is the time for one revolution. Using Equation 8-5 we have

$$\omega = \frac{v}{r} = (q/m)B$$

$$T = 2\pi/\omega = 2\pi \frac{(m/q)}{B} \qquad [8\text{-}6]$$

Equation 8-6 shows that the period T depends only on the mass-to-charge ratio (m/q) and on the flux density B; we notice that the period is *independent* of the particle's speed v and path radius r, provided the particle is nonrelativistic and the mass m is independent of the particle's speed.

THE CYCLOTRON The first and simplest cyclic machine is the *cyclotron*, invented by E. O. Lawrence and M. S. Livingston in 1932. In the cyclo-

tron, a charged particle is subject to a *constant* magnetic field, which bends
the particle into a circle, the particle being accelerated each half-cycle by
an electric field.

In this machine, positive ions, such as protons, deuterons, and α-par-
ticles, are injected into the central region (point C in Figure 8-16) between
two flat D-shaped hollow metal conductors (called "dees"). An alternating

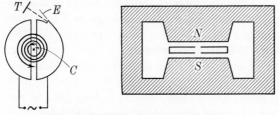

Figure 8-16 Cyclotron accelerator.

high-frequency voltage is applied to the dees, thereby producing an alternating electric field in the region between them. During the time when the
left dee is positive and the right dee is negative, the injected positive ions
are accelerated to the right by the electric field between the dees. Upon
entering the interior of the right dee, these ions are electrically shielded
from any electric field, and will therefore move in a semi-circle at a constant speed under the influence of the constant magnetic field. When ions
emerge from the right dee they will be further accelerated across the gap
if the left dee is now negative. This requires that the frequency of the alternating voltage applied to the dees be equal to the orbital frequency of
the ions. From Equation 8-6 it follows that the applied frequency must be

$$f = \frac{qB}{2\pi m} \qquad [8\text{-}7]$$

During each acceleration, the ions gain energy, move at a higher speed,
and travel in semi-circles of larger radii. As the ions spiral outward in the
dees they remain synchronized with, and in resonance with, the a-c source
of constant frequency, inasmuch as the time for an ion to move through
180° is independent of the ion's speed or radius, provided the ion mass m in
Equation 8-7 is constant. When the accelerated particles reach the circumference of the dees, they are deflected by the electric field of an ejector
plate E to strike the target T.

The final kinetic energy K of the particles is

$$K = \tfrac{1}{2} m v_{\max}^2 = \tfrac{1}{2} m \left(\frac{qBr_{\max}}{m} \right)^2$$

$$K = \frac{(q^2 B^2 r_{\max}^2)}{2m} \qquad [8\text{-}8]$$

We see from Equation 8-8 that the final energy of the particle depends on the square of the radius of the cyclotron dees and on the square of the magnetic flux density. To achieve the highest possible energies, B and r_{max} are made as large as possible. When the maximum attainable flux density ($\sim$ 2 weber/m^2) is used, the frequency f, as determined by Equation 8-7, is in the region of several megacycles per second (radio frequencies) for positive ions. The diameter of the dees, which is also the diameter of the electromagnet pole-faces, may be as large as 8 feet; such an electromagnet is enormous (400 tons of iron) and expensive. A typical alternating voltage across the dees is 200 kilovolts.

Heavy particles, such as protons, deuterons, and α-particles can be accelerated with a cyclotron to energies of 25 Mev. For all these particles the kinetic energy is much less than the rest energy (proton rest energy $\sim$ 1 Bev). Therefore, the mass does not increase appreciably, and the particles can remain in synchronism with the alternating voltage to energies of about 12 Mev for protons, and 25 Mev for deuterons. Because electrons can easily be accelerated to relativistic speeds (rest energy only $\frac{1}{2}$ Mev), they cannot be synchronized with the applied voltage and, therefore, cannot be accelerated to high energies by a cyclotron.

THE SYNCHROCYCLOTRON   The maximum energy obtainable with the ordinary fixed-frequency cyclotron is limited by the relativistic mass change. The orbital frequency of an accelerated particle, $qB/2\pi m$, Equation 8-7, decreases as the relativistic mass m increases. Therefore, as the particles spiral outward they fall increasingly behind the applied frequency, finally arriving at the gap between the dees so late that they are no longer accelerated by the electric field. This difficulty is overcome in the *synchrocyclotron*.

In this machine, the applied frequency is continuously *decreased* as the particles move to larger radii in such a way as to compensate for the increasing mass. The particles then remain in synchronism with the applied voltage and are accelerated to much higher energies than obtainable with a fixed-frequency cyclotron. Because the frequency is changed as the ions are accelerated and move in increasingly larger radii in the synchrocyclotron, this machine is sometimes known as an FM, or frequency-modulated, cyclotron.

A photograph of a 184-inch (dee diameter) synchrocyclotron, capable of accelerating protons to 720 Mev, is shown in Figure 8-18. The weight of the magnet alone in this machine is 4000 tons!

Although a synchrocyclotron accelerates particles to much higher energies than a cyclotron, the output beam current is much less (1 versus 10^3 microampere). This is so because only one pulse of particles can be accelerated in the machine at one time. Theoretically, there is no limit on the size, and therefore the energy, of a synchrocyclotron. However, it becomes

Figure 8-17. A 60-inch cyclotron which accelerates deuterons to 20 Mev. (Courtesy of Brookhaven National Laboratory).

Figure 8-18. The 184-inch synchro-cyclotron, which accelerates protons to 720 Mev. The lower pole of the cyclotron magnet is below floor surface and not visible. (Courtesy of the University of California Lawrence Radiation Laboratory.)

economically prohibitive to build such machines for accelerating particles to more than 1 Bev.

THE BETATRON Because of the appreciable mass increase of electrons at fairly low energies (10 per cent mass increase for 50 Kev electrons), cyclotrons and synchrocyclotrons cannot be used to produce high-energy electron beams. The most successful machine for accelerating electrons to energies of several Mev, that is to energies comparable to or exceeding the energies of β-particles emitted from nuclei, is the *betatron*, invented by D. W. Kerst in 1941.

The fundamental difference between the betatron and the other accelerators we have discussed is that, in the betatron, the charged particles are accelerated by an electric field produced, not by a potential difference, but by a changing magnetic field. In the betatron, electrons are maintained in a circular orbit by a magnetic field, and at the same time these electrons are given energy by an induced emf resulting from an increase in this magnetic field.

A schematic representation of the essential parts of a betatron are shown in Figure 8-19. The alternating magnetic field is produced by an electro-

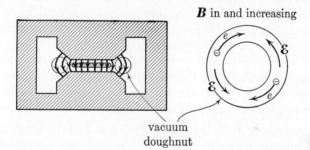

B in and increasing

vacuum
doughnut

Figure 8-19. Schematic diagram of a betatron accelerator.

magnet supplied by 60 cps alternating current. A pulse of electrons, having kinetic energies of about 50 Kev, are injected into the doughnut-shaped evacuated tube during that part of the cycle when the magnetic field is increasing (B is increasing into the paper in Figure 8-19). The increasing magnetic field induces an emf in the region of the doughnut and thus accelerates the electrons by electric lines of force which circle the doughnut. The betatron is, in effect, a transformer in which the secondary circuit is he electron beam.

In order that the electrons stay in a stable orbit with a *fixed* radius r, it s necessary that the magnetic field **B** at the electrons' orbit increase proportionately with the relativistic momentum mv, since from Equation 8-5,

$$r = mv/eB \qquad [8-9]$$

Therefore, by a proper design of the magnetic field and its variation in time, the electrons can gain momentum and kinetic energy continuously as they move in a circle of fixed radius.

The electrons must be ejected from the betatron when the magnetic field reaches its maximum value; otherwise, the electrons would be decelerated as the magnetic field decreases and the induced emf reverses direction. The electrons are deflected from their stable orbit by sending a pulse of current through an auxiliary coil. The high-energy beam of electrons can either be made to strike a target within the tube, thus producing an intense x-ray beam, or the electron beam can be removed through a "window."

Electron energies up to 340 Mev have been attained with betatrons. Much higher energies cannot be efficiently obtained because the electromagnetic radiation (*Bremsstrahlung*) continuously emitted by the circling, and therefore accelerating, electrons becomes increasingly important. Inasmuch as particles in a linear accelerator do not suffer such great accelerations and radiation losses (they move in straight lines, not circles), electrons can be accelerated to energies up to 1 Bev in a linear accelerator. Strictly, of course, an electron has a speed that is very closely equal to the speed of light after its energy has been raised to several Mev; therefore, the machines that bring electrons to hundreds of Mev energy might more properly be described as "energizers," or "ponderators," since they serve chiefly to increase the energy, or mass, of the electrons, but not the speed. The betatron, together with the cyclotron and synchrocyclotron, requires a very large electromagnet, extending over the entire area of the orbit, whose size and cost ultimately limit the maximum attainable energy.

THE SYNCHROTRON   Both the synchrocyclotron and the betatron accelerators must produce magnetic fields which extend over the entire region of particle orbits. The size of the electromagnet becomes enormous as the energy of the particles, and therefore, the orbit size increases, and it is economically infeasible to build machines of these types when the particle energies approach 1 Bev. Therefore, the *synchrotron*, which combines features of both the betatron and the synchrocyclotron, and which requires magnetic fields only at the fixed radius in which the particle moves, has been developed for accelerating (or energizing) particles to energies greater than 1 Bev. Our understanding of the fundamental particles of physics and their interactions at these very high energies has been profoundly affected by the development of synchrotron accelerating machines.

The principle of operation of the synchrotron is as follows: particles are confined, after injection at a relatively high energy, to an orbit of fixed radius by means of a magnetic field. Unlike the betatron or cyclotron, the magnetic field at the orbit is supplied by a ring-shaped magnet surrounding only the doughnut-shaped tube in which the particles move, and producing no appreciable field at the interior of the chosen orbit; therefore, far less iron is required for the electromagnet. Energy is supplied to the particles by a variable radio-frequency voltage source maintained in resonance with the frequency of the particles as they move in their orbit. The particles are injected into the synchrotron from a lower-energy accelerator, such as a Van de Graaff generator.

We will discuss the electron-synchrotrons and proton-synchrotrons separately, inasmuch as it is possible to inject electrons into the synchrotron at speeds very close to that of light, whereas protons are injected at much smaller speeds. In the electron-synchrotron of Figure 8-20, electrons enter the synchrotron orbit at point A with energies of about 2 Mev. The

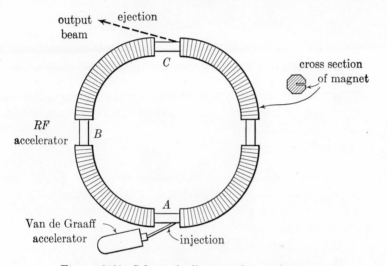

Figure 8-20. Schematic diagram of a synchrotron.

magnetic field maintains these highly-relativistic electrons in the fixed orbit of radius r; the frequency of the electrons in their orbital motion is $f = eB/2\pi m$. The 2 Mev electrons are injected at a speed of $0.98c$; therefore, during the "acceleration" within the synchrotron, the electron speed is closely c, and the frequency f is constant. If B is increased with time to compensate for the increase in the relativistic mass m, the frequency can be kept in resonance, or synchronism, with a constant-frequency oscillator. This oscillator at the point B gives energy to the electrons once each revolution. After making many revolutions, the electrons are ejected at point C with energies as high as 1 Bev. Radiation losses arising from the acceleration of the electrons will probably limit electron energies in a synchrotron to about 5 Bev.

The operation of the proton-synchrotron can also be explained in terms of Figure 8-20. A Van de Graaff generator injects 4 Mev protons at point

Figure 8-21. The Cosmotron, the 3-Bev proton synchrotron at Brookhaven National Laboratory. The Van de Graaff generator (left foreground) injects 3 Mev protons into the synchrotron; the magnet has an inside diameter of over 60 feet and weighs 2200 tons. (Courtesy Brookhaven National Laboratory.)

A. Unlike electrons of this energy, whose speed is close to c, 4 Mev protons are non-relativistic. Therefore, it is necessary to use a frequency-modulated oscillator which remains in resonance with the increasing frequency, $f = v/2\pi r$, of the proton's orbital motion until the protons become highly relativistic. Protons entering a synchrotron of 30-foot radius at 4 Mev can be accelerated to a final energy of 3 Bev, as the frequency of the accelerating field goes from 300 kilocycles/sec to 4.18 megacycles/sec. A photograph of a machine of this type, the Cosmotron at the Brookhaven National Laboratory, is shown in Figure 8-21.

8-12 Summary Nuclear detectors are devices which are sensitive to the passage of nuclear radiation (charged particles and photons) through the detector. *All* charged particles lose approximately 32 ev energy in creating a single ion pair in *any* gas. Nuclear radiation has the following general properties: it is more readily absorbed in a solid or liquid than in a gas; the degree of penetration *decreases*, for a given energy, in the following order, γ-rays, β-rays, and α-rays; and the range of charged particles increases with increasing energy.

DETECTORS OF NUCLEAR RADIATION

TYPE	REACTING MEDIUM	SPECIAL CHARACTERISTICS
Cloud chamber	*Supersaturated vapor*	*Particle tracks*
Bubble chamber	*Superheated liquid*	*Particle tracks (high density of absorber)*
Nuclear emulsion	*Photographic emulsion*	*Particle tracks (continuously sensitive)*
Ionization chamber	*Gas*	*Ionization proportional to radiation intensity*
Proportional counter	*Gas*	*Pulse size proportional to radiation intensity*
Geiger counter	*Gas*	*Same size pulse initiated by any type of ionizing radiation*
Scintillation counter	*Solid (or liquid)*	*Very short resolution time*

Radiation dose is measured in roentgen units: 1 roentgen unit (r) is that amount of radiation which creates 2.08×10^9 ion pairs in 1 cm³ of air (STP).

Measuring devices and selectors: (1) Velocity selector—crossed electric and magnetic fields, $v = \mathcal{E}/B$. (2) Momentum selector—uniform magnetic field, $mv = qBr$; an example of a momentum selector is the beta-ray spectrometer. (3) Mass spectrometer—at least one electric and one magnetic field; in its simplest form, a combination of a momentum and a velocity selector.

All charged-particle accelerators operate on the following principle: a charged particle can gain energy when it is accelerated by an electric field or by a changing magnetic field. Cyclic accelerators utilize a magnetic field to control the path of the charged particles so that the particles may gain energy repeatedly upon being recycled.

PARTICLE ACCELERATORS Listed below are the various types of high-energy particle accelerators. The nomenclature of the second column is: ϵ = electrons, p = protons, d = deuterons, and α = α-particles.

	PARTICLES ACCELERATED	MAXIMUM ENERGY (MEV)	AVAILABLE BEAM CURRENT (MICROAMPERES)
Linear Accelerators			
Voltage-multiplier	p, d, α	1	~1000
Van de Graaff	e, p, d, α	10	~1000
Waveguide	e, p, d, α and heavy ions up to Ar^{40}	1000	~1
Cyclic Accelerators			
Cyclotron	p, d, α	25	~1000
Synchrocyclotron	p, d, α	500	~1
Synchrotron	e, p	8000	~0.001
Betatron	e	340	~1

REFERENCES

Green, A. E. S., *Nuclear Physics*. New York: McGraw-Hill Book Company Inc., 1955. In Chapters 3 and 4 the problems of focusing, both for magnetic spectrometers and for particle accelerators, are treated. The design characteristics of a few large particle-accelerators are given in tabular form in Chapter 3.

Livingston, M. S., *High-Energy Accelerators*. New York: Interscience Publishers, Inc., 1954. This 152-page book discusses design problems that arise in particle accelerators and the basic equations governing the motion of high-energy particles.

Price, W. J., *Nuclear Radiation Detection*. New York: McGraw-Hill Book Company, Inc., 1958. The theory and fundamental basis for detecting nuclear radiation is discussed in Chapters 1 and 2; the remainder of this book is devoted to a detailed exposition of various detection instruments.

Segrè, E., ed., *Experimental Nuclear Physics*, Vol. I. New York: John Wiley & Sons, Inc., 1953. A comprehensive treatment of the instruments of nuclear physics.

Shamos, M. H., and G. M. Murphy, eds., *Recent Advances in Science*. New York: New York University Press, 1956. The chapter "Electro-nuclear Machines," by L. J. Haworth gives a qualitative review of the principal types of accelerating machines.

PROBLEMS

8-1 The range of a 5.00 Mev proton in air is 34 cm. Is the energy of such a proton which has passed through only 17 cm of air less or greater than 2.50 Mev?

8-2 The "thickness" of an absorber is often expressed in terms of the *areal density*, the product of the actual absorber thickness t and the absorber density ρ. It is found that the range of β-particles, given in terms of ρt, is essentially independent of the absorber material. If the range of 1.0 Mev β-particles in aluminum (density, 2.7 gm/cm³) is 420 milligrams/cm², (a) what thickness of aluminum stops 1.0 Mev electrons? (b) What thickness of wood (density, 0.6 gm/cm³) will stop 1.0 Mev electrons?

8-3 A certain G-M tube collects 10^7 electrons in each discharge. What is the average current in the tube when it registers 1000 counts/minute?

8-4 Compute the total electric charge of either sign collected in an ionization chamber if a 4.0 Mev α-particle loses its entire energy in the gas of the ionization chamber.

8-5 What is the average current in an ionization chamber resulting from the absorption of 200 5.0-Mev α-particles per minute?

8-6 Two hundred 4.2-Mev α-particles enter an ionization chamber each second. What is the value of a resistor that must be put in series with the ionization chamber to give a potential drop of 1.0 volt across the resistor?

8-7 * A Geiger tube is filled to a pressure of 10 cm Hg with a mixture of argon (90 per cent by weight) and ethyl alcohol (10 per cent). The volume of the tube is 40 cubic centimeters. If each pulse permanently dissociates 10^9 alcohol molecules, what is the useful life of the tube as measured by the total number of counts it can register. The molecular weights of argon and ethyl alcohol are 39.9 and 46.1, respectively.

8-8 * (a) Show that the electric field $\mathcal{E}$ between two concentric, oppositely charged cylinders at a distance r from the center is given by $\mathcal{E} = V/[r \ln (r_2/r_1)]$, where V is the potential difference between the cylinders, and r_1 and r_2 are the radii of the inner and outer cylinder respectively. (b) What is the force on an electron 1 mm from the central wire of a Geiger tube operated at 1000 volts and with a cathode of 2.0 cm diameter and a central wire of 0.005 cm?

8-9 A typical photomultiplier tube has 10 dynodes with a potential difference of 125 volts across each pair of adjoining dynodes, which are 8.0 mm apart. What is the total time elapsed between arrival of electrons at the first dynode and the appearance of a pulse at the last dynode, assuming that secondary electrons are released instantaneously and with negligible energy?

8-10 A cloud-chamber photograph reveals the path of a charged particle before and after its passage through a thin lead plate located 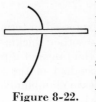 within the cloud chamber (see Figure 8-22). From the density of the droplets, it is established that the particle has the same mass as that of an electron. The radii of curvature of the particle above and below the lead plate are 7.0 cm and 10.0 cm, respectively. Assuming that the particle is traveling perpendicular to the constant magnetic field of flux density 1.0 weber/m² directed into the paper, (a) in which direction is the particle moving? (b) Is the particle a positron or an electron? (c) How much energy was lost by the particle in traversing the lead plate?

Figure 8-22.

8-11 What is the increase in the temperature of air exposed to a radiation dose of 1.0 roentgen for 1.0 hour? The specific heat of air is close to 1.0 joule/gm-°C, and its density is 1.3 kg/m³ at STP.

8-12 A *pocket dosimeter* is a small, well-insulated, air-filled electroscope, whose discharge by the ionizing effects of nuclear radiation can be used to measure radiation dosage. One such dosimeter, having a capacitance of 2.0 $\mu\mu f$ and an effective sensitive volume of .8 cm³, is charged initially to a potential difference of 150 volt. What potential difference will be read after the dosimeter has been exposed to a radiation dose of 100 milliroentgen?

8-13 * A certain radioactive material produces a radiation dose of 200 milliroentgen per minute at a distance of 3.0 ft from the source. If the emitted gamma rays have an energy of 1.33 Mev, for which energy the absorption coefficient in lead is 0.66 cm⁻¹, what thickness of lead absorber will permit a worker safely to spend 4.0 hours per week at a distance of 10 ft from the source?

8-14 Singly-charged ions having masses close to 14 and 15 amu are accelerated by a potential difference of 800 volts and then pass perpendicular to the lines of force of a uniform magnetic field of flux density 0.200 weber/m². What are the two radii of curvature?

8-15 What are the radii of curvature of (a) 1 Kev, (b) 1 Mev, and (c) 1 Bev protons in a magnetic field of flux density 1.0 weber/m²?

8-16 What are the radii of curvature of (a) 1 Kev, (b) 1 Mev, and (c) 1 Bev electrons in a magnetic field of flux density 1.0 weber/m²?

8-17   In a velocity selector, charged particles are sent through crossed electric and magnetic fields. What would be the motion of a charged particle sent through (a) two crossed electric fields and (b) two crossed magnetic fields?

8-18   A velocity selector employs a magnet that produces a flux density of 0.040 weber/m^2 and a parallel-plate capacitor with a plate separation of 0.010 m for the electric field. What potential difference must be applied to the capacitor in order to select charged particles having a speed of 4.00×10^6 m/sec?

8-19   A β-ray spectrograph bends electrons through 180° with a magnetic field of 0.0450 weber/m^2. What is the separation on the photographic plate of the lines produced by electrons having energies of 0.50 and 0.51 Mev?

8-20   A β-ray spectrometer records electrons which move in a radius of curvature of 0.15 m. What is the magnetic flux density required to measure β-particles having an energy of 4.0 Mev?

8-21   The collector in a mass spectrometer registers peak currents of 0.00138 and 0.00062 microamperes, respectively, when the two isotopes of copper, Cu^{63} and Cu^{65}, reach it. (a) What is the relative abundance of the two isotopes? (b) Compute the chemical atomic weight of copper.

8-22 * Singly-ionized mono-energetic atoms are deflected 5.00×10^{-4} m from their original path when they move through a parallel-plate capacitor (plates 0.0400 m long and separated by 0.00200 m) and the plates are charged to a 2500 volt potential difference. These same atoms are bent into a radius of curvature of 1.44 m if they pass into a uniform magnetic field of flux density 0.400 weber/m^2. (a) What is the charge-to-mass ratio of the ions? (b) Identify the atoms.

8-23 * A β-ray spectrometer is used to measure the energy of γ-ray photons by having the γ-rays strike a radiator in which photoelectric emission takes place. A source of radioactive cobalt, which emits γ-rays of 1.17 Mev and 1.33 Mev, is used with a lead radiator. The spectrometer detects electrons moving in a fixed radius of curvature when the electric current through a solenoid is changed. The photoelectric emission from lead involves the removal of a K electron, which is bound with an energy of 0.0891 Mev. If the spectrometer detects the photoelectron from the 1.17 Mev γ-ray when the current is set for 1.20 ampere, what is the current setting to detect the photoelectron from the 1.33 Mev γ-ray?

8-24   Protons of energy 1.2 Kev are sent through a mass spectrometer (see Figure 8-12). The magnetic flux density B_1 is 0.10 weber/m^2. By the proper adjustment of the electric field ε and the magnetic field B_2, the protons are made to fall on the photographic plate a distance of 0.08 m from the exit of the velocity selector. Find the values of ε and B_2 that will cause α-particles of the same energy as the protons to fall at the same place on the photographic plate.

8-25 * A narrow beam containing protons and deuterons, both of 1 Mev energy, is projected perpendicularly into a uniform magnetic field of flux density B. After being bent through 180°, the particles fall on a photographic plate. (a) Show that the separation on the plate of the two components is 1.2 $(K\, m_p)^{1/2}/eB$, where K is the kinetic energy of the particles and m_p is the proton mass, assumed to be one-half of the deuteron mass. (b) If $B = 0.50$ weber/m², what is the separation distance on the plate? (c) What is the minimum diameter of the magnet pole-pieces required?

8-26 A Van de Graaff generator can produce a potential difference of 8.0×10^6 v for accelerating electrons, protons, deuterons, and α-particles. (a) What are the final energies of the several particles in this machine? (b) Compute the speeds of the electrons, protons, and deuterons in terms of α-particle speed.

8-27 Electrons of 2.0 Mev energy are directed into a linear accelerator composed of 200 drift tubes connected alternately to a 3000 mega-cycle/sec oscillator. (a) If the electrons are to emerge with a final energy of 50 Mev, what must be the lengths of the second and the last drift tubes? (b) How many additional tubes would be needed to produce 100 Mev electrons in this accelerator? (c) What would be the over-all length of the 100 Mev linear accelerator, assuming the spacing between adjacent drift tubes to be negligible?

8-28 Protons of 4 Mev energy enter a linear accelerator which has 50 cylinders connected alternately to a 200-megacycle/sec oscillator. If the final energy of the protons is 40 Mev, what is the length of (a) the second cylinder and (b) the last cylinder?

8-29 A 20-lb satellite is fired from the Earth at a speed of 25,000 mph. Calculate the translational kinetic energy (in ev), per atom, deriving from this motion. (Assume that the satellite is composed entirely of iron atoms.)

8-30 A 350 Mev electron linear accelerator produces 60 pulses of 10^{12} electrons each second. What is (a) the average beam current and (b) the power output?

8-31 Calculate the energy of the following particles when accelerated by a 60-inch diameter cyclotron whose magnetic field has a flux density of 1.5 weber/m²: (a) protons, (b) deuterons, (c) α-particles.

8-32 (a) Show that a cyclotron which has been adjusted to accelerate deuterons can, without a change in the frequency or magnetic field, accelerate α-particles. (b) How must the frequency be changed, assuming the magnetic field is unchanged, to accelerate protons?

8-33 A cyclotron is to utilize a magnetic flux density of 2.0 weber/m²
and has a radius of 0.50 m. What is the frequency of the oscillator
for accelerating (a) protons, (b) deuterons, (c) α-particles. What
is the maximum energy of (d) protons, (e) deuterons, and (f)
α-particles?

8-34 A cyclotron accelerates deuterons when the flux density is B_p and
the frequency is f_p. (a) If the cyclotron is adjusted to accelerate
protons at the same frequency f_p, how must the magnetic flux
density be changed? (b) How will the proton energy compare
with the deuteron energy? (c) Repeat parts (a) and (b) for α-particles. (Assume that the proton mass is twice, and the α-particle
mass four times, that of the deuteron mass.)

8-35 * A cyclotron having a diameter of 5.0 ft accelerates α-particles to
an energy of 25 Mev. The maximum potential difference between
the dees is 200 kilovolts. (a) Assuming for simplicity that the
α-particles move, on the average, in a circle of 2.5 ft diameter,
compute the total distance traveled. (b) What is the total acceleration time?

8-36 A synchrocyclotron, having a magnetic flux density of 2.0
weber/m², accelerates 2 Mev protons to an energy of 250 Mev.
What is (a) the initial frequency, and (b) the final frequency of
the oscillator?

8-37 Electrons injected into a 100-cm radius betatron at 40 Kev, are
accelerated to a final energy of 50 Mev. If the induced voltage
per revolution is 300 volts, (a) what is the approximate total distance traveled by the electrons? (b) Calculate the acceleration
time of the electron pulses and compare this with the period of
the alternating magnetic field (60 cps).

8-38 * The proton-synchrotron, called the "Cosmotron," at Brookhaven
National Laboratory, accelerates protons to 3 Bev in one second.
Pulses containing 2×10^{10} protons are ejected at 5-second intervals. (a) What is the average output beam current? (b) What
would be the average force on the target if all particles were
brought to rest?

8-39 Electrons are injected into a synchrotron having a radius of 8.0 m
with an energy of 4.0 Mev and emerge, after acceleration, with an
energy of 1.0 Bev. (a) What are the initial and final frequencies
of the accelerating voltage? (b) If the electrons are accelerated
by a potential difference of 1000 volts during each revolution, for
what period of time are the electrons within the synchrotron? (c)
What total distance do the electrons travel within the synchrotron?

8-40 The 4 Bev protons from a synchrotron are to be deflected through
45° by a magnet having a maximum flux density of 2.0 weber/m².
What is the approximate diameter of the magnet pole-pieces?

N I N E

NUCLEAR STRUCTURE

Insofar as atomic structure is concerned, the atomic nucleus can be regarded essentially as a mass point and a point charge. The nucleus contains all of the positive charge and nearly all of the mass of an atom; it
provides, therefore, the center about which electron motion takes place.
Although the nucleus influences atomic structure primarily by its Coulomb
force of attraction with electrons, some rather subtle effects in atomic
spectra can be attributed to the nucleus. We recall that the Rydberg constant for a particular element is changed slightly by differences in the mass
of the isotopes. Furthermore, the phenomenon of hyperfine structure, the
occurrence of very-closely-spaced spectral lines in atomic spectra, has its
origin in the angular momentum and the very small magnetic moment of
the nucleus.

The fundamental α-particle scattering experiments of Rutherford established that, for distances greater than 10^{-14} m, the nucleus interacts
with other charged particles by the Coulomb electrostatic force. It was
found, however, that when the α-particles came closer than 10^{-14} m to the
nuclear center, the distribution of the scattered particles could not be

accounted for simply in terms of Coulomb's law. These experiments showed then that a totally new type of force, the nuclear force, acts at distances smaller than 10^{-14} m.

In this chapter we shall explore some of the simpler aspects of nuclear structure: the fundamental nuclear constituents, their interactions, the properties of stable nuclei, and the properties and decay characteristics of unstable nuclei. We shall see that, apart from the tremendous difference in their relative size, 10^{-10} m for atoms but less than 10^{-14} m for nuclei, nuclear structure is different from atomic structure in several significant respects.

Whereas atoms can be excited to emit their optical or x-ray spectra by gaining an energy never greater than 100 Kev, nuclei generally remain inert until they gain energies of the order of a few Mev. The simpler aspects of atomic structure can be understood on the basis of the Bohr model; no such simple model exists for nuclei. The primary force between the particles comprising the atom is the well-understood Coulomb force; the forces between the constituents of nuclei are only partially understood. An atom typically loses energy of excitation by emitting photons; an excited nucleus can lose its energy of excitation by emitting particles, as well as by emitting photons. Despite these differences there are a number of fundamental laws which are found to apply equally well to atoms and to nuclei; these are the rules of the quantum theory, and the conservation laws of mass-energy, linear momentum, angular momentum, and electric charge.

9-1   The nuclear constituents   The particles of which all nuclei are composed are the proton and the neutron. We will list here the fundamental properties—charge, mass, spin, and nuclear magnetic moment—of these particles.

CHARGE   The proton is the nucleus of the atom $_1H^1$, the light isotope of hydrogen; it carries a single positive charge, equal in magnitude to the charge of the electron.

The neutron is so named because it is electrically neutral. Because it carries no charge, the neutron shows only a feeble interaction with electrons, it produces no direct ionization effects, and it is, therefore, detected and identified only by indirect means. The existence of the neutron was not clearly established until 1932, when J. Chadwick demonstrated its properties in a series of classic experiments which will be discussed in Chapter 10.

MASS   We list below the masses of the proton (the bare nucleus of the $_1H^1$ atom) and the neutron, both in atomic mass units (see Section 2-10), together with the rest energies of these particles in units of Mev.

proton rest mass $= 1.007593 \pm 0.000003$ amu
proton rest energy $= 938.211 \pm 0.010$ Mev
neutron rest mass $= 1.008982 \pm 0.000003$ amu
neutron rest energy $= 939.505 \pm 0.010$ Mev

The proton and neutron have nearly the same mass, the neutron mass exceeding the proton mass by slightly less than 0.1 per cent. Both particles have rest energies of about 1 Bev. Because the proton carries an electric charge, its mass can be measured directly with high precision by the methods of mass spectrometry; electric and magnetic fields have virtually no effect on a neutron, and its mass is inferred indirectly from experiments to be described shortly.

SPIN   An important property of both the proton and the neutron is the intrinsic angular momentum, or the so-called *nuclear spin*. This angular momentum is intrinsic to the particles and does not depend on any orbital motion; the nuclear spin can be visualized, as in the case of electron spin, in terms of the spinning of the particle as a whole about some internal axis of rotation. The nuclear spin angular momentum is represented by the quantum number I, where the magnitude of the angular momentum is given by

$$P_I = \sqrt{I(I + 1)}\, \hbar \qquad\qquad [9\text{-}1]$$

in a fashion analogous to the spin angular momentum of the electron. See Equation 6-16.

The nuclear spin of both the proton and the neutron is $\frac{1}{2}$;

proton spin: $I = \frac{1}{2}$
neutron spin: $I = \frac{1}{2}$

The nuclear spin angular momentum is space-quantized by an external magnetic field, the permitted components along the direction of the magnetic field being $+\frac{1}{2}\hbar$ and $-\frac{1}{2}\hbar$, as shown in Figure 9-1. The magnitude of the angular momentum of a free proton or neutron, as well as the components of the angular momentum along the space-quantization direction are precisely the same as those for the electron spin.

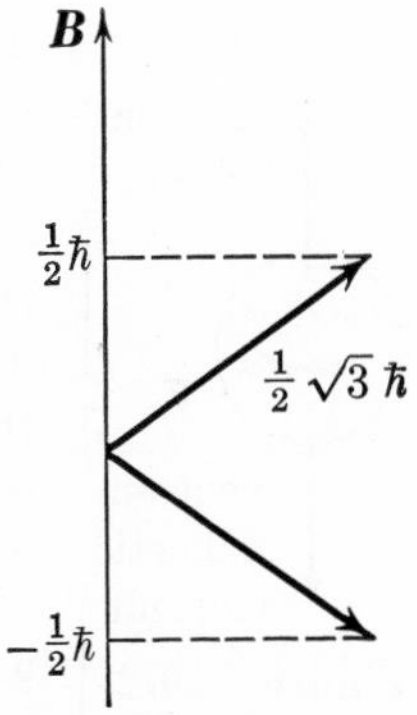

Figure 9-1. Space quantization of a proton spin or a neutron spin.

NUCLEAR MAGNETIC MOMENT   The component of the magnetic moment associated with electron spin along the direction of an external magnetic field is exactly one Bohr magneton (Section 6-5): $\beta = e\hbar/2m = .9273 \times 10^{-23}$ joule/(weber/m²). Because the electron has a negative electric charge, the electron-spin magnetic moment points in the opposite direction to that of the electron-spin angular momentum.

Let us now consider the magnetic moment associated with proton spin. The unit in which nuclear magnetic moments are measured is the *nuclear magneton* = β_I, which is defined as:

$$\text{nuclear magneton} = \beta_I = \frac{e\hbar}{2M_p} = 5.051 \times 10^{-27} \text{ joule/(weber/m}^2) \qquad [9\text{-}2]$$

where M_p, the proton mass, replaces the electron mass m in the Bohr magneton. Inasmuch as the proton mass is 1836 times that of the electron, the nuclear magneton is smaller than the Bohr magneton by this factor. The nuclear magnetic moment of the proton is found by experiment to be

$$\text{proton magnetic moment} = +2.7928 \text{ nuclear magneton}$$

The plus sign indicates that the magnetic moment of a proton points in the same direction as the proton's nuclear spin; the magnitude of the nuclear moment gives the *component* of the proton magnetic moment along the space-quantization direction in units of the nuclear magneton. It is significant that the size of the proton magnetic moment is *not* one nuclear magneton, but is, instead, nearly three times larger than what one might expect simply on the basis of the proton mass.

Despite the fact that the neutron, as a whole, carries no net electric charge, it does have a magnetic moment whose value is found to be

$$\text{neutron magnetic moment} = -1.9128 \text{ nuclear magnetons}$$

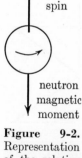

The negative sign indicates that the neutron magnetic moment is aligned *opposite* to the direction of the neutron angular momentum, as shown in Figure 9-2.

Because the proton moment is *not* one β_I and the neutron moment is *not* zero, the proton and neutron are more complicated entities than the electron. A successful theory of the fundamental nuclear particles must account for the apparently anomalous magnetic moments of these two particles.

Figure 9-2. Representation of the relative orientations of the neutron spin and magnetic moment.

9-2 The forces between nucleons

All nuclei consist of protons and neutrons bound together to form more or less stable systems; therefore, it is important to have some knowledge of the forces that act between these fundamental nuclear constituents. Consider first the force between two protons. The most direct way to examine this $(p-p)$ force is by a proton-proton scattering experiment. In such an experiment mono-energetic protons from a particle accelerator strike a target containing mostly hydrogen atoms, and therefore, protons. The angular distribution of the scattered protons is measured with a nuclear particle detector. From this distribution one can infer the force acting between the incident particles and the target particles—in this instance, both protons. The proton-proton scattering experiments show that the force can be represented by the potential curve shown in Figure 9-3. At large distances of separation the protons repel one another by the Coulomb electrostatic force. At a

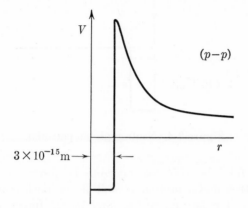

Figure 9-3. Proton-proton potential.

distance of approximately 3×10^{-15} m, a fairly sharp break in the potential curve occurs. This indicates the onset of the *nuclear force* between a pair of protons. The force is strongly *attractive* for smaller distances (although there is evidence for a repulsive "core" at very small distances). The "size" of the proton can be taken as the *range*, 3×10^{-15} m, of the nuclear $(p-p)$ force.

The force between a neutron and a proton can be investigated by neutron-proton scattering experiments. In these experiments, a mono-energetic neutron beam (from a nuclear reaction) bombards a target containing protons. Again, the distribution of the scattered neutrons is analyzed to deduce the $(n-p)$ force acting between the neutron and proton, or the potential whose (negative) derivative gives this force. The interaction between a neutron and proton can be represented by the potential curve shown in Figure 9-4. At large distances of separation there is *no* force between the two particles. But at a separation distance of about 2×10^{-15} m the neutron and proton attract one another by a strong nuclear force having a well-defined range. Clearly, this nuclear force is in no way dependent on the electric charge, inasmuch as the neutron is a neutral particle.

The nuclear force between two neutrons cannot be investigated by a neutron-neutron scattering experiment because it is impossible to prepare a target consisting of free neutrons. But a variety of indirect evidence indi-

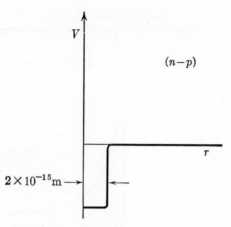

Figure 9-4. Neutron-proton potential.

cates that the force between two neutrons is closely equal to the force between a neutron and a proton, as well as the nuclear force between a pair of protons. The $(p-p)$, $(n-p)$, and $(n-n)$ forces are all approximately equal. Because a neutron and a proton are nearly equivalent in their interactions (apart from the Coulomb force between protons), it is customary to refer to a neutron *or* a proton as a *nucleon*. The term, nucleon, is used to designate either a proton or a neutron when the distinction between them is of little importance. This independence of the nuclear force on the particular participating nucleons is known as the charge-independence of the nuclear force. More sophisticated treatments of the proton-neutron interactions show that it is possible to consider the proton and neutron as two different states of the *same* particle.

9-3 The deuteron The simplest nucleus containing more than one particle is the *deuteron*, the nucleus of the deuterium atom. The deuteron consists of a proton and neutron bound together by the attractive $(n-p)$ nuclear force to form a stable system. The deuteron has a single positive charge, $+e$. Its mass is approximately twice that of the proton or neutron; more precisely,

$$\text{deuteron rest mass} = 2.014194 \text{ amu}$$

(It must be emphasized that the deuteron mass given above is that of the bare deuterium nucleus; the mass of the neutral deuterium atom exceeds that of the deuteron by the mass of an electron, 0.000549 amu, and is, therefore, 2.014743 amu.)

It is interesting to compare the mass of the deuteron, M_d, with the sum of the masses, M_p and M_n, of its constituents, the proton and neutron:

$$M_p = 1.007596 \text{ amu}$$
$$M_n = \underline{1.008986 \text{ amu}}$$
$$M_p + M_n = 2.016582 \text{ amu}$$
$$M_d = \underline{\underline{2.014194 \text{ amu}}}$$

$$M_p + M_n - M_d \text{ (mass difference)} = 0.002388 \text{ amu}$$

The total mass of the proton and neutron when separated *exceeds* the mass of the two particles when they are bound together to form a deuteron. This mass difference (sometimes misleadingly called mass defect) is easily interpreted on the basis of the relativistic conservation of mass-energy (see Section 2-9). When *any* two particles attract one another, the sum of their masses when separated exceeds that of the bound system, inasmuch as energy (or mass) must be added to the system to separate it into its component particles. This energy is called the binding energy; its value can be computed from the mass difference, using the mass-energy conversion factor, 1 amu = 931.14 Mev. Thus, the binding energy, E_b, of the neutron-proton to form a deuteron is given by

$$E_b + M_d c^2 = (M_p + M_n)c^2 \qquad \text{[9-3]}$$
$$E_b = (M_p + M_n - M_d)c^2$$

or $\qquad E_b = 0.002388 \text{ amu} \times 931.14 \text{ Mev/amu} = 2.224 \text{ Mev}$

If 2.224 Mev is added to a deuteron, the neutron and proton can be separated from one another, beyond the range of the nuclear force, both particles being left at rest and with no kinetic energy.

The mass difference with which we have been concerned here occurs for *any* system of bound particles, but for an atomic system, such as the hydrogen atom, this difference is so small, 1 part in 10^8 (see Section 2-9), that it cannot be measured directly. The binding energy is manifested as a measurable mass difference in nuclear systems because the nuclear force is very strong and the binding energy is very great. In fact, the binding energy (2.22 Mev) of two nucleons to form a deuteron is roughly a *million* times larger than the binding energy (13.6 ev) of a proton and an electron to form a hydrogen atom.

Let us recall that the binding energy of a hydrogen atom in the lowest, or ground, state, that is the ionization energy of the hydrogen atom, can be determined by knowing the energy of the photon whose absorption (photoelectric effect) frees the bound electron from the hydrogen nucleus. A completely analogous measurement can be made of the deuteron binding energy. In such an experiment deuterium gas is irradiated with a beam of high-energy γ-ray photons. If the energy of the mono-energetic photons is just equal to the binding energy of the deuteron, the photon absorption will produce a free neutron and a proton, at rest; and, if the photon energy exceeds the binding energy, the deuteron will be dissociated into a proton and neutron, each particle having kinetic energy. This *nuclear reaction* is written as follows:

$$\gamma + d \rightarrow p + n \qquad [9\text{-}4]$$

The conservation of mass-energy requires that

$$h\nu + M_d c^2 = M_p c^2 + M_n c^2 + K_p + K_n \qquad [9\text{-}5]$$

where K_p and K_n are the kinetic energies of the freed proton and neutron, respectively.

This process, whereby the proton and neutron are detached from one another by the absorption of a photon, is a nuclear photoelectric effect, or a nuclear *photodisintegration*. The threshold for the reaction occurs when $K_p = 0$ and $K_n = 0$; in this case,

$$h\nu_0 = (M_p + M_n - M_d)c^2 = E_b \qquad [9\text{-}6]$$

That is, the energy of the photon is equal to the binding energy of the deuteron. When the threshold photon energy $h\nu_0$ is measured and the values of M_p and M_d are known, the neutron mass can be computed from Equation 9-6. This is one of several ways in which the neutron mass can be measured.

The inverse reaction of the deuteron photodisintegration is that in which a neutron and proton, essentially at rest, combine to form a deuteron with the emission of a 2.22 Mev photon, as follows:

$$p + n \rightarrow d + \gamma \qquad [9\text{-}7]$$

Note that the nuclear reaction of Equation 9-7 is merely that of Equation 9-4 with the arrow reversed.

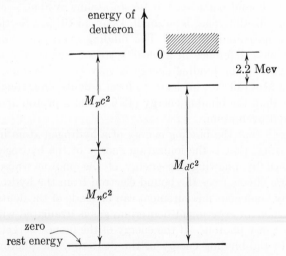

Figure 9-5. Energy-level diagram of the neutron-proton system, the deuteron.

The *nuclear energy-level diagram* of the deuteron is shown in Figure 9-5. Unlike all atoms and all other nuclei, the deuteron is found to have only *one* bound state. It can exist only in this, the ground, state; for the continuum of unbound states, the proton and neutron are free. The deuteron has no bound, excited states. To show the relationship between the rest masses and rest energies of the deuteron, proton, and neutron, these quantities are also shown (but not to scale) in the diagram. It is useful to compare this energy-level diagram for the simplest of all bound nuclear systems with that of the corresponding simplest two-particle atomic system, the hydrogen atom. The simplified (and grossly exaggerated) energy-level diagram of the bound proton-electron system is shown in Figure 9-6, where the masses of the electron and proton are also displayed.

The hydrogen atom has, of course, a whole series of possible excited

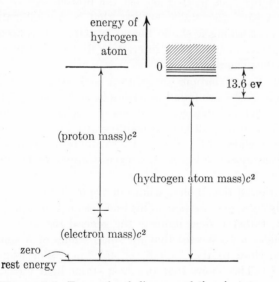

Figure 9-6. Energy-level diagram of the electron-proton system, the H[1] atom.

states; which is to say, the relativistic mass of the bound electron-proton system can assume any one of a large number of possible quantized values (compare Figure 9-6 with Figure 2-14). Because the binding energy for any one of the possible energy states of the hydrogen atom is small (less than 13.5 ev), it is not possible in practice to determine the quantized energies of the hydrogen atom by measuring its mass. But the large binding energies of nucleons do allow the binding energy of nuclei to be determined directly from the difference between the rest masses of the constituent particles and the mass of the bound nuclear system.

9-4 Stable nuclei We now consider the stable nuclei containing more than two nucleons. The number of protons in a nucleus is represented by the *atomic number Z*; the total number of nucleons is represented by the *mass number A*; and the number of neutrons is represented by N, where $N = A - Z$. The term *nuclide* is used to designate a particular species of nuclei with specific values of Z and of A. Nuclides having the same proton number Z are known as *isotopes*; nuclides having the same neutron number N are known as *isotones*; and nuclides having the same number of nucleons A are known as *isobars*. The nuclide, $_{17}Cl^{37}$, for example, has 17 protons, 20 neutrons, and 37 nucleons. The nuclides $_{17}Cl^{35}$ and $_{17}Cl^{37}$ are isotopes; the nuclides $_{17}Cl^{37}$ and $_{19}K^{39}$ are isotones; and the nuclides $_{17}Cl^{37}$ and $_{18}Ar^{37}$ are isobars.

The *stable* nuclides found in nature are shown in Figure 9-7, where the neutron number N is plotted against the proton number Z. Each point represents a particular stable nuclide, that is, a combination of protons and neutrons forming a stable bound system. By concentrating on its over-all features, a number of interesting and significant regularities can be seen in this diagram.

Only those combinations of protons and neutrons which appear as points in Figure 9-7 are found in nature as stable nuclides in their ground states; all other possible combinations of nucleons are, to some degree, unstable in that they decay, or disintegrate, into other nuclei. For example, the nuclides, $_8O^{16}$, $_8O^{17}$, and $_8O^{18}$, all isotopes of oxygen, exist as stable nuclear systems; but such oxygen isotopes as $_8O^{15}$ and $_8O^{19}$ are unstable.

The location of the stable nuclides in Figure 9-7 can be represented approximately by a *stability line*. This line does not, of course, pass through each point; rather it does indicate the general region in which the most stable nuclides fall. We see that for small values of N and Z, the stable nuclides lie close to the $N = Z$, or 45° line. (For example, $_8O^{16}$ has $N = Z = 8$.) This shows that the most stable light nuclides for a given mass number A are those for which the number of protons is closely equal to the number of neutrons. We can say that light nuclei prefer to have equal numbers of protons and neutrons because such aggregates are more stable than those for which there is a decided excess of protons or of neutrons.

For the heavier nuclides, the stability line of Figure 9-7 bends increasingly away from the 45° line; that is, for large A, $N > Z$. For example, the stable nuclide, $_{82}Pb^{208}$, has $Z = 82$ and $N = 126$. Thus heavy nuclides show a decided preference for neutrons over protons.

The neutron excess can be interpreted by taking into account the repulsive Coulomb force that acts between protons. If we start with a moderately heavy nucleus and attempt to construct from it a heavier

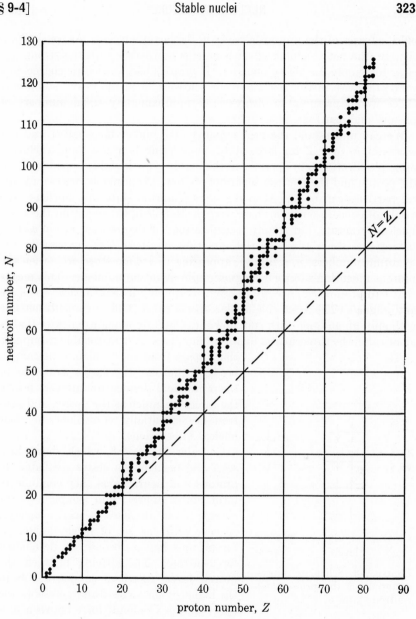

Figure 9-7. Neutron number versus proton number for the stable nuclides.

nucleus by adding one nucleon, the binding of an additional neutron will usually be greater than the binding of an additional proton. This is so because the neutron is attracted by the nuclear force, whereas the proton is both attracted by the nuclear force and repelled by the Coulomb repulsive forces of the protons already in the heavy nucleus. As we shall see shortly, the net Coulomb effect competes noticeably with the strong nuclear force only for heavy nuclides. If protons had no electric charge, but were otherwise distinguishable from neutrons, then the heavy stable nuclides would presumably always have approximately equal numbers of protons and neutrons.

We can understand the near equality of Z and N for small A and the excess of N over Z for large A by considering how the Pauli exclusion principle (Section 6-8) operates in the building up of stable nuclides. Both the proton and the neutron separately follow this principle: no two identical particles can be placed in the same quantum state in a nucleus. We need not concern ourselves here with the details of the quantum theory of nuclear structure, but simply recognize that, if two protons are in a state having the same three spatial quantum numbers (not necessarily the quantum numbers, n, l, and m_l appearing in atomic structure), then the protons must differ in their magnetic spin quantum numbers. This means that two protons can occupy the same state only if their nuclear spins are anti-aligned. The same rule applies to neutrons. Only two neutrons, one with spin up and one with spin down, can occupy a quantum state that is identical in the three spatial quantum numbers. It is reasonable to suppose that, apart from the Coulomb interaction between protons, the states available to the proton and the neutron are very nearly the same, inasmuch as the proton and neutron are essentially equivalent in their nuclear interactions.

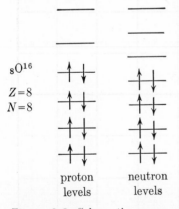

$_8O^{16}$

$Z = 8$
$N = 8$

proton   neutron
levels   levels

Figure 9-8. Schematic representation of the proton and neutron states of $_8O^{16}$.

Consider Figure 9-8, which shows in a *schematic* fashion the states available to protons and neutrons as they combine to form stable nuclei. For simplicity, the energy levels are shown nearly equally-spaced, with one set of levels designated for protons and a second set designated for neutrons. The spacing between the proton levels increases as one moves to the higher-lying levels; this represents the effect of the Coulomb force between protons. The neutron levels are shown as equally spaced for all values of N. We suppose that two protons, one with spin up, one with spin down, can be accommodated in each proton level, and that two neutrons can be placed into each neutron level.

The first proton level and the first neutron level are filled when a nucleus is formed with two protons and two neutrons. This is, of course, the very stable nuclide, $_2$He^4. In adding further nucleons to form heavy nuclides, one would expect the proton and neutron levels to be nearly equally populated to form the most stable nuclide for a given number of nucleons A. Thus, for small A, the stable nuclides will have $Z \simeq N$, in accord with observation. As the number of nucleons becomes larger, the most stable nuclides will again be formed when the lowest energy levels available to the protons and neutrons are filled. This requires that there be a neutron excess or, that $N > Z$ for large A. We see then that the general features of the stability line can be accounted for by applying the Pauli exclusion principle to the building up of stable nuclides, without knowing the details of the energy levels.

9-5 Nuclear radii There are a number of ways in which the radius of a nucleus can be defined and evaluated. The nuclear radius is given approximately by the results of the α-particle scattering experiments. Although the distribution of the scattered particles is accounted for perfectly by the Coulomb interaction for distances greater than 10^{-14} m, deviations from Coulomb's law occur when the α-particles come within a distance of about 10^{-14} m from the nuclear center. The nuclear radius can be defined, then, as the distance from the nuclear center at which the nuclear force becomes important.

The nuclear radius can, however, be inferred more directly and with higher precision from scattering experiments in which high-energy neutrons bombard target nuclei. Neutrons are not repelled by a Coulomb force; they are deviated from their incident directions or absorbed by the target nucleus only when they come within the range of the nuclear force of the bombarded nuclei. We can define the nuclear radius as being that distance from the center of the nucleus at which a neutron first feels the nuclear attractive force. Because the range of this nuclear force is quite definite, the nuclear interaction being effectively zero for greater separation distances, the neutron scattering experiments do not have the complicating effect of the Coulomb force to be subtracted out in analyzing the scattering data. In fact, a high-speed neutron will not be scattered at all unless it passes close enough to a target nucleus to be within the range of its nuclear force.

It is necessary to use high-speed neutrons because the neutron has, as do all particles, a de Broglie wavelength associated with it, and unless the wavelength of the neutron is small enough so that the neutron's position can be quite precisely specified, the neutron scattering experiments cannot be simply interpreted. Neutrons of 100 Mev energy have a wavelength of about 10^{-15} m. Many experiments have been carried out to measure the absorption of high-energy neutrons by a variety of targets. Some results are summarized in Figure 9-9, where the nuclear radius R is plotted against

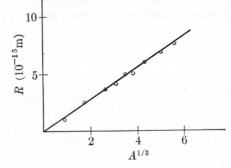

Figure 9-9. Nuclear radius as a function of the cube root of the mass number.

$A^{1/3}$, the cube root of the mass number. We see that the data are fitted by a straight line, so that the nuclear radius R can be represented by a simple formula,

$$R = r_0 A^{1/3}$$
$$\text{where } r_0 = 1.4 \times 10^{-15} \text{ m} \qquad \text{[9-8]}$$

The radius of any nucleus, defined in terms of the nuclear force with a neutron, can be computed using Equation 9-8. Even for so heavy a nucleus as $_{92}U^{238}$, R is found to be no larger than 9×10^{-15} m.

Scattering experiments with very-high-energy electrons provide a second method of measuring the nuclear radius. An electron with an energy of 500 Mev has a de Broglie wavelength of approximately 10^{-15} m, and such a well-localized particle is a suitable probe for studying nuclear radii. Unlike the neutron, an electron will not be appreciably affected by the nuclear force, but it will be scattered from its incident direction by the electric forces of protons within nuclei. As in the neutron experiments, the results of high-energy electron scattering experiments can be represented by the relation

$$R = r_0 A^{1/3} \qquad \text{[9-9]}$$

where $r_0 = 1.1 \times 10^{-15}$ m

Equation 9-9 is of precisely the same form as Equation 9-8; the value of r_0 for the nuclear-charge radius is, however, somewhat smaller than the value of r_0 for the nuclear-force radius.

The relationship for the nuclear radius leads to an important conclusion concerning the density of nuclear material. Cubing Equation 9-8 and multiplying by $4\pi/3$ we have

$$(4\pi R^3/3) = (4\pi r_0^3/3) A \qquad \text{[9-10]}$$

The quantity $(4\pi R^3/3)$ is the volume of the nucleus, assumed to be a sphere, or a near sphere; and we can interpret $4\pi r_0^3/3$ as the volume of a single nucleon. Therefore, Equation 9-10 becomes

(volume of the nucleus) = (volume of a nucleon) $\times$ (number of nucleons)

The total volume of a nucleus is merely the sum of the volumes of the several nucleons comprising it. Because this relation holds for all nuclei and because all nucleons have nearly the same mass, we conclude that *all* nuclei have the *same density* of nuclear matter. It is easily found that the density of nuclear material is 2×10^{17} kg/m^3, or about 10^9 tons/inch^3. (This extraordinarily high density is, of course, consistent with the fact that the radii of atoms are approximately 10^5 times that of nuclei.)

We are now in a position to discuss a question that has not been raised to this point; namely, why electrons do not exist as constituents of nuclei. Before the discovery of the neutron in 1932, it was thought that nuclei consisted of protons and electrons. On this basis such a nucleus as $_7$N^{14} would contain 14 protons and 7 electrons. The hypothesis that electrons existed in nuclei was strengthened by the observation that radioactive materials undergoing β^--decay actually emit electrons from their nuclei.

If an electron were confined and localized to within a nuclear dimension, say, 10^{-15} m, then the de Broglie wavelength of the electron could be no greater than this distance. But we have just seen that a 500 Mev electron has a wavelength of about 10^{-15} m. Therefore, if electrons were contained within nuclei, they would have kinetic energies of at least 500 Mev. But a 500 Mev electron can be bound and have a negative total energy, only if its potential energy is less than -500 Mev. Thus, an attractive potential of at least -500 Mev must exist for electrons if they are to be kept within the confines of a nucleus. There is no evidence whatsoever for the existence of so strong an attractive force on an electron; it must, therefore, be concluded that electrons cannot exist within a nucleus. We shall later discuss other equally compelling arguments against electrons as nuclear constituents. A nucleon (a proton or a neutron) can, however, be localized within a nuclear dimension when its kinetic energy is only a few Mev; for example, a 9 Mev nucleon has a wavelength of about 10^{-14} m.

9-6   The binding energy of stable nuclei   The nuclear force is so strong that the mass of a bound nuclear system is measurably smaller than the sum of the masses of the component nucleons. Thus, information on the binding energy of nuclear systems can be arrived at directly from a comparison of masses.

The masses of atoms can be measured with considerable precision (better than 1 part in 10^5) by the methods of mass spectrometry (Section 8-8). Appendix IV lists the masses of the *neutral* atoms in atomic mass units (amu). All measured atomic masses are very close to the integral mass number A; this is called the *whole-number rule*. An atom of $_8O^{16}$ has a mass of precisely 16 amu by definition.† The mass of the *nucleus* Z^A is the atomic mass less Z electron masses; because the binding energy of the atomic electrons to the nucleus is usually quite small compared with the atom's rest energy, we can take the neutral atom's mass to be the mass of the nucleus plus the masses of the electrons.

Consider the nucleus $_8O^{16}$ with 8 protons and 8 neutrons. We wish to calculate the *total* binding energy; that is, the energy required to separate an $_8O^{16}$ nucleus into its 16 component nucleons, each nucleon being at rest and effectively out of the range of the forces of the remaining 15 nucleons. In the following computation we shall use the mass of a *neutral hydrogen atom*, 1.00814 amu, along with the mass of an electron, 0.00055 amu.

$$
\begin{aligned}
8 \text{ protons} &= 8(1.00814 - 0.00055) \\
8 \text{ neutrons} &= 8(1.00899) \\
\hline
\text{total nucleon masses} &= 16.13704 - 8(0.00055) \\
_8O^{16} \text{ nuclear mass} &= 16.00000 - 8(0.00055) \\
\hline
\text{mass difference} &= 0.13704 \text{ amu} \\
\text{total binding energy} &= 0.13704 \text{ amu} \times 931.1 \text{ Mev/amu} = 127.5 \text{ Mev}
\end{aligned}
$$

Note that, because the electron masses cancel out, we can use the masses of the *neutral* atoms of hydrogen and oxygen rather than the masses of the proton and the oxygen nucleus.

We see that 127.5 Mev must be added to an $_8O^{16}$ nucleus to separate it completely into its constituent particles; and therefore, the 16 nucleons of $_8O^{16}$ are bound together with 127.5 Mev of binding energy to form the nucleus $_8O^{16}$ in its lowest energy state. The *average* binding energy per nucleon, E_b/A, is 127.5 Mev/16, or 7.97 Mev. This does *not* mean that each of the 16 nucleons is bound with an energy of precisely 7.97 Mev, but rather that 7.97 Mev is the energy with which each particle is bound, *on the average*.

We can write a general relationship giving the total binding energy E_b of any nucleus Z^A, having an atomic mass $M(Z^A)$, and comprised of Z protons of *atomic* mass M_H and $(A - Z)$ neutrons of mass M_n.

$$\boxed{E_b/c^2 = ZM_H + (A - Z)M_n - M(Z^A)} \qquad [9\text{-}10]$$

Let us compute the energy to remove *just one proton* from $_8O^{16}$, leaving a nucleus with 7 protons and 8 neutrons; namely, the stable nucleus, $_7N^{15}$. The binding energy, or *separation energy*, of the last proton to the remaining 15 nucleons in O^{16} can likewise be computed by using the rest masses of the particles. (We again use the masses of neutral atoms.)

---

† In *physical*, rather than chemical, *amu*.

$$\text{mass } {}_1H^1 = 1.00814$$
$$\text{mass } {}_7N^{15} = 15.00488$$
$$\text{mass of } {}_1H^1 + {}_7N^{15} = 16.01302$$
$$\text{mass } {}_8O^{16} = 16.00000$$
$$\text{mass difference} = 0.01302 \text{ amu}$$

The separation energy of the last proton in ${}_8O^{16}$ is 0.01302 amu $\times$ 931 Mev/amu = 12.1 Mev.

We see that the binding energy, 12.1 Mev, of one particular nucleon (the least-tightly-bound proton) in ${}_8O^{16}$, is *not* the same as the *average* binding energy of a nucleon, 7.97 Mev, in ${}_8O^{16}$, although they are of the same order of magnitude. (The separation energy of the least-tightly-bound *neutron* in ${}_8O^{16}$ is 15.6 Mev.)

The energy to remove the last proton in ${}_8O^{16}$ corresponds, in atomic structure, to the removal of the least-tightly-bound valence electron from an atom by ionization. We know that the ionization energy of an outer, valence electron in an atom is usually several orders of magnitude less than the ionization energy of the inner, tightly-bound electron (visible light for the former, x-rays for the latter); thus, we see that, unlike the case for electrons bound by electric interaction, the particles of a stable nucleus are all bound with at least approximately the same energy.

The average binding energy per nucleon, E_b/A, can be computed for all stable nuclei using Equation 9-10. When this is done, and the computed values of E_b/A are plotted against the corresponding values of the atomic

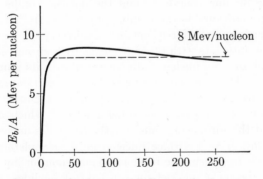

Figure 9-10. Average binding energy per nucleon as a function of mass number, for the stable nuclides.

mass A, we obtain the results shown in Figure 9-10. The points can be fitted fairly well with a smooth curve, although there are some points (for example, ${}_2He^4$) that fall somewhat off the curve.

The characteristics of the curve of E_b/A versus mass number A can be summarized as follows. The curve rises sharply going from the lightest stable nuclides to values of A in the vicinity of $A = 20$; for $A > 20$, the curve rises slowly, reaches a maximum close to the element, $_{26}Fe^{56}$, and then decreases slowly toward the heaviest stable nuclides.

The curve is approximately horizontal from $A = 20$ onward; and E_b/A is roughly a constant; namely, 8 Mev/nucleon.

$$\boxed{\text{for } A > 20. \quad E_b/A \simeq 8 \text{ Mev/nucleon}} \qquad [9\text{-}11]$$

Iron and nuclides close to it represent the most stable configurations of nucleons found in nature; in all elements lighter or heavier than iron, the typical nucleon is bound with a lesser energy.

9-7a   Nuclear models: the liquid-drop model   Excluding the initial rise in the curve of Figure 9-10, the value of E_b/A is nearly a constant; therefore, the total binding energy E_b is closely proportional to A, the total number of nucleons. This simple dependence of binding energy on the number of bound particles finds a simple interpretation in the nuclear model known as the *liquid-drop model*, proposed by Niels Bohr in 1936. In this model the binding between the nucleons in a nucleus is treated as similar to the binding of molecules in a liquid. We know that the total binding energy of a liquid is directly proportional to the mass of the liquid (for example, to boil 2 kg of water requires twice the energy to boil 1 kg.); and if the density of a liquid is constant, the total binding energy of a liquid is also proportional to the volume of the liquid. In many respects, nuclei show the same behavior. The binding energy E_b and the nuclear volume ($4\pi R^3/3$) are both directly proportional to the number of nucleons A. These regularities give us information on the forces between nucleons.

Each nucleon interacts with other nucleons by a strong, *short-range* nuclear force, and any one nucleon interacts only with its immediate neighboring nucleons. Thus, the nuclear force shows *saturation*: after a nucleon is completely surrounded by a full complement of neighbors with which it interacts, it can exert no force on other nucleons, which must be more distant. But at the surface of a nucleus, the force of a nucleon will be unsaturated, inasmuch as a surface nucleon is not completely surrounded. This corresponds to the surface-tension phenomenon for liquids. The surface effect will be most marked for the lightest nuclides, where a large fraction of the nucleons are to be found at the surface; for heavy nuclides, for which the surface to volume ratio is small and a small fraction of the nucleons are on the nuclear surface, the surface effect is less important.

Although the attractive nuclear force shows saturation, with each nucleon interacting only with its immediate neighbors, this is not true of the repulsive Coulomb force between protons in a nucleus. Each proton in a nucleus interacts with *every other* proton with an electrostatic potential energy E_e given by

$$E_e = ke^2/r \qquad [9\text{-}12]$$

where $k = 1/4\pi\epsilon_0$, e is the proton charge, and r is the distance between a pair of protons. Taking r to be 3×10^{-15} m as a typical, average distance between such a pair, we find from Equation 9-12 that $E_e \simeq \frac{1}{2}$ Mev. This energy is small compared to the value of 8 Mev for E_b/A, and the Coulomb energy is not an important influence in the lightest nuclei.

For the heavy nuclides (large Z), the Coulomb energy becomes quite significant. Each proton interacts with *every* other proton in the nucleus according to Equation 9-12, so that the *total* Coulomb energy depends on the number of proton pairs; that is, the total number of Z protons taken two at a time, $Z(Z - 1)/2!$. For a heavy nucleus the contribution of the Coulomb energy, which represents a tendency of the protons to disrupt the stable nucleon configuration, consequently varies approximately as Z^2.

According to the liquid-drop model, the total binding of nucleons is influenced principally by the following three effects:

(1) *The volume effect.* $E_b/A \simeq$ constant; therefore, the binding energy E_b is proportional to A, and to the nuclear volume.

(2) *The surface effect.* The *un*binding energy is proportional to the surface area of the nucleus, and therefore, to R^2, or $A^{2/3}$.

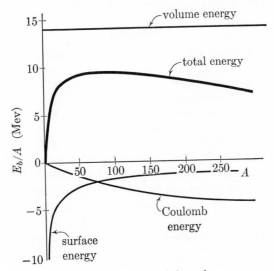

Figure 9-11. Contributions of the volume energy, surface energy, and Coulomb energy to the average binding energy per nucleon as a function of mass number.

(3) *The Coulomb effect.* This *un*binding energy is proportional to the total number of proton pairs, $Z(Z - 1)/2!$, and inversely proportional to the nuclear radius R; the total unbinding Coulomb energy is, therefore, proportional to $Z^2 A^{-1/3}$.

It is possible to fit the E_b/A curve of Figure 9-10 by combining the volume, surface, and Coulomb effects. The volume effect makes a *positive* contribution to the total binding energy; the surface and Coulomb effects make *negative* contributions. The three separate curves, and their algebraic sum, are shown in Figure 9-11. Because the liquid-drop model is capable of accounting so well for the principal features of the binding-energy-per-nucleon curve, the assumptions contained in this model must correspond, at least approximately, to the true nature of the interactions between nucleons.

9-7b   Nuclear models: the shell model   The liquid-drop model emphasizes the ways in which all nucleons are alike, inasmuch as it treats the proton and neutron as identical (apart from the Coulomb force between protons). This model is not successful, however, in accounting for some of the finer details of nuclear structure. The *shell model* is capable of explaining some of these observed nuclear properties. The number of known stable nuclides with even-even, even-odd, odd-even, and odd-odd proton-neutron numbers $(Z-N)$, are shown in Table 9-1.

Table 9-1

Nuclide type $(Z-N)$	Stable nuclides
Even-even	160
Even-odd	56
Odd-even	52
Odd-odd	4

Almost 60 per cent of the stable nuclides have an even number of protons and of neutrons. Clearly, nuclei have a greater stability when both the proton and neutron levels are fully occupied (see Figure 9-8) than if there is an odd number of protons or neutrons. In fact, the only examples of *stable* odd-odd nuclides are the lightest possible nuclides of this type, $_1H^2$, $_3Li^6$, $_5B^{10}$, and $_7N^{14}$, and all of these have $Z = N$.

The shell model is fairly successful in predicting the total nuclear angular momentum of stable nuclides. The total angular momentum of any nucleus consists of contributions from three sources; the intrinsic angular momentum, or nuclear spin, of the protons $\frac{1}{2} \hbar$; the intrinsic angular momentum, or nuclear spin, of the neutrons $\frac{1}{2} \hbar$; and the orbital angular momentum of the nucleons arising from their motion in the nucleus. These three contributions are combined by rules of vector addition to give the total, or resultant, nuclear angular momentum, misleadingly called the nuclear spin and represented by the symbol I.

The nuclear spins of the four categories of stable nuclides are given in Table 9-2.

Table 9-2

NUCLIDE TYPE $(Z - N)$	NUCLEAR SPIN
Even-even	0
Even-odd ⎱ Odd-even ⎰	$\frac{1}{2}, \frac{3}{2}, \frac{5}{2}, \frac{7}{2}, \ldots$
Odd-odd	1, 3

The fact that all even-even nuclides have zero total nuclear angular momentum is interpreted as follows. The protons fill the available proton levels in pairs of anti-aligned proton spins; the contribution of proton-spin angular momentum, therefore, is zero. Similarly, the neutrons are anti-aligned in pairs producing zero neutron-spin angular momentum. Finally, the angular orbital momentum of the protons and neutrons is zero, indicating that these nucleons are in closed shells analogous to the closed shells and subshells of atomic structure. Furthermore, all even-even nuclides have a zero nuclear magnetic moment. Just as the zero angular momentum and magnetic moment of inert-gas atoms is attributed to the pairing off of electron spins and orbital momenta, so too the zero nuclear angular momentum and the zero nuclear magnetic moment of even-even nuclides is attributed to the pairing off of neutron spins and proton spins·

In the even-odd or odd-even nuclides there is one odd nucleon, combining its half-integral intrinsic spin with the integral orbital angular momentum of the nucleus, thus giving half-integral I values. The nuclear magnetic moment of these nuclides is of the order of the nuclear magneton, Equation 9-2. This is the basis of another argument against electrons as nuclear constituents, for if electrons were to exist in the nucleus, the magnetic moment of the nucleus would be of the order of a Bohr magneton, which is roughly a thousand times larger than the observed nuclear magnetic moments.

The odd-odd nuclides have an odd proton and an odd neutron, each with spin $\frac{1}{2}$; therefore, the total nuclear spin I is integral. It is interesting to consider what the nuclear spin of an odd-odd nuclide such as the deuteron, $_1H^2$, would be if nuclei consisted of protons and electrons. On the proton-electron nuclear model, the deuteron would consist of 2 protons and 1 electron, each of the *three* particles with spin $\frac{1}{2}$. Therefore, this model predicts a *half*-integral value for I. On the other hand, the proton-neutron nuclear model, with *two* particles each of spin $\frac{1}{2}$, predicts that I be *integral*. The nuclear spin I of the deuteron is experimentally found to be 1. On this basis, the proton-electron nuclear model is again untenable.

Although we shall not pursue the shell model further, this model, emphasizing the quantum aspects of nuclear structure, is notably successful in describing many nuclear properties.

9-8   The decay of unstable nuclei   Thus far we have discussed only stable nuclei which, when left to themselves, will exist indefinitely without change.   Just as atoms can exist in any one of a number of quantized energy states, so too a nucleus has a set of discrete, quantized nuclear energy states.   Research in nuclear physics over the past years has accumulated voluminous experimental data on the energy states and decay schemes of somewhat more than 1200 nuclides; and although a complete theoretical understanding of nuclear structure is still wanting, many important aspects of nuclear structure have been established through a study of unstable nuclei.

In addition to the conservation laws of mass-energy, linear momentum, angular momentum, and charge, which hold for nuclear systems, there are other conservation laws that have been found to apply to nuclear transformations.   The only one of these which shall concern us at this time is the law of *conservation of nucleons*: the total number of protons and neutrons entering a reaction must equal the total number of nucleons leaving the reaction.   Therefore, the mass-energy, linear momentum, angular momentum, electric charge, and number of nucleons before a reaction or decay must all be equal to the respective quantities after the reaction or decay.

We will see that there are several important differences between unstable nuclei and unstable atoms:

(1) The spacing of nuclear energy levels is much greater than that of atomic energy levels.

(2) The time that an unstable nucleus spends on the average in an excited state can range from 10^{-14} sec to 10^{11} years, while atomic lifetimes are usually about 10^{-8} sec.

(3) Whereas excited atomic systems almost invariably emit photons upon de-excitation, an unstable nucleus may emit photons or particles of non-zero rest mass (for example, α-particles or β-particles).

All nuclear decays, whatever their differences in the particles emitted or the rates at which the disintegrations take place, follow a single law: the *law of radioactive decay*.   We shall call the initial unstable nucleus the parent and the nucleus into which the parent decays the daughter; the death of the parent gives birth to the daughter.   The probability that an unstable or excited nucleus will decay spontaneously into one or more particles having a lower energy is independent of the parent nucleus' past history, is the same for all nuclei of the same type, and is independent of external influences (temperature, pressure, etc.).

There is no way of predicting the time when any one nucleus will decay, and its survival is subject to the laws of chance. But during an infinitesimally small time interval, dt, the probability that a nucleus will decay is directly proportional to this time interval. Thus,

the probability that a nucleus *decays* in time $dt = \lambda\, dt$,

where the proportionality constant λ is called the *decay constant*, or the *disintegration constant*. Since the total probability that a nucleus will either survive or decay in the time dt is one (100 per cent),

the probability that a nucleus *survives* a time $dt = (1 - \lambda\, dt)$

and,   the probability that a nucleus *survives* a time $2\, dt$

$$= (1 - \lambda\, dt)(1 - \lambda\, dt) = (1 - \lambda\, dt)^2$$

Consider now the probability of the nucleus' surviving n time intervals, each of duration dt.

the probability that a nucleus survives a time $ndt = (1 - \lambda\, dt)^n$ [9-13]

Putting $ndt = t$, the total time elapsed, and remembering that, as $dt \to 0$, $n = t/dt \to \infty$, Equation 9-13 becomes

the probability that a nucleus survives a time $t = \mathop{\mathrm{Lim}}_{n \to \infty} (1 - \lambda t/n)^n$

[9-14]

Now, the definition of e^{-x}, where e is the base of the natural logarithms, is

$$e^{-x} \equiv \mathop{\mathrm{Lim}}_{n \to \infty} (1 - x/n)^n \qquad [9\text{-}15]$$

Comparing Equations 9-14 and 9-15, we see that

the probability that a nucleus survives a time $t = e^{-\lambda t}$ [9-16]

Equation 9-16 is the necessary consequence of the assumption that the decay of nuclei in unstable states is independent of the nucleus' present condition and past history. Although we cannot say precisely when a *single* nucleus will decay, we can predict the statistical decay of a *large* number of identical unstable nuclei. If there are initially N_0 unstable nuclei undergoing a decay process characterized by the decay constant λ, then the number N surviving after a period of time t has elapsed is merely N_0 times the probability that any one nucleus will have survived. Therefore, from Equation 9-16 we have

$$\boxed{N = N_0 e^{-\lambda t}} \qquad [9\text{-}17]$$

This exponential decay law holds, not only for unstable nuclei, but also for any unstable system (such as atoms in excited states) subject to decay by chance.

It is customary to measure the rapidity of decay in terms of the *half-life*, $T_{1/2}$, which is defined as the time in which one-half of the original unstable nuclei have decayed and one-half still survive. Therefore, $t = T_{1/2}$ when $N = \frac{1}{2} N_0$, and Equation 9-17 gives

$$\frac{1}{2} N_0 = N_0 e^{-\lambda T_{1/2}}$$

or,

$$\boxed{T_{1/2} = ln_e \, 2/\lambda = 0.693/\lambda} \qquad [9\text{-}18]$$

Thus, if a radioactive material decays with a half-life of 3 sec, after 3 sec one-half of the initial nuclei remain, after 6 sec one-quarter of the initial nuclei remain, and after 9 sec one-eighth of the initial nuclei remain. The decay constant λ has the units of reciprocal time (for example, sec^{-1}), which follows from its definition as the probability per unit time for decay.

The half-life is *not* the same as the *average lifetime*, or mean life, T_{av}, of an unstable nucleus; a straight-forward calculation (see Problem 9-18) shows that

$$T_{av} = 1/\lambda = T_{1/2}/ln_e 2 \qquad [9\text{-}19]$$

The decay of the parent and the concomitant growth of the daughter with time are shown in Figure 9-12. The number of daughter nuclei pro-

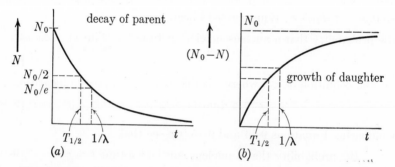

Figure 9-12. (a) Decay in the number of radioactive parent atoms as a function of time. (b) Growth in the number of (stable) daughter atoms as a function of time.

duced after a time t is $N_0 - N = N_0(1 - e^{-\lambda t})$, where N_0 is again the initial number of parent nuclei.

Another useful quantity describing radioactive decay is the *activity*, which is defined as the number of decays per second. It follows from Equation 9-17 that

$$dN/dt = -\lambda N_0 e^{-\lambda t} = -\lambda N$$

Therefore,

$$\boxed{\text{activity} = -dN/dt = \lambda N = (\lambda N_0)e^{-\lambda t}} \qquad [9\text{-}20]$$

The minus sign appearing in Equation 9-20 indicates that the number of unstable nuclei *decreases* with time. The activity λN, originally λN_0, falls off as $e^{-\lambda t}$. The activity of unstable nuclei which radiate particles or photons, that is the *radioactivity*, can be measured with a nuclear-radiation

counter over periods of time that are short compared to the half-life of
the decay. This provides then a simple and direct method of measuring λ
or $T_{1/2}$.

The common unit for measuring activity is the *curie*, which is defined
as 3.70×10^{10} disintegrations per second; a related unit, the millicurie
(mc) $= 3.70 \times 10^7$ sec^{-1}. Another unit sometimes used for activity is the
rutherford, defined as 10^6 decays per second.

9-9   Gamma decay   All stable nuclides are ordinarily in their lowest, or
ground, states. If such nuclei are excited and gain energy by photon or
particle bombardment, these nuclides may exist in any one of a number of
excited, quantized energy states. Furthermore, all radioactive nuclides
are initially in energy states from which they decay with the emission of
photons or particles. We shall be concerned here with the decay of a nucleus from an excited state by the emission of a photon. A photon emitted
from a nucleus in an excited state is
called a γ-ray.

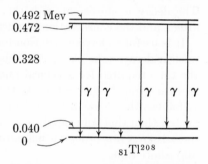

The nuclear energy levels of the
radioactive element thallium, $_{81}$Tl^{208}
(also known as thorium C''), are shown
in Figure 9-13, where the energy of the
nucleus in the ground state is chosen
as zero. The figure also shows the
transitions giving rise to γ-rays. The
spacings of nuclear energy levels range
from tens of Kev to a few Mev, in
contrast to the much smaller separa-
tions associated with atomic energy
levels. These nuclear energy levels
can be inferred from the γ-ray spec-

Figure 9-13. Nuclear energy-level
diagram and γ-ray transitions of
$_{81}$Tl^{208}.

trum emitted when excited nuclei make downward quantum jumps to
lower states.

Several methods are used in γ-ray spectroscopy to measure the energy
or wavelength of γ-rays: the crystal spectrometer (Section 4-3); the photoelectric effect (Section 3-2); the Compton effect (Section 3-4); pair production (Section 3-5); and absorption coefficient measurements (Section
3-7). In addition, a scintillation spectrometer (Section 8-6) can be used to
measure γ-ray energies, inasmuch as the output pulse is proportional to
the γ-ray energy.

Only those nuclear transitions occur for which the conservation laws are
satisfied. For the downward transition from an excited nuclear energy
state E_u to a lower-energy state E_l by the emission of a γ-ray photon of
energy $h\nu$, the conservation of energy requires:

$$\boxed{h\nu = E_u - E_l} \qquad [9\text{-}21]$$

The conservation of linear momentum demands that the total linear momentum following the γ-decay equal the linear momentum before the decay. If the decaying nucleus is originally at rest, it must recoil when the photon is emitted with a momentum equal to the momentum of the photon, $h\nu/c$. Thus,

$$\boxed{h\nu/c = (mv)} \qquad [9\text{-}22]$$

where mv is the momentum of the recoiling nucleus.

In γ-decay the γ-particle is created as the nucleus in the excited state $(Z^A)^*$ decays to a lower state, say, the ground state Z^A. We use here the conventional notation, in which a nucleus in an excited state is designated by an asterisk. Symbolically, we can write

$$\boxed{(Z^A)^* \to Z^A + \gamma}$$

The decay is consistent with the conservation of charge (Ze before, Ze after), and the conservation of nucleons (A before, A after).†

It is useful to have a criterion for judging the relative speed with which nuclear decays take place. For this purpose we define a *nuclear time*, t_n, as the time required for a typical nucleon, having an energy of several Mev and thus traveling at $\sim 0.1c$, to travel a nuclear distance, $\sim 3 \times 10^{-15}$ m. It follows that $t_n = (3 \times 10^{-15}$ m$)/(3 \times 10^7$ m/sec$) \simeq 10^{-22}$ sec. It is expected then that any rapid nuclear decay will have a half-life that is not more than a few orders of magnitude larger than 10^{-22} sec, an immeasurably short time.

The half-life for a typical γ-decay is predicted by theory to be about 10^{-14} sec; such a fast decay cannot be followed in time. But some γ-decays are so strongly forbidden that the half-life is greater than 10^{-6} sec, the minimum half-life that can be readily measured. Such nuclides having a measurably long half-life for γ-decay are called *isomers*. An isomer is not chemically distinguishable from the lower-energy nucleus into which it slowly decays. An extreme example of isomerism is that of niobium, $_{41}Nb^{91}$, which undergoes γ-decay with a half-life of 60 days!

The γ-decay of excited nuclei serves as a direct means of signaling the instability of the nuclei. An analysis of the γ-ray energies allows nuclear energy-level diagrams to be constructed. Any successful nuclear theory must, of course, be capable of predicting in detail the energy levels of nuclei; and although no complete theory presently exists, the energy-level

†Another process which competes with γ-decay is *internal conversion*. In this process a nucleus in an excited state converts its excitation energy internally (within the atom) to one of the inner atomic electrons, bound with an energy E_b; therefore, $E_u - E_l = E_b + K_e$, where K_e is the kinetic energy of the freed electron.

diagrams of many nuclides are partially understood on the basis of the quantum theory.

9-10 Alpha decay The study of γ-decay clearly demonstrates the discreteness of nuclear energy levels. Another observed decay mode of unstable nuclei, also verifying the quantum nature, is α-decay. Certain radioactive nuclei, those for which $Z > 82$, spontaneously decay, with the parent nucleus decaying into a daughter nucleus and a helium nucleus. Inasmuch as the α-particle is a very stable configuration of nucleons, it is perhaps not too surprising to imagine such a group of particles to exist within the parent nucleus prior to α-decay.

We now apply the conservation laws to α-decay. The conservation laws of charge and of nucleons require that

$$\alpha\text{-decay: } _{z}P^{A} \rightarrow _{z-2}D^{A-4} + _{2}\alpha^{4} \qquad [9\text{-}23]$$

where P and D refer to the parent and daughter nuclei, respectively. The pre-subscripts and the post-superscripts give the electric charge in units of e and nucleon numbers respectively; the conservation laws require that the respective sums on both sides of the reaction equation be equal. For example, bismuth 212 decays by α-emission into thallium 208 (in an older nomenclature these elements are called thorium C and thorium C''):

$$_{83}Bi^{212} \rightarrow _{81}Tl^{208} + _{2}\alpha^{4} \qquad [9\text{-}24]$$

Assuming that the radioactive parent is initially at rest, the conservation laws of energy and of linear momentum yield

$$M_{P}c^2 = (M_{D} + M_{a})c^2 + K_{D} + K_{\alpha} \qquad [9\text{-}25]$$

and $\qquad\qquad M_{D}v_{D} = M_{\alpha}v_{\alpha} \qquad\qquad\qquad\qquad [9\text{-}26]$

where M_{P}, M_{D}, and M_{α} are the rest (atomic) masses of the parent, daughter, and α-particle, and the K's and v's are the kinetic energies and velocities of the daughter and α-particle. Non-relativistic expressions for kinetic energy and momentum are used in Equations 9-25 and 9-26 because the energy released in α-decay is never greater than 10 Mev, whereas the α-particle rest energy is 4 Bev.

Obviously, the kinetic energies, K_{D} and K_{α}, can never be negative; therefore, α-decay is energetically possible, by Equation 9-25, only if

$$M_{P} > M_{D} + M_{\alpha} \qquad [9\text{-}27]$$

If the inequality of Equation 9-27 is not satisfied, α-decay simply cannot occur.

The energy released in the decay, $K_{D} + K_{\alpha}$, is called the *disintegration energy* and is represented by the symbol Q. Using Equation 9-25 we can write

$$Q = K_{D} + K_{\alpha} = (M_{P} - M_{D} - M_{\alpha})c^2 \qquad [9\text{-}28]$$

We see from Equations 9-27 and 9-28 that decay is energetically possible only for $Q > 0$.

When one observes an α-decay, it is the energy of the α-particle, K_α, which is usually measured; this can be done, for instance, by finding the range of the α-particle in a cloud chamber, or by measuring the radius of curvature in a magnetic field. Let us see how this measured energy is related to the disintegration energy Q. Squaring Equation 9-26 and multiplying by $\frac{1}{2}$ gives

$$M_D(\tfrac{1}{2} M_D v_D^2) = M_\alpha(\tfrac{1}{2} M_\alpha v_\alpha^2)$$

$$M_D K_D = M_\alpha K_\alpha \qquad [9\text{-}29]$$

The masses of the daughter and alpha particles are approximately $(A - 4)$ and 4 amu, respectively. Equation 9-29 then becomes

$$(A - 4)K_D = 4K_\alpha, \qquad \text{or} \qquad K_D = 4K_\alpha/(A - 4)$$

But, $\qquad Q = K_\alpha + K_D = K_\alpha\,[1 + 4/(A - 4)]$

Therefore, $\qquad K_\alpha = \left(\dfrac{A - 4}{A}\right) Q \qquad [9\text{-}30]$

Equation 9-30 shows that, for *two-particle* emission from an initially unstable nucleus at rest, the α-particle emerges with a *precisely-defined energy*: Q has a precise value, and therefore, so does K_α. The energy spectrum of the emitted α-particles from a radioactive substance in a simple α-decay is shown in Figure 9-14. The α-particles are *mono-energetic*.

Radioactive materials unstable to α-decay are heavy elements with $A \gg 4$. Equation 9-30 shows then that K_α is only slightly less than Q; for this reason, essentially all of the energy released in the decay is carried away as kinetic energy by the emitted α-particle.

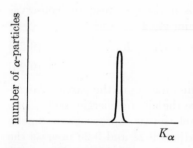

Figure 9-14. Energy spectrum of α-particles from a radioactive substance.

Most α-emitters show a group of discrete α-particle energies, rather than a single energy. This behavior is easily understood in terms of a nuclear energy-level diagram, as shown in Figure 9-15 for the decay of $_{83}\text{Bi}^{212}$ (see Equation 9-24). The parent can decay by α-emission to any one of a number of allowed energy states of the daughter, the ground state as well as excited states. The most energetic α-particles correspond to decay to the ground state of the daughter; the Q used in our analysis is defined for just this case, in that the mass of the daughter was taken to be its mass in the ground state.

A decay to an excited state of the daughter is followed by one or more
γ-emissions leading to the ground state. Because the half-life for γ-decay
is usually very short indeed, these γ-rays appear to be coincident in time
with the α-decays. The energies of the γ-rays are found to be completely
consistent with the differences in the energies of the emitted α-particles.
Both alpha and gamma decay prove conclusively that nuclei have quantized energy states.

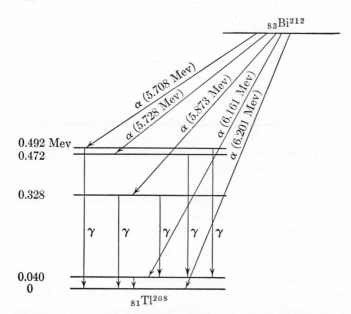

Figure 9-15. Nuclear energy-level diagram showing the decay
of $_{83}Bi^{212}$ by α-emission into the ground and excited states of
$_{81}Tl^{208}$.

At present about 160 α-emitters have been identified. The emitted
α-particles have discrete energies ranging from about 4 to 10 Mev, a
factor of 2, but with half-lives all the way from 10^{-6} sec to 10^{10} yr, a factor
of 10^{23}! The short-lived α-emitters have the highest energies, and conversely, as indicated by the three examples in Table 9-3.

Table 9-3

ALPHA EMITTER	K_α (MEV)	$T_{1/2}$	λ
$_{92}U^{238}$	4.18	4.49×10^9 yr	4.9×10^{-18} sec^{-1}
$_{83}Bi^{212}$	6.09	2.99 hr	6.4×10^{-4} sec^{-1}
$_{85}At^{215}$	8.00	10^{-4} sec	10^4 sec^{-1}

THEORY OF ALPHA DECAY   We wish to examine some of the details of the decay of uranium 238 into thorium 234 by α-emission. Consider Figure 9-16 which shows the potential energy of the daughter, $_{90}\text{Th}^{234}$, as seen by

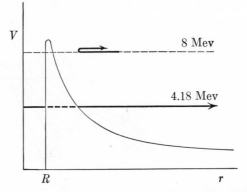

Figure 9-16. Potential energy of the nucleus $_{90}\text{Th}^{234}$ as seen by an α-particle.

an α-particle. When the α-particle is at a greater distance from the center of the nucleus than R, the range of the nuclear force (about 10^{-14} m), the force between the particles is given by Coulomb's law. This is established by α-particle scattering experiments, where α-particles having energies up to 8 Mev are scattered from the thorium nuclei by the Coulomb repulsive force. For distances less than R, an α-particle is subject to a strong attractive force which holds it within the nucleus. But this bound system, composed of the daughter nucleus $_{90}\text{Th}^{234}$ and an α-particle, is just the parent nucleus, $_{92}\text{U}^{238}$. It is assumed that two protons and two neutrons within the parent nucleus unite to form an α-particle, which exists for a time long compared with the nuclear time, 10^{-22} sec.

It is known from experiment that $_{92}\text{U}^{238}$ emits α-particles with an energy of 4.18 Mev, as shown in Figure 9-16. Inasmuch as the potential energy is zero when the α-particle is very far from the daughter nucleus, this kinetic energy, 4.18 Mev, also represents the *total* energy of the particle at any distance from the nucleus. Within the nucleus the total energy of the α-particle is again 4.18 Mev; this represents the algebraic sum of the potential energy (negative) and the kinetic energy (positive). Classically, if the α-particle is within the nuclear "walls," it must move back and forth between these walls indefinitely, striking them roughly 10^{21} times per second. The α-particle cannot penetrate the nuclear walls and escape, for it cannot have a negative kinetic energy. On this basis, it would be impossible for α-decay to take place!

Because α-decay does indeed occur, the classical argument is inapplicable. The explanation of α-decay is to be found in the remarkably successful quantum theory of α-decay, as first proposed by G. Gamow (1928) and R. W. Gurney and E. U. Condon (1928). We can merely indicate the main results here.

In quantum mechanics, the probability of locating the α-particle is given by its wave function $\psi(r)$ (see Section 4-5). The wave function for the potential of Figure 9-16 is that shown in Figure 9-17. It is oscillatory

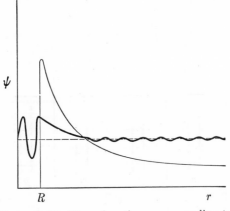

Figure 9-17. Wave function corresponding to the penetration of an α-particle through a nuclear barrier.

within the attractive potential well; ψ is drastically attenuated through the potential barrier; and the wave function is again oscillatory outside the nucleus, with a small, but finite, amplitude. This is interpreted to mean that there is a very small, but finite, probability that an α-particle originally within the nucleus will be found outside the nucleus. This unique, quantum phenomenon is known as the *tunnel effect*. The probability of tunneling through the barrier is strongly dependent on the height and thickness of the barrier, the probability of escape being greater when the α-particle has a higher energy.

One way of visualizing the decay is to imagine the α-particle to bounce between the nuclear walls until it finally escapes by penetrating the barrier. Let us compute the number of tries the α-particle must make before it breaks through the potential barrier. The half-life of $_{92}U^{238}$ is 4.5×10^9 yr $\simeq 10^{17}$ sec; on the average then, an α-particle must make 10^{21} tries per second for 10^{17} second, or, 10^{38} tries altogether, before it escapes.

The quantum theory accounts for α-decay in detail in that it gives theoretically the half-life in terms of fundamental properties of the decaying nucleus. A simplified and approximate formula, resulting from the quantum treatment, gives the decay constant λ in $\sec^{-1}$ in terms of the α-particle energy K_α in Mev as follows:

$$\log_{10} \lambda = 56 - 150/(K_\alpha)^{1/2} \qquad [9\text{-}31]$$

This relation, found by experiment long before the theory of α-decay was developed, is known as the *Geiger-Nuttall rule*.

9-11 Beta decay Beta decay can be defined as that radioactive decay process in which the charge of a nucleus is changed without changing the number of nucleons.

As an example of β instability, consider the three nuclides, $_5B^{12}$, $_6C^{12}$, and $_7N^{12}$, whose proton and neutron occupation levels are shown in Figure 9-18.

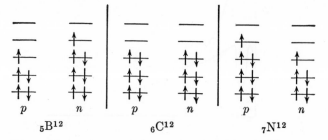

Figure 9-18. Proton and neutron occupation levels of $_5B^{12}$, $_6C^{12}$, and $_7N^{12}$.

These three nuclides are isobars, all having 12 nucleons, but they differ in the proton and neutron numbers, Z and N. Only the carbon nucleus, with 6 protons and 6 neutrons, is stable. The boron nucleus has too many neutrons, and the nitrogen nucleus has too many protons. The unstable $_5B^{12}$ nucleus decays to a lower-energy state by changing one of its nucleons from a neutron into a proton, the last neutron in $_5B^{12}$ (Figure 9-18) jumping to the lowest available proton level. In this process the $_5B^{12}$ nucleus has been transformed into the stable $_6C^{12}$ nucleus, and to preserve the conservation of electric charge, one unit of negative charge must be created. We know that an electron cannot exist *within* the nucleus; therefore, the created electron, or β^--particle, must be emitted from the decaying nucleus, according to the transformation

$$_5B^{12} \rightarrow {_6C^{12}} + \beta^-$$

The decay of $_7N^{12}$ is analogous. This isotope of nitrogen has too many protons, too few neutrons, or too much electric charge for the 12 nucleons. Therefore, $_7N^{12}$ decays to a lower-energy state by converting one of its nucleons from a proton into a neutron, with the last proton jumping to the lowest available neutron level (Figure 9-18). In this decay, the un stable $_7N^{12}$ nucleus is transformed into the stable $_6C^{12}$ nucleus, and charge conservation is preserved by creating a positron, or β^+-particle. Inasmuch as a positron cannot exist within a nucleus, it must be emitted. The decay of $_7N^{12}$ is given by

$$_7N^{12} \rightarrow {_6C^{12}} + \beta^+$$

These decay processes are also shown in Figure 9-19 on the N-versus-Z plot. The nucleus $_6C^{12}$ lies on the stability line, but $_5B^{12}$ lies above it and $_7N^{12}$ below it. The β-decay transformations occur along an isobaric ($-45°$) line in such a way as to bring the unstable nuclides closer to the stability line.

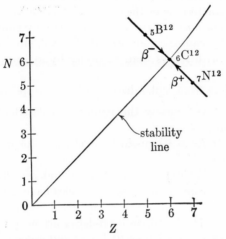

Figure 9-19. The β^+ decay of $_7N^{12}$ and the β^- decay of $_5B^{12}$ into $_6C^{12}$.

Another beta-decay process in which a proton is changed into a neutron within a nucleus is *electron capture* (EC). In electron capture, an atomic orbital electron combines with a proton of the nucleus to change it into a neutron. The number of nucleons is unchanged, but a nuclear proton is converted into a neutron, as in β^+-decay. The electrons of the atom have a finite probability of being found at the nucleus (see Figure 6-7); one of the innermost, or K, electrons has the highest probability of being captured within the nucleus, and the beta-decay process resulting from the nuclear capture of a K electron is often referred to as K *capture*.

No charged particle is emitted in the decay by electron capture. The absorption and annihilation of a particle is equivalent to the creation and emission of its anti-particle; in K capture an electron is absorbed, but in β^+-decay an electron anti-particle, or positron, is emitted. Both processes change a proton into a neutron. An example of an electron capture is that of the decay of unstable $_4Be^7$ into $_3Li^7$, where

$$e_K^- + {}_4\text{Be}^7 \rightarrow {}_3\text{Li}^7$$

The electron-capture process cannot, of course, be identified by an emitted charged particle. It may be inferred from the change in the chemical identity of the element undergoing the decay, or it may be detected by observing the x-*ray photons* emitted when the decay takes place. When a K electron is absorbed into the nucleus there is a hole, or vacancy, in the K shell; this vacancy is filled as higher-lying electrons in outer shells make quantum jumps to the inner vacancies, thereby emitting the characteristic x-ray spectrum. Because the x-ray emission must take place *after* the K electron vacancy is created, that is after the nuclear decay has occurred, the x-rays are characteristic of the *daughter* element, not the parent.

Many hundreds of nuclides are known to decay by emitting an electron, or a positron, or by capturing an orbital electron. In fact, essentially all unstable nuclides with Z less than 82 decay by at least one of the three β-decay processes.

Beta decay differs from alpha decay in several respects:

(1) In β-decay the parent and daughter have the same number of nucleons.

(2) As in γ-emission, the electron (or positron) is created at the time it is emitted.

(3) Whereas γ-ray photons and α-particles are emitted with a discrete spectrum of energies, β-particles have a continuous energy spectrum.

(4) The half-lives for β-decay are never less than 10^{-2} sec, in contrast to γ-decay (as small as 10^{-17} sec) and α-decay (as small as 10^{-7} sec).

β^--DECAY Let us consider β^--decay in somewhat more detail. By the conservation of electric charge and the conservation of nucleons, the β^--decay of a parent nucleus P into the daughter D can be represented by

$$ {}_zP^A \rightarrow {}_{z+1}D^A + {}_{-1}e^0 \qquad [9\text{-}32] $$

For example, boron 12 decays into carbon 12 and an electron with a half-life of 2.7×10^{-2} sec.

$$ {}_5\text{B}^{12} \rightarrow {}_6\text{C}^{12} + {}_{-1}e^0 $$

The conservation of mass-energy requires that the rest mass of the parent *nucleus*, $M_P - Zm_e$, exceed the rest masses of the daughter nucleus, $M_D - (Z + 1)m_e$, and the electron m_e; any excess mass-energy Q, appears as kinetic energy of the particles emerging from the decay. Therefore,

$$ [M_P - Zm_e] = [M_D - (Z + 1)m_e] + m_e + Q/c^2 \qquad [9\text{-}33] $$

or, for β^--decay: $\boxed{M_P = M_D + Q/c^2}$

where M_P and M_D are the neutral *atomic* masses of the parent and daughter respectively, and Q, the disintegration energy, is the energy released in the decay.

Equation 9-33 shows that β^--decay is energetically possible whenever the mass of the parent atom exceeds the mass of the daughter atom; that is, whenever $M_P > M_D$. Moreover, it is found that when β^--decay is energetically possible, it does occur, although the half-life may be extremely long.

The conservation of the linear momentum law requires that the sum of the linear momenta of the particles emerging from the decay equal zero if the decaying nucleus is initially at rest. Let us recall that for α-decay this implies that the α-particle and the daughter nucleus leave the site of the decay in opposite directions, the disintegration energy Q being divided between the kinetic energies of the two particles in such a way that the α-particle and the daughter nucleus each have the same magnitude of linear momentum. Thus, the α-particle and the recoiling daughter nucleus each have precisely defined and discrete energies.

If a similar situation were to apply in β^--decay, that is if the parent nucleus decayed into just *two* particles, the daughter nucleus and the electron, then the energies of the β^--particle and the daughter nucleus would both be precisely defined with $m_e v_e = M_D v_D$, and $Q = K_e + K_D$. Because the mass of the electron is at least several thousand times smaller than that of the daughter nucleus, essentially all of the released energy would be carried by the electron, the kinetic energy of the recoiling daughter nucleus being negligible by comparison. (The magnitudes of the momenta for the two particles would, of course, be equal.) This means then that if β^--decay were altogether analogous to α-decay, with a heavy and a light particle produced by the initially unstable nucleus, the electron would have a precisely defined kinetic energy, approximately equal to the disintegration energy Q. With $M_D \gg m_e$,

$$Q = K_e + K_D \simeq K_e$$

We can compute Q for the β^--decay of $_5B^{12}$ into $_6C^{12}$ directly from atomic masses using Equation 9-33

$$\begin{aligned} \text{mass } _5B^{12} &= 12.01819 \\ \text{mass } _6C^{12} &= \underline{12.00382} \\ M_P - M_D &= 0.01437 \text{ amu} \end{aligned}$$

$$Q = 0.01437 \text{ amu} \times 931 \text{ Mev/amu} = 13.4 \text{ Mev}$$

We might expect then to find *all* electrons emitted with a kinetic energy $K_e \simeq Q = 13.4$ Mev. Is this what is observed?

The distribution in energy of the emitted β^--particles from any particular radioactive element can be measured with a beta-ray spectrometer (see Section 8-8). The result for the $_5B^{12}$ decay is shown in Figure 9-20.

The emitted electrons are *not* mono-energetic! Instead, there is a distribution of electron energies from zero up to the maximum energy, K_{max} = 13.4 Mev. Those very few electrons having this maximum energy, and only those electrons, carry the kinetic energy expected on the basis of a two-particle decay; that is, the measurements show that

$$K_{max} = Q \qquad [9\text{-}34]$$

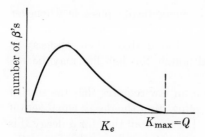

Figure 9-20. Distribution in energy of emitted β^- particles from $_5B^{12}$.

All other electrons, and this means almost all of the emitted electrons, would seem to have too little kinetic energy. In short, there is an apparent violation of the conservation of energy!

Furthermore, observations of individual β^--decays show that the electron and daughter do *not* necessarily leave the site of the disintegration in opposite directions. There is an apparent violation of the conservation of linear momentum! In addition, the angular momentum of the parent nucleus (integral spin, since A is even) can *not* equal the sum of the angular momenta of the daughter (integral spin) and the electron (half-integral spin). There is an apparent violation of the conservation of angular momentum!

We hasten to assure the reader that the fundamental conservation laws of energy, linear momentum and angular momentum are, in fact, *not* violated in β^--decay. This is so because our account of β^--decay has not included a third particle, the *neutrino* ("little neutral one"), also emitted in β-decay. The existence of the neutrino was first suggested by W. Pauli in 1930 as an alternative to abandoning the conservation principles; its existence was directly confirmed by experiment in 1956. It is now known with certainty that a radioactive nucleus decays in β^--emission into *three* particles: the daughter nucleus, the electron, and the neutrino.

Let us examine the properties of the neutrino. We shall see that all of the difficulties we have mentioned above disappear by virtue of the neutrino's participation in the β^--decay process.

THE NEUTRINO ν Electric charge = 0; rest mass = 0; linear momentum, p; and *total* relativistic energy, E; $E = pc$; intrinsic angular momentum, or spin = $\frac{1}{2}\hbar$.

The neutrino has zero electric charge. Charge is conserved in β^--decay *without* the neutrino. The neutrino cannot interact with matter by producing ionization. It interacts very, very weakly with nuclei, and is, therefore, essentially undetectable.

Energy *is* conserved for those very few electrons in β^--decay which are emitted with the maximum kinetic energy, $K_{max} = Q$. Therefore, the neutrino mass must be very small compared to the electron mass, and there are good theoretical reasons for taking the neutrino mass to be *exactly zero*. Since the neutrino has a zero rest mass and rest energy, it must, like a photon, always travel at the speed of light. Therefore, the total relativistic energy E of a neutrino is related to its relativistic momentum p by $E = pc$ (see Equation 2-46).

Let us consider again the conservation of energy and of linear momentum in β^--decay, assuming now that a neutrino, as well as an electron, is created in the decay and carries away energy and momentum. The conservation of energy requires that

$$Q = K_D + K_e + K_\nu \simeq K_e + K_\nu \qquad [9\text{-}35]$$

(The kinetic energy of the neutrino K_ν is also its total energy; $E_\nu = K_\nu$.) The conservation of linear momentum requires that the total *vector* linear momentum of the three particles add up to zero, as shown in Figure 9-21. It is no longer necessary that the particles leave the site of the decay along

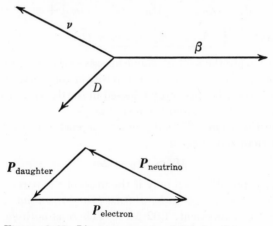

Figure 9-21. Linear momenta of the daughter nucleus, electron, and neutrino in β^- decay.

the *same* straight line; there are a variety of ways in which the separate momentum vectors can be arranged to add up to zero, satisfying Equation 9-35 in every instance. The electron and daughter nucleus will usually *not* move along the same straight line in opposite directions; but if they do, the neutrino momentum and energy can be zero, and from Equation 9-35, $K_e = K_{max} = Q$, in agreement with observation for the most energetic electrons. In all other decays, the essentially undetectable neutrino will carry energy and momentum and the electron will necessarily have a kinetic energy less than K_{max}. In two-particle decay, the emerging particles are mono-energetic; in three-particle decay, the particles are polyenergetic.

Consider the angular momentum, or spin, of the neutrino. It is $\frac{1}{2}$ in units of $\hbar$. In the β^--decay of $_5B^{12}$ into $_6C^{12}$, the parent and daughter nuclei both have integral nuclear spins; the electron has a spin of $\frac{1}{2}$. Therefore, when the neutrino's angular momentum is included, total angular momentum is conserved in β^--decay.

β^+-DECAY   The general relation giving the β^+-decay is

$$\boxed{{}_zP^A \rightarrow {}_{(Z-1)}D^A + {}_{+1}e^0 + \nu} \qquad [9\text{-}36]$$

A neutrino is emitted in β^+-decay, as in β^--decay. Positron decay can occur only if mass-energy is conserved. This means that the rest mass of the parent *nucleus* must exceed the sum of the rest masses of the daughter nucleus and the positron (the neutrino rest mass is zero). Any excess energy appears as kinetic energy of the three particles emerging from the decay. Therefore,

$$[M_P - Zm_e] = [M_D - (Z-1)m_e] + m_e + Q/c^2$$

or, for β^+-decay: $$\boxed{M_P = M_D + 2m_e + Q/c^2} \qquad [9\text{-}37]$$

where M_P and M_D are the *neutral atomic* masses of the parent and daughter, respectively, m_e is the rest mass of the positron (or electron), and Q, the disintegration energy, is the energy released in the decay and shared by the positron, daughter nucleus, and the neutrino.

We see from Equation 9-37 that β^+-decay is energetically possible, that is Q is greater than zero, only if

$$M_P > M_D + 2m_e \qquad [9\text{-}38]$$

Positron decay is possible then only if the mass of the parent atom *exceeds* the mass of the daughter atom *by at least two electron masses*, $2(0.00055)$ amu, or its energy equivalent, 1.02 Mev. (There is nothing fundamental in the appearance of the two electron masses here; this merely reflects the fact that neutral *atomic* masses rather than *nuclear* masses are used.)

Let us compute the Q of the positron decay of $_7N^{12}$ (half-life, 0.0125 sec) into $_6C^{12}$

$$\begin{aligned}
\text{mass } _7N^{12} &= 12.02290 \\
\text{mass } _6C^{12} &= \underline{12.00382} \\
M_P - M_D &= 0.01908 \\
2m_e &= \underline{0.00110} \\
Q/c^2 &= 0.01798 \text{ amu}
\end{aligned}$$

$$Q = 0.01798 \text{ amu} \times 931 \text{ Mev/amu} = 16.7 \text{ Mev}$$

This energy of 16.7 Mev is shared among the decay products, the positron, the neutrino, and the $_6C^{12}$ nucleus. When the positron energies are measured with a beta-ray spectrometer, it is found that there is a distribution of energies up to a maximum of 16.7 Mev, in agreement with the mass differences.

In actuality, the masses of short-lived radioactive electron- or positron-emitters cannot be easily measured; it is possible to compute the mass of a radioactive element by measuring the maximum energy of the emitted β-particle. The β^+-decay can be readily identified because an emitted positron will undergo annihilation with an electron and produce two annihilation photons, each with an energy of 0.51 Mev, the rest energy of an electron (or positron). Thus, β^+-decay is always characterized by the appearance of $\frac{1}{2}$ Mev annihilation quanta.

ELECTRON CAPTURE   The general relation for electron capture (EC) is written

$$_{-1}e^0 + {}_ZP^A \rightarrow {}_{Z-1}D^A + \nu \qquad [9\text{-}39]$$

an orbital electron is captured by the parent nucleus $_ZP^A$; the products of the decay are the daughter nucleus $_{(Z-1)}D^A$ and a neutrino.

Mass-energy is conserved when the energy released in the decay Q is equal to the sum of the rest masses entering the reaction less the sum of the rest masses leaving the reaction. Therefore,

$$m_e + [M_P - Zm_e] = [M_D - (Z-1)m_e] + Q/c^2$$

or, for electron capture $\boxed{M_P = M_D + Q/c^2}$ [9-40]

where M_P and M_D are again the neutral atomic masses of the parent and daughter respectively. Equation 9-40 shows that electron capture is energetically possible if the atomic mass of the parent exceeds that of the daughter.

Consider our earlier example of the decay of $_4Be^7$ (half-life, 54 days) into $_3Li^7$ by electron capture.

$$\text{mass } {}_4Be^7 = 7.01916$$
$$\text{mass } {}_3Li^7 = \overline{7.01823}$$
$$M_P - M_D = Q/c^2 = \overline{0.00093} \text{ amu}$$
$$Q = 0.00093 \text{ amu} \times 931 \text{ Mev/amu} = 0.85 \text{ Mev}$$

(Note that β^+-decay is energetically forbidden for $_4Be^7$.)

In this EC decay, 0.85 Mev is released. Where does it go? Unlike β^--decay and β^+-decay, electron capture produces only *two* particles. By the conservation of momentum, these two particles, the neutrino and the daughter nucleus, must move in opposite directions with the same magnitude of momentum, the sum of their kinetic energies being the disintegration energy, $Q = 0.85$ Mev. Because only two particles appear in electron capture, the neutrino and the recoiling nucleus each have precisely-defined energies. The neutrino's rest mass is zero; therefore, almost all of the energy is carried by the unobservable neutrino, and the nucleus recoils with an energy of only several electron volts. Some very delicate experiments have confirmed, nevertheless, that the recoiling nuclei are mono-energetic and that their energy is precisely the amount required to satisfy the conservation laws of momentum and of energy. Without the accompanying neutrino in electron capture, this decay process is completely inexplicable.

An energy-level diagram showing the decay of $_5B^{12}$ and $_7N^{12}$ into the stable nuclide $_6C^{12}$ is shown in Figure 9-22. By convention, a nuclide

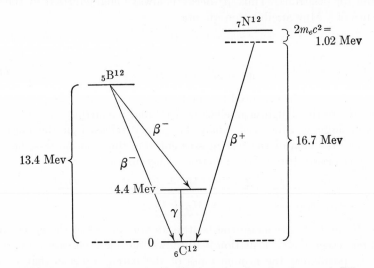

Figure 9-22. Energy-level diagram showing the decay of $_5B^{12}$ and $_7N^{12}$ into $_6C^{12}$.

undergoing β^--decay is shown on the left with respect to the daughter, and the nuclide undergoing β^+-decay or electron capture is placed on the right. (Z increases toward the right). We see that the decay of $_5B^{12}$ consists of electron emission to two states of $_6C^{12}$, the ground state and an excited state, 4.4 Mev above the ground state. Decay from the excited state to the ground state by the emission of a γ-ray photon is essentially simultaneous with the corresponding β^--decay. This near coincidence of the electron and photon emission can be verified experimentally by using two detectors, one for electrons and one for photons, and by noting that the pulses in the two detecting systems are coincident in time (within the resolving time of the detecting instruments).

An energy-level diagram for the radioactive element, $_{29}Cu^{64}$, which decays by β^--emission into $_{30}\text{Zn}^{64}$, and also by β^+-emission and electron capture into $_{28}\text{Ni}^{64}$, is shown in Figure 9-23.

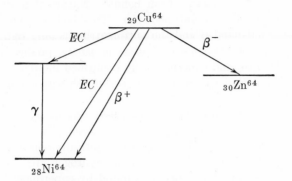

Figure 9-23. Energy-level diagram showing the decay of $_{29}\text{Cu}^{64}$ by β^- emission, β^+ emission, and electron capture.

It is interesting to note a general rule that applies to any two isobaric nuclides that differ in Z by one. The atomic mass of one must exceed that of the other. The nucleus of the heavier atom can, therefore, decay into the nucleus of the lighter atom either by β^--decay or by electron capture. Therefore, no two neighboring isobars can both be stable against β-decay, and this indeed is in accord with observation of the known nuclides (Figure 9-7).

The four basic reactions associated with β-decay are shown in Equation 9-41.

$$
\begin{array}{rl}
\beta^-\text{-decay:} & n \rightarrow p + e + \bar{\nu} \\
\beta^+\text{-decay:} & p \rightarrow n + \bar{e} + \nu \\
\text{electron capture:} & e + p \rightarrow n + \nu \\
\text{neutrino absorption:} & \nu + p \rightarrow n + \bar{e}
\end{array}
\qquad [9\text{-}41]
$$

The symbol e represents the electron (charge, -1); $\bar{e}$ represents the positron (charge, $+1$), the electron's anti-particle. The symbol ν designates a neutrino, and $\bar{\nu}$ designates an *anti-neutrino*.

We have, up to this point, recognized only one type of neutrino; there are, in reality, two neutrinos, one the anti-particle of the other. This distinction may seem to be completely formal, but it is not. It has been confirmed in subtle experiments, that the anti-neutrino, the neutrino's anti-particle, has the direction of its spin, or intrinsic angular momentum, along the direction of its linear momentum. For an anti-neutrino, the sense of rotation of the spin is clockwise, when viewed from behind, leading to a "right-handed" *helicity*, or *spirality*, associated with the anti-neutrino. On the other hand, the angular momentum and the linear momentum of the neutrino are in opposite directions, giving the neutrino a left-handed helicity, or spirality. The sense of rotation of the neutrino's spin is counter-clockwise when viewed from behind. Nature thus distinguishes between the neutrino and anti-neutrino. This lack of symmetry—the neutrino is *only* left-handed and the anti-neutrino is *only* right-handed—is a manifestation of the *non-conservation of parity*, predicted by C. N. Yang and T. D. Lee, and experimentally confirmed in 1957 by C. S. Wu *et al*. The principle that nature does *not* distinguish between left and right, the conservation of parity, is violated in β-decay.

The basic β^--decay shown in Equation 9-41, the decay of the neutron into a proton and electron, occurs for a *free neutron*, not merely for a neutron bound with a nucleus. The decay is energetically allowed because the mass of the neutron exceeds the mass of the hydrogen atom, $_1H^1$, Q being .78 Mev. The half-life of this decay is found by experiments of extreme difficulty to be 12.8 minutes. Because a free neutron is typically absorbed in less than 10^{-3} second when it passes through materials, the decay of the neutron is usually unimportant in situations involving free neutrons.

The basic β^+-decay process, in which a proton is converted into a neutron, positron, and a neutrino, is *not* permitted for a free proton, inasmuch as the mass on the left-hand side of the reaction is less than the mass on the right-hand side. Positron decay is possible only for protons bound within a nucleus.

The electron capture process is closely related to the β^+-decay process, of course. Note, in Equation 9-41, that the second reaction becomes the third reaction when the anti-electron is transferred to the left side, thereby becoming an electron. This follows from the general rule that the emission of a particle is equivalent to the absorption of an anti-particle, and conversely. Using this rule, together with the permitted reversal of the reaction arrow, it is seen that all four β reactions in Equation 9-41 are equivalent.

The final reaction in Equation 9-41 is that in which an anti-neutrino combines with a proton to become a neutron and a positron. Although the relative probability for neutrino capture in this process is extremely small, it has been observed directly with the very large neutrino flux from a nuclear reactor by C. L. Cowan and F. Reines in 1956, thus directly confirming the existence of the neutrino (strictly, the anti-neutrino). The anti-neutrino absorption process is identified by observing the neutron and the positron produced simultaneously when the anti-neutrino is captured by a proton. The neutron is observed by detecting the photon emitted from an excited nucleus that has absorbed the neutron; the positron is observed by detecting annihilation photons. The difficulty of this experiment can be appreciated from the fact that a neutrino or anti-neutrino has only 1 chance in 10^{12} of being captured while traveling completely through the Earth. Because neutrinos have such a very small probability of interacting with matter and being absorbed, a large fraction of the energy released in all β-decay processes is effectively lost.

9-12 Natural radioactivity We have discussed the three common modes of radioactive decay, α-, β-, and γ-decay, without concern as to how unstable nuclides are produced. It is customary to divide radioactive nuclides into two groups: (1) the unstable nuclides found in nature, which are said to exhibit *natural radioactivity*, and (2) the man-made unstable nuclides (usually produced by bombarding nuclei with particles) which are said to exhibit *artificial radioactivity*. To date, approximately 1000 artificially radioactive nuclides have been produced and identified. The number of identified isotopes for a given element varies: hydrogen has two stable and 1 unstable isotopes; xenon has 9 stable and 14 unstable isotopes. Nuclear reactions and the accompanying artificial radioactivity that can be produced by them will be discussed in Chapter 10. Here we discuss only the naturally radioactive nuclides.

It is believed that a cataclysmic cosmological event occurred about 5 billion years ago, at the time of the formation of the universe, in which *all* nuclides, stable and unstable, were formed in varying amounts. Those unstable nuclides with half-lives much less than 5×10^9 yr have long since decayed into stable nuclides; however, there are 14 unstable nuclides whose half-lives are comparable to, or longer than, the age of the universe, and which, therefore, are still found in measurable amounts in nature. These long-lived naturally radioactive materials are listed in Table 9-4.

Table 9-4. Naturally radioactive elements with half-lives comparable to or greater than the age of the universe

Element	Decay mode	Half-life in years	
$_{19}K^{40}$	β^-, EC	1.2×10^9	
$_{23}V^{50}$	EC	4×10^{14}	
$_{37}Rb^{87}$	β^-	6.2×10^{10}	
$_{49}In^{115}$	β^-	6×10^{14}	
$_{57}La^{138}$	β^-, EC	1.0×10^{11}	
$_{58}Ce^{142}$	α	5×10^{15}	All decay into stable daughters
$_{60}Nb^{144}$	α	3×10^{15}	
$_{62}Sm^{147}$	α	1.2×10^{11}	
$_{71}Lu^{176}$	β^-	5×10^{10}	
$_{75}Re^{187}$	β^-	4×10^{12}	
$_{78}Pt^{192}$	α	10^{15}	
$_{90}Th^{232}$	α	1.39×10^{10}	10 radioactive generations
$_{92}U^{235}$	α	7.07×10^9	11 radioactive generations
$_{92}U^{238}$	α	4.51×10^9	14 radioactive generations

The first 11 nuclides in Table 9-4 all decay into stable daughters. The last three nuclides in this table, thorium 232, uranium 235, and uranium 238, are all very heavy. They decay into daughters which are themselves radioactive and which decay in turn into still other radioactive daughters through several generations until finally a stable nuclide is reached. Thus, there are three naturally *radioactive series*, each series beginning with a very long-lived radioactive nuclide, whose half-life exceeds that of any of its descendants. The final stable nuclides into which these three series decay are isotopes of lead, $_{82}Pb^{208}$, $_{82}Pb^{207}$, and $_{82}Pb^{206}$. The age of the Earth can be estimated by measuring the relative amounts of the long-lived parent of the series and the appropriate stable lead isotope.

Figure 9-24 shows the complete decay scheme for all of the unstable nuclides of the so-called *thorium series*, plotted on a proton-versus-neutron

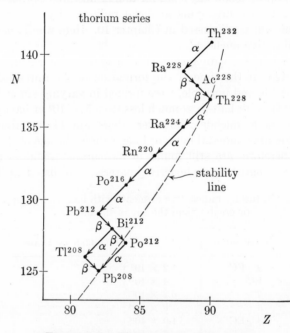

Figure 9-24. Thorium radioactive series.

diagram. A displacement on the figure in which Z and N are both decreased by 2 represents an α-decay, and a displacement in the figure in which Z increases by 1 and N decreases by 1 represents a β^--decay. The stability line in Figure 9-24 represents the least unstable nuclides for the particular value of A. Both α- and β^--decay often produce the daughter nuclei in excited nuclear states, which leads to subsequent γ-decay (see Figures 9-15 and 9-22).

All nuclides with $A > 209$ are unstable. We can say that all nuclides with more than 209 nucleons are too big; consequently, such heavy nuclides must lose nucleons to become more stable. The only decay mode in which a heavy naturally radioactive nucleus loses nucleons is α-emission, in which both Z and N are reduced by 2. But we see from Figure 9-24 that an α-decay tends to leave the daughter displaced to the left of the stability line. Decay by β^--emission is required to bring the daughters of α-emission closer to the stability line. Some nuclides, such as $_{83}Bi^{212}$ in the thorium series, are unstable to either α-decay or β^--decay, and a *branching* of the decay series occurs at such a nuclide.

The first nuclide in the thorium series has a mass number of 232, which is divisible by 4. All other nuclides in this series also have A values which are divisible by 4, inasmuch as the only decay mode that changes the number of nucleons is α-decay, and for α-emission the number of nucleons is reduced by 4. Therefore, the A values for any of the members of the thorium series can be written as $4n$, where n is an integer. The members of the *uranium series*, beginning with $_{92}U^{238}$, are characterized by A values which can be written as $4n + 2$; and the members of the third naturally radioactive series, the so-called *actinium series*, beginning with $_{92}U^{235}$, all have A values which are given by $4n + 3$.

There is a fourth radioactive series whose members have A values given by $4n + 1$, none of whose members have a half-life comparable to the age of the universe. Nuclides in this series do not, therefore, occur naturally. They can, however, be produced by nuclear reactions with the very heavy elements of the other series (for example, the capture of a neutron by $_{92}U^{236}$ followed by a β^--decay). The fourth, $4n + 1$ series is named the *neptunium series*, after the longest-lived nuclide in this series, $_{93}Np^{237}$, which decays with a half-life of 2.25×10^6 yr. The decay modes and half-lives of the neptunium, uranium, and actinium series are shown in Table 9-5.

The naturally radioactive materials show a tremendous range of half-lives, from $_{90}Th^{232}$ with $T_{1/2} = 1.39 \times 10^{10}$ yr to the nuclide $_{84}Po^{213}$ (polonium in the neptunium series) with $T_{1/2} = 4.2 \times 10^{-6}$ sec. How are such extraordinarily long or short half-lives measured? Clearly, it is impractical to measure $T_{1/2}$ by following in time the change in the activity of the decaying nuclei, and indirect methods must be used.

Consider first the measurement of very long half-lives. The fundamental radioactive decay law gives, following Equation 9-20,

$$\text{activity} = \lambda N = \lambda N_0 e^{-\lambda t} \qquad [9\text{-}42]$$

Equation 9-42 can be rearranged into

$$\lambda = \frac{\text{activity}}{N} \qquad [9\text{-}43]$$

Table 9-5

NEPTUNIUM $(4n + 1)$ SERIES

$$_{93}Np^{237} \xrightarrow[2.20 \times 10^6 y]{\alpha} {}_{91}Pa^{223} \xrightarrow[27.4d]{\beta} {}_{92}U^{233} \xrightarrow[1.63 \times 10^5 y]{\alpha} {}_{90}Th^{229} \xrightarrow[7000y]{\alpha} {}_{88}Ra^{225}$$

$$\xrightarrow[14.8d]{\beta} {}_{89}Ac^{225} \xrightarrow[10.0d]{\alpha} {}_{87}Fr^{221} \xrightarrow[4.8m]{\alpha} {}_{85}At^{217} \xrightarrow[.018s]{\alpha} {}_{83}Bi^{213}$$

URANIUM $(4n + 2)$ SERIES

$$_{92}U^{238} \xrightarrow[4.51 \times 10^9 y]{\alpha} {}_{90}Th^{234} \xrightarrow[24.5d]{\beta} {}_{91}Pa^{234} \xrightarrow[1.14m]{\beta} {}_{92}U^{234} \xrightarrow[2.48 \times 10^5 y]{\alpha} {}_{90}Th^{230}$$

$$\xrightarrow[8.0 \times 10^4 y]{\alpha} {}_{88}Ra^{226} \xrightarrow[1622y]{\alpha} {}_{86}Em^{222} \xrightarrow[3.825d]{\alpha} {}_{84}Po^{218}$$

ACTINIUM $(4n + 3)$ SERIES

$$_{92}U^{235} \xrightarrow[7.13 \times 10^8 y]{\alpha} {}_{90}Th^{231} \xrightarrow[24.6h]{\beta} {}_{91}Pa^{231} \xrightarrow[3.43 \times 10^4 y]{\alpha} {}_{89}Ac^{227}$$

The decay constant λ can be computed from Equation 9-43 if the activity and the number of surviving nuclei N are known. If $T_{1/2}$ is very long compared with t, the period of observation, $\lambda t \ll 1$, and $N = N_0 e^{-\lambda t} \simeq N_0$. That is, if a radioactive material decays very slowly, the number of atoms present is essentially constant over the period of observation.

For example, measurements show that a 1.0 milligram sample of uranium, $_{92}U^{238}$, emits 740 α-particles per minute. The atomic weight of this isotope is close to 238; therefore, the number of uranium atoms N in 1 milligram is $(10^{-3}\text{ gm})/(238)(1.66 \times 10^{-24}\text{ gm}) = 2.53 \times 10^{18}$ atoms. Using Equation 9-43 we have

$$\lambda = \frac{(740/60)\text{ disintegrations/sec}}{2.53 \times 10^{18}\text{ atoms}} = 4.87 \times 10^{-18}\text{ sec}^{-1}$$

and $$T_{1/2} = 0.693/\lambda = 4.51 \times 10^9\text{ yr}$$

Now consider the measurement of the half-life of a very short-lived member of a radioactive series. Assume that the radioactive descendants of the first member of the series remain with the original material. Then, after sufficient time has elapsed, *all* members of the series will be present together, nuclei of any particular member being formed by its parent while other nuclei of this member decay into its daughter. The relative amounts of the several member nuclides will be constant when the decay series reaches *radioactive equilibrium*, each nuclide decaying at the same rate at which it is formed. This means then that the activity, λN, of each member of the series is precisely the same as that of any other member.

$$(\text{activity})_1 = (\text{activity})_2 = (\text{activity})_3,\text{ etc.}$$

$$\lambda_1 N_1 = \lambda_2 N_2 = \lambda_3 N_3,\text{ etc.} \qquad [9\text{-}44]$$

or $$\frac{N_1}{(T_{1/2})_1} = \frac{N_2}{(T_{1/2})_2} = \frac{N_3}{(T_{1/2})_3},\text{ etc.} \qquad [9\text{-}45]$$

The half-life $(T_{1/2})_s$ of a very *short-lived* radio-nuclide in equilibrium with a *longer*-lived nuclide, $(T_{1/2})_l$, of the same series is given then by

$$(T_{1/2})_s = (T_{1/2})_l (N_s/N_l) \qquad [9\text{-}46]$$

where (N_s/N_l) is the relative numbers of these two members. Equation 9-45 shows that the relative numbers of atoms of the several members of a naturally radioactive series in equilibrium are directly proportional to their respective half-lives. Therefore, long-lived nuclides will be relatively abundant, and short-lived nuclides scarce.

A sample of a naturally radioactive substance emits α-, β-, and γ-rays simultaneously, because *all* members of the radioactive series are present and decaying. The α- and β-emissions involve changes in A and/or in Z; the γ-emissions result in decays from excited states. The early investigators in radioactivity distinguished among the three types of radiation emitted from radioactive substances by noting the deflection of the emitted rays in a magnetic field. The α-rays were found to be deflected in the same direction as positively-charged particles; the β-rays were deflected as negative particles; and the γ-rays were undeflected by the field. Furthermore, it was observed that the penetration of the radioactive emanations increased in the order α, β, and γ; and thus these rays were labeled by the first three letters of the Greek alphabet. All three types of nuclear radiation from naturally radioactive materials have energies up to several Mev; and until the development of high-energy particle accelerators in the early 1930's, these radioactive materials served as the only sources of high-energy nuclear particles for bombarding other nuclei.

Uranium 238 is the heaviest nuclide found in nature. Still heavier and relatively short-lived nuclides, called *transuranic elements*, are man-made in the sense that they can be produced by bombarding heavy elements with energetic particles. All transuranic elements up to nobelium 254, $_{102}No^{254}$, have been produced, at least momentarily, and identified.

The term, natural radioactivity, usually refers to those radioactive materials produced in the very distant past, and to the descendants of such long-lived nuclides. There are, however, radioactive materials *continuously* being produced in nature by nuclear collisions of high-energy particles from the cosmic rays (Section 10-9) with nuclei in the Earth's upper atmosphere. An example of this is the production of carbon 14, $_6C^{14}$, by the collision of neutrons with nitrogen nuclei, according to the reaction

$$_7N^{14} + {_0n^1} \rightarrow {_6C^{14}} + {_1p^1}$$

This radioactive isotope of carbon, or *radio-carbon*, decays by β^--emission with a half-life of 5760 years.

$$_6C^{14} \rightarrow {_7N^{14}} + \beta^-$$

A small fraction of the CO_2 molecules in the air will thus contain radioactive C^{14} atoms, in place of stable C^{12} atoms. Living organisms exchange CO_2 molecules with their surroundings, utilizing the carbon atoms, both C^{12} and C^{14}, in their structure. When the organisms die, their intake of C^{14} atoms ceases, and from that moment on the number of C^{14} atoms relative to the number of C^{12} atoms will decrease by virtue of the C^{14} decay, only half of the original C^{14} atoms being present after 5760 years. This offers a very sensitive method of determining the age of organic archeological objects, such as wood: by determining the relative fraction of C^{12} and C^{14} atoms. The number of C^{14} atoms is determined by measuring the radio-carbon activity and using Equation 9-43. This ingenious method of measuring the age of organic relics (to 25,000 years old) was originated by W. F. Libby in 1952 and is known as *radio-carbon dating*.

A second cosmic-ray nuclear reaction producing a naturally radioactive element continuously is

$$_7N^{14} + _0n^1 \rightarrow _6C^{12} + _1H^3$$

where $_1H^3$, called *tritium* (with a nucleus known as a *triton*), is a heavy radioactive isotope of hydrogen. Tritium decays into the stable helium isotope, $_2He^3$, with a half-life of 12.4 years by β^--emission.

$$_1H^3 \rightarrow _2He^3 + \beta^-$$

9-13 Summary

PROPERTIES OF THE NUCLEAR CONSTITUENTS

PROPERTY	PROTON	NEUTRON
Mass (amu)	1.007593	1.008982
Charge (units, e)	1	0
Spin (units, $\hbar$)	$\frac{1}{2}$	$\frac{1}{2}$
Magnetic moment (units, $e\hbar/2M_pc$)	2.79	-1.91

PROPERTIES OF THE NUCLEAR FORCE

Attractive and much stronger than the Coulomb force.

Short-range, $\sim 3 \times 10^{-15}$ m.

Charge independent; all three nucleon interactions, $n-p$, $p-p$, and $n-n$, are approximately equal.

NOMENCLATURE

Nucleon: proton or neutron

Atomic number, Z: number of protons

Neutron number, N: number of neutrons

Mass number, A: total number of nucleons $(Z + N)$

Nuclide: nucleus with a particular Z and a particular N

Isotopes: nuclides with same Z

Isotones: nuclides with same N

Isobars: nuclides with same A

PROPERTIES OF THE NUCLIDES

For stable nuclides, $N \simeq Z$ for small A, and $N > Z$ for large A.

The nuclear radius is given by, $R = r_0 A^{1/3}$, where $r_0 = 1.4 \times 10^{-15}$ m (neutron scattering) or $r_0 = 1.1 \times 10^{-15}$ m (electron scattering). All nuclei have the same nuclear density.

The total binding energy E_b of a nucleus Z^A is given by

$$E_b/c^2 = ZM_H + (A - Z)M_n - M(Z^A)$$

where all masses are those of the neutral atoms. For $A > 20$, $E_b/A \simeq 8$ Mev/nucleon.

In the liquid-drop model of the nucleus, the forces between nucleons are assumed to be analogous to the forces between molecules in a liquid; the important contributions to the binding energy are the volume energy, the surface energy (important for low A), and the Coulomb energy (important for high A). In the shell model of the nucleus, the quantum aspects of the nucleons are taken into account; the shell model is capable of accounting for nuclear spins and magnetic moments.

In the decay of all unstable nuclei, the conservation laws of electric charge, nucleons, mass-energy, and momentum are satisfied.

The law of radioactive decay is $N = N_0 e^{-\lambda t}$, where the decay constant, λ, the probability per unit time that any one nucleus will decay, is related to the half-life by $T_{1/2} = 0.693/\lambda$.

RADIOACTIVE DECAY MODES

	ALPHA	BETA	GAMMA
Particle	helium nucleus	electron, positron	photon
Half-lives	10^{-6} sec to 10^{10} yr	$> 10^{-2}$ sec	10^{-17} to 10^5 sec (isomer)
Energies	4 to 10 Mev	few Mev	Kev to few Mev
Decay mode	$_zP^A \rightarrow {}_{z-2}D^{A-4} + {}_2\alpha^4$	β^- $\quad _zP^A \rightarrow {}_{z+1}D^A + {}_{-1}e^0 + \bar{\nu}$ β^+ $\quad _zP^A \rightarrow {}_{z-1}D^A + {}_{+1}e^0 + \nu$ EC $\quad _{-1}e^0 + {}_zP^A \rightarrow {}_{z-1}D^A + \nu$	$(Z^A)^* \rightarrow Z^A + \gamma$
Disintegration energy equation (all neutral atom masses)	M_P $= M_D + M_\alpha + Q/c^2$	β^- $\quad M_P = M_D + Q/c^2$ β^+ $\quad M_P = M_D + 2m_e + Q/c^2$ EC $\quad M_P = M_D + Q/c^2$	$E_u = E_l + h\nu$
Energy distribution of decay products	Mono-energetic	β^- and β^+: poly-energetic EC: mono-energetic	Mono-energetic

Neutrino properties

$$\text{mass:} \quad 0$$
$$\text{charge:} \quad 0$$
$$\text{spin:} \quad \tfrac{1}{2}$$
$$\text{neutrino capture:} \quad \bar{\nu} + p \rightarrow n + \beta^{+}$$

Natural radioactivity

Series name	Longest-lived member	Type	Number of members
Thorium	$_{90}\text{Th}^{232}$	$4n$	13
Actinium	$_{92}\text{U}^{235}$	$4n + 3$	15
Uranium	$_{92}\text{U}^{238}$	$4n + 2$	18
Neptunium	$_{93}\text{Np}^{237}$	$4n + 1$	13

REFERENCES

Bitter, F., *Nuclear Physics*. Reading, Massachusetts: Addison-Wesley Publishing Company, Inc., 1950. This 200-page book is characterized by its emphasis on physical arguments.

Evans, R. D., *The Atomic Nucleus*. New York: McGraw-Hill Book Company, Inc., 1955. An advanced, rigorous, thorough, and heavy 972-page treatise.

Kaplan, I., *Nuclear Physics*. Reading, Massachusetts: Addison-Wesley Publishing Company, Inc., 1955. A comprehensive, yet elementary, book with many illustrative examples and tables of useful nuclear data.

Lapp, R. E., and H. L. Andrews, *Nuclear Radiation Physics* 2nd ed. Englewood Cliffs, New Jersey: Prentice-Hall, Inc., 1954. An elementary treatment of nuclear physics. A large portion of this book is devoted to the decay of unstable nuclei, their detection and measurement.

Leighton, R. B., *Principles of Modern Physics*. New York: McGraw-Hill Book Company, Inc., 1959. Chapters 15 and 16, on radioactivity and the systematics of stable nuclides, contain short but lucid accounts of these phenomena.

PROBLEMS

9-1 * (a) Show that the energy difference in joules between the two nuclear spin orientations of a proton is $(2.82 \times 10^{-26})B$ where B is the flux density of the external magnetic field in weber/m^2. (b) Compute the frequency of a photon whose absorption will reorient the proton spin for $B = 0.3$ weber/m^2. (c) In what region of the electromagnetic spectrum does this fall? This is the basis of the phenomenon known as *nuclear magnetic resonance*.

9-2 Show that in a p-p or an n-p elastic scattering experiment, no particles are scattered more than 90°. (Assume equal masses and low energy.)

9-3 Calculate the de Broglie wavelength of (a) a 100 Mev neutron; (b) a 500 Mev electron; (c) a 10 Mev nucleon.

9-4 A 5.22 Mev γ-ray photon collides with a deuteron initially at rest. The γ-ray is annihilated and the deuteron dissociates into a proton and neutron. The proton is observed to emerge from the collision at right angles to the incident photon direction. (a) Calculate the X- and Y-components of the emerging neutron's momentum. (b) What are the kinetic energies of the proton and neutron?

9-5 What nucleus (Z^A) has a radius that is one-half that of U^{238}?

9-6 Show that the density of a nucleus is approximately 2×10^{17} kg/m³ $\simeq 10^9$ tons/inch³.

9-7 Calculate the approximate ratio of the nuclear density to the atomic density of the hydrogen isotope $_1H^1$ in its ground state. (Use the Bohr radius as a measure of atomic size.)

9-8 Show that two protons separated by a distance equal to the classical electron radius (ke^2/m_ec^2) have a Coulomb energy equal to the rest energy of an electron, $\frac{1}{2}$ Mev.

9-9 Show that the quantity, ke^2/m_ec^2, the so-called classical electron radius, is closely equal to 2 r_0, where $r_0 = 1.4 \times 10^{-15}$ m and m_e is the electron mass.

9-10 Show that the separation energy of the least-tightly-bound neutron in $_8O^{16}$ is 15.6 Mev.

9-11 (a) Compute the separation energy of the least-tightly-bound nucleon in the stable nuclides $_{12}Mg^{24}$, $_{11}Na^{23}$, $_{10}Ne^{22}$, $_{10}Ne^{21}$, $_{10}Ne^{20}$, and $_9F^{19}$. (b) How does the separation energy for a nuclide with even A compare with that of the neighboring nuclides of odd A? (c) Explain this behavior in terms of the filling of proton and neutron levels.

9-12 (a) Show that an even-Z nuclide usually has many more stable isotopes than an odd-Z nuclide. (b) Between $_8O^{16}$ and $_{16}S^{32}$, there is one stable isotope for each odd-Z nuclide, and three stable isotopes for each even-Z nuclide. Explain this behavior in terms of the filling of neutron and proton shells.

9-13 * Why does hydrogen, $_1H^1$, show the greatest deviation from the whole-number rule of any element. (Hint: Examine Figure 9-10.)

9-14 * The equality of the n-n and p-p forces can be verified by considering *mirror nuclides*, such as $_6C^{13}$ ($Z = 6$ and $N = 7$) and $_7N^{13}$ ($Z = 7$ and $N = 6$). The total binding energies and masses of mirror nuclides would be identical if it were not for the neutron-proton mass difference and the difference in the Coulomb energy arising from the difference in Z. The average Coulomb energy for a *pair* of protons in a nucleus of mass number A is given by $6ke^2/5R$, where $R = (1.4 \times 10^{-15}$ m$)A^{1/3}$. (a) Calculate the expected mass difference between $_6C^{13}$ and $_7N^{13}$, and (b) confirm this result (and the charge independence of the nuclear force) by calculating the $_6C^{13} - _7N^{13}$ mass difference from the fact that the maximum energy of positrons emitted by $_7N^{13}$ is 1.20 Mev.

9-15 The binding energy per nucleon of the α-particle, 7.2 Mev, is particularly high as compared with that of neighboring nuclides. Is it reasonable that no stable nuclides exist with $A = 5$ and $A = 8$ ($_3Li^5$ and $_4Be^8$)?

9-16 Show that one gram of radium ($T_{1/2} = 1622$ yr) has an activity of 1.00 curie. (This was the basis of the original definition of the curie.)

9-17 The β-emitter, $_{82}Pb^{214}$, has a decay constant of 4.31×10^{-4} sec^{-1}. What amount of radioactive Pb^{214} will have an activity of one millicurie?

9-18 * Show that the mean life T_{av} of a radionuclide having a decay constant λ is given by $T_{av} = \int_{N_0}^0 t \, dN / \int_{N_0}^0 dN = 1/\lambda$.

9-19 The half-life of a radioactive nuclide is to be determined by measuring the activity of a sample at 10 minute intervals. If the measured activities are found to be: 7.3, 4.2, 2.5, 1.5, 0.9, 0.5, 0.3 millicuries, what is the half-life?

9-20 A radioactive element has a decay constant λ. Some one nucleus is known to have survived decay for a period of time of 10 half-lives. What is the probability that this particular nucleus will decay in a further time of one half-life?

9-21 Strontium 90 decays with a half-life of 28 years. How long must one wait until only 10 per cent of an original amount of Sr^{90} remains?

9-22 A certain radioactive material emits 12.0 microwatts of nuclear radiation at some instant of time and 1.5 microwatts of nuclear radiation after 3.6 hours. What is its half-life?

9-23 A certain radioactive material has an activity of 1.0 curie at one time and an activity of 1.0 rutherford 10 minutes later. What is the decay constant of the radionuclide?

9-24 * The half-life of C^{14} is 5760 years. Compute the radioactive power output per activity (in microwatts/millicurie) for this radioactive substance.

9-25 The range R in milligrams/cm^2 of β-particles is given for energies E greater than 2.5 Mev by the empirical relation, $R = 530E - 106$.

It is found that the β-particles emitted by Cu^{62} are absorbed by 5.33 mm of aluminum (density, 2-70 gm/cm³). What is (a) the maximum energy, and (b) the charge of the β-particles? (E is in Mev.)

9-26 Verify, by examining Figure 9-7, that adjacent stable isobaric pairs (differing in Z by 1) do not occur in nature.

9-27 Tritium ($_1H^3$) decays into helium 3 by β^--decay, the maximum kinetic energy of β^--particles being 19 Kev. No γ-rays are emitted. Compute the $_1H^3 - _2He^3$ mass difference, and compare this value with the values obtained from Appendix IV.

9-28 A $_{27}Co^{60}$ nucleus undergoes β^--decay into an excited state of $_{28}Ni^{60}$, with subsequent successive emissions of 1.17 Mev and 1.33 Mev γ-ray photons. What is the energy with which the Ni^{60} nucleus recoils when it emits the 1.33 Mev photon? (Note that the recoil energy is sufficiently large to break a typical chemical bond.)

9-29 Oxygen 14 decays by positron emission to an excited state of $_7N^{14}$, from which decay takes place to the ground state by γ-emission. The maximum kinetic energy of the observed positrons is 1.8 Mev, and the γ-ray energy is 2.3 Mev. Compute the mass of $_8O^{14}$ from the mass of $_7N^{14}$ and compare this value with the accepted value in Appendix IV.

9-30 * Beryllium 7 decays by electron capture into $_3Li^7$. Compute the (a) energy and (b) momentum of the neutrinos emitted in this decay. (c) With what energy does the $_3Li^7$ nucleus recoil?

9-31 The unstable nuclide $_{20}Ca^{41}$ decays by K-capture. (a) What fraction of the energy released is carried away by the neutrino? (b) Of what element are the x-rays emitted from a radioactive source of $_{20}Ca^{41}$ characteristic? The mass of $_{20}Ca^{41}$ is 40.975328 amu.

9-32 What is the minimum energy of an anti-neutrino for the reaction, $\bar{\nu} + p \rightarrow n + \bar{e}$, to occur?

9-33 * Some free 10 ev neutrons (half-life, 12.8 min) decay in flight into protons. (a) What is the maximum possible kinetic energy of the protons? (b) Through what distance do the neutrons travel so that the neutron flux is reduced to one-half?

9-34 A 2×10^{-4} gm sample of $_{90}Th^{230}$ has an activity of 3.86 μc. What is the half-life of $_{90}Th^{230}$?

9-35 * A $_6C^{14}$ nucleus, initially at rest, decays into a $_7N^{14}$ nucleus by β^--emission. (a) What is the maximum speed with which the $_7N^{14}$ nucleus can recoil? (b) If the recoil velocity of $_7N^{14}$ nucleus were zero, what fraction of the energy released in the $_6C^{14}$ decay would appear as kinetic energy of the electron?

9-36 (a) Determine whether it is energetically possible for a $_{19}K^{40}$ nuclide to decay by neutron emission, β^+-emission, electron capture, or β^--emission. (b) Calculate the energy released in the energetically possible decays of part (a). (c) Confirm your results in part (a) by reference to Figure 9-7.

9-37 Which of the following decay modes of $_7N^{16}$ are energetically possible: proton decay, neutron decay, α-decay, and β^--decay?

9-38 By what modes can $_7N^{13}$ decay?

9-39 What properties of a neutrino and a photon are (a) identical and (b) different?

9-40 Assume, for simplicity, that there were equal amounts of the radioactive nuclides $_{92}U^{238}$ and $_{93}Np^{237}$ at the time of the formation of the universe, and that no further production of these nuclides has taken place since that time (5 billion years). What would be the relative abundance of these two nuclides today?

9-41 * One gram of carbon from a living organism undergoes 12 disintegrations per minute. (a) What is the ratio of C^{14} to C^{12} atoms in such a material? (b) An organic relic having 220 grams of carbon is found to have a C^{14} activity of 930 disintegrations per minute. How long has this material been dead?

9-42 * Ignoring any γ-decay and assuming (for simplicity) that only a single particle is emitted by any member of a radioactive series in equilibrium, show that the total activity of the series is just the activity of any one member times the number of generations in the series.

9-43 Assume that an α-particle confined within a nucleus of 10^{-14} m has an energy of 4 Mev. How many collisions does it make per second with the nuclear walls?

9-44 Radium 226 is α-unstable and decays by emitting α-particle groups having energies of 4.777, 4.593, and 4.34 Mev. (a) Show the decay scheme on an energy-level diagram. (b) What γ-ray energies will be observed with the α-decays?

9-45 An α-emitter decays with the emission of two distinct groups of α-particles, having kinetic energies of K_1 and K_2. Show that γ-rays with an energy of $(K_1 - K_2)A/(A - 4)$ are emitted, where A is the mass number of the parent nucleus.

9-46 The energies of α-particles from $_{83}Bi^{212}$ are determined by noting that the α-particles are deflected by a magnetic field to move in a circle where $Br = 3.542 \times 10^{-1}$ weber/m. Compute (a) the α-particle energy, and (b) the disintegration energy.

9-47 The range R (cm) in air of an α-particle having an energy E between 4 and 8 Mev is given closely by the empirical formula, $R = 0.318 E^{3/2}$. The nuclide $_{83}Bi^{211}$ undergoes α-decay with the emission of two groups of α-particles, one with a range of 4.98 cm and the second with a range of 5.43 cm in air. Compute (a) th~

α-particle energies, (b) the disintegration Q and (c) the energy of the γ-ray expected to be found with this decay.

9-48 * In some naturally radioactive series there are as many as 3 successive α-decays. For example, in the neptunium series, Table 9-5, $_{89}Ac^{225}$ decays into $_{83}Bi^{213}$ by the successive emissions of 3 α-particles. (a) Show that the emission of a single $_6C^{12}$ nucleus by actinium-225 is energetically possible. (b) From the theory of α-decay (Figure 9-16) argue why the decay by C^{12}-emission is very improbable.

T E N

NUCLEAR REACTIONS

10-1 Low-energy nuclear reactions In the last chapter we saw that unstable nuclei will decay spontaneously, changing their nuclear structure without external influence. One can, however, induce a change in the structure of nuclei by bombarding them with energetic particles. Such a collision, in which the identity or characteristics of the struck particle are changed, is known as a *nuclear reaction*.

Thousands of nuclear reactions have been produced and identified since Rutherford observed the first reaction in 1919. The bombarding particles were, until the development of charged-particle accelerators in the 1930's, those emitted from radioactive substances. It is now possible to accelerate charged particles to energies of several Bev, and when such very-high-energy particles strike nuclei, they produce a severe disruption of the struck nuclei and may create such new and strange particles as *mesons* and *hyperons*.

We shall be concerned in this section with so-called *low-energy* nuclear reactions; that is, with reactions in which the incident particles have energies no greater than, say, 20 Mev. All low-energy nuclear reactions have several features in common:

(1) The bombarding particle is a light particle, such as an α-particle, γ-ray, proton, deuteron, or neutron.

(2) The reactions typically involve the emission of *one* of these light particles.

(3) No mesons or hyperons are created.

We shall illustrate several types of nuclear reactions with examples that have been important in the development of nuclear physics.

In the first observed nuclear reaction, Rutherford (1919) used 7.68 Mev α-particles from the naturally radioactive element, $_{84}Po^{214}$ (also known as radium C'). These α-particles were sent through a nitrogen gas, most of them being either undeflected by the nitrogen nuclei or elastically scattered in close encounters with the nuclei. Rutherford found, however, that in a few collisions (1 in 50,000), protons were produced, according to the nuclear reaction

$$_7N^{14} + {}_2He^4 \rightarrow {}_1H^1 + {}_8O^{17}$$

In this reaction an α-particle strikes a nitrogen 14 nucleus producing a proton and an oxygen 17 nucleus. Rutherford identified the emitted light particles as protons by measuring their range, which exceeded that of the incident α-particles (see Section 8-2). The identity of the emitted particles as protons for this reaction has since been established by measurements of the charge-to-mass ratio with a magnetic field. A cloud-chamber schematic of this nuclear reaction (Figure 10-1) shows the tracks of the incident α-particle, the emitted proton, and the recoiling oxygen nucleus. This reaction represents an *induced transmutation* of the element nitrogen into a stable isotope of oxygen; α- or β-radioactive decay represents, of course, a *spontaneous transmutation* of one element into another.

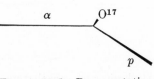

Figure 10-1. Representation of a cloud-chamber photograph of the reaction, $_7N^{14}$ (α, p) $_8O^{17}$.

The conservation laws of electric charge and of nucleons are satisfied in all nuclear reactions; therefore, the pre-subscripts, giving the electric charge of the particles, and the post-superscripts, giving the number of nucleons in each particle, each have the same sum on both sides of the reaction equation. The nuclear reaction can be written in abbreviated form as follows:

$$_7N^{14}(\alpha, p)_8O^{17}$$

where the light particles going into and out of the reaction are written in parentheses between the symbols for the target and product nuclei.

Until 1932 all nuclear reactions were produced by the relatively high-energy α-particles or γ-rays from naturally radioactive materials. In this year, J. D. Cockcroft and E. T. S. Walton, using a 500 Kev accelerator, observed the first nuclear reaction produced by artificially accelerated charged particles. They found that α-particles were emitted when a lithium target was struck by 500 Kev protons according to the reaction

$$_3\text{Li}^7 + _1\text{H}^1 \rightarrow _2\text{He}^4 + _2\text{He}^4$$

or,

$$_3\text{Li}^7(p, \alpha)_2\text{He}^4$$

The emitted α-particles each had an energy of 8.9 Mev; thus, an energy of 0.5 Mev had been put into the reaction, whereas 17.8 Mev was released as kinetic energy of the emerging particles. Here is a clear example of the release of nuclear energy. The total amount of energy released was trifling, of course, inasmuch as most of the nuclear collisions between the incident protons and lithium target nuclei did *not* result in nuclear disintegrations.

In the two reactions we have mentioned, the product nuclei were stable. The first nuclear reaction leading to an unstable, or radioactive, product nucleus was observed by I. Joliot-Curie and F. Joliot in 1934. In this reaction an aluminum target is struck by α-particles, leading to:

$$_{13}\text{Al}^{27} + _2\text{He}^4 \rightarrow _0n^1 + _{15}\text{P}^{30}$$

or,

$$_{13}\text{Al}^{27}(\alpha, n)_{15}\text{P}^{30}$$

The product nuclide, $_{15}\text{P}^{30}$, is not stable, but decays with a half-life of 2.5 minutes into a stable isotope of silicon by β^+-emission.

$$_{15}\text{P}^{30} \rightarrow _{14}\text{Si}^{30} + \beta^+$$

The production of unstable nuclides, which spontaneously disintegrate by the law of radioactive decay, is a feature of many nuclear reactions, and the artificially produced unstable nuclides are said to exhibit *artificial radioactivity*. Indeed, nuclear reactions provide the only means of obtaining artificial radioactive isotopes. Such materials, called *radio-isotopes*, are chemically identical to the element's stable isotopes. If a small amount of radio-isotope is added to stable nuclides of the same element, this unstable isotope can serve, through its radioactivity, as a *tracer* of this chemical element; that is, the presence and concentration of the element can be determined by measuring the radio-isotope's activity.

The discovery of the neutron came as a result of a nuclear reaction observed in 1930 by W. Bothe and H. Becker.

$$_4\text{Be}^9 + _2\text{He}^4 \rightarrow _0n^1 + _6\text{C}^{12}$$

or,

$$_4\text{Be}^9(\alpha, n)_6\text{C}^{12}$$

It was thought at first that the products of this nuclear reaction were a γ-ray and the stable nucleus, $_6\text{C}^{13}$, rather than a neutron and $_6\text{C}^{12}$, because an extremely penetrating radiation was found to result from the bombardment of beryllium by α-particles. Then, Curie and Joliot in 1932 found that when the penetrating radiation from this reaction fell on paraffin (which consists largely of hydrogen), protons with energies close to 6 Mev were emitted. This behavior was interpreted in terms of the Compton effect, in which a high-energy photon makes a Compton collision with a proton and ejects it from the paraffin. The photon energy required to transfer 6 Mev to protons is easily found from Equation 3-17 to be nearly 60 Mev. It is easy to show, from the conservation of mass-energy, that *less* than 60 Mev of energy would be released in the $_4Be^9(\alpha, \gamma)_6C^{13}$ reaction. Therefore, the photon hypothesis was untenable.

The proper interpretation of these experiments was given by J. Chadwick in 1932. Chadwick showed that all experimental results were consistent with the assumption that an uncharged, and therefore highly penetrating, particle having a mass close to that of the proton was emitted in the reaction of α-particles on beryllium. By the conservation of mass-energy, such a particle would be emitted with an energy of close to 6 Mev; and when the neutron struck a proton head-on, the neutron would come to rest, transferring its momentum and energy to the proton.

Neutrons are emitted in many nuclear reactions, and these emitted neutrons can themselves be used as bombarding particles. One of the important neutron-induced reactions is that in which a neutron is captured by a target nucleus and a γ-ray photon is radiated. This reaction is known as neutron *radiative capture*. For example,

$$_{13}Al^{27} + {}_0n^1 \rightarrow \gamma + {}_{13}Al^{28}$$

or,

$$_{13}Al^{27}(n, \gamma)_{13}Al^{28}$$

The product nucleus, $_{13}Al^{28}$ (an unstable isotope of the target nucleus) decays by β^--decay.

$$_{13}Al^{28} \rightarrow {}_{14}Si^{28} + \beta^-$$

Inasmuch as the neutron has no electric charge, the neutron-radiative-capture process can occur when a neutron of almost any energy strikes (almost) any nucleus; the heavier isotope thus produced is frequently radioactive, and the absorption of neutrons is, therefore, a common means of producing radio-isotopes.

Another important type of reaction resulting from neutron bombardment is that in which a charged particle, such as a proton or α-particle, is emitted. Such reactions offer a method of detecting neutrons because the emitted charged particles produce detectable ionization. One reaction frequently used in neutron detection is

$$_5B^{10} + {}_0n^1 \rightarrow {}_2He^4 + {}_3Li^7$$

or,

$$_5B^{10}(n, \alpha)_3Li^7$$

Photo-disintegration is the nuclear reaction in which the absorption of a

γ-ray photon results in the disintegration of the absorbing nucleus. One such reaction is

$$_{12}Mg^{25} + \gamma \rightarrow {}_1H^1 + {}_{11}Na^{24}$$

or,

$$_{12}Mg^{25}(\gamma, p)_{11}Na^{24}$$

followed by $_{11}Na^{24} \rightarrow {}_{12}Mg^{24} + \beta^-$.

We have listed as illustrative examples only a few of the many known nuclear reactions. One general statement concerning low-energy nuclear reactions, involving the light particles (p, n, d, α, and γ) either as bombarding particles or as emerging light particles, can be made: nuclear reactions with essentially all possible combinations of ingoing and outgoing light particles are found to occur.

One special type of low-energy nuclear reaction is that of *nuclear fission*. In this reaction, which we will discuss in more detail in Section 10-7, a low-energy neutron is captured by a very heavy nucleus, and the resulting aggregate splits into two moderately-heavy nuclei along with a few neutrons.

10-2 The energetics of nuclear reactions Consider the generalized nuclear reaction $X(x, y)Y$, where x is the bombarding particle, X the target nucleus, y is the emergent light particle, and Y is the product nucleus. The target nucleus is assumed to be at rest ($K_X = 0$) and the kinetic energies of x, y, and Y are designated by K_x, K_y, and K_Y, respectively.

The disintegration energy, or *Q-value*, of a radioactive decay was defined as the total energy released in the decay (Equation 9-28). In a similar way the Q-value of a nuclear reaction is defined as the total energy released in the reaction; that is, the kinetic energy out of the reaction less the kinetic energy into the reaction.

$$Q = (K_y + K_Y) - K_x \qquad [10\text{-}1]$$

The total relativistic energy of a particle is the sum of its rest energy and its kinetic energy. The conservation of mass-energy requires, then, that

$$(m_x c^2 + K_x) + (M_X c^2) = (m_y c^2 + K_y) + (M_Y c^2 + K_Y) \qquad [10\text{-}2]$$

where m_x, M_X, m_y, and M_Y are the *rest* masses. Combining Equations 10-1 and 10-2 gives

$$Q/c^2 = (m_x + M_X) - (m_y + M_Y) \qquad [10\text{-}3]$$

Equation 10-3 shows that Q/c^2, the mass equivalent of the energy released in the reaction, is simply the total rest mass into the reaction less the total rest mass out of the reaction. Thus, the nuclear energy released in a reaction can be computed directly from the masses of the participating particles; or, if one of the masses (most often that of the product heavy nucleus) is not known with precision, it can be computed if the Q-value is measured.

Nuclear energy is released in a reaction when $Q > 0$; such a reaction, in which mass is converted into the kinetic energy of the outgoing particles, is known as an *exothermic*, or *exo-ergic*, reaction. A reaction in which nuclear energy is absorbed, or consumed, with $Q < 0$, is called *endothermic*, or *endo-ergic*. An endothermic reaction can be thought of as an inelastic collision in which the identity of the colliding particles changes and kinetic energy is at least partially converted into mass.

A rather special sort of reaction is that in which the incoming and outgoing particles are identical; that is, $x = y$ and $X = Y$. If no kinetic energy is lost ($Q = 0$), the reaction is known as an elastic collision; if energy is lost ($Q < 0$), the reaction is an inelastic collision.

Let us compute the Q of the reaction, $_3\text{Li}^7(p, \alpha)_2\text{He}^4$. We may use the *neutral atomic* masses of the four particles because, in going from nuclear masses to atomic masses, an equal number of electron masses is added to both sides of Equation 10-2.

$$\text{mass } _1\text{H}^1 = 1.00814 \text{ amu}$$
$$\text{mass } _3\text{Li}^7 = \underline{7.01823}$$
$$m_x + M_X = 8.02637$$

$$\text{mass } _2\text{He}^4 = 4.00387$$
$$\text{mass } _2\text{He}^4 = \underline{4.00387}$$
$$m_y + M_Y = 8.00774$$

$$Q/c^2 = (m_x + M_X) - (m_y + M_Y) = 8.02637 - 8.00774 = 0.01863 \text{ amu}$$

$$Q = 0.01863 \text{ amu} \times (931 \text{ Mev/amu}) = 17.3 \text{ Mev}$$

This reaction is exothermic, with 17.3 Mev released. Thus, the total kinetic energy of the two outgoing α-particles will exceed the kinetic energy of the incoming proton by 17.3 Mev. In the original Cockcroft-Walton experiment, the incident protons had an energy of 0.50 Mev; therefore, the total energy carried by the two α-particles is $(17.3 + 0.5)$ Mev $= 17.8$ Mev, or about $\frac{1}{2}$ $(17.8) = 8.9$ Mev for each α-particle. The measured energy of the α-particles was in good agreement with this expectation. This, as well as all other reactions for which the masses and the kinetic energies of the participating particles are known, gives striking confirmation of the relativistic mass-energy equivalence.

The $_3\text{Li}^7(p, \alpha)_2\text{He}^4$ reaction that we have just discussed is atypical in that $y = Y$ in this reaction. Because of this, the two particles emerging from the site of the collision have nearly equal kinetic energies and momenta (magnitude). A more usual reaction is one in which $M_Y \gg m_y$, for which, then, $K_Y \ll K_y$.

Energy is released in an exothermic reaction; therefore, it is energetically possible for an exothermic reaction to occur even though the energy of the bombarding particle is nearly zero, although the probability for the occurrence of such a reaction may be very small. On the other hand, an endothermic reaction cannot occur unless the incident particle carries kinetic energy. At first thought, it might appear that an endothermic reaction with a Q of, say, -5 Mev would be energetically possible if 5 Mev of kinetic energy were carried into the collision by the bombarding particle; that is, if $K_x = 5$ Mev. But this is *not* true. The value of K_x must, in fact, exceed the magnitude of the Q-value for the reaction to go. The reason for this apparent anomaly is that linear momentum must be conserved in any nuclear reaction, and because of this, a fraction of the incident-particle energy is unavailable. These matters are resolved when we apply the conservation-of-momentum law to nuclear reactions.

10-3 The conservation of momentum in nuclear reactions The total linear momentum of any isolated system subject to no net external force is constant in magnitude and direction; this conservation law is found to hold for nuclear collisions and reactions, as well as for macroscopic systems. Therefore, the total momentum of particles interacting under the influence of their mutual forces in a nuclear reaction must be the same before and after the reaction.

If the target nucleus X is at rest in the laboratory, the total momentum of the system before the collision is simply $m_x v_x$, the momentum of the incident light particle. (We assume that the energies of the particles are never more than a few Mev, so that the classical expressions for momentum and kinetic energy apply.) The total linear momentum of particles y and Y emerging from the reaction must, therefore, add as vectors to yield a vector along the direction of the incident light particle whose magnitude is $m_x v_x$, as shown in Figure 10-2. This

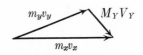
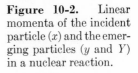

Figure 10-2. Linear momenta of the incident particle (x) and the emerging particles (y and Y) in a nuclear reaction.

means that it is impossible for a nuclear reaction to result in the particles y and Y both being at rest, for then the total momentum after the reaction would be zero.

It is convenient to state the conservation-of-momentum law in terms of the momentum of the center-of-mass of the system. The center-of-mass is at that point at which all of the mass of the system can be imagined to be concentrated. The momentum of the center-of-mass remains unchanged throughout a nuclear reaction; hence, the total momentum of the system, before and after the collision, is equal to the momentum of the center-of-mass. The momentum of the system before the reaction, $m_x v_x$, is equal to the momentum of the center-of-mass. After the reaction, the particles y and

Y leave the site at which they were formed in such a way that their total momentum is also equal to that of the center-of-mass.

Consider the motion of the center-of-mass as seen in the laboratory. Figure 10-3 shows the center-of-mass lying along a line connecting x and

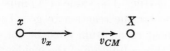

(a) before collision   (b) after collision

as seen in the laboratory

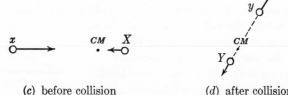

(c) before collision   (d) after collision

as seen from the center of mass

Figure 10-3. Nuclear collision or reaction as seen (a) and (b) in the laboratory, and (c) and (d) from the center of mass.

X, its distance to the respective particles being in the inverse ratio of their masses. As m_x moves to the right with a speed v_x, the center-of-mass moves to the right at a slower speed v_{cm}. The total momentum of the system before (or after) the reaction is therefore,

$$m_x v_x = (m_x + M_X)v_{cm}$$

and the speed of the center-of-mass with respect to the laboratory is

$$v_{cm} = v_x \left(\frac{m_x}{m_x + M_X} \right) \qquad [10\text{-}4]$$

Let us now view the collision of x and X from a coordinate frame attached to the center-of-mass. The situation is as shown in Figure 10-3, with both

x and X approaching the center-of-mass in opposite directions. Clearly, X approaches the center-of-mass with a speed v_{cm}. The momenta of x and X are equal and opposite: the *total momentum of x and X in the center-of-mass coordinate system is zero*. Therefore, the total momentum of y and Y, as viewed from the center-of-mass system, must also be zero; and the y and Y particles must always leave the origin in opposite directions with the same magnitude of momentum.

Consider a completely inelastic collision. In the center-of-mass system this would be a collision in which the x and X particles collide, unite, and then remain at rest at the center-of-mass. The total momentum before the collision is zero, and the total momentum after the collision is zero. In a coordinate system at rest in the laboratory the momentum before the collision is not zero, and the momentum after the collision, even when it is inelastic, cannot be zero. For this reason, not all of the energy that goes into the reaction, $K_x = \frac{1}{2} m_x v_x^2$ (as viewed in the laboratory system), can be used up in a collision in which energy is dissipated; that is, in an endothermic reaction. Only a portion of the K_x energy of the incident particle can be consumed in the reaction; the remainder represents energy that must be carried by the center-of-mass, which continues in its uniform motion. We label the energy carried by the center-of-mass K_{cm}. It is given by

$$K_{cm} = \frac{1}{2}(m_x + M_X)v_{cm}^2$$

Using Equation 10-4, we have

$$K_{cm} = \frac{1}{2}(m_x + M_X)\left[\frac{m_x v_x}{m_x + M_X}\right]^2 = \frac{1}{2} m_x v_x^2 \left(\frac{m_x}{m_x + M_X}\right)$$

$$K_{cm} = K_x\left(\frac{m_x}{m_x + M_X}\right) \qquad [10\text{-}5]$$

That portion of the energy carried by x which can be dissipated in the reaction is the *available energy*, E_a. Therefore, E_a is the total energy (in the laboratory) before the reaction (K_x) less the energy necessarily carried by the center-of-mass (K_{cm}).

$$E_a = K_x - K_{cm} = K_x\left[1 - \frac{m_x}{m_x + M_X}\right]$$

$$\boxed{E_a = K_x\left(\frac{M_X}{m_x + M_X}\right)} \qquad [10\text{-}6]$$

Equation 10-6 shows that only a fraction, $M_X/(m_x + M_X)$, of K_x is available to be dissipated, or consumed, in the reaction. If a reaction is endothermic, with $Q < 0$, the reaction cannot go unless an amount of energy K_x is supplied to the colliding particles. Thus, the reaction will occur only if the available energy, E_a, equals or exceeds the magnitude of Q. The value of K_x for which the reaction just becomes possible is called the *threshold energy* K_{th}; therefore, $K_x = K_{th}$ when $E_a = -Q$. Equation 10-6 then becomes

$$-Q = K_{th}\left(\frac{M_X}{m_x + M_X}\right)$$

or,

$$\boxed{K_{th} = -Q\left(\frac{M_X + m_x}{M_X}\right)} \qquad [10\text{-}7]$$

Consider the endothermic reaction, $_7N^{14}(\alpha, p)_8O^{17}$. From a comparison of the masses of the participating particles, it is easy to show that $Q = -1.19$ Mev. We can take M_X to be 14 and m_x to be 4. Therefore, the threshold energy, or the minimum α-particle energy required for this reaction to occur is, from Equation 10-7, $K_{th} = -(-1.19 \text{ Mev})(18/14) = 1.53$ Mev. This result is confirmed by experiment. When α-particles with energies less than 1.53 Mev strike nitrogen, no protons are released; after the threshold has been reached and exceeded, the reaction goes, as indicated by the appearance of protons (see Figure 10-4).

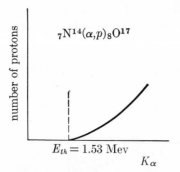

$E_{th} = 1.53$ Mev

K_α

Figure 10-4. Number of protons from the $_7N^{14}(\alpha, p)_8O^{17}$ reaction as a function of α-particle energy, indicating a threshold of 1.53 Mev.

The Q-values for endothermic reactions can, therefore, be determined quite directly by observing the threshold energy of the reaction and applying Equation 10-7. In the center-of-mass system, the y and Y particles leave the site of the collision in opposite directions. In the laboratory this is not true, however, inasmuch as the center-of-mass continues to the right after the reaction takes place, and there are a variety of directions in which y and Y can move and still have their respective momenta add up to that of the center-of-mass.

10-4 Cross section The decay of a radioactive material is characterized not only by the energy released in the decay products, but also by the half-life, or decay constant, for the disintegration process. For an initially unstable nucleus, the decay constant λ gives a measure of the probability for the occurrence of the decay in time. We have examined how the conservation laws of energy and of momentum apply to nuclear reactions, and we now wish to introduce a quantity that measures the probability of occurrence in space of a nuclear reaction. This quantity is called the reaction *cross section*. The conservation laws tell us whether the reaction is possible; the cross section will tell us whether the reaction is probable, and how probable.

Consider Figure 10-5, which shows a number of target nuclei (X) exposed to an incident beam of particles (x). Each X nucleus has associated with it an area σ, called the cross section, which is imagined to be oriented at right angles to the incident particles (regarded as mass points). The cross-sectional areas are assumed to be so small that, in a reasonably thin target material, no one nucleus is hidden from the incident particles by other nuclei. The area of the cross section is so chosen that, if an incident x particle strikes the area σ, the reaction $X(x, y)Y$ will take place; and if an incident particle misses the area σ, this particular reaction does not occur. The intrinsic probability for the occurrence of a nuclear reaction is, therefore, directly proportional to its cross section σ.

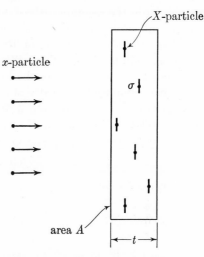

Figure 10-5. Target nuclei X in a target of thickness t, area A, and cross section σ being struck by x particles.

We take the number of incident particles in a thin foil of thickness t and area A to be n_i; and the number of these particles which undergo the nuclear reaction $X(x, y)Y$ is n_r. The quantity n_r is the number of y, or Y, particles produced. The number of target nuclei per unit volume is N, each with nuclear-reaction cross section σ. The total number of nuclei in the target foil is then $N(At)$, and the total exposed area resulting in reactions is σNAt. The ratio of the x particles undergoing reactions to the total number incident on the foil, namely n_r/n_i, must be equal to the ratio of the area σNAt to the total foil area A. Thus, $n_r/n_i = \sigma NAt/A$; or,

$$\boxed{n_r/n_i = \sigma Nt} \qquad [10\text{-}8]$$

(This derivation for reaction cross section is analogous to that for scattering cross section in Section 5-1.)

Equation 10-8 shows that the probability, n_r/n_i, that an incident particle will undergo a nuclear reaction is proportional to: the reaction cross section σ, the number of target nuclei per unit volume N, and the thickness of the target foil t. The most common unit for measuring nuclear cross sections is the *barn* $= 10^{-24}$ cm^2 $= 10^{-28}$ m^2. (Inasmuch as a nuclear cross section of 10^{-24} cm^2 is relatively large, to have an incident particle hit a nuclear target having a cross section of 1 barn is as easy as hitting the side of a barn.) The cross section varies from one reaction to another. Furthermore, σ is usually dependent on the energy of the bombarding particle.

Consider, as an example, the radiative capture of 500 Kev neutrons by aluminum in the reaction $_{13}Al^{27}(n, \gamma)_{13}Al^{28}$. The neutron-capture cross section for aluminum has been measured to be 2 millibarns $= 2 \times 10^{-31}$ m^2. Suppose that a neutron flux of 10^{10} neutrons/(cm^2-sec) is incident on an aluminum foil 0.20 mm thick. What is the number of neutrons captured per second in a 1 cm^2 area of the foil?

We can compute the density of aluminum atoms, or nuclei, N, from the ordinary density of aluminum, 2.70 gm/cm^3, Avogadro's number, and the atomic weight of aluminum, 27. Therefore,

$$N = (2.7 \text{ gm/cm}^3)(6.02 \times 10^{23} \text{ atoms/mole})/(27 \text{ gm/mole})$$
$$= 6.02 \times 10^{22} \text{ nuclei/cm}^3.$$

From Equation 10-8

$$n_r = n_i \sigma N t$$
$$n_r = (10^{10} \text{ neutrons/cm}^2\text{-sec})(2 \times 10^{-27} \text{ cm}^2)$$
$$(6.02 \times 10^{22} \text{ nuclei/cm}^3)(2 \times 10^{-2} \text{ cm})$$
$$n_r = 2.4 \times 10^4 \text{ neutrons/cm}^2\text{-sec}$$

Inasmuch as there were 10^{10} incident particles per cm^2 per second, only 2.4 out of every 10^6 neutrons striking the foil are captured in the reaction, $_{13}Al^{27}(n, \gamma)_{13}Al^{28}$.

Because the reaction cross section gives a measure of the probability that a certain nuclear reaction will occur, the measurement of reaction cross sections and their interpretation in the light of nuclear structure has been an important activity in nuclear physics. In reaction cross-section experiments, mono-energetic x particles from a charged-particle accelerator strike a target. (We will see shortly how mono-energetic neutron beams can be obtained.) The cross section can be measured by determining either the number of y particles or Y particles produced by a known number of x particles. The emerging light y particles can be counted by such devices as nuclear emulsions or counters. Alternatively, the heavy Y particles (often unstable) may be counted by measuring the radioactivity resulting from their decay; a much more difficult method, that of chemical quantitative analysis of the Y atoms, is seldom practicable because the concentration of Y atoms is typically very small.

There are a few general remarks that can be made concerning the behavior of reaction cross sections.

(1) For an endothermic reaction, which cannot proceed unless energy is added to the combining particles, the reaction cross section is necessarily zero until the threshold energy is exceeded.

(2) Reactions in which the incident particles are neutrons, particularly the neutron radiative-capture (n, γ) reactions, may exhibit high cross sections even when the energy of the bombarding particle is very small indeed. Unlike a charged particle, a neutron is undeflected by the electric charge of the bombarded nucleus, and a neutron can quite easily come within the range of the nuclear force and react with the target nucleus at slow speeds. A typical (n, γ) cross section is shown in Figure 10-6 as function of the neutron energy. It is seen that, apart from the

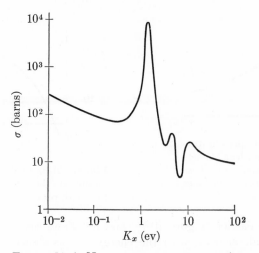

Figure 10-6. Neutron capture cross section for indium as a function of neutron energy.

quite pronounced peaks, the cross section increases as the energy or speed of the neutrons decreases. In fact, the cross section is found to be closely proportional to $1/v$, where v is the neutron speed. This $1/v$ law can be interpreted as follows: the probability that a neutron is captured is directly proportional to the time the neutron spends in the vicinity of any one bombarded nucleus or inversely proportional to its speed. The peaks in the cross-section curve are referred to as *resonances*; their interpretation, which will be discussed in Section 10-5, gives information on nuclear energy levels.

(3) When charged particles strike a target nucleus, the reaction cross section is influenced by the fact that the incident particles are repelled by the Coulomb electrostatic force. If it were not for the phenomenon of *barrier penetration*, discussed in Section 9-10 on α-decay, low-energy charged particles could not come within the range of the nuclear force of the target nucleus, the nuclear reaction could not occur, and the reaction cross section would be zero. But incident charged particles, even

with energies less than 1 Mev (much less than the height of the Coulomb potential barrier), are found to undergo nuclear reactions, thus indicating that the Coulomb barrier has been penetrated. The probability of barrier penetration depends very strongly on the barrier height and thickness. The more energetic the incident charged particle, the more easily it can

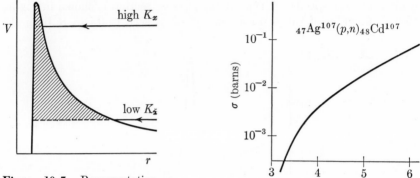

Figure 10-7. Representation of the relative nuclear barriers to be penetrated by low- and by high-energy incident charged particles.

Figure 10-8. Increase in the cross section of the $Ag^{107}(p,n)Cd^{107}$ reaction with proton energy.

penetrate the barrier (Figure 10-7), and it follows that the cross section will, in general, increase with the energy K_x, as shown in Figure 10-8.

10-5 The compound nucleus and nuclear energy levels As an introduction to the concept of the compound nucleus, let us compute the energy with which the "last" neutron in the stable cadmium nuclide, $_{48}Cd^{114}$, is bound to the remaining 113 nucleons. This is just the amount of energy that must be added to $_{48}Cd^{114}$ to separate the least-tightly-bound neutron, leaving the stable isotope, $_{48}Cd^{113}$. This separation energy E_s is found directly by comparing the masses of $(_0n^1 + {_{48}Cd^{113}})$ and $(_{48}Cd^{114})$;

$$E_s = [(1.00899 + 112.94036) - (113.93977) \text{ amu}] \times$$
$$931 \text{ Mev/amu} = 8.92 \text{ Mev}$$

Therefore, if 8.92 Mev of energy is absorbed by the $_{48}Cd^{114}$ nucleus, a neutron and a $_{48}Cd^{113}$ nucleus are formed, both particles being at rest. Symbolically,

$$_{48}Cd^{114} + 8.92 \text{ Mev} \rightarrow {_{48}Cd^{113}} + {_0n^1}$$

Now let us reverse the above procedure and bring together a neutron and a $_{48}Cd^{113}$ nucleus, both with zero kinetic energy, to form a $_{48}Cd^{114}$ nucleus. No energy need be added to the particles to have them amalgamate, inasmuch as the neutron will be attracted by the nuclear force of $_{48}Cd^{113}$

when it is sufficiently close to it. The $_{48}Cd^{114}$ nucleus thus formed will *not*, of course, be in its ground state; instead, it will be in an excited state, with an excitation energy of 8.92 Mev. The nucleus $_{48}Cd^{114}$* (the asterisk designates an excited state) is unstable and will quickly decay to its ground state by the emission of an 8.92 Mev, γ-ray photon. This over-all process can be written

$$_{48}Cd^{113} + _0n^1 \rightarrow {}_{48}Cd^{114*} \rightarrow \gamma(8.92 \text{ Mev}) + _{48}Cd^{114}$$

We have just described, in effect, the neutron radiative-capture reaction, $_{48}Cd^{113}(n, \gamma)_{48}Cd^{114}$. This reaction takes place in two stages: the amalgamation of the two original particles to form a single nucleus in an excited state; and the decay from this intermediate state, $_{48}Cd^{114}$*, into the products of the reaction. The energetics of this process are shown in Figure 10-9, where the total energies of the particles going into and coming out of the reaction are displayed.

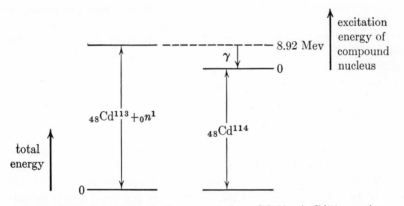

Figure 10-9. Energy-level diagram of the $_{48}Cd^{113}(n,\gamma)_{48}Cd^{114}$ reaction.

The neutron radiative-capture reaction illustrates a feature that is common to most low-energy nuclear reactions—the formation and decay of a *compound nucleus*. The existence of the compound nucleus as an intermediate stage in nuclear reactions was proposed by Niels Bohr in 1936. The assumptions are these:

(1) For the reaction, $X(x, y)Y$, the particles x and X combine to form the compound nucleus C, invariably in an excited state; $X + x \rightarrow C^*$. The energy carried into the reaction by x, is quickly shared among all of the nucleons in the compound nucleus.

(2) The compound nucleus C^* exists for a long time compared to the nuclear time ($\sim 10^{-21}$ sec), the time for a nucleon with a few Mev of energy to traverse a nuclear dimension. The average lifetime of a typical compound nucleus is at least 10^{-16} sec; this time, although long

compared with 10^{-21} sec, is so short that C^* is not directly observable. The compound nucleus lives for so long a time that it has no "memory" as to how it was formed. Because it does not "remember" its formation, there are a variety of (x and X) particles that can form the same nucleus C^* in the same excited state, as shown in Table 10-1.

Table 10-1

$$X + x \rightarrow C^* \rightarrow y + Y$$

$$
\left.\begin{array}{r}
{}_6C^{13} + p \\
{}_6C^{12} + d \\
{}_5B^{10} + \alpha
\end{array}\right\} {}_7N^{14*}
\left\{\begin{array}{l}
p + {}_6C^{13} \\
d + {}_6C^{12} \\
\alpha + {}_5B^{10} \\
n + {}_7N^{13} \\
\gamma + {}_7N^{14}
\end{array}\right.
$$

(3) The compound nucleus decays into the products of the reaction, $C^* \rightarrow y + Y$, as follows. After a fairly long time has elapsed (on a nuclear scale), the excitation energy of the compound nucleus, which was earlier distributed more or less evenly among the several nucleons, is finally concentrated on some one particle y, which is ejected, leaving the nucleus Y. A compound nucleus in some particular excited state may decay, then, by a variety of y and Y particles, as shown in Table 10-1. For a particular excited state of C^*, one particular decay mode will ordinarily dominate over all others.

(4) Inasmuch as the nuclear reaction is to be regarded as taking place in two distinct stages (the formation of C^* and the decay of C^*), the reaction cross section, which gives a measure of the probability that the *complete* reaction will take place, is proportional to *two* probabilities: the probability that x and X will amalgamate to form C^*, and the probability that C^* will decay into some particular y and Y particles.

Table 10-1 shows that there are a number of ways in which the compound nucleus, ${}_7N^{14*}$, can be formed, and a number of ways in which this nucleus can decay from an excited state. For example, when a 1 Mev proton combines with ${}_6C^{13}$, the most probable reaction is ${}_6C^{13}(p, \gamma){}_7N^{14}$; but when a 6 Mev proton strikes the same target and forms the same compound nucleus, the reaction ${}_6C^{13}(p, n){}_7N^{13}$ is most likely. In the first instance the excitation energy of ${}_7N^{14*}$ is about 8 Mev, and in the second one, about 13 Mev, as shown in Figure 10-10. Furthermore, when a 2 Mev α-particle combines with ${}_5B^{10}$, again forming ${}_7N^{14*}$ with an excitation of about 13 Mev, the observed reaction is ${}_5B^{10}(\alpha, n){}_7N^{13}$. The decay mode of the compound nucleus depends only on its excitation energy and *not* on the particles that form it. (Note that Q values for the reactions can be read directly from the differences in the rest energies in Figure 10-10.)

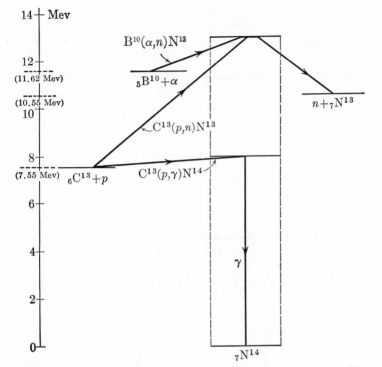

Figure 10-10. Energetics of several nuclear reactions for which $_7N^{14}$ is the compound nucleus. The scale gives energies with respect to the ground state of $_7N^{14}$ as zero.

Now consider the reaction, $_3Li^7(p, \alpha)_2He^4$. This reaction has as its compound nucleus, $_4Be^{8*}$, which does *not* exist as a stable nucleus in nature. In this reaction $_4Be^{8*}$ decays into two α-particles, each with about 8.8 Mev of kinetic energy. But another competing reaction is $_3Li^7(p, \gamma)_4Be^8$, for which the compound nucleus is again $_4Be^8$ in an excited state. In this second reaction, however, the compound nucleus decays by γ-emission to its ground state, emitting a very energetic, 17.6 Mev photon. Following this γ-decay, the unstable product nucleus, $_4Be^8$, now in its ground state, decays into two α-particles, each with a necessarily very-low energy. The energy-level diagram for these reactions is shown in Figure 10-11.

Let us return to the $_{48}Cd^{113}(n, \gamma)_{48}Cd^{114}$ reaction, now noting how the cross section for this reaction varies with the energy of the incident neutrons. The capture cross section for very-low neutron energies is shown in Figure 10-12. The cross section is well over 10 barns for all neutron energies, and there is a well-defined maximum, or *resonance*, for $K_x = 0.16$ ev. This means that there is a particularly high probability that the compound nucleus, $_{48}Cd^{114*}$, will be formed when its excitation energy is approxi-

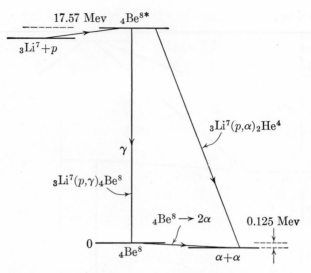

Figure 10-11. Nuclear energy-level diagram of the $_3Li^7(p,\alpha)_2He^4$ and $_3Li^7(p,\gamma)_4Be^8$ reactions.

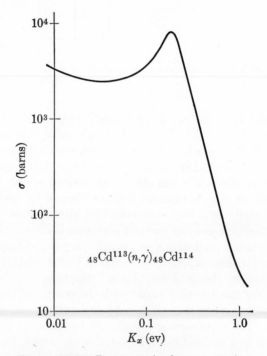

Figure 10-12. Resonance in the cross section of the $_{48}Cd^{113}(n,\gamma)_{48}Cd^{114}$ reaction.

mately† 0.16 ev higher than that which it has (8.92 Mev) for neutrons of zero kinetic energy. It indicates further that the nucleus $_{48}Cd^{114}$ has a well-defined quantized energy level when its excitation energy is just 0.16 ev higher than 8.92 Mev, as shown in Figure 10-13. This is just one of a number of discrete excited states for this nucleus. In general, the excited states

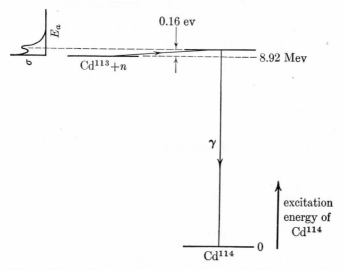

Figure 10-13. Resonance capture at an excited state of the compound nucleus for the reaction, $_{48}Cd^{113}(n,\gamma)_{48}Cd^{114}$. (The energy differences are *not* to scale.)

of the *compound nucleus* can be evaluated by observing resonances in the reaction cross section, and the existence of these well-defined resonances is strong evidence for the correctness of the compound-nucleus concept.

Nuclear-reaction data can also be used to deduce the excited energy levels of the *product nucleus*. For example, consider the reaction $_{13}Al^{27}(\alpha, p)_{14}Si^{30}$, having the compound nucleus, $_{15}P^{31*}$. This compound nucleus decays not only into the ground state of the product nucleus, $_{14}Si^{30}$, but also into several excited states of $_{14}Si^{30}$ followed by γ-decay to the ground state. The energy-level diagram is shown in Figure 10-14. The existence of these excited states is indicated experimentally, not only by the γ-rays emitted when $_{13}Al^{27}$ is bombarded by α-particles, but also by the fact that the protons observed (at some particular angle to the incident beam) have a spectrum of energies, as shown in Figure 10-15. The several *proton groups* correspond to the several possible decay modes of the com-

† Strictly, the additional energy added to the compound nucleus by a neutron with $K_x = 0.16$ is the *available* energy, $E_a = K_x[M_X/(M_X + m_x)] = (0.16)(113/114)$ ev $\simeq$ 0.16 ev, following Equation 10-6.

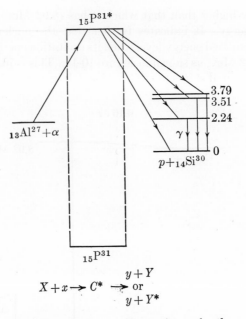

$$X + x \rightarrow C^* \rightarrow \begin{array}{l} y + Y \\ \text{or} \\ y + Y^* \end{array}$$

Figure 10-14.   Energetics of the $_{13}\text{Al}^{27}(\alpha, p)_{14}\text{Si}^{30}$ reaction, showing the excited states in the product nucleus.

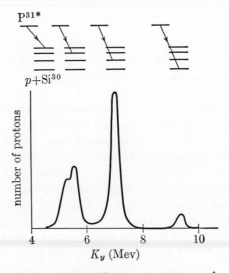

Figure 10-15. Energy spectrum of the proton groups from the reaction $_{13}\text{Al}^{27}(\alpha, p)_{14}\text{Si}^{30}$.

pound nucleus, and the excitation energies of $_{14}Si^{30*}$ can be evaluated by measuring the proton energies, K_y.

10-6   The production, detection, and moderation of neutrons   We have already discussed some important attributes of the neutron. In this section we wish to concentrate on some special problems that arise in producing and detecting neutrons, in measuring their energies, and in moderating neutrons in their passage through matter.

The most crucial property of the neutron is its electric charge—zero. Because of this, a neutron does not produce ionization directly, as do charged particles; it cannot be accelerated by electric fields; and it cannot be bent by magnetic fields. Its only means of communicating with other particles is through its strong nuclear interaction, unimpeded by any Coulomb repulsion.

NEUTRON PRODUCTION

(1) A common means of obtaining neutrons is by a *radium-beryllium source*, which depends on the reaction, $_4Be^9(\alpha, n)_6C^{12}$. The α-particles, coming from the radioactive decay of radium, collide with the beryllium with which the radium is mixed, and neutrons having a wide range of energies are emitted through the reaction $_4Be^9(\alpha, n)_6C^{12}$.

(2) A photodisintegration reaction can also be used as a source of neutrons. A simple example is $_4Be^9(\gamma, n)_4Be^8$. The photon energy must exceed 1.67 Mev for this reaction to occur; such γ-rays may be obtained from naturally or artificially radioactive materials.

(3) Accelerated charged particles can cause nuclear reactions in which neutrons are emitted. These reactions are particularly useful as neutron sources because the neutrons are mono-energetic. For example, the reaction, $_1H^3(d, n)_2He^4$, in which deuterons are accelerated to strike a target of tritium, has a Q of 17.6 Mev. Because energy and momentum must be conserved, the energy of the neutrons depends on the angle at which they leave the target with respect to the direction of the incident deuterons.

(4) Very-high-energy neutrons can be produced by a *stripping* reaction in which deuterons, having energies of several hundred Mev strike a target. The neutron is bound to the proton in a deuteron nucleus by an energy of only 2.2 Mev. When the very-high-energy deuteron strikes the target, the two particles may easily become separated, with the neutron continuing forward with about half of the incident deuteron energy.

(5) A still simpler means of obtaining very-high-energy neutrons utilizes a head-on collision of a very-high-energy proton with a single nucleon in a target nucleus. It is found, for example, that when

2 Bev protons strike a target, 2 Bev neutrons are knocked out in the forward direction, the proton transmitting its energy and momentum to an uncharged nucleon.

(6) The best source of a large flux of neutrons is a nuclear reactor, operating on the principle of nuclear fission. We shall discuss some properties of nuclear reactors in Section 10-7.

NEUTRON DETECTORS All common particle detectors depend in their operation upon the ionization produced by charged particles. The detection of neutrons must, therefore, take place in the following way: neutrons produce charged particles by some means, and the ionization due to these charged particles is detected.

(1) An ionization chamber or proportional counter is sensitive to neutrons, when filled with a gaseous boron compound, such as boron trifluoride (BF_3), or when lined with a solid boron compound. A neutron striking boron 10 can initiate the reaction, $_5B^{10}(n, \alpha)_3Li^7$, and the ionization of the α-particle is detected.

(2) The elastic collision of a neutron with a charged light particle, such as a proton, can be used as the basis for neutron detection. When a neutron makes a head-on, or *knock-on*, collision with a proton, the neutron is brought to rest and the proton moves forward with essentially the same energy as that of the original neutron. The energetic proton can be detected by its ionization in a gas-filled counter, or by its tracks in a nuclear emulsion, cloud-chamber, or bubble-chamber.

(3) Neutrons can be detected by the induced radioactivity that typically results from the neutron radiative-capture reaction. For example, when neutrons strike a silver foil, the silver 107 nuclei are *activated* by the reaction $_{47}Ag^{107}(n, \gamma)_{47}Ag^{108}$, and the product nuclei decay according to $_{47}Ag^{108} \rightarrow {}_{48}Cd^{108} + \beta^-$. The β^--activity may be detected and measured; and if the capture cross section, foil thickness, exposure time, and decay constant are known, the neutron flux may be computed.

MEASUREMENT OF NEUTRON ENERGY The kinetic energy of neutrons can be measured indirectly in some of the detection methods just described; for example, by measuring the energy of knock-on protons. There are, in addition, other methods with which neutron energies may be evaluated with considerable precision.

A particularly direct way of measuring a neutron's energy is to measure its speed. In this time-of-flight method, the procedure is to time the neutron's motion over a known distance; this can be done when neutrons are emitted in a reaction produced by pulsed charged particles from an accelerator, such as a cyclotron. If one measures the time delay between the reaction collisions and the arrival of neutrons at the neutron detector, a known distance from the target, the speed of the neutrons can be computed.

The speed of a neutron can also be determined with a mechanical device known as a *neutron velocity selector*, a simple form of which is shown in Figure 10-16. In order that the rotating discs be opaque to neutrons, the

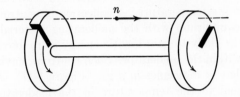

Figure 10-16. Schematic diagram of a neutron velocity selector.

discs are made of a material that strongly absorbs neutrons, such as cadmium. The neutron speed is evaluated by knowing the common angular speed of the two discs, their separation, and the angular displacement of the second slit relative to the first.

Another method of measuring neutron energies depends on the phenomenon of neutron diffraction (Section 4-3). A *neutron crystal-spectrometer* is a device for measuring the wavelength of neutrons by their diffraction through a crystal with a known crystalline structure. The wavelength λ of a neutron is related to its momentum mv by the de Broglie relation, $\lambda = h/mv$. Therefore, if the wavelength of a neutron is known, so is its momentum and kinetic energy. Neutron diffraction is feasible only for very-low-energy neutrons with energies of a fraction of an electron-volt or with wavelengths of several Ångstroms. This limitation arises from the fact that the interatomic spacing in crystalline solids is of the order of a few Ångstroms.

NEUTRON MODERATORS When neutrons pass through a material there are two types of interactions with the nuclei of the material that usually dominate over all others: neutron radiative-capture (n, γ); and *elastic collisions* between the neutrons and the nuclei of the material. For certain materials, the capture cross section is so small that the neutrons interact with the nuclei of the material primarily by elastic collisions. In passing through such materials, called *moderators*, the incident, originally energetic, neutrons are slowed down, or *moderated*.

In any elastic collision between two particles, the largest transfer of kinetic energy from one particle to the other occurs when the masses of the two particles are the same. Thus, a neutron can lose all of its kinetic energy when it makes a head-on collision with a proton, there being a lesser transfer of energy for oblique collisions. Therefore, hydrogenous materials, such as paraffin, are effective in moderating neutrons. When neutrons collide with heavier nuclei, they lose a smaller fraction of their kinetic energy, even in a head-on collision, and many more encounters are required to slow down the neutrons.

When neutrons are slowed down in a moderator they are never brought completely to rest, inasmuch as the nuclei in a material are in thermal motion at any finite temperature. A collection of neutrons can be said to be in thermal equilibrium with the moderating material when a typical neutron is just as likely to gain kinetic energy as to lose it upon colliding with a nucleus within the moderator. Such neutrons, having a distribution of speeds like that of molecules in a gas (Figure 1-17), can be assigned a temperature equal to the temperature of the moderator. The average kinetic energy of neutrons in equilibrium with a moderator at a temperature T is given by

$$\tfrac{1}{2} mv^2 = \tfrac{3}{2} kT$$

Neutrons in thermal equilibrium with a moderator at room temperature, 300°K, are said to be *thermal neutrons*. Their average kinetic energy is $\frac{1}{25}$ ev; their speed, 2200 m/sec; and their wavelength, 1.80 Å. A beam of high-energy (several Mev) neutrons incident on a typical moderator, such as graphite (carbon) or heavy water, is *thermalized* in less than 1 millisecond. The most probable fate of these moderated neutrons is capture by the nuclei of the moderator, inasmuch as the capture cross section increases drastically as the neutron energy falls. We recall that a free neutron is radioactive and decays into a proton and electron (and anti-neutrino) with a half-life of 12.8 minutes. This neutron decay, although possible, occurs very infrequently because it competes with the much faster process of neutron moderation and capture.

10-7 Nuclear fission A special type of nuclear reaction occurs in the very heavy nuclides. Unlike most low-energy nuclear reactions, in which a light particle and a heavy particle appear as products, this reaction results in the splitting, or fissioning, of the heavy nucleus into two parts of comparable mass; it is appropriately called *nuclear fission*. The identification of the nuclear-fission reaction was first made by O. Hahn and F. Strassman in 1939.

As an example of nuclear fission, consider the capture of a very-low-energy neutron, such as a thermal neutron, by the very heavy nucleus, uranium 235. The compound nucleus $_{92}U^{236}$ formed by this reaction is in an excited state with an excitation energy of 6.4 Mev. In almost all lighter nuclei, the excited compound nucleus formed by neutron capture decays by the emission of a γ-ray photon, the resulting heavy nucleus usually decaying by β^--emission. But an excited $_{92}U^{236*}$ nucleus can also decay by nuclear fission, splitting into two (and less frequently three or more) moderately heavy nuclei.

The behavior of a very-heavy, excited, compound nucleus, such as $_{92}U^{236*}$, can be understood from the point of view of the liquid-drop model. We recall that in this model a nucleus is regarded as analogous to a drop of liquid, with three important contributions to its total binding energy: the volume energy, the surface energy, and the Coulomb energy (see Section 9-7). The surface energy plays a similar role to that of the ordinary surface "tension" of a liquid in that it tends to minimize the surface area of the nucleus and thereby tends to make the nuclear shape spherical. The Coulomb energy represents a disruptive influence arising from the electrostatic force of repulsion between the protons.

Let us now suppose that, by virtue of a nuclear collision, a very heavy nucleus gains energy of excitation. The nucleus as a whole will oscillate, tending to distort its shape. One probable mode of deformation is shown in Figure 10-17, where the nucleus assumes, in turn, the shapes of: sphere,

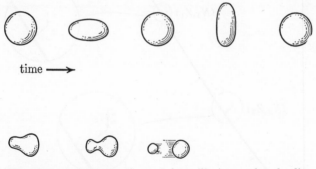

time ⟶

Figure 10-17. Deformations of an oscillating nucleus leading to nuclear fission.

prolate ellipsoid (cigar), sphere, oblate ellipsoid (pancake), etc. During these oscillations the nuclear volume is unchanged. The surface area does change, however, increasing for both the prolate and oblate distortions. The surface tension is manifest as a tendency for the nucleus to resume its spherical shape. On the other hand, the disruptive Coulomb energy is increased when in the prolate ellipsoidal shape, the positive charge at the two ends of the ellipsoid tending to increase the deformation. Thus, two competing influences are at work: the forces of surface tension, which tend to restore the deformed nucleus to its spherical shape; and the Coulomb forces, which tend to increase the nucleus' deformation.

If the degree of excitation of the compound nucleus is sufficiently large, the Coulomb forces succeed in deforming the nucleus into a dumbbell-like shape (Figure 10-17b). For so great a distortion of the nuclear shape, the surface tension is not strong enough to restore the nucleus to sphericity, and the Coulomb forces increase the separation between the ends until they split into two distinct nuclei, or *fission fragments*, usually of unequal size. The fission fragments repel each other by the Coulomb force, losing potential energy and gaining kinetic energy.

The transformations of this fission process are shown in Figure 10-18 on the N-versus-Z plot (Figure 9-7). Two typical fission fragments (Z_1, N_1)

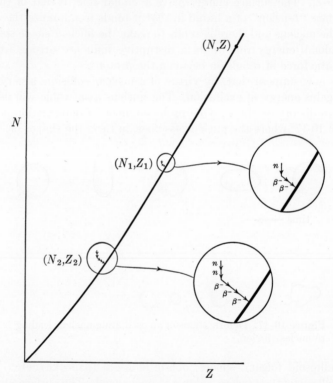

Figure 10-18. Transformations in the nuclear-fission reaction as they appear on a neutron-proton diagram.

and (Z_2, N_2), of a heavy compound nucleus share the protons and neutrons of the original compound nucleus (Z, N); that is, $Z = Z_1 + Z_2$ and $N = N_1 + N_2$. Both fragments fall to the *left* of the stability line; that is, both nuclei have too many neutrons to be stable. The neutron excess is so great that it is relieved almost instantaneously ($\sim 10^{-14}$ sec) by the release of two or three neutrons from the fission fragments. These nuclei still have too many neutrons, and they finally reach stability by changing neutrons into protons; that is, by β^--decay. These β^--decays are, of course, accompanied by γ-decay from excited nuclear states.

Two of the many known fission reactions resulting from neutron capture in uranium 235 are shown below, together with the subsequent β^--decays of the fission fragments.

$$_0n^1 + {}_{92}U^{235} \rightarrow {}_{92}U^{236*} \rightarrow {}_{56}Ba^{144} + {}_{36}Kr^{89} + 3{}_0n^1$$

where
$$_{56}Ba^{144} \xrightarrow{\beta^-} {}_{57}La^{144} \xrightarrow{\beta^-} {}_{58}Ce^{144} \xrightarrow{\beta^-} {}_{59}Pr^{144} \xrightarrow{\beta^-} {}_{60}Nd^{144}$$

and
$$_{36}Kr^{89} \xrightarrow{\beta^-} {}_{37}Rb^{89} \xrightarrow{\beta^-} {}_{38}Sr^{89} \xrightarrow{\beta^-} {}_{39}Y^{89}$$

Or,
$$_0n^1 + {}_{92}U^{235} \rightarrow {}_{92}U^{236*} \rightarrow {}_{54}Xe^{140} + {}_{38}Sr^{94} + 2{}_0n^1$$

where
$$_{54}Xe^{140} \xrightarrow{\beta^-} {}_{55}Cs^{140} \xrightarrow{\beta^-} {}_{56}Ba^{140} \xrightarrow{\beta^-} {}_{57}La^{140} \xrightarrow{\beta^-} {}_{58}Ce^{140}$$

and
$$_{38}Sr^{94} \xrightarrow{\beta^-} {}_{39}Y^{94} \xrightarrow{\beta^-} {}_{40}Zr^{94}$$

The basic requirement for the occurrence of fission in the very heaviest nuclides is that the compound nucleus be formed with sufficient excitation energy so that it splits. Neutron capture is just one of the several ways in which nuclear fission can be induced. Fission can also result from the bombardment of heavy nuclei by protons, deuterons, α-particles, and γ-rays (*photofission*).

Let us compute the total energy released in a typical fission process. We see from Figure 9-10 that the very heavy elements ($A \simeq 240$) have an average binding energy per nucleon, E_b/A, of approximately 7.6 Mev/nucleon and that moderately heavy nuclei ($A \simeq 120$), for which an average nucleon is more tightly bound, have an E_b/A value of approximately 8.5 Mev/nucleon. Thus, if we take the mass number A of the original nucleus to be roughly 240, the total energy released in the fission process is about (240 nucleons) $\times$ (8.5 − 7.6) Mev/nucleon $\simeq$ 200 Mev. The total energy released in a fission reaction is very large indeed, as compared with the few Mev of energy released in a typical low-energy exothermic nuclear reaction.

The nuclear-fission reaction is characterized by: the decay of the compound nucleus into two moderately heavy nuclei; the emission of a few neutrons; and the β^--decay of the radioactive fission fragments. In an average fission reaction about 200 Mev is released and distributed approximately as follows: the kinetic energy of the fission fragments, 170 Mev; the kinetic energy of the fission neutrons, 5 Mev; the energy of β^-- and γ-rays, 15 Mev; and the energy of the neutrinos associated with the β^--decay, 10 Mev.

The light isotope of uranium, $_{92}U^{235}$, undergoes fission with thermal neutrons, the excitation energy gained by the compound nucleus $_{92}U^{236*}$ in slow-neutron capture being great enough to cause the fission to occur. (Uranium 235 is the only *natural* nuclide which undergoes fission with slow neutrons.) The much more abundant (99.3 per cent) heavy isotope of uranium, $_{92}U^{238}$, will undergo fission, but only if fast neutrons, having an initial kinetic energy exceeding 1 Mev, are used. Low-energy neutrons are captured by $_{92}U^{238}$, but the excited compound nucleus, $_{92}U^{239*}$, has too little excitation energy to decay by fission and it decays instead by γ-emission.

Uranium 235 is fissionable for both low- and high-energy neutrons, the (n, γ) reaction being less probable than fission for all energies; on the other hand, uranium 238 is fissile for high-energy neutrons only, low-energy neutrons being captured without fission. Clearly, the compound nucleus, $_{92}U^{239*}$, does not gain enough excitation energy from the capture of low-energy neutrons to decay by fission, whereas the compound nucleus, $_{92}U^{236*}$, is sufficiently excited to undergo fission. This difference is attributable to the fact that U^{236}, being an even-even nuclide, with the last neutron relatively tightly bound (6.4 Mev), gains more excitation energy in capturing a zero-energy neutron, than does the even-odd nuclide, $_{92}U^{239}$, for which the last odd neutron is relatively weakly bound (4.9 Mev).

The fact that nuclear fission with uranium 235 can be initiated by low-energy neutrons and that, on the average, 2.5 neutrons are released in the fission process, makes it possible to extract useful energy from uranium. The energy released in exothermic nuclear reactions produced by particle bombardment from accelerators cannot be utilized in a practical way because the number of reactions is typically very small. The total energy released in the reactions is much less than the total energy supplied to the many accelerated particles, of which only a small fraction cause reactions. The fission process can, on the other hand, be made much more efficient by the possibility for a *self-sustaining chain reaction*.

In essence, the neutrons from one fission reaction may initiate other fission reactions with the further release of fission energy continuing ideally until all of the nuclear fuel, or fissionable material, is consumed. For the fission reactions to continue, once initiated, a number of conditions must be fulfilled. These conditions are achieved in a *nuclear reactor*. The engineering problems connected with nuclear reactors lie in the area of nuclear technology, and we shall merely outline the physical principles on which reactor operation is based.

NUCLEAR REACTORS The first self-sustaining, nuclear-fission chain reaction was achieved by E. Fermi in 1942. The reactor used natural uranium (0.7 per cent U^{235} and 99.3 per cent U^{238}) as the fuel, and graphite as the neutron moderator. Although there are many different types, we shall illustrate the basic features of nuclear reactors with a simple reactor using natural uranium, a graphite moderator, and based on the fission of uranium 235 by slow neutrons.

For a fission chain reaction to be self-sustaining, it is required that, for each uranium atom split, there be at least one neutron which will split another uranium atom. In the fission of uranium, each decay produces, on the average, 2.5 neutrons. Therefore, no more than 1.5 neutrons can be lost without stopping the chain reaction. The important ways in which neutrons become unavailable for uranium 235 fission are: capture without fission by U^{238} (and, to a lesser extent, by U^{235}); capture by other materials; and leakage from the interior of the reactor to the outside.

Let us first consider the problem of neutron leakage. If the reactor (the fuel elements and the moderator) is very small, a large fraction of the neutrons produced in some initial fission reactions leak out of the reactor before inducing further fission reactions. As the size of the reactor is made larger, the number of fission reactions increases. The fission-reaction production rate is roughly proportional to the volume of the reactor, and the leakage rate is roughly proportional to the surface area of the reactor. Therefore, by increasing the reactor size, neutron losses due to leakage are reduced.

The fission cross section in uranium 235 increases as the neutron energy decreases, reaching 550 barns for thermal neutrons. On the other hand, both the fission cross section and the neutron-capture cross section by uranium 238 are very small for thermal neutrons. The capture cross section in uranium 238 is fairly large, however, for larger-energy neutrons. Therefore, the problem in operating a reactor with natural uranium is to slow the high-energy (few Mev) neutrons emitted in the U^{235} fission to thermal energies, where further fission reactions in U^{235} are more probable, without losing the neutrons by capture in U^{238} on the way down. These conditions are met by using a moderator to slow down (but not capture) the neutrons, and by properly arranging uranium fuel blocks within the moderator.

The function of the moderator is to slow down neutrons without capturing them. Although hydrogen atoms, with a mass essentially equal to that of the neutron, cause the greatest fractional loss in the kinetic energy of a neutron, hydrogen is unsuitable as a moderator because of the relatively high probability of the neutron-capture reaction, $_1H^1(n, \gamma)_1H^2$. The lightest usable moderator materials are heavy water (D_2O), beryllium (Be), and graphite (C^{12}). Most other light elements are not usable because of their large neutron-capture cross sections. The fuel elements are arranged as blocks in the medium of the moderator. Under ideal conditions, a fast neutron from a fuel element escapes into the moderator and is slowed down, thereby avoiding capture (without fission) in U^{238}. Then the thermal neutron enters another fuel element causing another fission in U^{235}, the whole process taking place in less than a millisecond.

When all sources of neutron loss have been minimized, it is possible for the reactor to go *critical*, with each fission reaction leading to exactly one more fission reaction. The power level of the reactor, or the rate at which fission reactions occur, can be controlled by the insertion into the reactor of materials, such as cadmium, for which the neutron-capture cross section is very high. Such materials which strongly absorb neutrons are usually in the form of rods, called *control rods*. A reactor is said to be *subcritical* if, on the average, each fission reaction produces *less* than one further fission; the fission reaction is then not self-sustaining. On the other hand, if each fission reaction produces more than one further fission, the reactor is said to be *supercritical*; an extreme example of a supercritical fission reaction is an atom bomb.

The control of reactors by mechanically-actuated control rods would be virtually impossible if the only neutrons available were the *prompt neutrons* released at the instant of fission. But there are, in addition, *delayed neutrons* (0.7 per cent), which are emitted by a few of the fission fragments, usually *after* one of more β^--decays have occurred. A delayed neutron can cause a further fission after (10 seconds on the average) the fission which released this delayed neutron. This is in contrast to a prompt neutron which causes a further fission in less than a millisecond.

One example of a fission-fragment decay leading to a delayed neutron is:

$$_{35}\mathrm{Br}^{87} \Big\langle \begin{array}{l} \nearrow \;_{36}\mathrm{Kr}^{87*} \rightarrow \;_{36}\mathrm{Kr}^{86} + \;_{0}n^{1} \\ \searrow \;_{36}\mathrm{Kr}^{87} \xrightarrow{\beta^-} \;_{37}\mathrm{Rb}^{87} \xrightarrow{\beta^-} \;_{38}\mathrm{Sr}^{87} \end{array}$$

(56 sec)

There are many designs of nuclear reactors. They may differ in the following respects: the fuel (natural uranium, uranium enriched in U^{235}, other artificially produced fissionable materials), the moderator (water, graphite, beryllium); the distribution of fuel within the moderator (homogeneous, heterogeneous); the energy of neutrons producing fission (fast, intermediate, and slow neutrons); and the heat exchanger (gas, water, liquid metals). Figure 10-19 is a schematic of a nuclear reactor.

Nuclear reactors can also be classified according to their use: power,

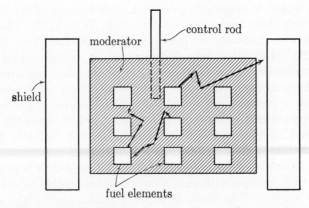

Figure 10-19. Simple elements of a nuclear reactor.

neutron source, production of radio-isotopes, and production of fissionable material.

(1) The large kinetic energy of the fission fragments in a nuclear reactor is a source of energy in the form of heat, which can be extracted through a heat exchanger to do useful work, such as generating electric energy.

(2) The interior of a reactor is a region in which the neutron flux can be as high as 10^{19} neutrons/m²-sec. Such a high neutron flux can be used to perform experiments in physics, or to irradiate materials, thereby producing radio-isotopes with (n, γ) reactions.

(3) Such materials as uranium 238 and thorium 232, which do *not* undergo fission with low-energy neutrons, can be converted in a nuclear reactor into nuclides which are fissile by thermal neutrons. Two such reactions are:

$$_0n^1 + {}_{92}U^{238} \rightarrow {}_{92}U^{239*} \xrightarrow[\text{(23 min)}]{\beta^-} {}_{93}Np^{239} \xrightarrow[\text{(2.3 days)}]{\beta^-} {}_{94}Pu^{239}_{\text{(24,000 yrs)}}$$

$$_0n^1 + {}_{90}Th^{232} \rightarrow {}_{90}Th^{233*} \xrightarrow[\text{(23 min)}]{\beta^-} {}_{91}Pa^{233} \xrightarrow[\text{(27 days)}]{\beta^-} {}_{92}U^{233}_{\text{(1.6}\times10^5\text{ yr)}}$$

Both of the nuclides, uranium 238 and thorium 232, are *not* fissile by thermal neutrons. But when these materials capture neutrons, the reactions lead to the materials, plutonium 239 and uranium 233, both fissile by thermal neutrons.

These two reactions lead to the possibility of a *breeder reactor*. In such a reactor there are two fuel materials, one which is fissionable (such as plutonium 239) and the other which is fertile (such as uranium 238), in that it can be converted in the reactor into fissionable material. In the fission of Pu^{239} there are, on the average, 3 neutrons released. Of these three neutrons, 1 neutron must sustain the reaction producing the fission of a Pu^{239} nucleus; of the remaining 2 neutrons, at least one must be captured by U^{238}, thus leading to Pu^{239}, to maintain the same amount of fissionable fuel in the reactor. When more than 1 of these 2 neutrons is captured by U^{238}, the reactor can breed fissionable Pu^{239}; that is, more fissionable material is produced than consumed.

10-8   Nuclear fusion   The origin of energy radiated from the Sun and other stars is a series of exothermic nuclear reactions. The particles participating in such reactions in the interior of the star are completely ionized, all electrons having been removed from the atoms. Such a collection of electrons and bare nuclei is called a *plasma*. The nuclear particles are at a very high temperature (up to 10^8 °K) and make frequent collisions with one another. At such high temperatures the Coulomb repulsion between positively charged nuclei may be overcome in these collisions, thus enabling the particles to approach one another closely enough so that they interact by nuclear forces, and reactions take place with high probability. A nuclear reaction which occurs by virtue of the high temperature, or thermal motion, of the interacting particles is called a *thermonuclear reaction*.

The two most common cycles of thermonuclear reactions releasing energy in stars are the *carbon-cycle* and the *proton-proton cycle*.

Carbon Cycle

$$H^1 + C^{12} \rightarrow N^{13} + \gamma$$
$$N^{13} \rightarrow C^{13} + \beta^+ + \nu$$
$$H^1 + C^{13} \rightarrow N^{14} + \gamma$$
$$H^1 + N^{14} \rightarrow O^{15} + \gamma$$
$$O^{15} \rightarrow N^{15} + \beta^+ + \nu$$
$$H^1 + N^{15} \rightarrow C^{12} + He^4$$

In this cycle, the carbon acts merely as a catalyst, inasmuch as one begins with one C^{12} nucleus and ends with one C^{12} nucleus. In the process, however, 4 protons are, in effect, converted, or fused, into one α-particle and two positrons, with a net energy release of about 25 Mev.

Proton-Proton Cycle

$$H^1 + H^1 \rightarrow H^2 + \beta^+ + \nu$$
$$H^1 + H^2 \rightarrow He^3 + \gamma$$
$$He^3 + He^3 \rightarrow He^4 + 2H^1$$

This cycle again fuses 4 protons into an α-particle plus 2 positrons. The first reaction in this proton-proton cycle, in which a positron is created in the collision of two protons, has a very small cross section. The over-all Q for this cycle is again about 25 Mev.

We see from Figure 9-10 why both fission of the heaviest elements and fusion of the lightest elements result in highly exothermic reactions: both fission and fusion reactions lead to more-tightly-bound nuclear configurations. The fractional conversion of mass into energy is greater for a fusion reaction (0.66 per cent) than for a fission reaction (0.09 per cent).

Much interest has been aroused by the possibility of producing controlled thermonuclear fusion reactions with the resultant large energy release. Such an energy source has significant advantages over that of a fission reactor: there is a virtually unlimited supply of fuel; the reactions do *not* result in radioactive wastes; and there exists the possibility of generating electric energy more directly than through heat-exchangers and turbines.

There are, however, formidable difficulties in achieving a power source based on controlled nuclear fusion. Extraordinarily high temperatures are required to overcome the Coulomb repulsion between the interacting nuclei, and the (high-temperature) plasma must be confined for long periods of time so that many collisions between the plasma particles take place (ordinary containers are unsuitable because they would chill the plasma below the spontaneous-fusion temperature).

Possible fusion reactions using isotopes of hydrogen (H^2, deuterium; H^3, tritium) are:

$$_1H^2 + {}_1H^3 \rightarrow He^4 + n + 17.6 \text{ Mev}$$
$$_1H^2 + {}_1H^2 \rightarrow He^3 + n + 3.3 \text{ Mev}$$
$$_1H^2 + {}_1H^2 \rightarrow H^3 + H^1 + 4.0 \text{ Mev}$$

Because the Coulomb barrier between charged particles increases with increasing nuclear charge, a thermonuclear reaction with hydrogen requires lower temperatures than other elements. Deuterium is particularly attractive as fusion fuel because it is readily available in almost unlimited quantities; for example, in sea water (one D_2O for every 6000 H_2O molecules).

The high-temperature plasma in a thermonuclear fusion reactor can be prevented from striking the walls of its container by means of magnetic fields, and the design of such "magnetic bottles" is under active development. Obviously, a plasma having a temperature of millions of degrees will exert an uncontainable pressure unless the density of the plasma is very low indeed; the pressure of the unheated plasma must, therefore, be about 10^{-4} atmosphere.

A self-sustaining, controlled, thermonuclear fusion reactor has not yet been achieved. The only man-made thermonuclear reactions have used a supercritical fission reaction (atomic bomb) to achieve the high temperatures required to initiate a supercritical fusion reaction (hydrogen bomb).

10-9　Cosmic radiation　The study of the cosmic rays has been a most challenging field of research in high-energy nuclear physics. It is concerned with the origin and identification of very-high-energy particles continuously entering the Earth's upper atmosphere and the complex phenomena occurring when these high-energy particles collide with nuclei in the atmosphere.

The study of cosmic rays began with the discovery, by J. Elster and H. Geitel (1899) and by C. T. R. Wilson (1900), that even the best electrically-insulated charged electroscopes slowly lose their charge. This effect was correctly attributed to some type of radiation which ionized air molecules in the electroscope, thus rendering the air at least slightly conducting. The apparently obvious explanation, that the ionizing radiation resulted entirely from the minute traces of naturally radioactive elements always present in laboratory materials, was disproved by V. F. Hess (1911). Sending charged electroscopes aloft in balloons, Hess found that the ionizing radiation *increased* with altitude, and therefore proposed that this penetrating radiation comes from outer space. Because the intensity striking the Earth was observed to be essentially constant, independent of the time of day or of the year, Hess suggested that this extra-terrestrial radiation had a cosmic origin from without our solar system.

Cosmic-ray phenomena are unusually complicated because the primary cosmic-ray particles entering the Earth's atmosphere initiate a variety of high-energy nuclear reactions. The products of these reactions can themselves generate other reactions lower in the Earth's atmosphere. Furthermore, many of the particles emitted in these reactions are radioactive and decay as they travel toward the Earth. Finally, new and strange particles, such as mesons and hyperons, are created in cosmic-ray nuclear collisions. Thus, the secondary cosmic-radiation observed at sea-level contains contributions from many origins. Only the primary events, high in the Earth's atmosphere, are relatively simple to analyze. For this reason, we shall not follow the historical development of cosmic radiation, based in the early years mostly on observations at sea level. Instead, we shall first discuss (1) the identification and (possible) origin of the *primary* cosmic-ray particles, (2) the interaction of this primary radiation with nuclei in the Earth's upper atmosphere, and finally (3) the *secondary* radiation.

THE PRIMARY COSMIC RADIATION It is now well established that the primary cosmic rays are very-high-energy protons and, to a lesser extent, heavier nuclei, which rain upon the Earth in all directions from outer space.

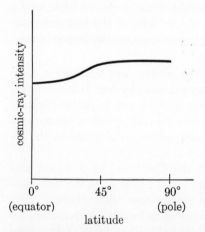

Figure 10-20. Variation in the cosmic-ray intensity at 2000 m above sea level as a function of latitude.

That the primary cosmic rays are *charged* particles, rather than uncharged particles or photons, is established by the *latitude effect*. This effect, based on the observation that the cosmic-ray intensity is less near the Equator than at higher latitudes, is intimately related to the Earth's magnetic field. If it is assumed that the cosmic rays approach the Earth uniformly in all directions, no such variation with latitude would be found unless the particles were charged. The Earth's magnetic field is weak, but it extends, of course, far out past the Earth's atmosphere; therefore it can influence the paths of incoming charged particles even at great distances from the Earth. Charged particles approaching normal to the Equator are strongly deflected by the Earth's magnetic field, the amount of the deflection depending on the momentum and charge of the particles. In fact, at the Equator all particles with energies less than 10 Bev are so strongly deflected that they completely miss the Earth's atmosphere. On the other hand, particles approaching the Earth along its magnetic axis travel parallel to the Earth's magnetic lines of force and are therefore undeflected. Primary cosmic-ray particles of *any* energy can enter the Earth's atmosphere in the vicinity of the magnetic poles. The increase in cosmic-ray intensity with latitude, as shown in Figure 10-20, is thus accounted for.

Experimental determinations of the energy spectra and identity of the cosmic rays are made in a variety of ways, using such instruments as nuclear emulsions, cloud chambers, ionization chambers, and Geiger counters. It has been ascertained that almost all primary cosmic-ray particles have energies greater than $\frac{1}{2}$ Bev, some very few have energies as high as 10^{19} ev, but only 15 per cent have energies over 15 Bev. The intensity of cosmic radiation incident on the Earth is about 10^{-5} watt/m², which is approximately the same as the intensity of visible photons from outer space; namely, starlight.

That the primary cosmic-ray particles carry a *positive* charge is indicated by the *East-West effect*. Consider the path of particles, assumed to be positively-charged, approaching the Earth at the equator, as shown in Figure 10-21. It is clear that, if the particles are positively-charged, more

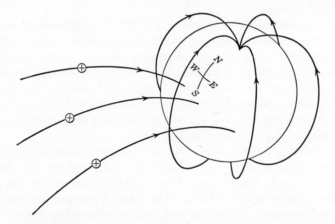

Figure 10-21. Paths of incident, positively-charged cosmic ray particles under the influence of the Earth's magnetic field.

will arrive from the West than from the East. Negative primary charged particles would, of course, show an assymmetry in the opposite sense. Measurements show that there *is* an East-West effect, with an excess of particles arriving from the west; therefore, the primary cosmic-ray particles are positive. These results are also confirmed by high-altitude measurements.

The primary cosmic radiation approaching the Earth's atmosphere is found to consist mainly of protons (77 per cent), together with about 21 per cent of α-particles and 2 per cent of still heavier nuclei. The relative abundances of the various nuclides present in the primary cosmic rays are close to the relative abundances of the elements in the Earth's crust, as well as in the stars.

The origin and acceleration mechanism of the primary cosmic rays are still only partially understood. It is presently believed that most of the charged particles are emitted by stars within our own galaxy (the Milky Way). These particles become trapped by the weak magnetic fields known to exist in interstellar regions; and if these magnetic fields change with time, it is possible for the trapped charged particles to be accelerated by the changing magnetic fields. After a long period of entrapment and continued acceleration, some of these particles leak out of the confining magnetic fields, travel through space, and eventually collide with objects such as the Earth.

THE SECONDARY RADIATION   Three general types of events occur when high-energy primary cosmic-ray particles collide with nuclei (mostly nitrogen and oxygen) in the upper atmosphere: *star events; cascade showers;* and *meson showers.*

When a nucleus in the atmosphere is struck by a high-energy cosmic-ray particle, the nucleus can gain an enormous amount of energy, and split or disintegrate into a large number of smaller nuclides. These emerging particles, mostly protons and neutrons, leave the site of the collision with high energies. Such an event is known as a *star event* because of the appearance of the tracks in a photographic emulsion, as shown in Figure 10-22. Many of the particles from such a star event have very high energies and may themselves produce still other star events.

The collision between a cosmic-ray particle and a nucleus may also result in the production of high-energy γ-rays (we shall soon see how these photons are produced). Such a γ-ray may annihilate into an electron-positron pair when passing near a nucleus. The created charged particles, having large kinetic energies, will collide with and be deflected by still other nuclei they encounter. By virtue of their acceleration, these particles will radiate high-energy photons by the *Bremsstrahlung* process. If the secondary photons have energies greater than 1 Mev, they can produce further electron pairs. Thus, by the repeated occurrence of the pair-production and *Bremsstrahlung* processes, a *cascade shower* of electrons, positrons, and photons is produced, the energy of the original photon having been degraded and spread among many particles. The shower is effectively extinguished after pair-production becomes energetically impossible. **Figure**

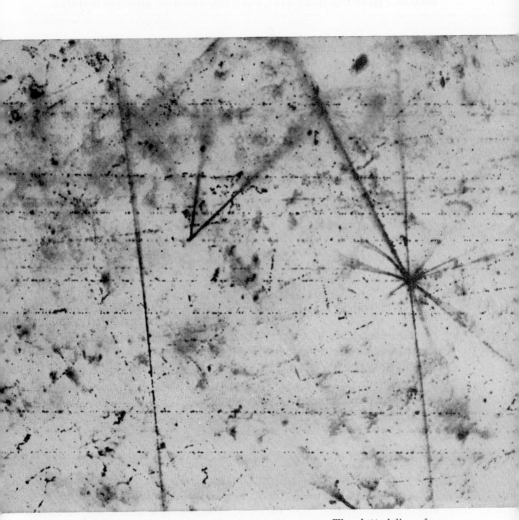

Figure 10-22. A nuclear emulsion showing a star event. The dotted lines from left to right were made by 2 Bev protons. A star, representing the disintegration of a bromine or silver atom in the emulsion, appears on the right. The prong, left of center, is an event in which an incoming particle caused two heavy particles to be ejected from a nucleus. (Courtesy of Brookhaven National Laboratory.)

10-23 shows a cascade shower schematically; cloud-chamber photographs, such as Figure 8-2 (page 277), have confirmed its principal features.

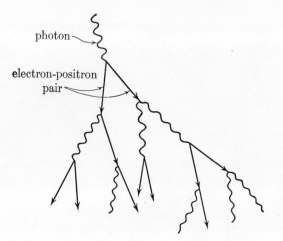

photon

electron-positron
pair

Figure 10-23. Schematic representation of a cascade shower.

A third common type of high-energy cosmic-ray collision is that in which entirely new kinds of particles emerge from the collision. Some of these particles are *mesons* (Greek *mesos*, "middle"). They are so called because their masses are intermediate between the mass of the electron, m_e, and the mass of a proton, 1836 m_e. Large numbers of mesons may be produced in high-energy collisions. These created mesons are called pi-*mesons*, or *pions*. Heavier mesons have also been observed from cosmic-ray collisions. In fact, particles having a *greater* mass than that of a nucleon (proton or neutron) are found; such particles are given the general name, *hyperons* (*hyper*, "over").

The road to understanding the properties of the several types of mesons and hyperons, their interactions, and especially their role in physical theory, has been a long and arduous one. Until the 1940's, the high-energy reactions required for the creation of mesons were available only in the uncontrolled and infrequent events initiated by cosmic radiation. But when accelerators of several hundred Mev and higher energies were developed, large numbers of high-energy meson events could be produced under controlled conditions in the laboratory, and our knowledge of these particles increased correspondingly.

THE π- AND μ-MESONS   Three types of π-mesons are now known to exist: two charged π's, π^+ and $\overline{\pi^-}$, with charges $+e$ and $-e$, and a mass of 273 m_e; and an uncharged π-meson, π^0, with a mass of 264 m_e. (As before, a bar above the symbol indicates an *anti*-particle.) Although the charged π-mesons were first observed in cosmic radiation in 1947 and the uncharged π^0 meson in 1950, the existence of particles having a mass intermediate between that of an electron and a nucleon had been theoretically predicted in 1935 by H. Yukawa.

Yukawa hypothesized that the nuclear force between nucleons could be thought of as an exchange of quanta, or particles, associated with the nuclear-force field, just as the electric force between charged particles may be described in terms of the exchange of electromagnetic quanta (photons) between charged particles. Whereas an electromagnetic force, such as the Coulomb force, extends throughout all space (long-range force), the nuclear force is zero for distances greater than about 1.4×10^{-15} m (short-range force). A simplified argument, based on the uncertainty principle, illustrates the essential aspects of mesons as agents of the nuclear force, and leads to a value for the rest mass in close agreement with that of the π-meson.

Assume that a nucleon creates and emits a π-meson, and in a short time Δt a second nucleon with which the first interacts absorbs this meson. During the time-of-flight of the meson, the conservation-of-energy principle is violated by an amount $\Delta E = m_\pi c^2$, the rest energy of the created, but unobservable (*virtual*), meson. The time duration of this energy violation is, by the uncertainty principle,† limited to

$$\Delta t \simeq \hbar/\Delta E \qquad [10\text{-}10]$$

When the meson is absorbed by the second nucleon, the energy of the system is restored. If the meson travels essentially at the speed of light, c, during the time Δt, its maximum distance of travel R must be, from Equation 10-10,

$$R = c \, \Delta t = c(\hbar/\Delta E) = \hbar/m_\pi c$$

Therefore, $$m_\pi \simeq \frac{\hbar}{Rc} \qquad [10\text{-}11]$$

Taking the distance between interacting nucleons to be 1.4×10^{-15} m, Equation 10-11 gives

$$m_\pi \simeq 274 \, m_e \qquad [10\text{-}12]$$

This predicted mass is very close—fortuitously, because of this oversimplified argument—to that observed for the π-mesons. By virtue of their exchange between nucleons, the π-mesons are the quanta, or agents, of the nuclear-force field. The three different π-mesons (π^+, $\overline{\pi^-}$, and π^0) allow for the charge independence of the $(p\text{-}p)$, $(p\text{-}n)$, and $(n\text{-}n)$ nuclear forces.

The properties of the π-mesons are listed in Table 10-2. Like a photon, (spin, 1) the π-meson has an integral spin, (0). The π-mesons are all radio-

† The uncertainty principle is sometimes written with an $\hbar$ rather than h.

Table 10-2.   Properties of π-mesons.

	π^+	$\overline{\pi^-}$	π^0
Mass	273.23 m_e	273.23 m_e	264.4 m_e
Charge	$+e$	$-e$	0
Spin	0	0	0
Magnetic moment	0	0	0
Mean-life	2.54 $\times$ 10^{-8} sec	2.54 $\times$ 10^{-8} sec	$\sim$10^{-15} sec
Decay mode	$\pi^+ \rightarrow \overline{\mu^+} + \nu$ (99.99%) $\pi^+ \rightarrow \overline{e^+} + \nu$ (0.01%)	$\overline{\pi^-} \rightarrow \mu^- + \bar{\nu}$	$\pi^0 \rightarrow \gamma + \gamma$ (98.8%) $\pi^0 \rightarrow e^+ + e^- + \gamma$ (1.2%)

active and decay (when free) according to the schemes shown in Table 10-2. The dominant decay mode of the charged π's is into a neutrino and a particle with a mass of 207m_e, called a μ-*meson* (*muon*). The uncharged, π^0-meson decays most often into two high-energy γ-rays. As one would expect, the π-mesons interact very strongly with nuclei. Figure 10-24 shows

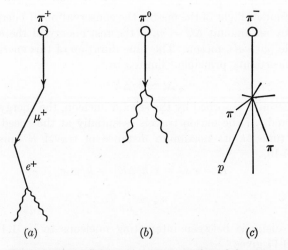

Figure 10-24. Schematic representation of the most probable fates of π^+, π^0, and π^- mesons.

the most probable fate of the three π-mesons. The charged pions lose their energy by ionization, coming nearly to rest. The π^+ is repelled by the Coulomb force of any nucleus and decays into $\overline{\mu^+} + \nu$. The $\overline{\pi^-}$, attracted to a nucleus, can be absorbed by it. Upon its capture by a nucleus, this π-meson releases its rest energy (140 Mev) to the nucleons of the nucleus, which disintegrates violently, producing a star.

The properties of the lighter mesons, the μ-mesons, are summarized in

Table 10-3. We notice that there are only two different μ-mesons, μ^- and $\overline{\mu^+}$, analogous to the electron-positron pair. Their spin is also half-integral, $\frac{1}{2}$. The muon interacts with matter solely through its charge. It is not an agent of the nuclear force, and therefore interacts very weakly with nucleons. As a consequence, μ-mesons are very penetrating.

Table 10-3.   Properties of the μ-mesons.

	$\overline{\mu^+}$	μ^-
Mass	$206.84 \, m_e$	$206.84 \, m_e$
Charge	$+e$	$-e$
Spin	$\frac{1}{2}\hbar$	$\frac{1}{2}\hbar$
Magnetic moment	$1.0026 \, (m_e/m_\mu)\beta$	$-1.0026 \, (m_e/m_\mu)\beta$
Mean-life	2.22×10^{-6} sec	2.22×10^{-6} sec
Decay mode	$\overline{\mu^+} \rightarrow \bar{e}^+ + \nu + \bar{\nu}$	$\mu^- \rightarrow e^- + \nu + \bar{\nu}$

We are now in a position to interpret the observed measurements of cosmic-ray experiments both at high altitude and at sea level. Upon entering the upper regions of the Earth's atmosphere, the primary cosmic-ray particles (mainly protons) collide with nuclei, and many π-mesons are created. The π^0-mesons decay very quickly into high-energy γ-rays, which are then dissipated in cascade showers. The charged π-mesons decay into μ-mesons and (undetectable) neutrinos. These μ-mesons interact only weakly with matter in the Earth's atmosphere, penetrating it easily. However, the μ-mesons are themselves unstable and decay into positrons or electrons, depending on their charge. Thus, at sea level one finds a very penetrating, or *hard, component* in the μ-mesons, and an easily absorbed, or *soft, component* in the electrons and positrons. About 75 per cent of the secondary cosmic radiation at sea level is composed of μ-mesons, about 20 per cent is electrons and positrons, and only 1 per cent is π-mesons, most of the π's not having survived decay from the top of the Earth's atmosphere to sea level.

An interesting confirmation of the time-dilatation phenomenon, predicted by the theory of special relativity and discussed in Section 2-7, is found in the decay of high-energy and thus, high-speed μ-mesons. The *mean*-life of the μ-meson (Table 10-3) is 2.22×10^{-6} sec and the *half*-life is 1.54×10^{-6} sec, both as measured in the inertial system in which the μ-meson is at rest. Therefore, if 1000 μ-mesons are at rest with respect to an observer, he will find only 500 μ-mesons surviving after 1.54×10^{-6} sec has elapsed (according to *his* clock).

If, on the other hand, the mesons move with respect to an observer, the decay time will be dilated and the mesons will appear to live longer for this observer. According to Equation 2-31, he will measure a half-life T given by $T = T_0/[1 - (v/c)^2]^{1/2}$, where T_0 is the half-life in the inertial system of the mesons, and v is the relative velocity between the mesons and the observer. Inasmuch as mesons approach an observer on the Earth's surface with speeds very close to c, time dilatation is appreciable and readily observed.

EXAMPLE　Consider 1000 μ-mesons approaching the Earth's surface at a speed of 0.98c from an altitude $L = 2260$ m, this height being measured by the observer on Earth. The half-life as measured by the observer on Earth is $T = T_0/[1 - (v/c)^2]^{1/2} = 5T_0 = 7.70 \times 10^{-6}$ sec. The flight time of the μ-mesons as measured by the ground observer is $L/(0.98\,c) = (2260\,\text{m})/(0.98 \times 3.0 \times 10^8\,\text{m/sec}) = 7.70 \times 10^{-6}$ sec. (The flight time is equal to the half-life because of our choice of L.) Therefore,

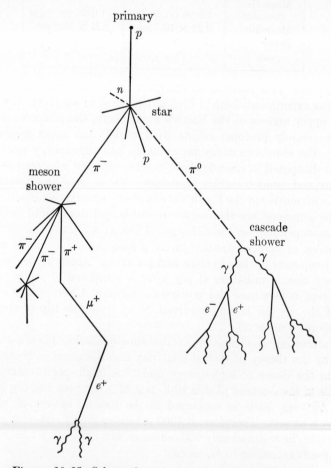

Figure 10-25. Schematic representation of typical events initiated by a high-energy primary cosmic-ray particle.

of the original 1000 μ-mesons at 2260 m, only 500 will have survived decay upon reaching the Earth's surface.

How do the μ-meson decays appear to an observer moving with the decaying mesons? He finds that half the mesons decay in a time of $T_0 = 1.54 \times 10^{-6}$ sec. This observer sees the Earth approach him at a speed of $0.98c$; and, because of the space-contraction phenomenon, the Earth's distance from him is contracted. At that time when this observer measures, $(2260 \text{ m})[1 - (0.98)^2]^{1/2} = 452$ m. Thus, the time of flight, as determined by this observer moving with the decaying mesons, is $(452 \text{ m})/(0.98 \times 3 \times 10^8 \text{ m/sec}) = 1.54 \times 10^{-6}$ sec. This is just the decay half-life in the rest system of the mesons; and therefore, an observer in this system will also find 500 of the original 1000 mesons surviving when the Earth reaches the mesons. Although the two observers, one on the Earth and one with the mesons, disagree on the measurements of time and of length, they both agree that 500 mesons survive when the Earth and mesons meet.

The three principal types of events initiated by high-energy primary cosmic-ray particles—star events, cascade showers, and meson showers—are shown schematically in Figure 10-25.

THE STRANGE PARTICLES   Other new particles besides the π-mesons are sometimes created in high-energy collisions. Inasmuch as these particles seem to play no altogether clear role in the present theory of matter, they are appropriately called *strange particles*. The strange particles may be classified into two general groups:

(1) The *K-mesons*, which have masses between those of the π-mesons and the nucleons and with zero spin (like π mesons).

(2) The *hyperons*, all of which have masses greater than those of the nucleons and with spin, $\frac{1}{2}$ (like nucleons).

We list in Tables 10-4 and 10-5 the presently known K-mesons and hyperons, together with their anti-particles. Although only one of the anti-hyperons has been experimentally observed ($\overline{\Lambda^0}$), theoretical arguments justify inclusion of the other anti-hyperons. The symbols in parentheses represent the anti-particles (always designated with a bar above the symbol). The charge is shown by the post-superscript in units of the electron charge, e; and m_e is the electron mass.

Table 10-4.   The K-mesons.

SYMBOL	MASS	SPIN	MEAN-LIFE (SEC)	PRINCIPAL MODES OF DECAY
$K^+(\overline{K^-})$	$967\ m_e$	0	1.2×10^{-8}	$K^+ \rightarrow 2\pi^+ + \overline{\pi^-}$ $\overline{K^-} \rightarrow 2\overline{\pi^-} + \pi^+$
$K^0(\overline{K^0})$	$966\ m_e$	0	$\sim 10^{-10}$	$K^0 \rightarrow \pi^+ + \overline{\pi^-}$

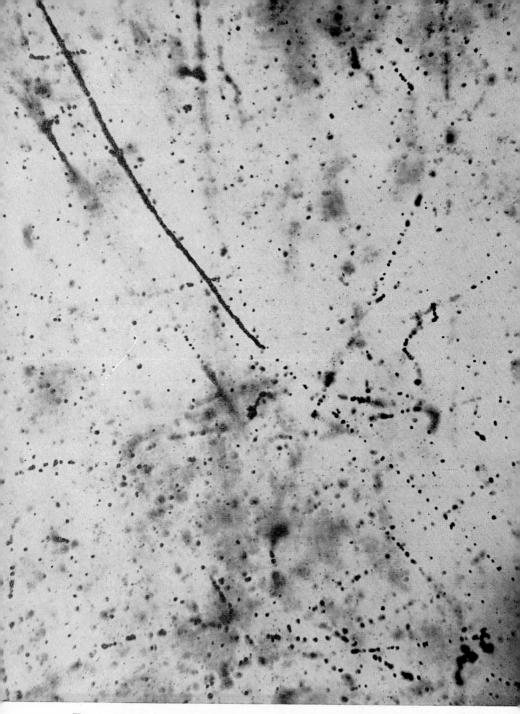

Figure 10-26. A nuclear emulsion showing the decay in flight of a K-meson into a π^+ meson and π^0 meson. The π^0 meson does not produce a track; its identity is established by applying conservation of energy and of momentum to the decay event. (Courtesy of Brookhaven National Laboratory.)

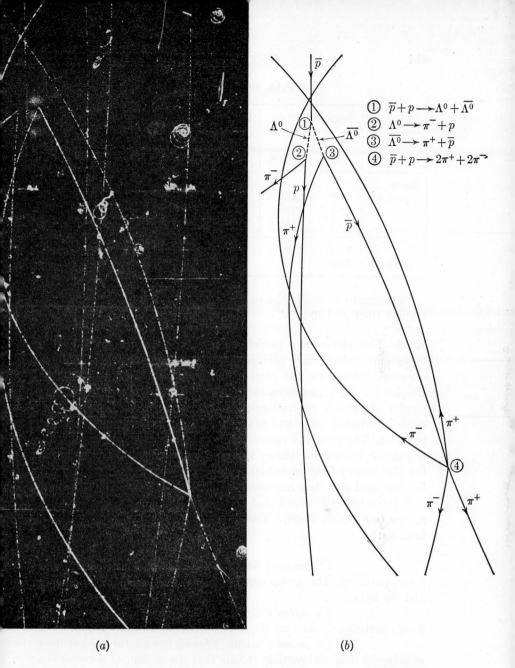

Figure 10-27. (a) A liquid-hydrogen bubble-chamber photograph establishing the existence of the anti-lambda particle (L. W. Alvarez *et al.*, 1959). (b) Identification of the series of events in which a $\overline{\Lambda^0}$ particle appears. An anti-proton enters from the top, unites with a proton, and these decay into a lambda and anti-lambda. (Courtesy of the University of California Lawrence Radiation Laboratory.)

Table 10-5. The Hyperons.

Particle	Symbol	Mass	Mean-life (sec)	Principal modes of decay
Lambda particle	$\Lambda^0(\overline{\Lambda^0})$	2182 m_e	2.6×10^{-10}	$\Lambda^0 \rightarrow p + \pi^-$ or, $n + \pi^0$
Sigma particles	$\Sigma^0(\overline{\Sigma^0})$ $\Sigma^+(\overline{\Sigma^-})$ $\Sigma^-(\overline{\Sigma^+})$	2326 2328 2342	$\sim 10^{-18}$ 8×10^{-11} 1.7×10^{-10}	$\Sigma^0 \rightarrow \Lambda^0 + \gamma$ $\Sigma^+ \rightarrow p + \pi^0$ or, $n + \pi^+$ $\Sigma^- \rightarrow n + \pi^-$
Xi particles	$\Xi^-(\overline{\Xi^+})$ $\Xi^0(\overline{\Xi^0})$	2585 ?	$\sim 10^{-10}$?	$\Xi^- \rightarrow \Lambda^0 + \pi^-$ $\Xi^0 \rightarrow \Lambda^0 + \pi^0$

Photographs in which K-mesons and hyperons participate are shown in Figures 10-26 and 10-27.

10-10 The elementary particles Considering the multiplicity of the (apparently) elementary particles found in modern physics, the recurring question, "What are the elementary particles of nature?" arises again. Despite the incompleteness, tentative character, and extraordinary complexity of present elementary-particle theory, much progress has been made in bringing order and simplicity into the relationships among these particles. The recondite nature of these questions precludes our giving, in any detail, the current theory of elementary particles. We can, however, list the known (and probable) particles, show their grouping into four families, and give their important means of interaction.

It is currently believed that 30 distinct elementary particles occur in nature (see Figure 10-28). These particles fall rather naturally into four families:

The baryon ("heavy") family is composed of 8 particles and 8 anti-particles. This group comprises the heaviest particles, the nucleons and the hyperons.

The lepton ("light") family is composed of 3 particles and 3 anti-particles. Note that the μ-*meson* is placed in this group.

The meson ("middle") family has a total of 7 particles, the π^0 being its own anti-particle. (Note that the misnamed μ-meson is *not* in this family.)

The photon family consists of just one type of particle, the photon, which is its own anti-particle.

The following properties of the elementary particles are now established (or inferred, where experimental observation is still incomplete):

(1) Each of the thirty elementary particles has a charge of $+e$, $-e$, or zero. In any reaction, or decay, the conservation of charge holds.

(2) Mass-energy, linear momentum, and angular momentum are conserved in any reaction.

(3) The intrinsic angular momentum, or spin, of any elementary particle is quantized, the possible values being half-integral or integral multiples of $\hbar$. Those particles with half-integral spin obey the Pauli exclusion principle; those with integral spin, do not. All particles in the baryon family and in the lepton family have spin $\frac{1}{2}$ (see Figure 10-28). All particles in the meson family have spin zero (integral spin) and the particle of the photon family has spin 1 (integral spin).

(4) In any reaction, the number of baryons is conserved, and the number of leptons is conserved. For these conservation laws, the

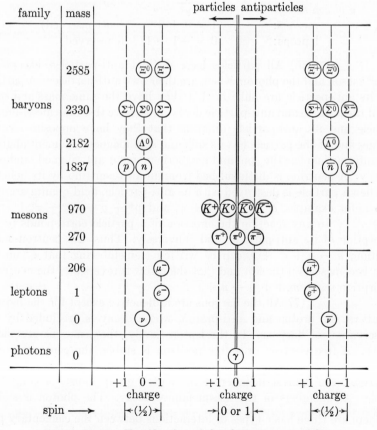

Figure 10-28. Classification of the elementary particles.

anti-particles are taken to be negative. We note that the baryons and leptons are all spin $\frac{1}{2}$ particles. An example of these two conservation laws is neutron decay (see Equation 9-41).

$$n \rightarrow e^- + p + \bar{\nu}$$

baryons: $1 = 0 + 1 + 0$ (conserved)

leptons: $0 = 1 + 0 - 1$ (conserved)

(5) The photon (spin, 1) and the mesons (spin, 0) are the quanta of the electromagnetic field and nuclear field respectively. For field-quantum particles, there is *no* conservation law restricting the total number of particles. They can be created or annihilated in a reaction; for example,

Pair annihilation: $e^- + \overline{e^+} \rightarrow \gamma + \gamma$

photons: $0 + 0 \neq 1 + 1$ (*not* conserved)

leptons: $1 - 1 = 0 + 0$ (conserved)

π-meson decay: $\pi^+ \rightarrow \overline{\mu^+} + \nu$

mesons: $1 \neq 0 + 0$ (*not* conserved)

leptons: $0 = -1 + 1$ (conserved)

(6) All particles have distinct anti-particles, except for the π^0-meson and the photon, which are their own anti-particles. A particle and its anti-particle are alike in that they have the same mass and spin; and if the particle and anti-particle decay, both have the same mean-life. A particle and its anti-particle differ in that they have opposite electric charge; and, if the particle has its spin and spin magnetic moment aligned, the anti-particle has the spin and magnetic moment anti-aligned, and conversely. A neutrino is distinguished from an anti-neutrino by its helicity.

When a particle is destroyed with its anti-particle, field quanta are produced; for example, $e^- + \overline{e^+} \rightarrow \gamma + \gamma$, and $p^+ + \overline{p^-} \rightarrow \pi^0 + \pi^0 + \pi^- + \pi^+ + \ldots$. In any reaction the emission of a particle corresponds to the absorption of its anti-particle, and conversely. Thus, the neutron-decay reaction, $n \rightarrow p + e^- + \bar{\nu}$, can be written (remembering that e^- on the right becomes $\overline{e^+}$ on the left and that the arrow is reversible) the neutrino-absorption reaction, $\bar{\nu} + p \rightarrow n + \overline{e^+}$.

(7) All the baryons are radioactive except for the lightest members, the proton and anti-proton, whose decay is precluded by the conservation of baryons. In the lepton family, the μ-mesons are radioactive. A free electron or a free positron is stable, the conservation of charge preventing their decay into neutrinos. A free neutrino (or anti-neutrino), the lightest member of the lepton group, must, of course, be stable. All members of the meson family decay. The photon is stable.

There are three basic types of interactions between the elementary particles of physics:

(1) *The nuclear force.* It is a strong, short-range interaction between baryons, and the quanta associated with this force are the π-mesons.

(2) *The electromagnetic force.* This is a long-range force having a strength approximately 1/100th the nuclear interaction. This force exists between all *charged* particles, and is transmitted by photons.

(3) *The weak interaction.* Its strength is only 10^{-14} the nuclear interaction. The most common example of the weak interaction is the (very slow) process of β-decay. All the radioactive elementary particles, except for the Σ^0 and the π^0, decay by means of the weak interaction, as indicated by the relatively long half-lives. The Σ^0 and π^0 decay by a strong interaction, and thus have very short half-lives, 10^{-18} sec and 10^{-15} sec, respectively.

The very weak force of universal gravitation, which emerges as a necessary consequence of the properties of space and time in the general theory of relativity, is not as yet included in the scheme of elementary particles. It is only 10^{-41} that of the nuclear interaction.

The gravitational force, although very weak, is dominant in macroscopic systems, such as the solar system; the electromagnetic forces dominate in microscopic systems, such as atoms; and the very strong nuclear forces dominate in submicroscopic systems, such as nuclei.

10-11   Summary

Most low-energy nuclear reactions, or transmutations, are of the general form, $X(x,y)Y$, where x is a light incident particle, X is the target nucleus, y is the emerging light particle, and Y is the product nucleus (often radioactive). The number of nucleons and the electric charge are conserved in a nuclear reaction. The reaction Q is defined by

$$Q = [(M_X + m_x) - (M_Y + m_y)]c^2 = (K_y + K_Y) - K_x$$

where the target nucleus is at rest in the laboratory. When $Q > 0$, the reaction is exothermic and nuclear energy is released; when $Q < 0$, the reaction is endothermic and energy is consumed. Reactions in which the x or y particles are photons are known respectively as photodisintegration and radiative capture.

Because momentum must be conserved, only a portion of K_x is available for a nuclear reaction; the available energy is $K_x M_X/(m_x + M_X)$. The threshold for an endothermic reaction is given by $K_{th} = -Q(M_X + m_x)/M_X$.

The reaction cross section σ gives a measure of the intrinsic probability that a nuclear reaction will occur. The fractional number of x particles undergoing a reaction in a thin foil of thickness t and containing N X-particles per unit volume is $n_r/n_i = \sigma N t$.

According to the concept of the compound nucleus, nuclear reactions take place in two distinct stages: the formation of the compound nucleus and the decay of the compound nucleus

$$X + x \rightarrow C^* \rightarrow y + Y$$

The occurrence of peaks, or resonances, in the reaction cross section is a manifestation of the compound nucleus in quantized excited states.

Neutrons are detected indirectly by the ionization effects of charged particles produced by neutron-initiated reactions or by neutron collisions. Neutron energies can be evaluated by measuring their time-of-flight, by a mechanical velocity selector, or by neutron-diffraction in crystals. Neutrons are moderated when they lose kinetic energy in elastic collisions with a material. Neutrons in equilibrium with a moderator at temperature T have an average kinetic energy of $\frac{3}{2} kT$.

In a nuclear-fission reaction, an excited heavy nucleus, such as uranium 235, splits into fission fragments and several neutrons with the release of about 200 Mev, mostly in the form of kinetic energy of the fragments. The fragments are unstable and decay by β^--emission. The process of fission is best understood in terms of the liquid-drop model, with a competition between surface-tension and Coulomb-repulsion effects in the deformation of the heavy nucleus. The principal elements of a nuclear reactor operating on a self-sustaining nuclear-fission reaction are the fuel elements, moderator, and control rods. A reactor can be used as a source of heat or of neutrons, or as a means to render fertile materials fissile.

A thermonuclear-fusion reaction is a highly exothermic reaction of relatively light particles in a very-high-temperature plasma. The common stellar thermonuclear reactions occur by means of the carbon cycle and the proton-proton cycle.

The primary cosmic rays come from outer space and produce high-energy nuclear reactions in the Earth's upper atmosphere. This incoming radiation consists of nuclei, mostly protons, with an average energy of 20 Bev; the secondary cosmic radiation is produced by reactions initiated by the primary particles. Important cosmic-ray events include: star events, in which a nucleus is exploded by an energetic incident particle; cascade showers, in which a photon is degraded by the *Bremsstrahlung* and pair-production processes into many electrons, positrons, and lower-energy photons; and meson showers, in which many π-mesons are emitted.

The π-mesons are the agents of the nuclear force; the μ-mesons, the decay products of π-mesons, interact weakly with nuclei and comprise the hard component of the cosmic radiation.

The elementary particles are organized into the following families (Figure 10-28): the baryons, consisting of hyperons (Ξ, Σ, and Λ particles) and nucleons (p and n); the mesons (K and π); the leptons (μ-mesons, electrons, positrons, neutrinos); and the photons. In all nuclear reactions the total number of baryons and the total number of leptons are conserved.

REFERENCES

All of the nuclear-physics texts listed in the references for Chapter 9 (page 363) contain material pertaining also to the topics of this chapter.

Bishop, A. S., *Project Sherwood, The U. S. Program in Controlled Fusion.* Reading, Massachusetts: Addison-Wesley Publishing Company, Inc., 1958. The elementary theory of controlled nuclear fusion and some devices in which controlled fusion has been attempted are discussed in this book. The multi-colored diagrams illustrating the confinement of charged particles by magnetic fields are especially noteworthy.

Halliday, D., *Introductory Nuclear Physics*, 2nd ed. New York: John Wiley & Sons, Inc., 1955. Chapters 9 and 13 contain excellent discussions of neutrons and nuclear reactions, respectively.

Murray, R. L., *Introduction to Nuclear Engineering.* Englewood Cliffs, New Jersey: Prentice-Hall, Inc., 1954. This text treats the elementary engineering aspects of nuclear reactors.

Semat, H., *Introduction to Atomic and Nuclear Physics.* New York: Rinehart & Company, Inc., 1954. Many nuclear reactions are listed and described in Chapter 11.

Smyth, H. DeW., *Atomic Energy for Military Purposes.* Princeton, New Jersey: Princeton University Press, 1945. The Smyth report gives an historical account of the remarkable scientific and technological developments leading to the first large-scale release of nuclear energy.

PROBLEMS

(The values of atomic masses are given in Appendix IV.)

10-1 (a) Indicate why unstable products of a (n, γ) reaction are likely to decay by β^--emission. (b) Indicate why unstable products of a (p, n) reaction are likely to decay by β^+-emission or electron capture.

10-2 Write at least three nuclear reactions which can be used to produce the radioactive nuclide $_{15}P^{32}$ using stable targets.

10-3 * Before the neutron was properly identified by Chadwick, it was thought that the bombardment of $_4Be^9$ by α-particles led to the reaction $_4Be^9(\alpha, \gamma)_6C^{13}$. When the penetrating radiation from the beryllium bombardment struck paraffin, protons were ejected with an energy of 5.7 Mev. Show that if the proton is assumed to have been energized in a Compton collision, the γ-ray photon must have an energy of 55 Mev.

10-4 Protons are emitted when a target of $_{16}S^{32}$ is struck with α-particles. (a) What is the reaction Q? (b) What is the minimum α-particle energy that will permit the reaction to occur?

10-5 What is the minimum energy with which α-particles must strike a $_4\mathrm{Be}^9$ target to produce a reaction with proton emission?

10-6 The threshold energy of the reaction $_{29}\mathrm{Cu}^{65}$ $(p, n)_{30}\mathrm{Zn}^{65}$ is found to be 2.16 Mev. (a) What is the reaction Q? (b) What is the atomic mass of $_{30}\mathrm{Zn}^{65}$? (c) By what modes can $_{30}\mathrm{Zn}^{65}$ decay?

10-7 A cyclotron accelerates protons to 4 Mev, deuterons to 2 Mev, and α-particles to 4 Mev. What nuclear reactions might be expected to occur when these particles bombard a $_6\mathrm{C}^{12}$ target?

10-8* A 4.0 Mev photon produces the photodisintegration of a deuteron. If the proton is observed to move at 90° to the direction of the initial photon, what is (a) the kinetic energy of the proton, (b) the kinetic energy of the neutron, and (c) the direction of the neutron with respect to the direction of the incident photon?

10-9 The threshold for the reaction $_3\mathrm{Li}^7(p, n)_4\mathrm{Be}^7$, is found to be 1.882 Mev. Compute the neutron-proton mass difference, using the masses of Li^7 and Be^7.

10-10 * (a) What is the minimum kinetic energy of a α-particle beam striking a stationary helium target that will permit the reaction $_2\mathrm{He}^4(\alpha, p)_3\mathrm{Li}^7$ to go? (b) Suppose now that two α-particle beams of equal energy are made to collide head-on. What is the minimum kinetic energy of particles in either beam that will permit the same reaction to go?

10-11 * The Q of a nuclear reaction can be evaluated by measuring the energies, K_x and K_y, of the incident and emerging light particles and by knowing the direction of the y particles with respect to the incident beam. Show that, for y particles observed at 90° to the incident beam, the Q is given by

$$Q = K_y[1 + (m_y/M_Y)] - K_x[1 - (m_x/M_Y)]$$

10-12 When 7.30 Mev α-particles strike an aluminum target, the reaction $_{13}\mathrm{Al}^{27}(\alpha,p)_{14}\mathrm{Si}^{30}$ occurs. The protons observed at 90° are found to have an energy of 9.34 Mev. (a) What is the Q of this reaction using the relation in Problem 10-11? (b) What is the Q of this reaction as computed from the masses of the particles? (c) What would be the energy of protons observed in the forward direction?

10-13 A proton from the $_7\mathrm{N}^{14}(\alpha, p)_8\mathrm{O}^{17}$ reaction moves in the same direction as the incident α-particle. (a) What is the energy of the proton if the α-particle energy is 5.00 Mev? (b) What is the energy of the $_8\mathrm{O}^{17}$ nucleus?

10-14 A 1.00 Mev proton initiates the reaction $_3\mathrm{Li}^7(p, \alpha)_2\mathrm{He}^4$. What is (a) the energy available for the reaction, (b) the energy carried by the center-of-mass, (c) the energy with which an α-particle leaves the center-of-mass, (d) the energy of an α-particle in the laboratory, assuming that both α-particles have the same energy?

10-15 Suppose that the reaction $A(a, b)B$ has a positive Q. If b particles from the $A(a, b)B$ reaction are used to bombard a target of material B, and the reaction $B(b, a)A$ occurs, show that the final emerging a particles will have a smaller kinetic energy than the a particles striking the first target.

10-16 * A 0.0050 cm gold foil is irradiated with a beam of neutrons. The beam contains 2.0×10^{11} neutrons/sec. The radioactive nuclide $_{79}Au^{198}$ is produced by the (n, γ) reaction. The neutron-capture cross section of gold is 99 barns; the density of gold is 19.3 gm/cm³. (a) How many $_{79}Au^{198}$ atoms are formed if the target is irradiated for 1000 sec? (b) The nuclide $_{79}Au^{198}$ decays with a half-life of 2.7 days. What is the activity of the foil at the end of the bombardment period, assuming that a negligible number of decays occur during irradiation? (Conversely, by measuring the activity, one can deduce the neutron flux.)

10-17 The neutron radiative-capture reaction (n, γ) usually has, for a given target material, a much larger cross section than such competing reactions as (n, p), (n, α), or (n, d). Why is this to be expected?

10-18 (a) What is the height of the Coulomb barrier encountered by a proton striking a $_3Li^7$ nucleus? (Use the relation $R = (1.4 \times 10^{-15}m)A^{1/3}$ for computing the range of the nuclear force.) (b) The reaction $_3Li^7(p, \alpha)_2He^4$ takes place with protons having energies as small as 0.500 Mev. What inference can be drawn concerning penetration of the Coulomb barrier?

10-19 The thermal-neutron capture cross section of $_{79}Au^{197}$ is 99 barns. (a) Assuming that the $1/v$ law holds, what is the cross section for 1.0 ev neutrons? (b) What thickness of gold (density, 19.3 gm/cm³) foil will absorb only 10 per cent of 1.0 ev neutrons incident on it?

10-20 The deuteron beam from a cyclotron has an energy of 8.0 Mev, and the beam current is 1.0 microampere. (a) What is the power output (watt) of the cyclotron beam? The deuteron beam strikes a target of $_5B^{10}$ and neutrons are released in the reaction. (b) How much energy is released in each reaction? If the target is thick, it is found that there are one and one-half $_5B^{10}(d, n)_6C^{11}$ reactions for every million deuterons striking the target. (c) What is the nuclear energy released per second (watt)? (d) What is the ratio of the output nuclear power from this reaction to the input power from the cyclotron beam?

10-21 Write down as many reactions as possible leading to $_{14}Si^{28}$ as the compound nucleus.

10-22 * (a) What is the fractional loss in the kinetic energy of a neutron making a head-on elastic collision with a C^{12} nucleus? (b) How many such collisions are required to thermalize a 5.0 Mev neutron?

10-23 What is the (a) speed and (b) energy of neutrons which pass through a mechanical velocity selector of the sort shown in Figure 10-16, if the discs are separated by 1.0 m, the angular displacement is 90°, and the rotation rate is 32,000 rpm?

10-24 Neutrons are produced in a cyclotron by a (d, n) reaction. The cyclotron is pulsed, with deuterons striking the target in pulses having a duration of 2.0 microseconds and repeated 1000 times per second. A BF_3 neutron detector is located 5.0 m from the cyclotron target. What is the speed and energy of neutrons that reach the neutron detector 1200 microseconds after the initiation of the cyclotron pulse?

10-25 A neutron crystal spectrometer utilizes Bragg planes separated by 0.7323 Å in a beryllium crystal. What must be the angle between the incident poly-energetic neutron beam and the Bragg reflection planes so that the reflected neutron beam will be mono-energetic with an energy of 4.00 ev?

10-26 The nucleus $_{92}U^{236}$ decays into the fission fragments $_{56}Ba^{146}$ and $_{36}Kr^{90}$. (a) Assuming that the two fragments have a spherical shape and are just touching after their formation, compute their electrostatic potential energy in Mev. (b) How does this energy compare with energy released in fission?

10-27 A nuclear reaction uses nuclear fuel at the rate of 10^6 watt. (a) What is the rate at which mass is converted into energy? (b) What is the power lost in neutrinos? (c) If the reactor uses uranium, how much of U^{235} fuel is consumed in a one-day period? (d) For a conventional power plant operating at 10^6 watt, what is the rate at which mass is converted into energy?

10-28 What would be the temperature of a "gas" of fission fragments? Assume that two fragments of equal size are produced per fission and that each fragment has a kinetic energy of 70 Mev.

10-29 (a) What is the electrostatic potential energy of two deuterons separated by 2.0×10^{-15} m? (b) What must the temperature of a gas of deuterons be in order that the particles with an average kinetic energy approach one another within the range of the nuclear force?

10-30 Confirm that the energy released in the carbon-cycle and proton-proton cycle is 24.7 Mev in both cases.

10-31 The total amount of water on Earth is about 10^{21} kg. There is one D_2O molecule for every 6000 H_2O molecules. (a) What is the total amount of energy that can be extracted from the deuterium in water assuming that it is used in the fusion reaction, $_1H^2(d, p)_1H^3$? (b) How many kilograms of fissionable uranium have the same nuclear energy content?

10-32 What is the maximum energy of neutrons obtained from the reaction $H^3(d, n)He^4$ if the deuteron energy is 1.0 Mev?

10-33 * A 10^{18} ev primary cosmic-ray proton approaches the Earth. (a) What is the thickness of the Earth (8000 mi) as viewed by an observer at rest with respect to the proton? (b) When the proton is 10^4 km away from the Earth as measured by an observer on Earth, how long does it take the proton to reach the Earth as measured by an observer traveling with the proton?

10-34 A cascade shower is initiated by a photon having an energy of 250 Mev. What is the maximum number of electrons or positrons that it can produce?

10-35 * The mean-life of a Σ^- hyperon is 1.7×10^{-10} sec. The entire life history of such a particle, from creation to decay, is shown in a cloud-chamber photograph. (a) If the energy of the particle is 500 Mev and a negligible amount of energy is lost in forming the cloud-chamber track, how long is the Σ^- track assuming that this hyperon decays in its mean-life? (b) If a very large number of 500 Mev Σ^- tracks are observed, will all tracks have the same length? Why?

10-36 * Show that the μ-mesons emitted by decaying π-mesons at rest are mono-energetic and compute the μ-meson kinetic energy.

10-37 A μ^--meson is captured by an iron nucleus forming a mesic-atom. (a) What is the radius of the first Bohr orbit? (b) What is the ionization energy of this mesic-atom? (c) What is the nuclear radius for iron?

10-38 * Show that, when a π^0-meson decays in flight into two photons each making an angle θ with respect to the direction of the original meson, $\sin \theta = (E_0/E)$, where E_0 and E are the rest and total relativistic energies of the meson, respectively.

10-39 A π^0-meson is found to decay in flight into two photons both of which make an angle of 30° with respect to the direction of the original π^0-meson. Show that the kinetic energy of such a π^0-meson is equal to its rest energy.

10-40 What is the energy of the photon produced by the decay of Σ^0 particle at rest into a Λ^0 particle?

10-41 Write the decay reactions leading from a Ξ^- particle into finally intrinsically stable particles.

10-42 π-mesons can be created in the high-energy reaction $p(\gamma, \pi^+)n$. What is the minimum photon energy that will permit the reaction to go?

ELEVEN

MOLECULAR AND SOLID-STATE PHYSICS

The simpler aspects of atomic and nuclear structure can be understood on the basis of a few fundamental principles of the quantum theory, often without recourse to the formal mathematical procedures of wave mechanics. This happy situation does not, however, obtain when one deals with more complicated systems, such as molecules or solids, which consist of many interacting particles. These many-body systems can be treated adequately only by a thorough-going wave-mechanical analysis, together with the statistical methods for handling very large numbers of particles. For this reason our discussion of molecular structure and the physics of the solid state will necessarily be of a somewhat qualitative nature, and certain results of wave mechanics and statistical mechanics will be given without proof. We will illustrate the nature of some fundamental problems in molecular and solid-state physics with a few simple examples: molecular binding, rotational and vibrational molecular spectra, the classical and quantum distribution laws, the specific heats of gases and solids, black-body radiation, the free-electron theory of metals, and the properties of conductors, semiconductors, and insulators.

11-1   Molecular binding   Let us consider the ways in which a simple diatomic molecule, consisting of two atoms held together by an attractive force, may be bound. If the molecule is to exist as a stable bound system the energy of the two atoms when close together must be *less* than the energy of the two atoms when separated by a great distance; and to show that molecular binding occurs, it is merely necessary to establish that the total energy of the atoms is reduced when they are brought sufficiently close together.

There are two important ways in which molecules are bound: *ionic*, or *heteropolar*, *binding*; and *covalent*, or *homopolar*, *binding*. An example of a molecule bound almost completely by ionic binding is sodium chloride, NaCl, and an example of covalent binding is the hydrogen molecule, H_2.

IONIC BINDING   We wish to show that NaCl can exist as a stable molecule, the total energy of a NaCl system being less when the two atoms are brought together than the total energy of the atoms when separated. The element sodium, $_{11}$Na, is in the alkali-metal group, the first column of the periodic table (see Figure 6-30); as such, it has one electron outside a closed subshell. The electron configuration of $_{11}$Na in the ground state is $1s^2 2s^2 2p^6 3s^1$. The single $3s$ electron is relatively weakly bound to the atom, and can be removed by adding 5.1 ev, thereby ionizing the atom and leaving it with a net electric charge of $+1e$. The alkali metals are said to be *electropositive* because they are easily ionized to form positive ions, with a resulting electron configuration consisting of closed electron shells, like those of the inert gases.

The element chlorine, $_{17}$Cl, is a halogen element, falling in the seventh column of the periodic table; all elements in this column lack one electron of completing a closed p subshell. The electron configuration of $_{17}$Cl in the ground state is $1s^2 2s^2 2p^6 3s^2 3p^5$. The neutral chlorine atom lacks one electron of filling a tightly-bound, complete $3p$ subshell; and indeed the energy of the chlorine atom is *lowered* by 3.8 ev when an electron is added to it, thus forming a negative ion of charge $-1e$. The halogen elements are said to be *electronegative*, and the *electron affinity energy* of $_{17}$Cl is 3.8 ev. It follows that 3.8 ev of energy must be added to the Cl$^-$ ion to remove the last electron, leaving the neutral Cl atom.

Suppose that one begins with a neutral Na atom and a neutral Cl atom, infinitely separated. Now to remove one electron from Na, thereby forming Na$^+$, costs 5.1 ev. But when this electron is transferred to Cl, thereby forming Cl$^-$, 3.8 ev of this energy is repaid, the over-all energy required being only $(5.1 - 3.8)$ ev $= 1.3$ ev. The energy differences are shown in Figure 11-1. We now have a positive ion and a negative ion, still separated. These ions will attract one another by the Coulomb electrostatic force of attraction, the Coulomb potential energy being $-ke^2/r$, where r is the distance between the centers of the two ions. When the Na$^+$ and Cl$^-$ ions are brought together, the total energy of the system decreases, inasmuch as

5.1 ev

1.3 ev
_____

0
_____   - - - - - - -   - - - - - - -   - - - - - - -

−3.8 ev
_____

(Na+Cl)   (Na⁺+Cl)   (Na+Cl⁻)   (Na⁺+Cl⁻)

Figure 11-1. Energy differences of sodium and chlorine atoms and ions.

the force is attractive and the potential energy is negative; and if r is chosen as, say, 4.0 Å—which is larger than the sum of the radii of the closed subshells of the respective ions—the Coulomb energy is easily found to be −1.8 ev. Thus, when the ions are separated by 4.0 Å, the total energy, $(1.3 - 1.8) = -0.5$ ev, is clearly less than that of a Na atom and Cl atom infinitely separated. The net cost of forming two ions is more than repaid by the electrostatic attraction of the ions thus formed.

We have seen that Na and Cl atoms will attract one another when their nuclei are separated by less than about 4 Å. When the separation between the nuclei is decreased still further, a repulsive force begins to act between the ions. The ions repel one another when the electron clouds of the two ions, each of which can be regarded as spherical, begin to overlap. The Pauli exclusion principle governs the number of electrons that can be fitted into any given atomic electron shell. Both the Na⁺ and Cl⁻ ions have their full electron quotas, and no further electrons can be accommodated unless they occupy relatively high-lying energy states. Consequently, as the interatomic distance is reduced, the electron shells are prevented from overlapping by the Pauli exclusion principle. The electrons must, therefore, go to higher-lying available states, and the total energy of the molecule is thereby increased. Because the atoms attract

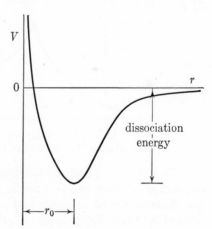

Figure 11-2. Molecular potential as a function of interatomic distance.

one another for larger separation distances but repel one another for sufficiently small separation distances, there exists an equilibrium interatomic separation distance r_0 at which the total potential energy of the system is a minimum, as shown in Figure 11-2.

The *dissociation energy* of a molecule is its binding energy; that is, the energy that must be added to a molecule in its lowest energy state to separate it into its component atoms. The molecule NaCl has a dissociation energy of 4.24 ev, and an equilibrium separation distance r_0 of 2.36 Å. For a molecule such as NaCl, held together by ionic binding, the end of the molecule containing the Na nucleus represents a region of positive electric charge, and the end of the molecule with the Cl atom represents a region of negative electric charge. Thus, an ionic molecule is a *polar molecule*, with a permanent *electric dipole moment*, and for this reason, ionic binding is also known as *heteropolar binding*.

COVALENT BINDING   A simple example of covalent binding is the hydrogen molecule, H_2. Before discussing this molecule, however, we shall first consider a somewhat simpler system, the hydrogen molecule ion, H_2^+. We can imagine an H_2^+ molecule to be formed when a neutral hydrogen atom, that is an electron bound to a proton, is brought together with an ionized hydrogen atom; that is, a bare proton. When the protons are far apart, the electron will move in hydrogen orbits about the one proton alone; but when the protons are separated by a distance of approximately 1 Å, the electron can jump from one proton to the other, making orbits about one or the other of the protons, or about both.

It is possible to solve in detail for the wave function ψ of the single electron, the probability for finding the electron at any position being given by ψ^2. The wave-mechanical results are shown in Figure 11-3. We see that

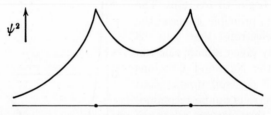

Figure 11-3. Quantum-mechanical probability for finding the electron in the hydrogen molecule ion.

there is a relatively high probability for the single electron to be found in the region between the two nuclei, as compared to the probability that the electron will be at either end of the molecule. When the electron is located between the protons, it attracts both protons by the Coulomb electrostatic force; when the electron is at some exterior location, the protons will be less strongly attracted to the electron and their mutual repulsion becomes more important. Because the mutual repulsive force between the two nuclei increases as r (their separation) decreases, the H_2^+ molecule has a minimum in the potential energy. It is found that $r_0 = 1.06$ Å, and the dissociation energy = 2.65 ev.

Now consider the hydrogen molecule, H_2, formed when two neutral hydrogen atoms are brought close together. The binding together of two neutral, identical atoms is completely inexplicable in classical terms, and the binding of the hydrogen molecule can be understood fully only on the basis of the quantum theory and the Pauli exclusion principle. When the two hydrogen atoms are separated by a distance that is large compared to the size of the first Bohr orbit (0.5 Å), each electron will be clearly identified with its own parent nucleus. But when the separation of the atoms is comparable to the size of the orbits, one must recognize that there is absolutely no way of distinguishing between the two electrons, and it is no longer possible to associate one electron with one particular nucleus.

There are, however, two distinct ways in which the two hydrogen atoms can be brought together: with the two electron spins aligned in the same direction, or with the electron spins anti-aligned, or in opposite directions. The wave mechanics shows that, when the hydrogen atoms are brought together, the total energy of the system increases when the electron spins are aligned, but decreases when the spins are anti-aligned, as shown in Figure 11-4. This difference in energy arises from the Pauli exclusion principle.

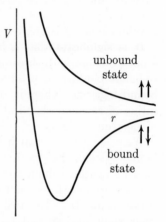

Figure 11-4. Potential between two hydrogen atoms with aligned and anti-aligned spins.

The force binding the (two-electron) neutral hydrogen molecule, which is not present in the (one-electron) hydrogen molecule ion, is associated with the so-called exchange energy. The occurrence of *electron exchange* is a strictly wave-mechanical phenomenon, for which there is no classical analog; and its features can be examined in detail only through mathematical analysis. Nevertheless, there is a simple way of visualizing the exchange of two identical, indistinguishable electrons; namely, through the behavior of their spins. We have noted that the hydrogen molecule is bound, with a sharing of the two electrons by the two nuclei, only when the electrons have their spins anti-aligned. But it is possible for this bound state to persist if the two electrons exchange their spin orientations; that is, if the electron with spin up becomes an electron with spin down, and the electron with spin originally down simultaneously becomes an electron with spin up. This exchange of electron spins is the principal contribution to the binding of the molecule; inasmuch as two anti-aligned electrons can both be simultaneously in the region between the protons, they can exert attractive forces on both protons. We can say, roughly, that the two nuclei of the hydrogen molecule share the two electrons so that each nucleus, at least for a part of the time, can have both electrons filling a closed shell about it.

Because of the exchange phenomenon, the total energy of the two hydrogen atoms is reduced as the atoms approach one another with anti-aligned spins; but eventually the mutual repulsion of the nuclei will exceed the binding due to exchange. Thus, a minimum energy exists, corresponding to the equilibrium configuration of the molecule. For the hydrogen molecule, $r_0 = 0.74$ Å, and the dissociation energy $= 4.48$ ev.

The chemical binding of two identical, non-polar atoms is called *covalent binding*, inasmuch as it is the sharing of valence electrons that is responsible for the attractive force. A molecule such as H_2 has no permanent electric dipole moment, and is said to be non-polar. For this reason the covalent binding is also referred to as *homopolar binding*.

It is significant that a bound hydrogen molecule can exist with two atoms, but not with three or more; that is, the valence forces operating in covalent binding show *saturation*, a limited number of the atoms being bound together. There is, of course, no classical explanation for the saturation of covalent chemical forces because it arises from the phenomenon of electron exchange.

We wish to show that it is impossible for the molecule H_3 to exist as a bound system. Let us first consider, for simplicity, the forces acting between a neutral helium atom and a neutral hydrogen atom. When helium is in its ground state, the two electrons complete the $1s$ shell, the configuration being $1s^2$. The two electrons must, by the Pauli principle, have their spins anti-aligned ($\downarrow \uparrow$), the total spin of the atom being zero. Now suppose that a hydrogen atom (with electron spin $\frac{1}{2}$) approaches a helium atom (with total electron spin 0). If there is to be chemical binding between the two atoms it must arise from the exchange of electron spins. There are only two possible ways in which the single electron ($\uparrow$) in the hydrogen atom can interact with the helium atom: by exchanging spins with the one electron ($\uparrow$), or with the second electron ($\downarrow$). Suppose that the hydrogen electron ($\uparrow$) exchanges spins with the ($\uparrow$) helium electron. The two spins are aligned ($\uparrow \uparrow$), and the exchange force between the atoms is, as in the case of the hydrogen molecule, repulsive. Suppose now that the hydrogen electron ($\uparrow$) exchanges spins with the ($\downarrow$) helium electron; if this were to happen, the electron spins of the helium atom would be aligned ($\uparrow \uparrow$), which is prohibited by the Pauli exclusion principle. In short, this exchange is impossible. Therefore, the only possible exchange force between a hydrogen atom and a helium atom is repulsive, and the molecule HHe does not exist.

We can easily extend this argument to the interaction between a hydrogen molecule, H_2, and a hydrogen atom H_1. The molecule H_2, like the He

atom, has its electron spins anti-aligned. The only possible electron exchange between H_2 and H_1 leads to a repulsive force, and the chemical
homopolar bond between hydrogen atoms is saturated with two electrons.
The molecule H_3 is not formed. The homopolar binding of atoms can be
extended to more complicated structures, and it is found that the binding
in all organic molecules is of this type. Most inorganic molecules are, however, bound by the action of both ionic and covalent binding, one or the
other type typically dominating.

A third type of force, the only one operating in the interaction between
closed-shell atoms of inert gases, is the *Van der Waals force*. This force is
responsible for the cohesion in the liquid and solid state of rare gases.
When two such atoms approach one another, the "center" of the negative
charge is displaced from the positive nucleus. The atoms then weakly
interact through the electric dipoles induced by the charge displacement.

The ionic and homopolar binding processes can hold atoms together to
form crystalline solids. An example of an ionic crystal is NaCl (shown in
Figure 4-1), an alkali halide, in which the Na^+ and Cl^- ions are found
alternately on the corners of a cubic lattice. An example of a covalent
crystal is diamond (carbon; C), where in the elementary cell structure there are carbon nuclei at the center and corners of a tetra-
hedron. A third type of crystalline binding, for which there is no counter-
part in molecules, is metallic binding. This type of binding, arising from
the Coulomb interaction between the fixed positive ions and the free
electrons of the metal, is a quantum-mechanical phenomenon.

11-2 Molecular rotation and vibration

MOLECULAR ROTATION   A simple motion that a diatomic molecule as a
whole can execute is that of pure
rotation about its center-of-mass.
The molecule closely resembles a
dumbbell structure, with atoms of
masses m_1 and m_2 at distances r_1
and r_2, respectively, from the center-
of-mass of the molecule, as shown
in Figure 11-5. From the definition
of the center-of-mass,

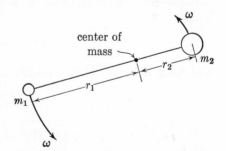

$$m_1 r_1 = m_2 r_2 \qquad [11\text{-}1]$$

Figure 11-5. A rotating diatomic molecule.

The moment of inertia I of the
molecule about an axis of rotation
perpendicular to the interatomic axis and passing through the center-of-
mass is

$$I = m_1 r_1^2 + m_2 r_2^2 \qquad\qquad [11\text{-}2]$$

(The molecule's moment of inertia *along* the interatomic axis is negligible because of the small size of the nuclei and the small mass of the electrons.)

Using Equation 11-1, Equation 11-2 can be written in the form

$$I = \left(\frac{m_1 m_2}{m_1 + m_2}\right)(r_1 + r_2)^2 \qquad [11\text{-}3]$$

We recognize the first factor in Equation 11-3 as the reduced mass μ (Equation 5-30) and the second factor as the interatomic separation distance between the two masses, $r_0 = r_1 + r_2$. Therefore, Equation 11-3 can be written in the simple form,

$$I = \mu r_0^2 \qquad [11\text{-}4]$$

Thus, the motion of the two masses, both rotating at the same angular speed ω about the center-of-mass, can be described in terms of the equivalent rotation of a single particle having a mass μ, at a distance r_0 from the axis of rotation, and rotating at the same angular speed ω.

The angular momentum P_r of molecular rotation is

$$P_r = I\omega \qquad [11\text{-}5]$$

The quantum theory shows that the orbital angular momentum of an atom P_θ is given by $\sqrt{l(l+1)}\hbar$, where l is the orbital angular momentum quantum number (Equation 6-1). In a similar fashion, the rotational angular momentum P_r is given by

$$\boxed{P_r = \sqrt{J(J+1)}\hbar} \qquad [11\text{-}6]$$

where the *rotational quantum number* J has the possible integral values, $0, 1, 2, 3, \ldots$.

The total energy E_r of the free rotation of the rigidly fixed nuclei in a diatomic molecule is given by the rotational kinetic energy of the reduced mass:

$$E_r = \tfrac{1}{2} I\omega^2 = \frac{(I\omega)^2}{2I} = \frac{P_r^2}{2I} \qquad [11\text{-}7]$$

and using Equation 11-6 in Equation 11-7 gives

$$\boxed{E_r = \frac{J(J+1)\hbar^2}{2I}} \qquad [11\text{-}8]$$

We see from Equation 11-8 that the rotational kinetic energy of a molecule is quantized, the possible values being $0, 2(\hbar^2/2I), 6(\hbar^2/2I), 12(\hbar^2/2I)$, $\ldots$, as shown in Figure 11-6.

Let us consider now a transition between a lower and an upper rotational state with energies E_l and E_u. Such a molecular transition results from the absorption of a photon of energy $h\nu$.

$$h\nu = E_u - E_l = (\hbar^2/2I)[J_u(J_u + 1) - J_l(J_l + 1)] \qquad [11\text{-}9]$$

Photon absorption, arising from a transition in the rotational states, is possible only if the molecule is polar with a permanent electric dipole moment. Thus, a rotational spectrum is found for an ionically bound molecule, such as HCl, but not for a molecule such as H_2, with homopolar binding.

The quantum theory shows that the permitted transitions are those for which the rotational quantum number J changes by one unit; that is, the selection rule for rotational transitions is

$$\boxed{\Delta J = \pm 1} \qquad \text{[11-10]}$$

Therefore, the only possible transitions are those between adjacent rotational levels. We take the quantum number of the upper state J_u to be $J_l + 1$, and Equation 11-9 becomes

$$h\nu = (\hbar^2/2I)(2)(J_l + 1)$$

or, $$\nu = \frac{\hbar}{2\pi I}(J_l + 1) \qquad \text{[11-11]}$$

We see from Equation 11-11 that the *pure rotational spectrum*, resulting from changes in the rotational state of a molecule, consists of equally-spaced lines. The lines fall in the far-infrared and microwave regions of the electromagnetic spectrum, with wavelengths ranging from 10^{-3} cm to 10 cm. The observation of the rotational spectrum permits the moment

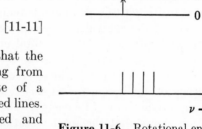

Figure 11-6. Rotational energy-level diagram of a diatomic molecule, together with the pure rotational spectrum.

of inertia, $I = \mu r_0^2$, of the molecule to be determined, and therefore, also the interatomic distance r_0. For example, it is found that the molecule $I^{127}Cl^{35}$ shows rotational lines at the microwave wavelengths 4.30 cm $(J = 0 \rightarrow 1)$ and 1.43 cm $(J = 2 \rightarrow 3)$; using Equations 11-11 and 11-4 it is found that $I = 2.40 \times 10^{-45}$ kg-m^2, and $r_0 = 2.32$ Å.

The rotational spectra of polyatomic polar molecules are complicated by the fact that such molecules must be characterized by more than one moment of inertia. The spectra have more rotational lines in the far-infrared or microwave regions, and an analysis of these spectra yields information on interatomic distances and bond angles.

MOLECULAR VIBRATION   We have seen that the equilibrium configuration of a diatomic molecule is that in which the total energy of the molecule is a minimum; for $r < r_0$ the atoms repel one another, and for $r > r_0$ the atoms attract one another. When the separation distance differs from the equilibrium distance r_0, the atoms will tend to be restored to the equilibrium position, and thus the atoms will vibrate along the interatomic axis.

In the vicinity of the minimum, the potential of a diatomic molecule (Figure 11-2) can be represented quite well by the equation for a parabola:

$$V = V_0 + \tfrac{1}{2} k(r - r_0)^2$$

To find the force on either of the two nuclei, we take the negative derivative of V with respect to the interatomic distance r.

$$F = -dV/dr = -k(r - r_0)$$

Therefore, a restoring force, proportional to the displacement from the equilibrium position, acts on each nucleus. This is, of course, the requirement for simple harmonic motion. Inasmuch as the molecular potential is parabolic near the equilibrium position, the atoms will vibrate in simple harmonic motion around r_0. In still simpler fashion, the reduced mass μ undergoes simple harmonic motion.

Let us consider the results of the quantum theory as applied to a particle which is restricted to a parabolic potential, $V = \tfrac{1}{2} kx^2$, where x is the displacement of the equivalent classical simple harmonic oscillator. At the amplitude position, $x = A$, the total energy of the oscillator is entirely potential; and therefore, $\tfrac{1}{2} kA^2$ is the total energy of the oscillator for *any* displacement. For a particular force constant k and mass μ, the classical frequency of oscillation, $f = (1/2\pi)(k/\mu)^{1/2}$, is independent of the oscillator's amplitude A, or of its energy. The solution of the Schrödinger wave equation shows that the total vibrational energy E_v of the oscillator is quantized, the permitted energies being given by

$$\boxed{E_v = (v + \tfrac{1}{2})h(1/2\pi)\sqrt{k/\mu} = (v + \tfrac{1}{2})hf} \qquad \text{[11-12]}$$

where v, the *vibrational quantum number*, is restricted to the integral values, $v = 0, 1, 2, 3, \ldots$ (See Appendix III.)

The energy-level diagram for a simple harmonic oscillator is shown in Figure 11-7, the permitted energies of the oscillator being equally spaced. It is especially noteworthy that for the lowest energy level, $v = 0$, the total energy is *not* zero, but rather $E_0 = \tfrac{1}{2} hf$. This must be interpreted to mean that an oscillator can never be completely at rest, and that a *zero-point-vibration* takes place even for the ground state. This is quite unlike the classical case, where the oscillator may, of course, be exactly at rest. The occurrence of the zero-point-vibration is a particularly striking manifestation of the uncertainty principle, which requires that the product of the uncertainties in the momentum and position be of the order of h, $\Delta p_x \Delta x \geq h$. If the oscillator were to be at rest, both the position ($x = 0$)

and the momentum ($p_x = 0$) would be known with complete precision, which the uncertainty principle forbids.

We now turn to the spectrum of photons absorbed when a simple harmonic oscillator undergoes a quantum transition from a lower state (v_l) to an upper state (v_u). As in the case of rotation, the absorption (or emission) of photons by a vibrating diatomic molecule is possible only if it has an electric dipole moment, which changes as the molecule vibrates. The photon energy is

$$h\nu = E_u - E_l = hf[(v_u + \tfrac{1}{2}) - (v_l + \tfrac{1}{2})] \qquad [11\text{-}13]$$

The selection rule governing the possible changes in the vibrational quantum number v is

$$\boxed{\Delta v = \pm 1}$$

That is, the only permitted transitions are those between adjacent energy levels, for which v changes by 1. Putting $v_u = v_l + 1$ in Equation 11-13 gives then

$$\boxed{h\nu = hf} \qquad [11\text{-}14]$$

The photons absorbed are of a *single* frequency. The simple harmonic oscillator is atypical in that the photon frequency ν is precisely equal to the classical frequency of vibration f. The allowed transitions and the resulting single-line spectrum are shown in Figure 11-7.

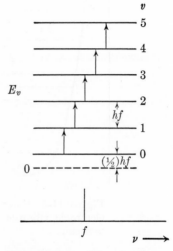

Figure 11-7. Vibrational energy-level diagram of a diatomic molecule, together with the vibrational spectrum.

The vibrational spectrum of a polar diatomic molecule differs from that of an ideal harmonic oscillator (for which the potential is strictly parabolic for all displacements) because the molecular potential shows departure from the parabolic shape for high degrees of vibrational excitation above the ground state. For this reason, the vibrational energy levels are not equally spaced for high quantum numbers; instead, they crowd together near the dissociation limit. For such an *an*harmonic oscillator, transitions for which v changes by 2, 3, . . . , as well as 1, are allowed.

The vibrational frequencies of diatomic molecules fall in the infrared portion of the electromagnetic spectrum, at roughly 100 times the rotational frequencies of molecules. The differences in energy between vibrational energy states are, therefore, approximately 100 times larger than the differences in energy between rotational states. Thus, for each vibrational state there is a whole set of possible rotational states, as shown in

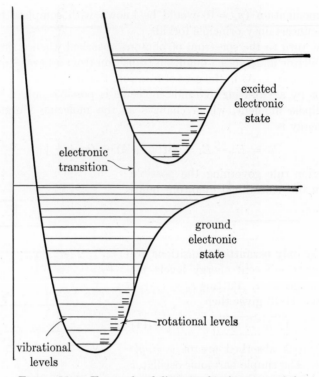

Figure 11-8. Energy-level diagram showing some of the electronic, vibrational, and rotational levels of a diatomic molecule.

Figure 11-8. Because of the rotational fine structure of the vibrational states, the absorption or emission spectra from diatomic molecules consist, not of a single vibrational line, but rather of a group of very-closely-spaced rotational lines crowded near the frequency of the vibrational transition. The infrared spectra of diatomic molecules reflect changes both in the vibrational and rotational states of the molecule and are called *vibrational-rotational spectra*. The vibrational-rotational lines of the molecule, HCl, for example, are found to be centered about 3.3×10^{-4} cm wavelength.

In our discussion of molecular vibration and rotation we have been primarily concerned with the motion of the nuclei of the atoms, and not with electron motion. It is proper to assume that there is no change in the state of the electrons in a molecule during vibration or rotation because these molecular motions take place so slowly, as compared with the motion of electrons, that the electrons are able to follow the nuclei. But changes in the electronic structure of a molecule can take place when an electron in the molecule changes its state or orbit. Such an electronic transition re-

Figure 11-9. A portion of the band spectrum of cyanogen (CN) in the ultra-violet. (Courtesy of RCA Laboratories, Princeton, New Jersey.)

sults in the emission or absorption of a molecular *electronic spectrum* in the visible or ultraviolet portion of the electromagnetic spectrum.

The lowest electronic state and an excited electronic state of a diatomic molecule are shown in Figure 11-8. The potential curve of the molecule in an excited electronic state is displaced upward by an amount that is large compared to either the vibrational or rotational energy differences. The molecular electronic excited state can be thought of as that in which one of the electrons of the atom has been raised to an excited state. As a consequence, the equilibrium position r_0 may be displaced.

The electronic spectra of molecules are enormously complicated because, for a single electronic transition, there are a large number of possible rotational and vibrational states between which transitions can occur. The molecular spectrum in the visible or ultraviolet region consists then, not of a relatively small number of sharply defined lines, as for atomic spectra, but rather of many groups of very closely spaced lines, which with moderate resolution appear as nearly continuous *bands*, as shown in Figure 11-9.

11-3   The statistical distribution laws   Many problems arising in physics are concerned with the behavior of systems comprised of a very large number of weakly interacting, identical particles. The statistical methods for handling large numbers of particles, whose mechanics is known, are called *statistical mechanics*.

A familiar example of a system that can be treated by the methods of statistical mechanics is that of an ideal gas composed of a large number of identical point-particles obeying Newton's laws of motion. Although it is possible, in principle, to describe in detail the motion of every particle of this system, the problem is so mathematically formidable as to be virtually beyond solution. Actually, what is of interest is *not* the *detailed behavior* of every particle of the system, but rather the *average behavior* of the microscopic particles and their influence on macroscopic measured quantities. Thus, it is possible to predict the pressure (a macroscopic quantity) of a gas on its container in terms of the mass and average speed of the molecules (microscopic quantities) as discussed in Section 1-5. Furthermore, it is possible to relate another macroscopic quantity, the absolute temperature T, to a microscopic quantity, the average kinetic energy $\bar{\epsilon}$ of the molecules $(\bar{\epsilon} = \frac{3}{2} kT)$.

In such a system as a gas of particles, interacting only weakly with one another, an equilibrium distribution results. The molecules interact with one another and with the walls by collisions, whose duration is short compared to the time between collisions. Some molecules will have a small kinetic energy, others will have a large kinetic energy; in short, there will be a distribution of the particle energies (and, thus, of the speeds) over a considerable range (see Figure 1-17). If the molecules are in equilibrium and their number is very large, the relative number of molecules of any particular energy will be essentially constant, although the energy of any one molecule will, of course, change with time through collisions.

Statistical mechanics allows one to determine the energy distribution of *any* system of weakly-interacting particles in thermal equilibrium, whether the particles obey classical or quantum mechanics. Although a rigorous development of statistical mechanics is far beyond the scope of this book, we will state and briefly discuss some important results. We will then apply these results to several physical problems of interest in molecular and solid-state physics.

It is assumed that each particle in a system composed of a very large number of identical, weakly-interacting particles has available to it a discrete set of quantum states, the energy of each state i being designated by ϵ_i. For the classical situation, there is a continuum of allowed energies, with the separation between adjacent energies taken to be zero. Inasmuch as the particles are imagined to interact only weakly with one another, each particle has its own set of states; and if the particles are all identical they will all have an identical set of states available to them for occupancy.

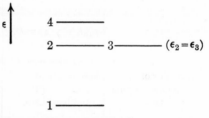

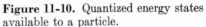

Figure 11-10. Quantized energy states available to a particle.

A very simple hypothetical example is shown in Figure 11-10, where each particle must, at any one time, exist in one of the four possible states. We notice that the states 2 and 3 have identical energies, that is, $\epsilon_2 = \epsilon_3$. Whenever two or more states, such as 2 and 3 above, have the same energy, the states are said to be *degenerate*. This degeneracy can, however, be removed by some external disturbance; for example, the Zeeman degeneracy is removed by a magnetic field (Section 6-5). Now the question is, "How are the particles of the system distributed among the various available states?"

Statistical mechanics predicts the most probable distribution of the particles among the various states for the three kinds of statistical systems met in physical problems. Because most systems of interest have a large number of particles, the most probable distribution becomes overwhelmingly more probable than any other distribution, and thus represents (with almost complete certainty) the distribution of an actual system. In Table 11-1 are listed the three types of probability distributions: Maxwell-Boltzmann, Bose-Einstein, and Fermi-Dirac statistics. Also included in the table are the characteristic properties defining the statistical behavior, and some examples of physical systems which obey the various distributions.

Table 11-1

	Maxwell-Boltzmann	Bose-Einstein	Fermi-Dirac
Characteristics determining the statistics	Identical, but distinguishable particles	Identical, indistinguishable particles of integral spin	Identical, indistinguishable particles of half-integral spin and obeying Pauli exclusion principle
Distribution function $f(\epsilon_i)$	$f_{\text{MB}}(\epsilon_i) = Ae^{-\epsilon_i/kT}$	$f_{\text{BE}}(\epsilon_i) = \dfrac{1}{e^{\alpha}e^{\epsilon_i/kT} - 1}$	$f_{\text{FD}}(\epsilon_i)$ $= \dfrac{1}{e^{(\epsilon_i - \epsilon_F)/kT} + 1}$
Examples of systems obeying statistics	Essentially all gases at all temperatures	Liquid helium (spin = 0) Photon gas (spin = 1) Phonon gas (spin = 0)	Electron gas (spin = $\frac{1}{2}$)

The *distribution function* $f(\epsilon_i)$ in Table 11-1 *represents the average number of particles in the state i.* Inasmuch as $f(\epsilon_i)$ depends only on the energy of the state, the average number of particles in states having the same energy (e.g., states 2 and 3 in Figure 11-10) will be the same (e.g., $f(\epsilon_2) = f(\epsilon_3)$). The distribution functions are derived on the basis of the *fundamental postulate of statistical mechanics: any particular distribution of particles among the various available states of the system is just as likely as any other distribution.* Any particular distribution must, of course, be consistent with the characteristics of the particles, and with the conservation laws of energy and of particles.

(1) *Maxwell-Boltzmann distribution.* This classical distribution applies to a system of *identical particles* which are, nevertheless, *distinguishable* from one another. (For example, a collection of billiard balls of identical mass and diameter, but painted red, blue, etc.) The average number of particles, f_{MB} in a state i with energy ϵ_i is

$$f_{\text{MB}}(\epsilon_i) = Ae^{-\epsilon_i/kT} \qquad [11\text{-}15]$$

where A is a constant, k is the Boltzmann constant, 1.38×10^{-23} joule/°K, and T is the absolute temperature of the system of particles, always assumed to be in equilibrium. The average behavior of a gas of atoms or molecules is adequately described by the Maxwell-Boltzmann distribution law. This distribution function $f_{MB}(\epsilon_i)$, is plotted in Figure 11-11 for two different temperatures.

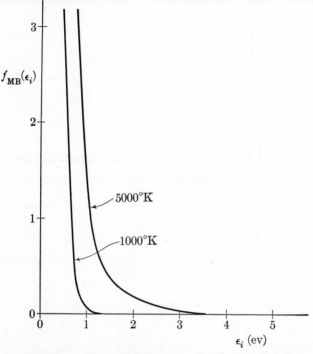

Figure 11-11. Maxwell-Boltzmann distribution function.

(2) *Bose-Einstein distribution.* This distribution law applies to a system of *identical particles* which are *indistinguishable*, each particle having an *integral spin.* Such particles are called *bosons.* The average number of particles occupying a particular state i of energy ϵ_i is given by

$$f_{BE}(\epsilon_i) = \frac{1}{e^{\alpha}e^{\epsilon_i/kT} - 1} \qquad [11\text{-}16]$$

The Bose-Einstein distribution is plotted in Figure 11-12 for $\alpha = 0$. It can be shown that for a system of photons or for a system of phonons (to be defined in Section 11-7), the constant α must always be zero (thus, $e^{\alpha} = 1$). This arises from the fact that the total number of photons (or phonons) in a system is not conserved. It can be seen from both Equation (11-16) and Figure 11-12 that the distribution function $f_{BE}(\epsilon_i)$ approaches the Maxwell-Boltzmann distribution $f_{MB}(\epsilon_i)$ when $\epsilon_i \gg kT$ (with $\alpha = 0$).

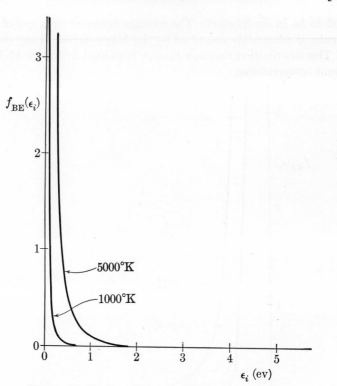

Figure 11-12. Bose-Einstein distribution function.

At low energies, $\epsilon_i \ll kT$, the term (-1) in the denominator of Equation 11-16 becomes important and has the effect of making $f_{BE}(\epsilon_i)$ much larger than $f_{MB}(\epsilon_i)$ for the same energy.

(3) *Fermi-Dirac distribution.* The Fermi-Dirac distribution applies to a system of *indistinguishable, identical particles*, each particle having an *half-integral spin*. Particles of half-integral spin (*fermions*), such as the electron, proton, or neutron, obey the Pauli exclusion principle. This excludes more than one particle from existing in the same state at the same time. The Pauli principle represents, so to speak, a very strong interaction between identical Fermi particles precluding any two such particles from occupying the same state. Although the Maxwell-Boltzmann and Bose-Einstein distribution laws impose no restriction on the number of particles that can occupy the same state, the Fermi-Dirac statistics allows, at most, only one particle in a particular state. The average number of particles in a particular quantum state i, whose energy is ϵ_i, is given by

$$f_{FD}(\epsilon_i) = \frac{1}{e^{(\epsilon_i - \epsilon_F)/kT} + 1}$$

[11-17]

The quantity, ϵ_F, is called the *Fermi energy* and is a constant for many problems of interest, being nearly independent of temperature.

The physical meaning of the Fermi energy can be seen from Equation 11-17. The average number of particles in a state for which $\epsilon_i = \epsilon_F$, the Fermi energy, is $\frac{1}{2}$; that is, the probability that a state of energy ϵ_F is occupied is just $\frac{1}{2}$. For those states with energies much less than ϵ_F, the exponential term in the denominator of Equation 11-17 is essentially zero, and $f_{FD} = 1$; thus, all such states have their full quota of particles, one per state, and are filled. For states with energies much greater than the Fermi energy, the exponential term becomes much greater than the $+1$ term, and f_{FD} reduces to the Maxwell-Boltzmann distribution, Equation 11-15. At the absolute zero of temperature, the Fermi-Dirac distribution (see

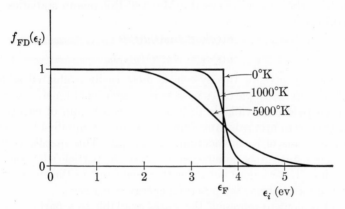

Figure 11-13. Fermi-Dirac distribution function.

Figure 11-13) f_{FD} is 1 for all states up to ϵ_F, and zero for all states with energies greater than ϵ_F.

Now, it must be recalled that, for all three types, the distribution function $f(\epsilon_i)$ gives only the *average* number of particles occupying a *state i* of energy ϵ_i, and *not* the number of particles, $n(\epsilon_i)$, *with the energy* ϵ_i. This is so because there may be two or more states with the same energy. Therefore, we introduce the quantity, $g(\epsilon_i)$, called the *statistical weight*, which gives the number of states with the same energy ϵ_i. (Thus, in Figure 11-10, $g(\epsilon_1) = 1$, $g(\epsilon_2) = 2$, and $g(\epsilon_4) = 1$.) It follows then that

$$\boxed{n(\epsilon_i) = f(\epsilon_i)g(\epsilon_i)}\qquad\text{[11-18]}$$

In many situations the energy levels are so closely spaced as to be regarded as continuous, and one then wishes to know the number of particles, $n(\epsilon)d\epsilon$, having energies between ϵ and $\epsilon + d\epsilon$. Equation 11-18 then becomes

$$\boxed{n(\epsilon)d\epsilon = f(\epsilon)g(\epsilon)d\epsilon}\qquad\text{[11-19]}$$

where $g(\epsilon)$, the *density of states*, gives the number of states per unit energy.

Equation 11-19 will form the foundation of all our later discussions in this chapter; for if one knows the applicable distribution function $f(\epsilon)$, and if the density of states $g(\epsilon)$ can be computed for a particular system, then one will know the (most probable) number of particles $n(\epsilon)d\epsilon$ within the range ϵ to $\epsilon + d\epsilon$. With this knowledge, such properties of the system as its average energy, specific heat, etc., can be computed.

11-4 Application of Maxwell-Boltzmann statistics to an ideal gas

Consider a classical, ideal gas composed of N identical atoms or molecules, assumed to be point particles obeying Newton's laws of motion. We wish to calculate the number of atoms, $n(\epsilon)d\epsilon$, within the energy range $d\epsilon$. Inasmuch as this system obeys the Maxwell-Boltzmann statistics, Equation 11-19 becomes

$$n(\epsilon)d\epsilon = f_{\mathrm{MB}}(\epsilon)g(\epsilon)d\epsilon$$
$$n(\epsilon)d\epsilon = Ae^{-\epsilon/kT}g(\epsilon)d\epsilon \qquad [11\text{-}20]$$

The quantity $g(\epsilon)$ for this system is most easily evaluated as follows. The only energy these particles possess is translational kinetic energy, and any one particle has available to it a whole continuum of energy states, ranging from zero upward. Any state of an atom is specified by giving the three components of its momentum, (p_x, p_y, p_z). This specification of the free particle's state gives no information as to the particle's location within the container, but we need not know this, inasmuch as the energy depends *only* on the momentum, the potential energy being zero.

It is convenient to represent the states available to a particle by points in *momentum space*, where the coordinates in this space are the three components (p_x, p_y, p_z) of the momentum. Each point in classical momentum space thus corresponds to a possible available state, all points in momentum space being allowed.

We now have a means of finding the number of states $g(\epsilon)d\epsilon$ in the energy range $d\epsilon$, where ϵ represents the total energy (which here also equals the kinetic energy) of any atom. We write

$$\epsilon = \tfrac{1}{2} mv^2 = p^2/2m \qquad \text{and} \qquad d\epsilon = p\, dp/m \qquad [11\text{-}21]$$

where the magnitude of the momentum p is

$$p = [p_x^2 + p_y^2 + p_z^2]^{1/2}$$

Furthermore, $\qquad\qquad g(\epsilon)d\epsilon = g(p)dp \qquad\qquad [11\text{-}22]$

where $g(p)dp$ gives the number of states with a magnitude of momentum between p and $p + dp$. This number is proportional to the volume of a spherical shell in momentum space; that is, to $4\pi p^2 dp$ as shown in Figure 11-14. Thus,

$$g(p)dp \propto p^2 dp \qquad [11\text{-}23]$$

Using Equations 11-21 and 11-22 in Equation 11-23,

$$g(\epsilon)d\epsilon \propto (\epsilon)(d\epsilon/\sqrt{\epsilon})$$

or, $$g(\epsilon)d\epsilon \propto \epsilon^{1/2}d\epsilon \qquad [11\text{-}24]$$

Finally, the most probable number of particles with energies between ϵ and $\epsilon + d\epsilon$ is, from Equation 11-20,

$$\boxed{n(\epsilon)d\epsilon = Ce^{-\epsilon/kT}\epsilon^{1/2}d\epsilon} \qquad [11\text{-}25]$$

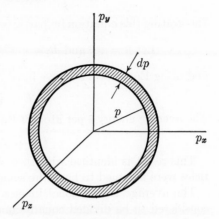

Figure 11-14. Momentum states between p and $(p + dp)$.

C being a proportionality constant. The distribution of the particles as a function of energy is shown in Figure 11-15. The constant C may be evaluated by applying the conservation of particles, their total number N being fixed.

$$N = \int_0^\infty n(\epsilon)d\epsilon = C \int_0^\infty \epsilon^{1/2}e^{-\epsilon/kT}d\epsilon \qquad [11\text{-}26]$$

Each of the $n(\epsilon)d\epsilon$ particles within $d\epsilon$ has an energy ϵ; therefore, the total energy E of the gas is

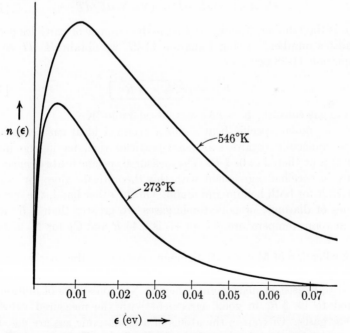

Figure 11-15. Energy distribution of the molecules of an ideal gas.

$$E = \int_0^\infty \epsilon n(\epsilon) d\epsilon = C \int_0^\infty \epsilon^{3/2} e^{-\epsilon/kT} d\epsilon$$

Integrating this equation by parts,

$$\left(\text{Let } u = \epsilon^{3/2} \text{ and } dv = e^{-\epsilon/kT} d\epsilon \text{ in } \int u \, dv = uv - \int v \, du. \right)$$

and using Equation 11-26, we have

$$E = \tfrac{3}{2} NkT$$

The average energy $\bar\epsilon$ per atom is then

$$\bar\epsilon = E/N = \tfrac{3}{2} kT \qquad [11\text{-}27]$$

This result is identical with that obtained in Section 1-5, where all particles were assumed to have the same speed or kinetic energy.

The average translational kinetic energy per atom, $\bar\epsilon = \tfrac{3}{2} kT$, may be considered to be divided equally among the three translational *degrees of freedom*. A degree of freedom is defined as one of the independent coordinates needed to specify the position of the particle. Therefore, the average energy per degree of freedom may be taken to be $\tfrac{1}{2} kT$.

The molar specific heat C_V of a gas at a constant volume is defined as the energy necessary to increase the temperature of one mole of gas 1°K, the volume of the gas being fixed; that is,

$$C_V \equiv (1/n) dE/dT = (N_0/N) dE/dT \qquad [11\text{-}28]$$

where n is the number of moles and N_0 is the number of particles per mole (Avogadro's number). Using Equation 11-27, we obtain $dE/dT = \tfrac{3}{2} kN$, and Equation 11-28 becomes

$$\boxed{C_V = \tfrac{3}{2} kN_0 = \tfrac{3}{2} R} \qquad [11\text{-}29]$$

where the gas constant $R = kN_0 = 1.99$ cal/mole-°K.

Thus, the molar specific heat C_V of a classical ideal gas comprised of atoms or molecules regarded as point-particles (i.e., having no internal structure) is predicted to be $\tfrac{3}{2} R$. The measured specific heats of *monatomic* gases are in excellent agreement with this theoretical value; for example, $C_V = 1.50 R$ for both helium and argon. On the other hand, the measured C_V values of diatomic or polyatomic gases are greater than $\tfrac{3}{2} R$; for example, at room temperature, C_V for H_2 is 2.45 R and C_V for N_2 is 2.50 R.

11-5   Application of Maxwell-Boltzmann statistics to the specific heat of a diatomic gas   In the last section the specific heat of a classical ideal gas, imagined to consist of mass-points obeying Newton's laws of motion, was computed to be $\tfrac{3}{2} R$, in good agreement with the measured values for *monatomic* gases. Of course, the atoms of a monatomic gas are *not* simple mass points; they have a complicated internal structure and follow the laws of the quantum theory. But their translational kinetic energy is *not* quantized; and for temperatures less than thousands of degrees, we have seen that the electron configuration is always that of the ground state. Thus, the atoms of a monatomic gas behave (at moderate temperatures) *as if* they were mass points.

Now let us consider the specific heat of a gas of diatomic molecules. Here there are three contributions to the total energy of such molecules: the (unquantized) *translational kinetic energy* of the center of mass; the (quantized) *rotational kinetic energy* of the molecule as a whole about the center of mass; and the (quantized) *vibrational energy* of the atoms of the molecule. The molecules' translational kinetic energy will contribute to the total energy of the gas at all finite temperatures. But there will be contributions to the total energy from rotation and vibration only if there are molecules existing in excited rotational or vibrational states.

If all of the molecules were to be found in the lowest rotational state ($J = 0$), the rotational kinetic energy would be zero; and if all of the molecules were to be found in the lowest vibrational state ($v = 0$), there would be no contribution from vibrational energy (except for the ever-present zero-point vibration). We can determine the temperatures at which rotation and vibration make significant contributions to the total energy of a diatomic gas by applying the Maxwell-Boltzmann statistics.

The number of molecules $n(E_r)$, each having a rotational kinetic energy E_r, can be found by using Equation 11-18 and the Maxwell-Boltzmann distribution function, Equation 11-15. The density of states $g(E_r)$ for pure rotation is $(2J + 1)$ where J is the rotational (angular momentum) quantum number. [Recall that the Zeeman degeneracy for the orbital angular momentum quantum number l is $(2l + 1)$.] Therefore,

$$n(E_r) = (2J + 1)Ae^{-E_r/kT} \qquad [11\text{-}30]$$

where
$$E_r = J(J + 1)\hbar^2/2I$$

from Equation 11-8. The number of molecules occupying some particular rotational state J depends only on the moment of inertia of the molecules, I, and the temperature of the gas, T.

Let us consider the population of the rotational states for the hydrogen molecule, H_2, which has a relatively small moment of inertia, $I = 4.64 \times 10^{-48}$ kg-m². Figure 11-16 shows the relative number of molecules in the first several rotational states for the temperatures 56°K, 170°K, and 340°K, as computed from Equation 11-30. It is clear that at 56°K, most of the hydrogen molecules are in the lowest ($J = 0$) rotational state; at this temperature most of the molecules do not rotate, and there is very little contribution to the total energy of the gas from molecular rotation. On the other hand, for a temperature of 340°K, a large fraction of the molecules are rotating. In short, there is a large increase in the rotational

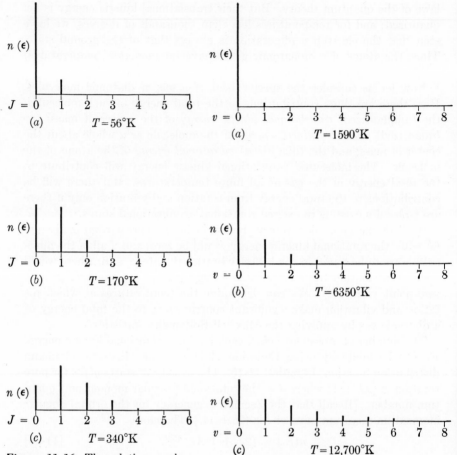

Figure 11-16. The relative number of H_2 molecules occupying rotational states at three different temperatures.

Figure 11-17. The relative number of H_2 molecules occupying vibrational states at three different temperatures.

energy of this gas in going from 56°K to 340°K, because the higher rotational states become more populated as the temperature of the gas is raised.

We can make a similar computation for the relative population of the vibrational states. For vibration $g(E_v) = 1$, and the number of molecules $n(E_v)$, each having a vibration energy E_v, is then given by

$$n(E_v) = Ae^{-E_v/kT} \qquad [11\text{-}31]$$

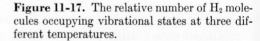

where $\qquad E_v = (v + \tfrac{1}{2})hf \qquad$ and $\qquad f = \dfrac{1}{2\pi}(k/\mu)^{1/2}$

from Equation 11-12. For the hydrogen molecule, H_2, $f = 1.32 \times 10^{14}$ sec^{-1}. Figure 11-17 shows the relative number of hydrogen molecules in the lowest several vibrational states for the temperatures 1590°K, 6350°K and 12,700°K, computed from Equation 11-31.

It is clear that for temperatures less than about 1600°K, essentially all molecules are in the ground state ($v = 0$ vibrational state); and that for temperatures of 10,000°K or more, a substantial fraction of the molecules are in excited vibrational states. Thus, for $T < 1600°$K appreciable vibration does not occur; for $T > 10,000°$K, the hydrogen gas has a significant contribution to its total energy from molecular vibration.

We have seen that there is an energy of $\frac{1}{2} kT$ per molecule associated with each of the three degrees of freedom for translation, the total translational kinetic energy being $\frac{3}{2} kT$ per molecule. There are *two* degrees of freedom associated with rotation, one for each of the two mutually perpendicular directions with respect to the interatomic axis about which rotation can take place. Thus, at temperatures for which a substantial fraction of the molecules are rotating, the rotational energy per molecule is $2 \times \frac{1}{2} kT = kT$. For the single vibrational degree of freedom there are two contributions to the vibrational energy, one for the kinetic energy and one for the potential energy, each $\frac{1}{2}kT$; therefore, again for sufficiently high temperatures, the vibrational energy per molecule is also kT.

We have seen that for $T < 50°$K only translation can occur in a hydrogen gas; therefore, the total energy per molecule is $\frac{3}{2} kT$ in this region, as was the case for a gas of monatomic molecules. Molecular rotation of H_2 becomes significant for temperatures of a few hundred degrees and the total energy per molecule is then $[(3/2) + (2/2)]kT = (5/2)kT$. Finally, for a few thousand degrees, molecular vibration, as well as rotation and translation, occurs, and the total energy per molecule is then

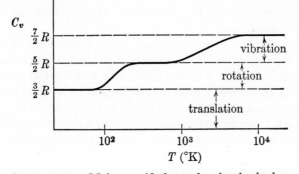

Figure 11-18. Molar specific heat of molecular hydrogen (H_2) as a function of temperature, showing the specific-heat contributions arising in consequence of translation, rotation, and vibration.

$[(3/2) + (2/2) + (2/2)]kT = (7/2)kT$. In a fashion exactly analogous to that used in Section 11-4, it follows that the molar specific heats (constant volume) for the three temperature regions are $(3/2)R$, $(5/2)R$, and $(7/2)R$, respectively. The observed variation in the specific heat of a hydrogen gas is shown in Figure 11-18. The observed values of the specific heat, which clearly demonstrates the quantum nature of molecular vibration and rotation, are in complete accord with theoretical expectation.

11-6 Blackbody radiation As an example of a physical system of particles obeying the Bose-Einstein distribution law, Equation 11-16, we choose that of *blackbody radiation*. Actually, it was the successful theoretical explanation of the electromagnetic radiation from a solid by Max Planck in 1900 that marked the beginning of the quantum theory. We have, however, postponed until now a discussion of this phenomenon because an interpretation of the radiation from solids involves not only the quantum theory, but also the statistical distribution of particles in a many-particle system.

All substances at a finite temperature radiate electromagnetic waves. The radiation spectra from atomic gases, where the atoms are far apart and interact only feebly with one another, consist of discrete frequencies or wavelengths. The spectra of molecules, with contributions from rotational and vibrational, as well as electronic transitions, again consist of discrete lines. The molecular lines in the visible region appear, on casual observation, as continuous bands. A solid represents a still more complex radiator or absorber, and it may be regarded in some ways as an enormous molecule with a correspondingly increased number of degrees of freedom. The radiation emitted by solids consists of a *continuous spectrum*, all frequencies or wavelengths being radiated. An adequate theory of blackbody radiation must explain how the radiation is distributed among the various frequency components and how this radiation varies with the temperature of the emitting surface.

Consider first what is meant by the term, blackbody. Any solid will absorb a certain fraction of the radiation incident on its surface, the remainder being reflected. An ideal blackbody is defined as a material that absorbs *all* of the incident radiation, reflecting none. From the point of view of the quantum theory, a blackbody is then a material which has so many quantized energy levels, spaced over so wide a range of energy differences, that *any* photon, whatever its energy or frequency, is absorbed when incident on the material. Inasmuch as the energy absorbed by a material would increase its temperature if no energy were emitted, a perfect absorber, or blackbody, is also a perfect emitter.

A very good approximation to a blackbody which can be achieved in the laboratory is a hollow container, completely closed, except for a small hole through which radiation can enter or leave the interior. Any radiation entering the container through the hole has a very small probability of being immediately reflected out again. Instead, the radiation is absorbed or reflected repeatedly at the inner walls, and effectively all radiation incident through the hole is absorbed in the container. By the same token, the radiation leaking out through the hole is representative of the radiation in the interior.

When the container is maintained at some fixed temperature T, the inner walls will emit and absorb photons at the same rate. Under these conditions, the electromagnetic radiation can be said to be in thermal equilibrium with the inner walls; in different language, the *photon gas* can be said to be in thermal equilibrium with the system of particles (in the walls) creating and absorbing the photons.

The observed frequency distribution of the radiation from a blackbody, analyzed by measuring the radiation escaping through the hole, is shown in Figure 11-19 for two fixed temperatures. Some general features of blackbody radiation can be seen from the figure.

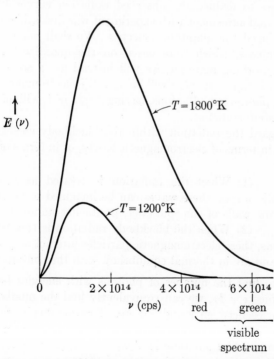

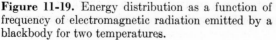

Figure 11-19. Energy distribution as a function of frequency of electromagnetic radiation emitted by a blackbody for two temperatures.

(1) For a fixed temperature, the energy $E(\nu)\, d\nu$, emitted in the small frequency range $d\nu$ between the frequencies ν and $\nu + d\nu$, first increases with frequency, reaches a maximum, and then decreases for still higher frequencies.

(2) $E(\nu)\, d\nu$ increases with the temperature T for any frequency; consequently the total energy

$$E_T = \int_0^\infty E(\nu)\, d\nu,$$

increases with T. Before Planck's development of the theory of blackbody radiation, E_T was known to vary as T^4, the so-called *Stefan-Boltzmann law*.

(3) A larger fraction of the emitted radiation is carried by the higher-frequency components as the temperature of the radiating body is increased. The wavelength corresponding to the peak in the radiation spectrum was found to be inversely proportional to the absolute temperature, this relation being known as the *Wien displacement law*.

(4) It is found that the radiation spectrum of the blackbody is independent of the material of which the radiator is constructed.

All attempts to deduce the observed radiation curves from classical theory failed, and agreement with experiment was first achieved only when Planck introduced the quantum concepts. We shall not follow Planck's original arguments, which were based on the quantized nature of the emitting or absorbing material; we will instead use a somewhat simpler approach, in which our concern will be with the electromagnetic radiation, regarded as a photon gas. Photons, having a spin of 1, will, of course, obey the Bose-Einstein statistics.

We can regard the radiation within the blackbody enclosure in either of two ways: in terms of electromagnetic waves, or in terms of particle-like photons.

(1) When the radiation is treated as a collection of electromagnetic waves, these waves can be imagined to be repeatedly reflected from the walls of the container, producing standing waves.

(2) When the blackbody radiation is treated as a collection of photons, these electromagnetic particles interact only with the container walls and are in thermal equilibrium with the container.

We wish to find the number of photons with energies between ϵ and $\epsilon + d\epsilon$; or, since $\epsilon = h\nu$, we can equivalently find the number of photons with frequencies between ν and $\nu + d\nu$. This number is the product of the number of available energy states $g(\epsilon)$ between ϵ and $\epsilon + d\epsilon$ times the Bose-Einstein distribution function, $f_{BE}(\epsilon) = 1/(e^{\epsilon/kT} - 1)$, the average number of photons in a particular state of energy ϵ.

The strategy for finding the number of available photon states—which is to say, the number of possible electromagnetic waves—is this: we imagine plane electromagnetic waves to be confined to a cube (for simplicity) of edge L; we then count up the number of possible stationary, or standing, wave patterns that can exist within the cube. This procedure is *not* unduly restrictive, for the cube may be arbitrarily large, so that even the longest waves can be accommodated.

The state of a photon is completely specified by giving the three components of its linear momentum, p_x, p_y, and p_z, and one of its two possible polarization directions. Thus, there are two states for each particular set of p_x, p_y, and p_z values. The procedure for fitting stationary electromagnetic waves within a three-dimensional enclosure is analogous to that used in Section 4-7 for finding the permitted quantum states of a particle confined to a uni-dimensional potential box. Only certain values of p_x, p_y, and p_z will lead to stationary states.

Figure 11-20 shows one particular electromagnetic wave, traveling obliquely with respect to the sides of the box, its direction of propagation being given by the momentum vector p. The wave fronts, to which p is perpendicular, are shown one-half wavelengths apart, where $\lambda = h/p$. For stationary waves to exist within the cubical box, the projection of any side along the direction of propagation must be an integral number of half wavelengths. Thus, for the side parallel to the p_x direction we must have

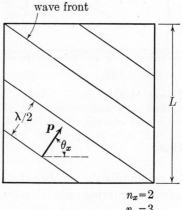

Figure 11-20. An allowed stationary plane electromagnetic wave in a cubical box.

$$L \cos \theta_x = n_x(\lambda/2) \qquad [11\text{-}32]$$

where n_x is an integer and θ_x is the angle between p and the p_x-axis.

For photons, $p = h/\lambda$, and therefore,

$$p_x = h \cos \theta_x/\lambda \qquad [11\text{-}33]$$

Combining Equations 11-32 and 11-33 we find for the permitted values of the p_x-components of p,

$$p_x = (h/2L)n_x$$

Similarly, $$[11\text{-}34]$$

$$p_y = (h/2L)n_y \qquad \text{and} \qquad p_z = (h/2L)n_z$$

Figure 11-21 shows the allowed values of (p_x, p_y, p_z) in momentum space. Each point actually represents two possible states because of the two possible polarization directions. The

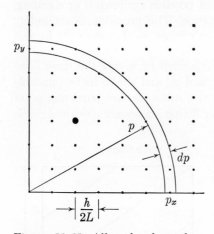

heavy dot in Figure 11-21 corresponds to the state illustrated in Figure 11-20. For a macroscopic length L, the separation between the adjacent points, $(h/2L)$, is very small as compared with the momentum of all photons except those of the longest wavelengths. For example, if the cube is as small as 5 cm, electromagnetic waves with wavelengths less than 10 cm (in the microwave radio region) can exist within the cube, and essentially all wavelengths in the visible region ($\sim 10^{-7}$ m) can be accommodated within the box.

Figure 11-21. Allowed values of p_x, p_y, and p_z in momentum space. The heavy dot corresponds to the state $(n_x = 2, \ n_y = 3)$ illustrated in Fig. 11-20.

We are interested in the number of states within the small energy range ϵ to $\epsilon + d\epsilon$, where $\epsilon = pc$ and $p = (p_x^2 + p_y^2 + p_z^2)^{1/2}$. This can be found by computing the number of states within a shell of radius p and of thickness dp, counting only the positive values of p_x, p_y, and p_z. Thus,

$$g(p)dp = \frac{2(\frac{1}{8})(4\pi \ p^2 \ dp)}{(h/2L)^3} \qquad [11\text{-}35]$$

where the factor 2 accounts for the two polarization directions, the factor $(\frac{1}{8})$ is introduced so that only one octant of the spherical shell is included (only *positive* values of n_x, n_y, and n_z), and the factor $(h/2L)^3$ is the volume associated with each point in momentum space.

Using $p = \epsilon/c$ and $L^3 = V$, the total volume of the box, Equation 11-35 becomes

$$g(\epsilon)d\epsilon = \frac{8\pi V \epsilon^2 d\epsilon}{h^3 c^3} \qquad [11\text{-}36]$$

The quantity $g(\epsilon)d\epsilon$ represents the number of states available for occupancy by photons in the energy range ϵ to $\epsilon + d\epsilon$. Multiplying $g(\epsilon)d\epsilon$ by the average number of photons per state [that is, by $1/(e^{\epsilon/kT} - 1)$] will give the number of photons in the infinitesimal energy range $d\epsilon$. Since each photon has an energy $\epsilon = h\nu$, we obtain for $E(\nu) \ d\nu$, the radiation energy per unit volume within the frequency range $d\nu = d\epsilon/h$,

$$E(\nu)\,d\nu = \frac{h\nu\,g(\epsilon)\,d\epsilon}{V(e^{\epsilon/kT}-1)} = \frac{(h\nu)8\pi V(h\nu)^2(h\,d\nu)}{Vh^3c^3(e^{\epsilon/kT}-1)}$$

$$\boxed{E(\nu)\,d\nu = \frac{8\pi\,h\nu^3}{c^3}\cdot\frac{1}{e^{h\nu/kT}-1}\,d\nu}\qquad\text{[11-37]}$$

This is the *Planck radiation equation*, giving the blackbody radiation spectrum. It is in complete agreement with the observed experimental curves, such as Figure 11-19.

It is of interest to note that the Planck equation reduces for low frequencies, $h\nu/kT \ll 1$, to the classical *Rayleigh-Jeans radiation formula*:

For low ν: $\qquad E(\nu)\,d\nu = (8\pi\,\nu^2\,kT/c^3)d\nu$ $\qquad$ [11-38]

The classical Rayleigh-Jeans relation fails, of course, in the high-frequency region because Equation 11-38 predicts an infinite value for $E(\nu)$ as ν approaches infinity. This failure is known as the *ultraviolet catastrophe*.

For high frequencies, $h\nu/kT \gg 1$, the Planck relation, Equation 11-37, reduces to the *Wien formula*.

For high ν: $\qquad E(\nu)\,d\nu = \left(\frac{8\pi\,h\nu^3}{c^3}\,e^{-h\nu/kT}\right)d\nu$ $\qquad$ [11-39]

This relation, Equation 11-39, fails, of course, for low frequencies.

Finally, by setting $dE(\nu)/d\nu = 0$ in Equation 11-37 one finds that the peak in the radiation distribution follows the *Wien displacement law*:

$$kT/h\nu_{\max} = \text{constant}$$

Or, $\qquad\qquad \lambda_{\max}T = 2.898 \times 10^7\ \text{Å} - °\text{K}$

where $\nu_{\max}$ and $\lambda_{\max}$ are the frequency and wavelength corresponding to the maximum $E(\nu)$ in the radiation spectrum. The Wien displacement relation shows that when the temperature of a blackbody is changed, the peak in the radiation spectrum is displaced inversely as the absolute temperature, a blackbody becoming progressively red, white, and blue, as its temperature is raised.

The total energy E_T radiated from a blackbody is obtained by integrating $E(\nu)\,d\nu$ in Equation 11-37 over the entire range of emitted frequencies

$$E_T = \int_0^\infty E(\nu)\,d\nu = CT^4$$

where C is a constant. The power P radiated from a unit area of a black body is given by $P = \sigma T^4$, where σ is 5.67×10^{-8} watt/m$^2 - °$K^4.

11-7 The theory of the specific heats of solids Another application of the quantum statistics is to the specific heat of solids. In this section we shall present the partially successful classical theory and then give the quantum theory, utilizing again the Bose-Einstein statistics.

Consider a crystalline solid composed of N atoms, each atom being bound in the crystal lattice by forces arising from its neighboring atoms. When any one atom is displaced from its equilibrium position, it is subject, in a first approximation, to a restoring force proportional to its displacement. Thus, any atom displaced from its equilibrium position will undergo simple harmonic motion. But, when one atom is displaced from its equilibrium position, so too are the neighbors with which it is coupled by interatomic binding forces. Consequently, if one atom undergoes simple harmonic motion, it causes neighboring atoms also to oscillate, the disturbance or deformation being propagated through the crystal as an elastic wave.

At temperatures below the melting point, the total energy content of the solid consists of the following contributions from each atom: the kinetic energy of the essentially free, outer valence electrons; and the energy of vibration of the remainder of the atom—namely, the nucleus plus the tightly-bound, inner electrons. At all moderate temperatures there is no change in the quantum state of any of the bound electrons. For this reason, the nucleus-plus-bound-electrons may be treated as a single, inert, vibrating particle. If the energy content of the solid changes, so too does the temperature, the change in the energy content of the crystal per unit change in temperature being the specific heat of the solid. The total specific heat of the solid consists of the *electronic specific heat* and the *lattice* (vibrational) *specific heat*. For all except the very lowest temperatures, the electronic specific heat is negligible (Section 11-8), and in this section we shall discuss only the contributions arising from the lattice vibrations.

We first compute the specific heat of a solid using the classical theory, and attribute the lattice-energy content to simple harmonic oscillators. It can be shown that, for each degree of freedom of a simple harmonic oscillator, there is $\frac{1}{2}kT$ of energy associated with potential energy, and $\frac{1}{2}kT$ associated with kinetic energy (see also Section 11-5). Therefore, the total vibrational energy E is the number of degrees of freedom, $3N$, times the energy per degree of freedom, kT.

$$E = (3N)(kT) = 3NkT$$

and the classical lattice specific heat per mole, C_v, is

$$C_v = \frac{1}{n}\frac{dE}{dT} = 3\left(\frac{N}{n}\right)k = 3N_0 k = 3R \qquad [11\text{-}40]$$

$$C_v = 5.96 \text{ cal/}°\text{K-mole}$$

where n is the number of moles, N_0 is Avogadro's number, and R is the constant of the general-gas law. This classical relation is known as the *Dulong-Petit law*.

Equation 11-40, predicting that the molar specific heat of any solid is a constant, independent of the material and of the temperature, is in agreement with experiment at *high* temperatures. This simple classical theory is, however, incapable of explaining the observed decrease in the specific heat for low temperatures, as shown in Figure 11-22.

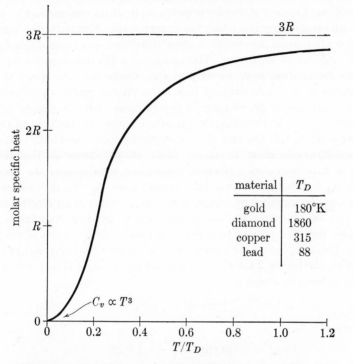

material	T_D
gold	180°K
diamond	1860
copper	315
lead	88

Figure 11-22. Observed molar specific heat of solids as a function of temperature. The temperatures are given in units of T_D, the Debye temperature (Eq. 11-45).

The first successful theoretical treatment of the lattice specific heat for *all* temperatures was given by Einstein in 1906. This early quantum treatment was improved upon by P. Debye in 1912, and is usually referred to as the *Debye theory of specific heats.*

The essential quantum feature of lattice vibrations is the quantization of the atoms' vibrational energies. In this view, any one atom gains or loses energy in discrete amounts and transfers energy to neighboring atoms in discrete amounts, the amount of the mechanical energy transferred being hf, where f is the classical frequency of vibration of the atom about its equilibrium position. Because the energy propagated through the crystal lattice in the form of elastic deformations is quantized, one can speak of the propagation of particle-like quanta of vibrational energy, called *phonons*. Phonons are created and absorbed by quantized lattice vibrators upon changing their quantum states, just as photons are created or absorbed by particles of a blackbody. The phonons represent the thermal energy content of a crystalline lattice, just as the photons represent the electromagnetic radiation content of a blackbody.

The Bose-Einstein statistics imposes no limit on the number of photons that can occupy any one available energy state. Similarly, the number of possible phonons is unrestricted, their distribution being governed also by the Bose-Einstein statistics. The energy of a phonon is given by $\epsilon = hf$, and its momentum is given by $p = \epsilon/v_s$, where v_s is the speed of sound.

Because of the close analogy between a photon gas in equilibrium with a blackbody and a *phonon gas* in equilibrium with the simple harmonic oscillators of an elastic solid, the quantum lattice specific heat of the solid is closely related to the radiation distribution of a blackbody. Appropriate modifications must be made, however, by virtue of the difference between a photon and a phonon. Again we wish to find the number of states $g(\epsilon)d\epsilon$ available within the energy range $d\epsilon$. To do this, we must again find the number of waves, now elastic rather than electromagnetic, that can be fitted as stationary waves between the boundaries of the medium, now the crystal boundaries rather than the walls of the blackbody. Thus, replacing the speed c by the speed v_s in Equation 11-36, and removing the factor 2 for the two photon polarization directions, we have for the density of states

$$g(\epsilon)d\epsilon = \frac{4\pi V \epsilon^2 d\epsilon}{h^3 v_s^3}$$

or, since $\epsilon = hf$,

$$g(f)df = \left(\frac{4\pi V}{v_s^3}\right) f^2 \, df$$

For elastic waves, there are two distinct types of waves propagated through the medium, a transverse wave traveling at a speed v_t having two possible, mutually perpendicular polarization directions, and a longitudinal wave traveling at a speed v_l. The number of elastic modes of vibration, or the total number of available phonon states, in the frequency range df is

$$g(f)df = 4\pi V \left[\frac{2}{v_t^3} + \frac{1}{v_l^3}\right] f^2 \, df \qquad [11\text{-}41]$$

The total number of vibrational modes is limited, however, to the total number of degrees of freedom of the crystal, $3N$. One of Debye's contributions consisted in recognizing this restriction. The elastic vibrations are cut off at the frequency f_D, called the *Debye frequency*, as follows:

$$\text{Total number} \atop \text{of modes} = \int_0^{f_D} g(f)df = 4\pi V \left[\frac{2}{v_t^3} + \frac{1}{v_l^3}\right] \int_0^{f_D} f^2 \, df = 3N$$

Therefore,
$$f_D^3 = \frac{9N}{4\pi V}\left[\frac{2}{v_t^3} + \frac{1}{v_l^3}\right]^{-1}$$
[11-42]

Equation 11-41 can now be written in terms of the Debye frequency f_D

$$g(f)df = \frac{9N}{f_D^3} f^2\, df$$
[11-43]

Because the elastic vibrations are quantized, but the number of phonons in any particular state is not restricted by the exclusion principle, the phonon distribution is given by the Bose-Einstein statistics, where again $\alpha = 0$ in Equation 11-16. Therefore, the number of phonons in the frequency range between f and $f + df$ is

$$n(f)\, df = g(f)\, \frac{1}{e^{hf/kT} - 1}\, df$$

Inasmuch as the energy per phonon is hf, the total vibrational energy content of the crystal is

$$E = \int_0^{f_D} hf\, \frac{g(f)\, df}{e^{hf/kT} - 1} = 9N\left(\frac{kT}{hf_D}\right)^3 kT \int_0^{x_m} \frac{x^3\, dx}{e^x - 1}$$
[11-44]

where we have used Equation 11-43 for $g(f)$, and defined $x \equiv hf/kT$ and $x_m \equiv hf_D/kT$.

It is convenient to define a characteristic temperature, called the *Debye temperature*, T_D, as that temperature for which $hf_D = kT_D$. Therefore,

$$T_D = hf_D/k \quad \text{and} \quad x_m = T_D/T$$
[11-45]

Equation 11-44 can then be written

$$E = 9N\left(\frac{T}{T_D}\right)^3 kT \int_0^{x_m} \frac{x^3\, dx}{e^x - 1}$$
[11-46]

The integral in Equation 11-46 must be evaluated numerically.

The molar specific heat, C_v, immediately follows from the definition, $C_v = (1/n)(dE/dT)$. Although one cannot easily evaluate C_v for all temperatures (because x and x_m in Equation 11-46 are functions of T), it is possible to find expressions for E and C_v for high temperatures and low temperatures.

In the high-temperature limit, $kT \gg hf_D$, and $x \ll 1$, and therefore $e^x \simeq 1 + x$. Equation 11-46 becomes, at high temperatures,

$$E \simeq 9N\left(\frac{T}{T_D}\right)^3 kT \int_0^{x_m} x^2\, dx = 9N\left(\frac{T}{T_D}\right)^3 (kT)\left(\frac{T_D^3}{3T^3}\right)$$

$$E = 3NkT$$

and
$$C_v = \frac{1}{n}\frac{dE}{dT} = 3N_0 k = 3R$$

In the high-temperature region, the quantum theory of lattice specific heat gives exactly the same result, $C_v = 3R$, as the classical theory, Equation 11-40.

Consider now the low-temperature region, where $kT \ll hf_D$ and $x_m \to \infty$. The integral in Equation 11-46 can be evaluated in closed form to yield

$$\int_0^\infty \frac{x^3 \, dx}{e^x - 1} = \frac{\pi^4}{15}$$

and Equation 11-46 becomes, at low temperatures,

$$E \simeq \tfrac{3}{5} \pi^4 N k T \left(\frac{T}{T_D} \right)^3$$

and

$$C_v = \frac{1}{n} \frac{dE}{dT} = \left(\frac{12\pi^4 R}{5T_D^3} \right) T^3$$

The lattice specific heat is thus seen to vary as T^3 in the low-temperature region, again in accord with observation. See Figure 11-22.

The observed temperature dependence of the specific heats for solids, whether *insulators or conductors*, are found to be in good agreement with the Debye theory, as shown in Figure 11-22. This is, at first sight, rather surprising, inasmuch as the Debye theory takes into account the internal energy arising from the lattice vibrations, but *not* the contribution to the specific heat of the conduction electrons. An electric or thermal insulator is a material in which there are essentially no free electrons. It is to be expected then that the specific heat of an insulator would have contributions from the lattice vibrations alone, in agreement with experiment and the Debye theory, but that a conductor would have, in addition, a contribution to the specific heat from the free electrons.

A good conductor is imagined to have a large number of unbound, free electrons, which can wander throughout the material. (These conduction electrons will show a net flow in one direction when a temperature gradient or external electric field is applied, thus accounting qualitatively for the high thermal and electrical conductivities of metals.) Let us compute the electronic specific heat of a solid under the simple assumption that each of the N atoms of the solid has one free, valence electron, and that these N free electrons can be regarded as classical particles of a Maxwell-Boltzmann gas. Three degrees of freedom are associated with each particle, and if these free electrons are regarded as classical particles wandering throughout the solid conductor, much as molecules of a gas, then the total electronic energy is $E_e = N(\tfrac{3}{2}kT)$. The electronic contribution to the molar specific heat would be $C_{ve} = (1/n)(dE_e/dT) = \tfrac{3}{2}(Nk/n) = \tfrac{3}{2}R$. At the high-temperature, classical limit of the Debye theory, the lattice specific heat is $3R$. Thus, if the valence electrons of a metallic conductor were to behave as classical free particles, the *total* molar specific heat of a conductor at relatively high temperatures would be $3R + \frac{3}{2}R$, or $\frac{9}{2}R$. This is an increase of 50 per cent over the observed value of C_v for conductors. (Note that in Figure 11-22, the observed values of C_v for insulators and conductors are the same, $\sim 3R$ for $T > T_D$.)

This simple treatment of electronic specific heat is clearly untenable. It fails to take into account the fact that a gas of free electrons is *not* a collection of classical particles, but rather a system of particles obeying the Pauli exclusion principle and the Fermi-Dirac statistics. The almost negligible electronic specific heat of metallic conductors, which is inexplicable with the classical free-electron theory, is understood on the basis of the quantum free-electron theory of a metal, to which we now turn.

11-8 The free-electron theory of metals The free-electron model of a metal, first developed by W. Pauli and A. Sommerfeld in 1927, is an example of a system of particles subject to the Pauli exclusion principle and obeying the Fermi-Dirac statistics.

In the free-electron model, a metallic crystal is imagined to consist of two components: the nuclei together with their tightly-bound electrons; and the weakly-bound valence electrons which may be considered to belong to the entire crystalline solid rather than to any one particular atom. The valence electrons are assumed to be free in the sense that any one such electron experiences no net force from the remaining valence electrons or from the nuclei and bound electrons of the lattice. Therefore, it is assumed that, in the interior of the solid, each valence electron has a constant electrostatic potential energy, $-E_i$, which is independent of its location within the crystal. The electrostatic potential rises markedly at the boundaries of the crystal to zero potential; this occurs by virtue of the net electrostatic attraction acting on a valence electron at the boundary. A plot of the potential energy of a free electron is shown in Figure 11-23.

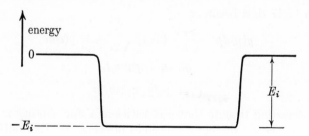

Figure 11-23. Average potential energy of a free electron in a conducting solid.

In the free-electron model of a metal one has, therefore, a collection of a large number of free particles confined to a box; that is, to the interior of the metal. The *electron gas* cannot, however, be regarded as equivalent to an ordinary gas, whose distribution is given by the Maxwell-Boltzmann statistics; rather, electrons exist in states restricted by the Pauli exclusion principle, and thus their distribution among available states is governed by the Fermi-Dirac statistics.

We wish to find the number of free electrons, $n(\epsilon)d\epsilon$, with energies in the range ϵ to $\epsilon + d\epsilon$, where $n(\epsilon) = g(\epsilon)f_{FD}(\epsilon)$. With a knowledge of the energy distribution of the free electrons, we will be able to interpret some aspects of the macroscopic behavior of a metal.

The density of states, $g(p)dp$, in the momentum range p to $p + dp$ is computed by considering the total number of ways in which N free electrons, regarded as waves, can be fitted within a three-dimensional box of side L. This problem is exactly analogous to that of finding the total number of ways in which N photons, regarded as electromagnetic waves, can be arranged within boundaries to form stationary wave patterns. Therefore, Equation 11-35, derived in Section 11-6 (blackbody radiation), can be used directly:

$$g(p)dp = \frac{8\pi V p^2 dp}{h^3} \qquad [11\text{-}47]$$

The factor 2, which was earlier introduced to account for the two possible polarization directions, is retained; this factor now accounts for the two electrons, one with spin up, the other with spin down, which have the same momentum components. Let us, for simplicity, measure electron energies upward from the constant potential in the interior of the metallic crystal, now taking the potential energy to be zero in the interior of the metal. Then the total energy ϵ of a free electron is purely kinetic, and we can write

$$\epsilon = \tfrac{1}{2} m v^2 = p^2/2m$$
$$d\epsilon = (p/m)dp = (\sqrt{2m\epsilon}/m)\, dp$$

Equation 11-47 then becomes

$$g(p)dp = \frac{8\pi V}{h^3}\,(2m\epsilon)\,\sqrt{\frac{m}{2\epsilon}}\, d\epsilon = g(\epsilon)d\epsilon$$

$$g(\epsilon)d\epsilon = C\epsilon^{1/2}\, d\epsilon \qquad [11\text{-}48]$$

where $\qquad\qquad C = 8\sqrt{2}\pi V m^{3/2}/h^3 \qquad\qquad\qquad [11\text{-}49]$

It is interesting to note that $g(\epsilon)$ varies as ϵ^2 for photons and phonons (bosons), but as $\epsilon^{1/2}$ for both molecules obeying the Maxwell-Boltzmann statistics and for electrons (fermions).

The distribution function, Equation 11-17, for a collection of fermions is given by

$$f_{FD} = \frac{1}{e^{(\epsilon - \epsilon_F)/kT} + 1} \qquad [11\text{-}50]$$

Equation 11-19 then becomes, using Equations 11-48 and 11-50,

$$n(\epsilon)d\epsilon = f_{FD}(\epsilon)g(\epsilon)d\epsilon$$

$$\boxed{n(\epsilon)d\epsilon = \frac{C\epsilon^{1/2}\,d\epsilon}{e^{(\epsilon-\epsilon_F)/kT}+1}} \qquad\qquad [11\text{-}51]$$

Equation 11-51 gives the energy distribution of free electrons in equilibrium with a material at a temperature T. The meaning of the quantity ϵ_F, called the Fermi energy, is best seen by considering the distribution of electron energies for a metal at the absolute zero of temperature.

METALS AT ABSOLUTE-ZERO TEMPERATURE   A plot of the energy distribution for an electron gas, Equation 11-51, is shown in Figure 11-24 for $T = 0$. The plot is based on the product of two energy-dependent terms: the factor $\epsilon^{1/2}$, which accounts for the parabolic rise in $n(\epsilon)$ from

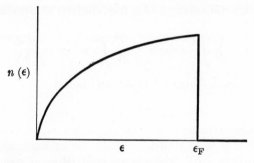

Figure 11-24. Energy distribution of free electrons in a metal at $T = 0°K$.

$\epsilon = 0$ upward; and the factor, $1/[e^{(\epsilon-\epsilon_F)/kT}+1]$, the Fermi-Dirac probability distribution function (Figure 11-13), which, for $T = 0$, has the value of 1 from $\epsilon = 0$ to $\epsilon = \epsilon_F$, and zero for $\epsilon \geq \epsilon_F$.

Figure 11-24 shows that the free electrons do *not* all have zero kinetic energy at the absolute zero of temperature, as would the particles in a classical gas; rather, there are electrons with finite energies up to a maximum energy, ϵ_F, the Fermi energy. The reason why the electrons have finite energies and are in motion, even at $T = 0$, is found in the Pauli exclusion principle, to which electrons are subject. No more than two electrons, one for each of the two possible electron spin orientations, are permitted in some one particular energy state; hence, all of the lowest states become filled, until one reaches the most energetic electrons with $\epsilon = \epsilon_F$. At absolute zero, the Fermi energy is the kinetic energy of the most energetic electrons, all states of lesser energy being filled, and all states of greater energy being empty.

The value of ϵ_F, the Fermi energy, can be computed quite directly. The total number N of free electrons is

$$N = \int_0^{\epsilon_F} n(\epsilon)d\epsilon = C \int_0^{\epsilon_F} \epsilon^{1/2}\, d\epsilon = \tfrac{2}{3}\, C\epsilon_F^{3/2} \qquad [11\text{-}52]$$

Using the value of C from Equation 11-49, we have

$$\epsilon_F = \frac{h^2}{2m}\left(\frac{3n}{8\pi}\right)^{2/3} \qquad [11\text{-}53]$$

where n is the number of free electrons per unit volume and m is the electron mass. The Fermi energy of copper, which has a valence of 1, and one free electron per atom, is computed from Equation 11-53 to be 7.0 ev; for the conductor, sodium, ϵ_F is 3.1 ev. The values of ϵ_F are typically of the order of a few ev. Thus, the most energetic electrons of a conductor have a kinetic energy of several ev, even at the lowest possible temperature.

The average kinetic energy $\bar{\epsilon}$ of a free electron at absolute zero can be found directly.

$$\bar{\epsilon} = \frac{1}{N}\int_0^{\epsilon_F} \epsilon n(\epsilon)d\epsilon = \frac{C}{N}\int_0^{\epsilon_F} \epsilon^{3/2}\, d\epsilon = \frac{2C\epsilon_F^{5/2}}{5N}$$

But $C\epsilon_F^{3/2} = \tfrac{3}{2}N$, from Equation 11-52, and hence

$$\bar{\epsilon} = \tfrac{3}{5}\epsilon_F \qquad [11\text{-}54]$$

The relatively high average kinetic energy, a few ev, of a free electron in a metal at $T = 0°\text{K}$ is to be contrasted with the average kinetic energy per classical free particle, $\tfrac{3}{2}kT$, which is a mere 0.04 ev at room temperature, and zero, of course, at $T = 0°\text{K}$. This extraordinary behavior, in which electrons of a material at $T = 0°\text{K}$ have a sizable kinetic energy is, like the zero-point-vibration of a simple harmonic oscillator, strictly a quantum phenomenon.

An energy-level diagram showing the occupied energies for the free

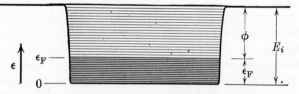

Figure 11-25. Occupation of the energy levels by the free electrons of a metal at $T = 0°\text{K}$.

electrons of a metal at absolute zero is shown in Figure 11-25. The electrons occupy energy states continuously up to the Fermi energy; all higher states are unfilled. The binding energy of the least-tightly-bound electrons of the metal (those at the Fermi surface) is, of course, the work function ϕ (Section 3-2); hence,

$$E_i = \epsilon_\mathrm{F} + \phi \qquad [11\text{-}55]$$

Inasmuch as all three quantities in Equation 11-55 can be determined, the simple features of the quantum free-electron model can be verified. The work function ϕ can be measured by experiments involving the photoelectric effect. The Fermi energy ϵ_F is evaluated using Equation 11-53. The difference in potential energy E_i between the interior and exterior of the crystal produces a change in the speed of an electron upon entering the crystal. This results in a refraction of electrons at the surface and this refraction is manifest in the *diffraction* of electrons by the crystal lattice. In this way E_i can be measured and Equation 11-55 verified. For example, the conductor lithium, with one valence electron, has $E_i = 6.9$ ev, $\epsilon_\mathrm{F} = 4.7$ ev, and $\phi = 2.2$ ev; these values are in complete agreement with Equation 11-55. We shall see that the values of ϵ_F and E_i are essentially independent of the temperature, and that Equation 11-53 holds for all moderate temperatures.

METALS AT A FINITE TEMPERATURE   We now inquire into the changes that do occur when the temperature of the conductor is raised. A plot of the energy distribution for a finite temperature is shown in Figure 11-26,

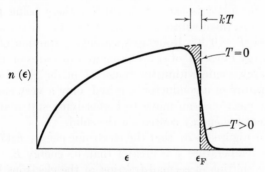

Figure 11-26. Energy distribution of free electrons in a metal at a finite temperature.

following Equation 11-51. The energy distribution at a moderate temperature is quite similar to that for $T = 0°\mathrm{K}$; the only significant difference is that the corners of the plot, at $\epsilon = \epsilon_\mathrm{F}$, are now slightly rounded. This rounding of the corners at the Fermi energy is a result which follows from the change in the Fermi-Dirac distribution function, $f_\mathrm{FD}(\epsilon)$, shown in Figure 11-13.

At room temperature, $kT = 0.03$ ev, an energy which is much less than the typical Fermi energy, ϵ_F, always a few electron volts. Therefore, the

Fermi-Dirac distribution function, $f_{FD}(\epsilon)$, is unchanged from $f_{FD}(\epsilon)$ at $T = 0$, except for an electron energy ϵ close to the Fermi energy, ϵ_F. For $kT \ll \epsilon_F$,

$$f_{FD}(\epsilon) = \frac{1}{e^{(\epsilon - \epsilon_F)/kT} + 1} \simeq 1 \quad \text{when} \quad \epsilon \ll \epsilon_F,$$

$$f_{FD}(\epsilon) \simeq 0 \quad \text{when} \quad \epsilon \gg \epsilon_F$$

There is a difference between the energy distributions, $n(\epsilon)$ at a finite temperature T and at $T = 0°$K, only within the small energy range for which $|\epsilon - \epsilon_F| \simeq kT$; that is, there is a significant difference only within the range kT of the Fermi energy. We see from Equation 11-50 that, when $\epsilon = \epsilon_F$, $f_{FD}(\epsilon) = \frac{1}{2}$; thus, the Fermi energy corresponds to that energy for which there is a 50-50 chance that the state will be occupied. It can be shown that ϵ_F can be assumed to be a constant, independent of the temperature, for temperatures less than a few thousand degrees.

The distribution in energy of the free electrons of a metal at a finite temperature (Figure 11-26) has an interesting interpretation. Those electrons whose energy is much less than the Fermi energy, remain in the low-lying energy states that they occupy at $T = 0°$K. Only the most energetic electrons, those within the range kT of the Fermi energy, have available to them unoccupied higher-energy states to which they can be excited by thermal excitation. The lower-lying electrons are locked to their energy states when the metal is excited thermally, there being no unoccupied states available to them within an energy range kT, either above or below their states for $T = 0°$K. Roughly speaking, a fraction of the electrons within the energy range kT of ϵ_F are promoted to states on the high side of the Fermi energy, again within the range kT of the Fermi energy. Thus, as the temperature of a conductor is raised, only a very small fraction of all of the free electrons can move to higher-lying states and thereby increase the electronic energy content of the solid.

We see from Figure 11-26, that the electronic energy $E_e(T)$ of the metal at some finite temperature T is greater than its energy $E_e(0)$ at $T = 0°$K because of the shifting of a small fraction of the electrons (shaded areas) from below to above the Fermi energy. The number of electrons promoted to higher energies is proportional to kT. Furthermore, the average increase in the energy of each promoted electron is approximately kT. Therefore,

$$E_e(T) = E_e(0) + A(kT)^2$$

where A is a constant. The molar *electronic specific heat*, C_{ve}, is then given by

$$\boxed{C_{ve} = (1/n)[dE_e(T)/dT] = \gamma T} \qquad [11\text{-}56]$$

where γ is a constant. The quantum free-electron theory thus predicts that the electronic contribution to the specific heat of a conductor is directly proportional to the absolute temperature. A more detailed analysis shows that $\gamma = (\pi^2/2)z(k/\epsilon_F)R$, where z is the number of valence electrons per atom. Therefore, Equation 11-56 becomes

$$C_{ve} = \left(\frac{\pi^2}{2}\right) z \left(\frac{kT}{\epsilon_F}\right) R \qquad [11\text{-}57]$$

In copper, a typical conductor, $z = 1$, $\epsilon_F = 7.0$ ev, and $kT = 0.03$ ev for room temperature. Using Equation 11-57 we find that $C_{ve} \simeq 0.02R$ for copper at room temperature. The *lattice* molar specific heat for copper is close to $3R$ at this temperature, so that the electronic contribution to the specific heat is negligible at moderate temperatures.

The electronic specific heat is comparable to the lattice specific heat only at the very lowest temperatures (few degrees K). That the electronic contribution is very small follows from the fact that only a very small fraction of the free, or valence, electrons are able to participate in an energy change when the temperature of a metal is changed.

If the temperature of a conductor is raised to several thousand degrees so that kT becomes comparable to the work function of the metal, some of the free electrons will have enough energy to escape from the metal surface, their kinetic energy ϵ equaling or exceeding the internal potential energy, E_i (see Figure 11-25). Thus, severe thermal excitation of free electrons can result in their emission from the metal. This process is known as *thermionic emission*.

Although the quantum free-electron theory of metals is capable of accounting for such properties as the electronic specific heat and thermionic emission, it is not able to account for other important properties of solids, properties which depend on the fact that the electrons, even the valence electrons, in a metal are *not* completely free. In the next section we will discuss a more realistic model of solids, where the electrons move through a non-constant potential within the metal.

11-9 The band theory of solids: conductors, insulators and semiconductors The *band theory* of solids is the basis for understanding such phenomena as electrical and thermal conductivities and the distinction between conductors, insulators, and semiconductors. The band theory is able to account for the tremendous range in electrical resistivity from a good insulator to a good conductor, for which the ratio of the resistivities may be as large as 10^{30}. Although a detailed, quantitative treatment of the band theory of solids involves the rigorous application of wave mechanics, it is possible to understand some of the important qualitative features of this highly successful theory without mathematical analysis. There are two approaches to the band theory:

(1) The theory of F. Bloch (1928) emphasizes the fact that a valence electron in a metal does *not* see a constant potential in its motion through the crystal, but rather experiences a periodic potential, whose periodicity is that of the crystalline structure.

(2) The theory of W. Heitler and F. London (1927) considers the effects on the electron orbits when isolated atoms are brought close together to form a crystalline solid.

Consider first a perfect crystal with nuclei located at fixed lattice positions (called *sites*) within the crystal. These nuclei form a geometrically ordered array, with small groups of nuclei being repeated throughout the crystal. Associated with these nuclei are the electrons, whose total number is such that the crystal as a whole is electrically neutral. An inner electron is tightly bound to an individual nucleus, and must, therefore, remain at all times with this nucleus. On the other hand, an outer, or valence, electron is weakly bound to any one nucleus and may wander from one nucleus to another. Such a valence electron, according to the Bloch theory, is considered to belong to the entire crystal rather than to any one nucleus. Furthermore, a valence electron sees a periodic potential, due to the fixed nuclei and the remaining electrons.

A representation of a simple periodic potential seen by an electron in a crystal is shown in Figure 11-27. Now the problem of determining the

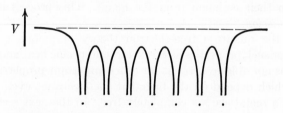

Figure 11-27. Periodic potential seen by an electron in a crystalline solid.

allowed states and energies of valence electrons is that of determining what electron wavelengths are possible within the crystal. Thus, one is confronted with the difficult problem of finding what electron wavelengths can be fitted to the periodic potential which characterizes the interaction between a valence electron and the remainder of the crystal, a very complicated problem indeed.

A second approach to the band theory of solids, that of Heitler and London, lends itself more easily to a qualitative description. Let us consider N identical isolated (non-interacting) atoms. Each atom has its own particular set of energy levels, and the permitted states of some one atom are identical with those of any other atom. As an example, the energy-level diagram of a lithium atom is shown in Figure 11-28a, together with the number of available states for each allowed energy. Inasmuch as the energy-level diagrams for all atoms are identical, the combined energy-

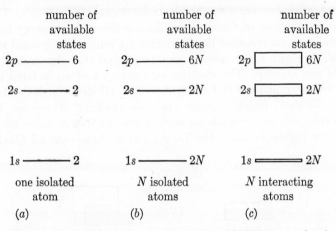

Figure 11-28. Schematic representation of the energy levels and states available to (a) one isolated lithium atom, (b) N isolated lithium atoms, and (c) N interacting lithium atoms.

level diagram for the N atoms, all far separated from one another, is simply that of the single atom with, however, the number of available states for each energy level now increased by a factor N. For a single atom, one can accommodate two electrons in an s energy level and six electrons in a p energy level (see Section 6-8); but with N atoms, there is room for $2N$ electrons in the s energy level, and $6N$ electrons in the p energy level, as shown in Figure 11-28b.

When the N atoms are brought together so that the separation between adjacent atoms is comparable to the separation of the atoms in a crystalline solid, the atoms interact fairly strongly with one another. The consequence of this interaction is to broaden the energy levels of the system such that those states which were earlier degenerate, having the same energy, now have a slightly different energy. (We saw an example of this splitting in Section 11-1, Figure 11-4, for the case of two hydrogen atoms being brought together to form H_2.) The effect of bringing together a very large number of the originally isolated atoms of Figure 11-28b to form a bound system is shown schematically in Figure 11-28c. The $2N$ available states for the $1s$ energy level are no longer coincident but are spread (essentially) continuously throughout the $1s$ *energy band*. In a similar fashion, there are $2N$ available states in the $2s$ energy band, and $6N$ available states in the $2p$ energy band. The regions between the available energy bands cannot be occupied by *any* electron, and these regions are known as *forbidden bands*. The width and separation of the energy bands depends, of course, on the particular crystalline material with which they are associated.

We have, up to this time, discussed only the *available* states in the energy bands of a solid, but not how the electrons occupy these levels. To illustrate the distribution of the electrons among the various energy bands of the crystal we first consider the conductor sodium in the ground state at $T = 0°K$. We will, for simplicity, assume at first that the several energy bands do not overlap. The electron configuration of an isolated sodium atom in the ground state is $1s^2 2s^2 2p^6 3s^1$; thus, all electron shells are filled up to the $3s$ shell, which contains only one electron. Therefore, the sodium crystal has energy bands for each of the electron shells of the atom, as shown in Figure 11-29a. The $1s$, $2s$, and $2p$ bands are all filled, with

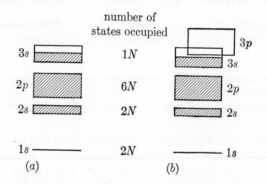

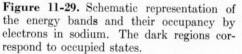

Figure 11-29. Schematic representation of the energy bands and their occupancy by electrons in sodium. The dark regions correspond to occupied states.

$2N$, $2N$, and $6N$ electrons respectively. The $3s$ band, which has $2N$ available states, is only half filled with N $3s$ electrons. We note that the $1s$ band, corresponding to inner electrons in small electron shells, is quite narrow compared with higher-lying bands; this band is narrow because the inner electrons are strongly attracted to the individual nuclei, and are less influenced by neighboring electrons and nuclei.

It is the fact that the uppermost energy band of a conductor, such as sodium, is only partially filled that is responsible for the high electrical conductivity of these materials. Consider what happens to the occupation of the energy bands when an electric field is applied to the metal. It is possible then for all of the electrons in the partially filled band to gain small amounts of energy by the action of an external electric field, and thus be promoted to the continuum of available states lying immediately above.

In a similar way, one can account for the high thermal conductivity of metallic crystals.

The distribution of the electrons among the available states at some finite temperature differs only slightly from the distribution at the absolute zero of temperature. The shift in occupation of electrons is controlled here, as in the simple free-electron theory, by the Fermi-Dirac statistics. Consequently, the significant change in the distribution occurs for those very few electrons which lie within a region of energy kT about the uppermost filled level at $T = 0°$K.

The energy bands shown for sodium in Figure 11-29a have not included the unoccupied $3p$ band which comes immediately after the $3s$ band. Not only is the $3p$ band quite broad, but, in addition, it overlaps the $3s$ band, as shown in Figure 11-29b. Thus, the number of unoccupied levels available to the electrons in the $3s$ shell is increased, leading to a very high electrical conductivity.

The very low electrical conductivity of an insulator, such as diamond ($_6$C), can also be understood on the basis of the band theory. The electron configuration of carbon in its ground state is $1s^2 2s^2 2p^2$. Because the $2p$ energy band is only partially filled, with $6N$ available states but only $2N$ electrons, it might at first appear that diamond would be an electrical conductor. There are, however, two $2p$ energy bands, separated from each other by a forbidden region of 6 ev, as shown in Figure 11-30. This separation of the $2p$ band arises from the nature of the crystalline structure of diamond. Therefore, the lower $2p$ band is filled completely with $2N$ elec-

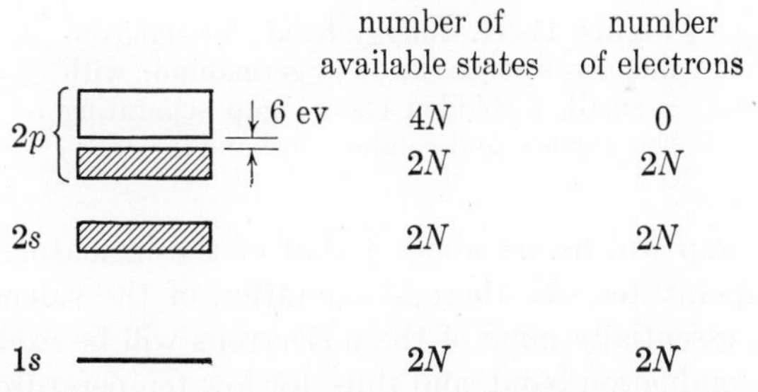

		number of available states	number of electrons
$2p$ $\Big\{$	$\downarrow$ 6 ev $\uparrow$	$4N$	0
		$2N$	$2N$
$2s$		$2N$	$2N$
$1s$	——	$2N$	$2N$

Figure 11-30. Schematic representation of the energy bands and their occupancy in diamond. Note the sizeable forbidden region between the two $2p$ bands.

trons in the $2N$ available states. At room temperature $kT \simeq 0.03$ ev, and the gap width for diamond is so much greater than the thermal excitation energy kT that there are virtually no electrons in the upper $2p$ band, even at this temperature. When an external electric field is applied it is impossible for electrons to gain enough energy to be promoted to the upper unoccupied $2p$ band, and thus an external field cannot cause a net electron flow, or electric current. In short, a substance, such as diamond, is a good insulator in that there is a sizable energy gap between a filled band, called the *valence band*, and the next empty (but available) energy band, called the *conduction band*.

Similarly, the electrons in the valence band of an insulator cannot have their energies raised by the absorption of photons whose energy is less than the gap width. For diamond, this means that all visible-light photons are transmitted through the crystal without absorption; which is to say, diamond is perfectly transparent to visible light. By the same token, conductors are opaque; this follows from the fact that there is a continuum of unfilled energy states immediately above the filled states to which electrons in a conductor can be promoted by the absorption of photons over a continuum of wavelengths.

Some crystalline solids, such as silicon and germanium, have a filled valence band and an empty conduction band, like diamond, but with a much *smaller* forbidden region separating the bands, as shown in Figure 11-31. The energy gap for silicon is 1.1 ev, and for germanium, 0.70 ev;

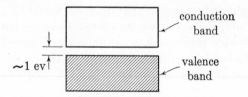

Figure 11-31. Energy bands for semiconductors, such as silicon or germanium, with a small, forbidden energy gap separating the valence and conduction bands.

both of these gap widths are about $\frac{1}{6}$ that of the insulator, diamond. At very low temperatures, the thermal excitation of the valence electrons is so small that essentially none of these electrons will be excited to occupy states in the conduction band, and thus, for low temperatures, these materials behave as insulators.

Consider now, however, the occupation of states in the conduction band at higher temperatures. If the gap width is small, there will be some electrons occupying available states in the conduction band; and under the influence of an external electric field, the few electrons in the conduction band can participate in a net electron flow through the material. At the same time, the unfilled states in the valence band, called *holes*, will also contribute to the electric current. The conductivity of materials such as these lies between the very low values of insulators and the very high values of conductors, and thus they are known as *semiconductors*. The type of semiconductor described above, which consists strictly of atoms of a single type and which depends for its semiconductivity on the electrons in the conduction band which have been thermally excited across the energy gap, are known as *intrinsic semiconductors*.

A second type of semiconductor, called an *extrinsic*, or *impurity*, *semiconductor*, depends for its semiconductivity upon the presence within a semiconducting crystal of a few atoms, called *impurity atoms*, of a type different from those of the crystal. Before we discuss the influence of the impurity atoms on the energy-band structure, we must first examine the bonding of atoms and impurity atoms within the crystal. Silicon and germanium both lie in the fourth column of the periodic table, their outermost shells containing $3s^2 3p^2$ and $4s^2 4p^2$ electron configurations, respectively. Each silicon or germanium atom in the crystalline solid is bound to its four nearest neighbors by covalent bonds, each saturated bond representing the sharing of two valence electrons, as shown in Figure 11-32. This struc-

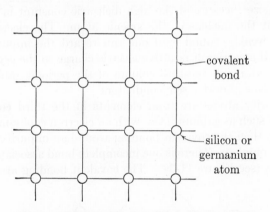

covalent bond

silicon or germanium atom

Figure 11-32. Covalent bonding of the atoms in a crystalline solid of silicon or germanium. Each line between nearest neighboring atoms represents a saturated covalent bond, with the sharing of two valence electrons.

ture is called the diamond structure since it is like that of the diamond crystal ($_6$C with $2s^2 2p^2$ electrons).

Now let us consider the effect of a few atoms of arsenic in a silicon crystal. The element, $_{33}$As, lies in the fifth column of the periodic table, its ground state electron configuration being $4s^2 4p^3$. Thus, a neutral atom of arsenic has one more electron than a neutral atom of silicon; and when an arsenic atom replaces a silicon atom in the crystal, there is one additional

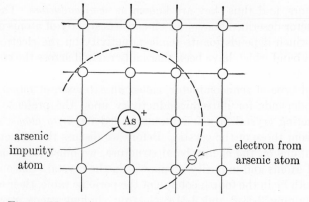

Figure 11-33. An example of a donor impurity atom in silicon, producing n-type semiconductivity.

electron which is not bonded covalently (see Figure 11-33). The unpaired electron from the impurity atom can be imagined to move in Bohr orbits of very great size (because of the high dielectric constant in the crystal's interior) about the nucleus of the arsenic atom. The electron is consequently very weakly bound, and one can regard the impurity atom as having donated a carrier of negative electric charge to the crystal. Therefore, impurity atoms in the fifth column of the periodic table are *donors* and lead to *n-type* impurity semiconductors.

If the impurity atoms are from elements in the third column of the periodic table, such as gallium, $_{31}Ga$, with an electron configuration of only *one p* electron, then the covalent bonding around an impurity atom in the crystal is not complete. There is one incomplete bond associated with each impurity atom (see Figure 11-34). The covalent bonding around such an

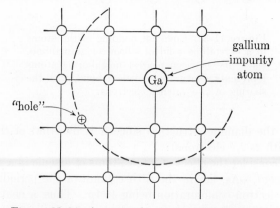

Figure 11-34. An example of an acceptor impurity atom in silicon, producing p-type semiconductivity.

impurity atom is completed by the impurity atom's accepting an electron from the valence band, thereby producing one vacancy, or hole, in the valence band. This hole moves in Bohr orbits about the now negatively charged impurity ion. When one speaks of the motion of a hole, imagined as an equivalent positive charge, one is, in effect, describing the motion in the opposite direction of electrons. An impurity atom, which accepts an electron to complete bonding, is known as an *acceptor*; and a semiconductor containing such impurities (and hence, holes), is known as a *p-type impurity semiconductor*.

The effect on the energy-band structure of impurities in silicon is shown in Figure 11-35. The *n*-type impurities introduces additional, closely

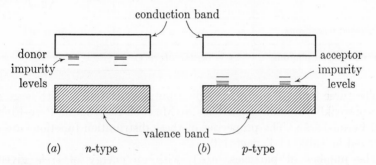

Figure 11-35. Energy-level diagram of a silicon crystal modified by (a) donor impurities and (b) acceptor impurities.

spaced energy levels lying just below the top of the conduction band, these levels being occupied by the weakly-bound electrons contributed by the donor atom. The electrons in these discrete energy levels can move to the available states in the conduction band, which lie immediately above. The *p*-type impurities introduce closely-spaced levels just above the valence band. Therefore, electrons from the valence band, just below, can move upward to occupy these available impurity states, thus accounting for the slight conductivity of the semiconductor. The conductivity of an impurity semiconductor can be controlled by the relative concentration of the impurity atoms. Impurity semiconductors have a number of significant technological applications when used, singly or in combination, to form rectifiers, amplifiers, detectors, transistors, and other solid-state devices.

11-10 Summary The two principal types of molecular binding are ionic (heteropolar) binding, resulting from the electrostatic attraction of ions; and covalent (homopolar) binding, resulting from the sharing of valence electrons.

The rotational and vibrational energies of a diatomic molecule are quantized. The photons absorbed or emitted in transitions between rotational states or vibrational states are in the far-infrared and infrared regions of the electromagnetic spectrum, respectively. The application of quantum mechanics to the rotational and vibrational motions of a diatomic molecule leads to the following results.

	ROTATION	VIBRATION
Allowed energies of molecule	$E_r = \dfrac{J(J+1)\hbar^2}{2I}$ where $J = 0, 1, 2, \ldots$ and $I = \mu r_0^2$	$E_v = (v + \tfrac{1}{2})hf$ where $v = 0, 1, 2, \ldots$ and $f = \left(\dfrac{1}{2\pi}\right)\sqrt{k/\mu}$
Allowed transitions	$\Delta J = \pm 1$	$\Delta v = \pm 1$
Frequency of photons	$\nu = (\hbar/2\pi I)(J_l + 1)$	$\nu = f$

The three kinds of probability distributions for dealing with large numbers of weakly-interacting particles are Maxwell-Boltzmann, Bose-Einstein, and Fermi-Dirac. The properties of these distribution functions are summarized in Table 11-1, page 440.

The number of particles, $n(\epsilon_i)$, with an energy of ϵ_i is given by $n(\epsilon_i) = f(\epsilon_i)g(\epsilon_i)$ where $f(\epsilon_i)$, the distribution function, is the average number of particles in the state i, and $g(\epsilon_i)$ is the number of states with the energy ϵ_i. For very closely-spaced energy levels one can write $n(\epsilon)d\epsilon = f(\epsilon)g(\epsilon)d\epsilon$ where $g(\epsilon)$, the density of states, gives the number of states per unit of energy, and $n(\epsilon)$ is the number of particles per unit energy.

Some applications of Maxwell-Boltzmann, Bose-Einstein, and Fermi-Dirac statistics are summarized in the table on the opposite page.

Characteristics of various solids:

> Conductors: The uppermost band containing electrons is only partially occupied.

> Insulators: The uppermost band holding electrons is completely filled; the next available, higher-lying (conduction) band is separated from the filled (valence) band by a forbidden gap of a few ev.

> Intrinsic semiconductors: The conduction and valence bands are separated by a narrow forbidden gap, and semiconductivity arises from the electrons thermally excited to the conduction band.

	$f(\epsilon)$	PARTICLES	$g(\epsilon)$	RESULTS
Ideal gas	f_{MB}	Point particles	$g(\epsilon) \propto \epsilon^{1/2}$ $(\epsilon = p^2/2m)$	$\bar{\epsilon} = \frac{3}{2}kT$ $C_v = \frac{3}{2}R$
Diatomic gas	f_{MB}	Diatomic molecules	$g(E_r) = (2J+1)$ $g(E_v) = 1$	Low temperatures: $\bar{\epsilon} \simeq \bar{\epsilon}_{tr} = \frac{3}{2}kT$ $C_v \simeq (C_v)_{tr} = \frac{3}{2}R$
				Intermediate temperatures: $\bar{\epsilon} \simeq \bar{\epsilon}_{tr} + \bar{\epsilon}_r = \frac{5}{2}kT$ $C_v \simeq (C_v)_{tr} + (C_v)_r = \frac{5}{2}R$
				High temperatures: $\bar{\epsilon} \simeq \bar{\epsilon}_{tr} + \bar{\epsilon}_r + \bar{\epsilon}_v = \frac{7}{2}kT$ $C_v \simeq (C_v)_{tr} + (C_v)_r + (C_v)_v = \frac{7}{2}R$
Blackbody radiation	f_{BE}	Photons	$g(\epsilon) \propto \epsilon^2$ $(\epsilon = pc)$	Planck radiation equation $E(\nu)d\nu = \left(\dfrac{8\pi h\nu^3}{c^3}\right)\left(\dfrac{1}{e^{h\nu/kT}-1}\right)d\nu$
Lattice specific heat of solids	f_{BE}	Phonons	$g(\epsilon) \propto \epsilon^2$ $(\epsilon = pv_s)$	$E(f)df = \left(\dfrac{9Nhf^3}{f_D^3}\right)\left(\dfrac{1}{e^{hf/kT}-1}\right)df$ $E = 9N\left(\dfrac{T}{T_D}\right)^3(kT)\displaystyle\int_0^{x_m}\dfrac{x^3 dx}{e^x - 1}$
				For high temperatures: $T \gg T_D$ $(C_v)_{\text{lattice}} = 3R$
				For low temperatures: $T \ll T_D$ $(C_v)_{\text{lattice}} \propto T^3$
Electronic specific heat	f_{FD}	Electrons	$g(\epsilon) \propto \epsilon^{1/2}$ $(\epsilon = p^2/2m)$	Fermi energy $\epsilon_F = \dfrac{h^2}{2m}\left(\dfrac{3n}{8\pi}\right)^{2/3}$ $(C_v)_{\text{electronic}} \propto T$
Band theory of solids	f_{FD}	Electrons	Not calculated. $(\epsilon = p^2/2m + V)$ where V is the periodic potential energy between an electron and the crystalline lattice	Available states for electron occupancy are bands (1s, 2s, 2p band, etc.) with possible forbidden gaps between the bands

Impurity
semiconductors: Traces of impurities introduce available states into
the region of the forbidden gap. The donor impurity
atoms of an n-type semiconductor introduce discrete
states lying just below the conduction band, and semiconductivity results from electron transport. The acceptor impurity atoms of a p-type semiconductor introduce discrete states lying just above the valence
band and semiconductivity results from the transport
of holes.

REFERENCES

Dekker, A. J., *Solid State Physics*. Englewood Cliffs, New Jersey: Prentice-
Hall, Inc., 1957. A fairly comprehensive treatment of topics in
solid-state physics is given here.

Kittel, C., *Introduction to Solid State Physics*, 2nd ed. New York: John
Wiley & Sons, Inc., 1956. This authoritative introductory solid-
state physics text contains detailed references to original papers
and the numerical values of many physical properties of solids.

Leighton, R. B., *Principles of Modern Physics*. New York: McGraw-Hill
Book Company, Inc., 1959. Chapter 9 summarizes molecular
binding and spectra in some detail. Succinct derivations of the
three statistical distribution laws are given in Chapter 10.

Sproull, R. L., *Modern Physics*. New York: John Wiley & Sons, Inc., 1956.
A clear exposition of solid-state physics, based on elementary wave
mechanics, is presented in this text for engineers.

PROBLEMS

11-1   At what separation distance will a Na^+ ion and a Cl^- ion have
the same electrostatic potential energy as a neutral Na atom and
a neutral Cl atom, infinitely separated?

11-2   The equilibrium separation of the HCl^{35} molecule is 1.27 Å. (a)
Calculate the reduced mass and moment of inertia of this molecule.
(b) What are the three lowest values of the angular momentum
and rotational energy of this molecule?

11-3   (a) Calculate the energy (ev) and wavelength of the photon absorbed when a $Hg^{200}Cl^{35}$ molecule ($r_0 = 2.23$ Å) makes the rotational transitions, $J = 0 \rightarrow 1$ and $J = 1 \rightarrow 2$. (b) In what region
of the electromagnetic spectrum are these lines found?

11-4   The adjacent lines in the pure rotational spectrum of $Cl^{35}F^{19}$ are
separated by a frequency of 1.12×10^{10} sec^{-1}. What is the interatomic distance of this molecule?

11-5   Show that the frequency of a photon emitted by a rigid rotator in a quantum transition equals the mechanical rotational frequency in the limit of very large rotational quantum numbers.

11-6   It is found from a study of the rotational spectrum that the moment of inertia of a diatomic molecule increases as the molecule occupies states with increasingly larger J values. Explain this effect qualitatively.

11-7   Show that the permanent electric dipole moments of polar molecules are of the order of 10^{-29} coulomb-m.

11-8   Photons of 34,800 Å wavelength are absorbed by the diatomic gas HCl when the molecules make vibrational transitions. What is the total zero-point vibrational energy (joules) of 1 mole of this gas at the absolute-zero temperature?

11-9   (a) What is the zero-point vibrational energy of a pendulum 1.0 m long? (b) What is the vibrational quantum number when the pendulum oscillates with an amplitude of 1.0 mm? The mass of the pendulum is 1.0 kg.

11-10   Calculate the ratio of the vibrational frequencies of H^1Cl^{35} and H^2Cl^{37} molecules, assuming that the force constant is the same for both molecules.

11-11   An electronic transition in the molecule CO produces bands of lines in the visible region (6000 Å) of the spectrum. What is the approximate separation in wavelength between adjacent rotational lines of the bands, if the interatomic distance for CO is 1.128 Å? This illustrates the apparently continuous band spectrum of molecules in the visible region.

11-12   Show that the high-energy limits ($h\nu \gg kT$) of the Bose-Einstein and Fermi-Dirac distribution functions approach the Maxwell-Boltzmann distribution.

11-13 *   Show that the Maxwell-Boltzmann velocity distribution of an ideal gas of classical point particles is given by

$$n(v) = Av^2 e^{-mv^2/2kT}$$

where $A = (4/\sqrt{\pi})(m/2kT)^{3/2}$. (*Hint:* $n(v)\, dv = n(\epsilon)d\epsilon$.)

11-14   Show that the molar specific heat of a monatomic gas at *constant pressure* is $\frac{5}{2}R$.

11-15   How much energy is required to raise the temperature of 1.0 mole of helium by 50°C at constant volume?

11-16   At what temperature will 2 per cent of the molecules of a CO gas be found in the first rotational state, the remainder being in the zero-th rotational state? The interatomic distance of carbon monoxide is 1.128 Å.

11-17   Compute the relative number of hydrogen molecules occupying the first several rotational states at 170°K, and compare your results with Figure 11-16.

11-18   Compute the relative number of hydrogen molecules occupying the first several vibrational states at 6350°K, and compare your results with Figure 11-17.

11-19   Show that the molar specific heat at constant *pressure* of H_2 is $\frac{5}{2}R$, $\frac{7}{2}R$, and $\frac{9}{2}R$ for the three temperature regions shown in Figure 11-18.

11-20   The vibrational frequency f of the molecule KCl is 8.40×10^{12} sec^{-1}. The dissociation energy of KCl is 4.42 ev. (a) Assuming, for simplicity, that *all* of the vibrational energy levels are equally spaced, compute the vibrational quantum number corresponding to dissociation of the molecule. (b) At what approximate temperature will KCl dissociate by thermal excitation?

11-21   What is the number of modes of electromagnetic waves between the wavelengths of 4000 Å and 4100 Å in a black box 5 cm on a side?

11-22 *   (a) Calculate the momentum and energy of photons in a box, 10 cm on a side, in the available state for which n_x, n_y, and n_z are $(1, 0, 0)$ and $(1, 1, 0)$. (b) What is the difference in energies of these two states compared with the energy of the lower state? (c) Repeat parts (a) and (b) for the states $(100, 0, 0)$ and $(100, 1, 0)$.

11-23   A blackbody is at a temperature of 1000°K. (a) What is the radiation energy per unit volume in the (visible) region from 4000 Å to 4400 Å? (Assume that $E(\nu)$ is constant over this range of wavelengths.) (b) What is the radiation energy per unit volume in the same wavelength band at the radio frequency, 10 megacycles/sec?

11-24   (a) Show that the Rayleigh-Jeans radiation formula, Equation 11-38, is the low-frequency approximation of the Planck radiation formula. (b) Show that the Wien formula, Equation 11-39, is the high-frequency approximation of the Planck radiation formula.

11-25 *   Verify that the Stefan-Boltzmann radiation law follows from the Planck radiation relation.

11-26 *   Verify that the Wien displacement law is a consequence of the Planck radiation relation.

11-27   The Debye temperature of diamond is 1860°K. What is the molar specific heat of diamond at room temperature? (Use Figure 11-22.)

11-28   The molar specific heats of copper and beryllium are 2.8 R and 1.7 R, respectively, at room temperature. Which material has the higher Debye temperature?

11-29 The element gold has a Debye temperature of 180°K. What is the specific heat of gold (in cal/gm-°K) at room temperature?

11-30 The Debye temperatures for the materials sodium, potassium, diamond, and beryllium are 150°K, 100°K, 1860°K, and 1000°K, respectively. Estimate from Figure 11-22, the molar specific heats for each of these elements at room temperature.

11-31 * Show that if the atoms of a lattice are assumed to be arranged in a cubical array (as in Figure 4-1), the Debye frequency corresponds to an elastic wave for which the distance between adjacent atoms is approximately one-half wavelength. (Assume, for simplicity, that $v_t = v_l$.)

It is assumed in the Debye theory that the elastic waves are propagated through an essentially continuous medium, for which the wavelength is long compared to the interatomic distance. The Debye cut-off occurs when the wavelength is so short that elastic waves cannot be propagated.

11-32 The density of aluminum is 2.700 gm/cm³, and the number of free electrons per atom is 3. Compute the Fermi energy for this metal.

11-33 (a) Compute the total energy (joules) of the valence electrons in one gram-mole of copper (valence, 1; Fermi energy, 7.0 ev) at absolute-zero temperature, according to the free-electron model. (b) Can this energy be extracted from the material?

11-34 * Show that the fraction of the total number of conduction electrons in copper (valence 1), which, at room temperature, have energies greater than the Fermi energy, is of the order of 10^{-3}.

11-35 Verify that the electronic specific heat of a typical conductor is very small compared to the lattice specific heat at room temperature.

11-36 For the element copper, the Debye temperature is 315°K, the Fermi energy is 7.0 ev, and the number of valence electrons per atom is 1. At what temperature are the contributions from the lattice and the electronic specific heats equal?

11-37 * For barium, the work function is 2.51 ev, the internal potential energy seen by a free electron is 6.31 ev, the atomic weight is 138, and the density is 3.78 gm/cm³. Calculate the number of free electrons per atom in barium.

11-38 Explain why the work function of a metal is essentially independent of its temperature.

11-39 * The electron configuration of magnesium, a divalent metal, is $1s^2 2s^2 2p^6 3s^2$. Magnesium is a good conductor. What inference can one draw concerning the relative positions of the $3s$ and $3p$ energy bands?

11-40 When a fifth-column impurity atom, such as arsenic, replaces a silicon atom in a silicon crystal, there is an unbonded electron which sees a charge of $+e$ located at the arsenic ion. This electron can be imagined to move in Bohr orbits about the positive charge in the silicon medium characterized by a dielectric constant of 12. (a) Compute the radius of the first Bohr orbit, and compare it with the distance between nearest-neighbors in silicon, 2.35 Å. (b) Compute the binding energy of this electron to the arsenic ion, and compare this with kT for room temperature.

11-41 Make the computations called for in Problem 11-40, but with an arsenic impurity atom now in a germanium crystal with a dielectric constant of 16. The distance between nearest neighbors for germanium is 2.44 Å.

ELECTROMAGNETIC WAVES

The following analysis of a rectangular electromagnetic pulse illustrates some of the properties of electromagnetic waves. Although this derivation is based on rectangular pulses, the results are of a general character, as is confirmed from a rigorous solution of Maxwell's equations.

Let us assume that a rectangular pulse of magnetic field with flux density **B** is moving along the positive X-axis at a speed v, as shown in Figure I-1.

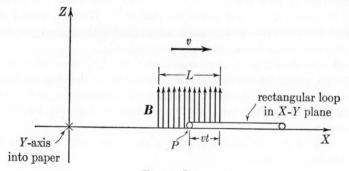

Figure I-1.

This field is moving through empty space, far removed from any electric charges and currents. The flux density has a constant magnitude B along the Z-axis in the interval L, and is zero outside this interval. We wish to show that an electric field $\mathcal{E}$ must necessarily accompany this **B**.

It will be helpful to consider a rectangular loop, lying in the X-Y plane, whose length in the Y-direction is w, and whose length in the X-direction is indefinitely long. Then, if at the time $t = 0$, the leading edge of the magnetic pulse is just arriving at the left end of the loop, this moving edge after a time t will have progressed a distance vt into this loop. By Faraday's law of electromagnetic induction we can evaluate the electric field induced in the loop by the passing magnetic pulse.

$$\text{voltage induced in loop} = \int_{\substack{\text{around} \\ \text{loop}}} \mathcal{E} \cos \theta \, ds = -d\phi/dt \qquad [\text{I-1}]$$

where ϕ is the magnetic flux, $\mathcal{E}$ is the induced electric field, and ds is an element of length along the loop. The flux at the time t is

$$\phi = \int B \cos \theta \, dA = BA = B(wvt)$$

Therefore,
$$d\phi/dt = Bwv \qquad [\text{I-2}]$$

The only contribution to the integral of Equation I-1 around the loop is along the left side, where the integral becomes $(\mathcal{E}w)$. Using this result, along with Equation I-2 in Equation I-1, we have (magnitudes only)

$$(\mathcal{E}w) = (Bwv)$$

or,
$$\mathcal{E} = Bv \qquad \text{[I-3]}$$

The direction of $\mathcal{E}$ is easily found by Lenz's law: as the magnetic pulse passes into the loop, the magnetic flux through the loop increases in the $+Z$-direction. Consequently, an induced emf is set up to oppose the increase in the magnetic flux; this requires that the direction of $\mathcal{E}$ along the left side of the loop be inward; i.e., in the $+Y$-direction. Thus, while the magnetic pulse passes the point P, there will be an electric field $\mathcal{E}$ in the $+Y$-direction of magnitude, $\mathcal{E} = Bv$. At any time when the pulse is *not* passing the point P, there is no electric field at P. We find, using Faraday's law, that a transverse electric pulse accompanies the transverse magnetic pulse, the two fields being mutually perpendicular.

Before considering an analogous situation in which we start with a moving electric pulse, let us digress for a moment to develop an important relation. This relation shows how a changing electric field produces a magnetic field, just as Faraday's law shows how a changing magnetic field produces an electric field.

We know that a current i creates a magnetic field of flux density $\boldsymbol{B}$ given by Ampere's law

$$\int_{\substack{\text{closed} \\ \text{loop}}} B \cos \theta \, ds = \mu_0 i \qquad \text{[I-4]}$$

As was indicated in Section 1-4, i must represent *both* the real current i_r, and also Maxwell's so-called displacement current i_d, which arises from a changing electric field. For the situation discussed here there are no real currents. The displacement current can be determined by considering a parallel-plate capacitor of area A, charge q, and surface charge density $\sigma = q/A$ (see Figure 1-8). Then, $i_d = dq/dt = A \, d\sigma/dt$. For a parallel-plate capacitor $\mathcal{E} = \sigma/\epsilon_0$, and we have

$$i_d = A\epsilon_0 \, d\mathcal{E}/dt \qquad \text{[I-5]}$$

or,
$$i_d = \epsilon_0 \, d/dt \, (A\mathcal{E}) \qquad \text{[I-6]}$$

Equation I-6 represents the more general form for i_d than does Equation I-5. A displacement current can thus be produced either by a changing area A, or by a changing electric field $\mathcal{E}$, or both.

Applying Ampere's law, Equation I-4, under the condition that $i_r = 0$, and using Equation I-6, for i_d, gives

$$\int_{\substack{\text{closed} \\ \text{loop}}} B \cos \theta \, ds = \mu_0 i_d = \mu_0 \epsilon_0 \, d/dt \, (A\mathcal{E}) \qquad \text{[I-7]}$$

In a fashion similar to that employed for the magnetic pulse, we now consider a rectangular pulse of electric field, as shown in Figure I-2. A rec-

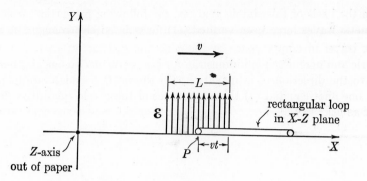

Figure I-2.

tangular loop lies in the X-Z plane, whose length in the Z-direction is b. Equation I-7 then becomes

$$\int_{\substack{\text{closed} \\ \text{loop}}} B \cos \theta \, ds = Bb = \mu_0 \epsilon_0 \, d/dt \, (bvt\mathcal{E})$$

or,
$$B = \mu_0 \epsilon_0 v \mathcal{E} \qquad \qquad \text{[I-8]}$$

The direction of $\boldsymbol{B}$ at the point P is out of the paper (in $+Z$-direction), inasmuch as the displacement current i_d through the loop is in the $+Y$-direction, and the accompanying magnetic field surrounding this current follows from the right-hand rule.

In Figure I-1 we began with a moving magnetic pulse and found the electric field produced by this pulse. In Figure I-2 we began with a moving electric pulse and found the accompanying magnetic field. We see that the relative directions of $\mathcal{E}$, $\boldsymbol{B}$, and the direction of propagation of the field pulses are self-consistent in Figures I-1 and I-2. In addition, the magnitudes of $\mathcal{E}$ and $\boldsymbol{B}$, specified by Equations I-3 and I-8, must also be self-consistent.

Substituting the value of $\mathcal{E}$ from Equation I-3 into Equation I-8 immediately gives

$$v = 1/\sqrt{\epsilon_0 \mu_0} \qquad \qquad \text{[I-9]}$$

Therefore, if this electromagnetic pulse is to exist it must travel at the constant speed $1/\sqrt{\epsilon_0 \mu_0}$, which is just equal to c, the speed of light! Using the value of v from Equation I-9 in Equation I-8, we have

$$B = \sqrt{\mu_0 \epsilon_0} \mathcal{E} \qquad \qquad \text{[I-10]}$$

Recalling that, for empty space, the magnetic intensity $\mathcal{3C}$ is $\mathcal{3C} = \boldsymbol{B}/\mu_0$. Equation I-10 can be written in the familiar form

$$\mathfrak{K}/\mathcal{E} = \sqrt{\epsilon_0/\mu_0}$$

or, $$\mu_0\mathfrak{K}^2 = \epsilon_0\mathcal{E}^2$$

On the basis of this simple analysis, the following properties of electromagnetic waves have been verified: (1) an isolated electromagnetic wave must travel in empty space at the constant speed $c = 1/\sqrt{\epsilon_0\mu_0}$; (2) the electric and magnetic field intensities are perpendicular to one another and also to the direction of propagation; (3) the fields $\mathcal{E}$ and $\mathfrak{K}$ are in phase with one another; and (4) the magnitudes of these field quantities are related as $\mu_0\mathfrak{K}^2 = \epsilon_0\mathcal{E}^2$.

WAVE PACKETS

The representation of a monochromatic wave, traveling along the X-axis with a velocity $v = \nu\lambda$, is given by the equation for a running wave

$$A = A_0 \cos 2\pi(x/\lambda - \nu t) \qquad \text{[II-1]}$$

where λ and ν represent the wavelength and frequency respectively. The wave disturbance A is given in Equation II-1 as a function both of position x and time t, and has the maximum value A_0. For an electromagnetic wave A stands for the electric or magnetic field intensity; for a sound wave through air A represents the pressure; and for a transverse wave on a string, A represents the transverse displacement.

Using the definition

$$k = 2\pi/\lambda \qquad \text{[II-2]}$$

where, for simplicity, k will be called the *wave number*, Equation II-1 can be written in the form

$$A = A_0 \cos k(x - vt) \qquad \text{[II-3]}$$

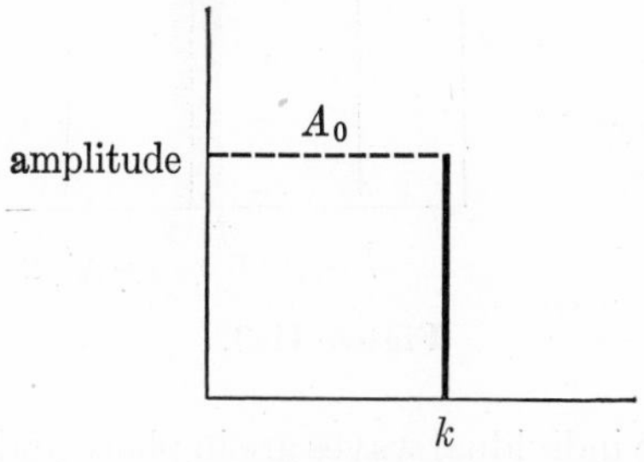

Figure II-1.

Figure II-1 shows the amplitude of the single monochromatic wave having a wave number, k, or wavelength $2\pi/k$; and Figure II-2 shows A as a function of x at the particular time, $t = 0$.

Let us now consider a collection, or packet, of monochromatic waves, all traveling at the same speed v (showing no dispersion) in the $+X$-direction. For convenience we imagine that all waves have the same amplitude A_0 and that the wave packet includes all wave numbers running from $k - \Delta k/2$ to $k + \Delta k/2$. Therefore, all waves lie within the band of width Δk, as shown in Figure II-3. If $\Delta k = 0$, the band of waves becomes the single,

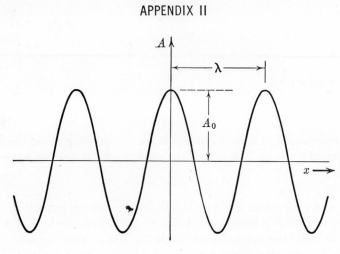

time, $t = 0$

Figure II-2.

monochromatic wave of Figure II-1. Figure II-4, showing the spatial extent of the wave packet at time $t = 0$, corresponds to Figure II-2.

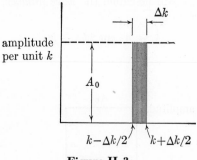

Figure II-3.

At the origin all the individual waves are in phase and add constructively, giving, therefore, a large resultant amplitude at this point. As we leave the origin in either direction, the waves become increasingly out-of-phase, and the algebraic addition of the individual waves gives a resultant amplitude A which rapidly approaches zero.

Using the principle of superposition, we now compute mathematically the resultant amplitude A at any point x and any time t, comprised of contributions of all monochromatic waves within the band Δk. We sum the individual contributions $A_k dk$ from $(k - \Delta k/2)$ to $(k + \Delta k/2)$. It is convenient to let $x - vt = x'$. Thus, Equation II-3 becomes

$$A_k = A_0 \cos kx'$$

where A_0 is now the amplitude per unit k.

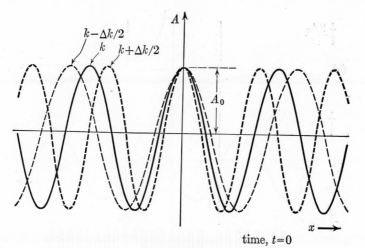

time, $t=0$

Figure II-4.

The resultant displacement is given by

$$A = \int_{k-\Delta k/2}^{k+\Delta k/2} A_k dk = A_0 \int_{k-\Delta k/2}^{k+\Delta k/2} \cos kx' \, dk = (A_0/x') \sin kx' \Big]_{k-\Delta k/2}^{k+\Delta k/2}$$

$$A = \frac{A_0}{x'} \left[\sin x' \left(k + \frac{\Delta k}{2} \right) - \sin x' \left(k - \frac{\Delta k}{2} \right) \right] \qquad \text{[II-4]}$$

We can simplify Equation II-4 by using the trigonometric identity

$$\sin (a + b) - \sin (a - b) = 2 \sin b \cos a$$

Equation II-4 then becomes

$$A = (2A_0/x') \sin (x' \, \Delta k/2) \cos x'k \qquad \text{[II-5]}$$

Figure II-5a,b,c show the separate factors of Equation II-5 plotted against x'; and Figure II-5d,e show the resultant wave A and the envelope of A^2 as a function of x'. Inasmuch as $x' = x - vt$, Figure II-5e is a "snapshot" of the intensity (proportional to the square of the amplitude) of the wave.

More than one-half of the total energy of the wave packet is within the region $2\pi/\Delta k$. (The shaded area of Figure II-5e is approximately 75 per cent of the total area under the curve.) The uncertainty Δx in the width of the wave packet (at any instant of time) is at least as large as $2\pi/\Delta k$. Therefore,

$$\Delta x \geq 2\pi/\Delta k \qquad \text{[II-6]}$$

We have introduced the wave number k merely for mathematical convenience in evaluating the integrals. Now let us rewrite Equation II-6 in terms of the more familiar wave property, λ. By definition

$$k = 2\pi/\lambda$$

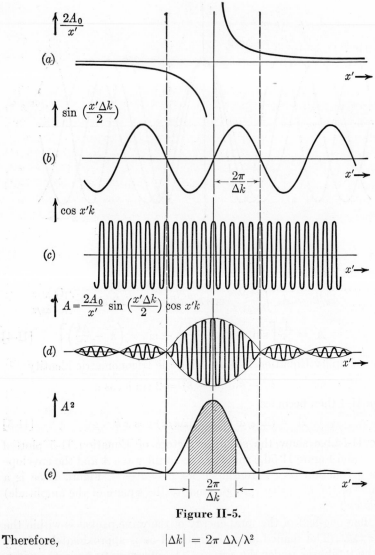

Figure II-5.

Therefore, $|\Delta k| = 2\pi \, \Delta\lambda/\lambda^2$

Equation II-6 then becomes

$$\Delta x \, \Delta\lambda \geq \lambda^2 \qquad\qquad [\text{II-7}]$$

Equation II-7 shows that if $\Delta\lambda$ is small (that is, if the wave packet consists of almost monochromatic waves) then the spatial extent of the wave packet, Δx, becomes very large. On the other hand, if a wave disturbance is to be confined to a very small region in space Δx, then $\Delta\lambda$ must be very large; that is, one must add together monochromatic waves over a wide range of wavelengths.

THE SCHRÖDINGER EQUATION

A very brief introduction to the Schrödinger wave theory of a material particle is given here. The one-dimensional, time-independent Schrödinger wave equation is developed and illustrated by an example.

First consider a free particle of mass m moving with a speed v in the positive X-direction. The particle's total energy E is $E = K = \frac{1}{2} mv^2 = p^2/2m$, and its linear momentum is $\boldsymbol{p} = m\boldsymbol{v}$. Inasmuch as the energy and momentum are well-defined, and $E = h\nu$ and $p = h/\lambda$, the wave associated with this free particle is monochromatic, having a precise frequency ν and wavelength λ. The traveling wave can be represented by the function $\Psi_1 = \Psi_1(x, t)$ of the form

$$\Psi_1 = A_1 \cos 2\pi[(x/\lambda) - \nu t]$$

A similar monochromatic wave traveling to the left can be represented by

$$\Psi_2 = A_1 \cos 2\pi[(x/\lambda) + \nu t]$$

The superposition of these traveling waves results in standing, or stationary, waves of the form

$$\Psi = \Psi_1 + \Psi_2 = A \cos (2\pi x/\lambda) \cos (2\pi \nu t) \qquad \text{[III-1]}$$

where $\Psi = \Psi(x, t)$ is now seen to be the product of a spatial-dependent term $\psi(x) = A \cos 2\pi x/\lambda$, and a time-dependent term $f(t) = \cos 2\pi\nu t$. We will limit our discussion to the spatial-dependent term $\psi(x)$.

Taking the partial derivation of $\Psi(x, t)$ twice with respect to x, we obtain

$$\frac{\partial^2 \Psi}{\partial x^2} = -(2\pi/\lambda)^2 \Psi$$

Inasmuch as $\Psi(x, t) = \psi(x) f(t)$, we have

$$d^2\psi/dx^2 = -(2\pi/\lambda)^2\psi = -(p^2/\hbar^2)\psi \qquad \text{[III-2]}$$

Although the second-order differential equation for $\psi(x)$, Equation III-2, has been developed for a free particle, Schrödinger (1926) assumed that this equation also describes the wave behavior of a particle under the influence of a conservative force. For a particle whose potential energy is $V = V(x)$, the total energy is

$$E = K + V = (p^2/2m) + V$$

Therefore, $\qquad\qquad p^2 = 2m(E - V)$

and Equation III-2 becomes

$$\frac{\hbar^2}{2m}\frac{d^2\psi}{dx^2} + (E - V)\psi = 0 \qquad\qquad \text{[III-4]}$$

Equation III-4 is the one-dimensional, time-independent Schrödinger equation. To obtain the three-dimensional Schrödinger equation one merely replaces $d^2\psi/dx^2$ in Equation III-4 by $(d^2\psi/dx^2 + d^2\psi/dy^2 + d^2\psi/dz^2)$.

Solutions of the Schrödinger equation for a "particle" under the influence of a known force (having a known potential energy $V(x)$ as a function of its position) and consistent with the boundary conditions will yield the permitted energies E_n (called *eigenvalues*) and the corresponding wave functions ψ_n (called *eigenfunctions*). From a knowledge of ψ one can determine the probability of finding the particle at any position. Except for very simple situations, the solution of the second-order Schrödinger differential equation is mathematically complicated.

For all bound systems, a general remark can be made which is independent of the particular form of $V(x)$. Regarding the "particle" as a wave, the traveling wave can be imagined to be reflected back and forth within the confines of the bound system, forming stationary, or standing, waves. Only those solutions of the Schrödinger equation leading to stationary waves are meaningful. Indeed, it is the fitting of stationary waves to satisfy the boundary conditions that leads to the allowed (quantized) energies following

$$\lambda = h/p = h/\sqrt{2m(E - V)} \qquad\qquad \text{[III-5]}$$

from Equation III-3.

A simple problem that is easily solved by the Schrödinger equation is that of a particle confined to a one-dimensional box of length L. The potential energy $V(x)$ within the box is constant (taken to be zero) and rises sharply to infinity at the boundaries, $x = 0$ and $x = L$. The Schrödinger equation, Equation III-4, for this potential then becomes

$$\frac{\hbar^2}{2m}\frac{d^2\psi}{dx^2} + E\psi = 0$$

for $0 < x < L$, or

$$\frac{d^2\psi}{dx^2} = -B^2\psi \qquad\qquad \text{[III-6]}$$

where $\qquad\qquad B^2 \equiv 2mE/\hbar^2$

The particle cannot be outside the box. Recalling (Section 4-7) that $\psi^2(x)dx$ is the probability of finding the particle in the length dx, $\psi(x) = 0$ for $x \leq 0$ and $x \geq L$. The solution of Equation III-6 consistent with the boundary conditions, $\psi(0) = \psi(L) = 0$, is the sine function

$$\psi(x) = A \sin Bx$$

as is verified by substitution into Equation III-6. The boundary condition, $\psi(0) = 0$, is satisfied; but the second boundary condition, $\psi(L) = 0$, is satisfied only if $BL = n\pi$ (sin $n\pi = 0$ for all integral n). Substituting for B we

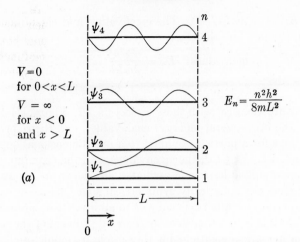

$V = 0$
for $0 < x < L$
$V = \infty$
for $x < 0$
and $x > L$

(a)

$$E_n = \frac{n^2 h^2}{8mL^2}$$

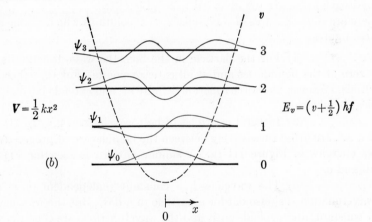

$V = \frac{1}{2}kx^2$

(b)

$$E_v = \left(v + \tfrac{1}{2}\right) hf$$

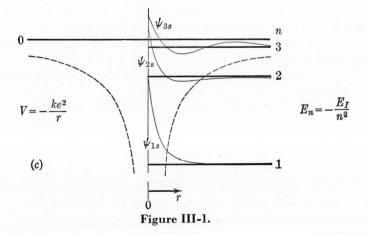

$V = -\dfrac{ke^2}{r}$

(c)

$$E_n = -\frac{E_I}{n^2}$$

Figure III-1.

obtain $n\pi = BL = \sqrt{2mE}(L/\hbar)$. The eigenvalues and eigenfunctions for the particle in the box are, therefore,

$$\text{eigenvalues: } E_n = \frac{n^2\pi^2\hbar^2}{2mL^2}$$

$$\text{eigenfunctions: } \psi_n(x) = A_n \sin (n\pi x/L)$$

These results are identical with those of Section 4-7. Figure III-1a shows the potential, the first several (energy) eigenvalues, and their corresponding eigenfunctions for a particle confined to a one-dimensional box.

The potential, $V(x) = \frac{1}{2} kx^2$, the eigenvalues, and the eigenfunctions for a one-dimensional simple harmonic oscillator are shown in Figure III-1b. Figure III-1c gives the spherically symmetrical solutions (s-states) of the Schrödinger equation for a particle subject to an inverse-square force with $V(r) = -k\,e^2/r$; this is just the wave-mechanical problem of the hydrogen atom. The mathematical solutions for the simple harmonic oscillator and the hydrogen atom are too involved to be given here.

There are some features of wave-mechanical solutions which can be seen in these figures:

(1) For the particle in the box, the wave function is exactly zero at the boundaries and at all exterior points. For the other two potentials, however, the wave function is finite at, and extends beyond, the classical boundaries.

(2) Integral multiples of half-wavelengths $1(\lambda/2)$, $2(\lambda/2)$, $3(\lambda/2)$, . . . are fitted successively between the boundaries for progressively higher energies in Figure III-1a. A similar pattern is seen for Figures III-1b and c.

(3) The wavelength is constant (independent of x) and the wavefunction is sinusoidal for a particle in a box. But the wavelength is not constant, and depends on x, and the wavefunctions are consequently non-sinusoidal for the simple-harmonic oscillator and the hydrogen atom. From Equation III-5, $\lambda = h/[2m(E - V)]^{1/2}$, it follows that the wavelength depends on the potential energy $V(x)$ and therefore on x. For this reason, the wavelength is shorter at the center than at the boundaries in Figures III-1b and c.

(4) In all three examples the lowest, or ground-state, energy is *not* zero.

(5) For the particle in the box, the energy levels are crowded at the bottom; for the simple-harmonic oscillator, they are equally-spaced; and for the hydrogen atom, they are crowded at the top. This behavior is related to the shape of the potential curve. In Figure III-1a the potential is bent toward the vertical with respect to the simple-harmonic oscillator potential, whereas in Figure III-1c the potential is bent toward the horizontal with respect to that of the simple-harmonic oscillator.

THE ATOMIC MASSES

Given here are the masses of the neutral atoms of all stable nuclides and a few of the unstable nuclides. The radioactive nuclides are indicated by an asterisk following the mass number A. Errors in the listed masses are in the last significant figure only. Mass values enclosed in parentheses are suspected to be unreliable.

These data are taken from W. H. Johnson, K. S. Quisenberry, and A. O. Nier, Part 9, Chapter 2, p. 55, *Handbook of Physics*, ed. by E. U. Condon and H. Odishaw, McGraw-Hill Book Company, Inc., New York, 1958.

Element	A	Atomic mass amu	Element	A	Atomic mass amu
$_0$n	1	1.008 9860	$_8$O	14*	14.013 06
$_1$H	1	1.008 1451		15*	15.007 784
	2	2.014 7425		16	16.000 000
	3*	3.017 005		17	17.004 5374
$_2$He	3	3.016 986		18	18.004 8847
	4	4.003 874		19*	19.009 620
	5*	5.013 89	$_9$F	17*	17.007 512
	6*	6.020 79		18*	18.006 676
$_3$Li	5*	5.014 0		19	19.004 443
	6	6.017 034		20*	20.006 35
	7	7.018 232	$_{10}$Ne	18*	18.011 2
	8*	8.025 020		19*	19.007 940
$_4$Be	7*	7.019 158		20	19.998 7979
	8*	8.007 849		21	21.000 5241
	9	9.015 060		22	21.998 3771
	10*	10.016 725		23*	23.001 61
$_5$B	8*	8.026 7	$_{11}$Na	23	22.997 091
	9*	9.016 208	$_{12}$Mg	24	23.992 669
	10	10.016 124		25	24.993 782
	11	11.012 808		26	25.990 854
	12*	12.018 185	$_{13}$Al	27	26.990 111
$_6$C	10*	10.020 2	$_{14}$Si	28	27.985 826
	11*	11.014 939		29	28.985 71
	12	12.003 8156		30	29.983 290
	13	13.007 4900	$_{15}$P	31	30.983 6126
	14*	14.007 692	$_{16}$S	32	31.982 238
	15*	15.014 4		33	32.981 947
$_7$N	12*	12.022 9		34	33.978 664
	13*	13.009 877		36	35.978 525
	14	14.007 5257	$_{17}$Cl	35	34.979 972
	15	15.004 8783		37	36.977 657
	16*	16.011 20			
	17*	17.014 00			

Element	A	Atomic mass amu	Element	A	Atomic mass amu
$_{18}$Ar	36	35.978 983	$_{34}$Se	74	73.946 05
	38	37.974 802		76	75.943 42
	40	39.975 093		77	76.944 44
$_{19}$K	39	38.976 100		78	77.942 17
	40*	39.976 709		80	79.941 88
	41	40.974 856		82	81.942 68
$_{20}$Ca	40	39.975 293	$_{35}$Br	79	78.943 49
	42	41.971 967		81	80.942 15
	43	42.972 444	$_{36}$Kr	78	77.944 97
	44	43.969 471		80	79.942 00
	46	45.968 297		82	81.939 50
	48	47.967 766		83	82.940 42
$_{21}$Sc	45	44.970 211		84	83.938 19
$_{22}$Ti	46	45.967 241		86	85.938 11
	47	46.966 685	$_{37}$Rb	85	84.939 02
	48	47.963 190		87*	86.936 9
	49	48.963 429	$_{38}$Sr	84	83.939 9
	50	49.960 669		86	85.936 7
$_{23}$V	50*	49.963 045		87	86.936 62
	51	50.960 175		88	87.934 0
$_{24}$Cr	50	49.961 931	$_{39}$Y	89	88.934 0
	52	51.957 026	$_{40}$Zr	90	89.932 9
	53	52.957 482		91	90.934 2
	54	53.956 023		92	91.933 9
$_{25}$Mn	55	54.955 523		94	(93.937 5)
$_{26}$Fe	54	53.956 759		96	95.939 8
	56	55.952 725	$_{41}$Nb	93	92.935 20
	57	56.953 511	$_{42}$Mo	92	(91.937 8)
	58	57.951 736		94	(93.935 8)
$_{27}$Co	59	58.951 920		95	94.934 6
$_{28}$Ni	58	57.953 772		96	95.935 4
	60	59.949 824		97	96.937 0
	61	60.950 462		98	97.937 1
	62	61.948 029		100	99.938 3
	64	63.948 284	$_{44}$Ru	96	95.939 2
$_{29}$Cu	63	62.949 604		98	(97.985)
	65	64.948 426		99	98.937 5
$_{30}$Zn	64	63.949 471		100	
	66	65.947 013		101	
	67	66.948 419		102	101.936 1
	68	67.946 458		104	103.937 3
	70	69.947 576	$_{45}$Rh	103	102.937 3
$_{31}$Ga	69	68.947 63	$_{46}$Pd	102	101.937 26
	71	70.947 37		104	103.936 3
$_{32}$Ge	70	69.946 23		105	104.938 2
	72	71.944 46		106	105.936 6
	73	72.946 53		108	107.937 8
	74	73.944 51		110	109.939 4
	76	75.945 43	$_{47}$Ag	107	106.938 8
$_{33}$As	75	74.945 54		109	108.939 2

Element	A	Atomic mass amu	Element	A	Atomic mass amu
$_{48}$Cd	106	105.939 6	$_{58}$Ce	136	135.950 3
	108	107.938 4		138	137.949 9
	110	109.938 3		140	139.949 77
	111	110.939 54		142	141.954 42
	112	111.938 6	$_{59}$Pr	141	140.952 28
	113	112.940 36	$_{60}$Nd	142	141.952 60
	114	113.939 77		143	142.955 0
	116	115.941 8		144*	143.955 56
$_{49}$In	113	112.940 2		145	144.958 1
	115*	114.940 2		146	145.959 09
$_{50}$Sn	112	111.940 7		148	147.963 49
	114	113.940 1		150	149.968 49
	115	114.939 9	$_{62}$Sm	144	143.957 41
	116	115.939 0		147*	146.961 20
	117	116.940 27		148	147.961 5
	118	117.939 5		149	148.964 2
	119	118.941 0		150	149.964 57
	120	119.940 3		152	151.967 7
	122	121.942 2		154	153.970 9
	124	123.944 6	$_{63}$Eu	151	150.967 5
$_{51}$Sb	121	120.942 15		153	152.969 2
	123	122.943 3	$_{64}$Gd	152	
$_{52}$Te	120	119.942 6		154	153.969 9
	122	121.941 66		155	154.972 0
	123	122.943 4		156	155.971 8
	124	123.942 5		157	159.674 0
	125	124.944 3		158	157.974 6
	126	125.943 90		160	159.978 1
	128	127.946 2	$_{65}$Tb	159	
	130	129.948 26	$_{66}$Dy	156	
$_{53}$I	127	126.945 0		158	
$_{54}$Xe	124	123.945 52		160	159.974 8
	126	125.944 5		161	160.977
	128	127.944 18		162	161.977 1
	129	128.945 7		163	162.980
	130	129.944 81		164	163.980 6
	131	130.946 70	$_{67}$Ho	165	164.981 5
	132	131.946 10	$_{68}$Er	162	
	134	133.947 99		164	163.981 9
	136	135.950 42		166	165.981 5
$_{55}$Cs	133	132.947 39		167	166.983 6
$_{56}$Ba	130	129.947 54		168	167.984 2
	132	131.947 1		170	169.989 6
	134	133.946 83	$_{69}$Tm	169	
	135	134.948 5	$_{70}$Yb	168	
	136	135.947 58		170	
	137	136.949 06		171	
	138	137.948 73		172	(171.980)
$_{57}$La	138*	137.950 6		173	
	139	138.950 20		174	(173.981)
				176	

Element	A	Atomic mass amu	Element	A	Atomic mass amu
$_{71}$Lu	175		$_{78}$Pt	190*	
	176*	175.997 8		192	(192.025 8)
$_{72}$Hf	174			194	(194.025 0)
	176	175.996 7		195	(195.028 2)
	177	176.998 8		196	(196.029 4)
	178	177.999 6		198	198.027 6
	179	179.002 1	$_{79}$Au	197	197.029 8
	180	180.003 3	$_{80}$Hg	196	196.028 7
$_{73}$Ta	181	181.003 3		198	198.030 3
$_{74}$W	180	180.001 7		199	(199.031 5)
	182	182.003 9		200	(200.031 9)
	183	183.006 3		201	201.034 0
	184	184.007 4		202	
	186	186.009 8		204	204.039 1
$_{75}$Re	185		$_{81}$Tl	203	203.036
	187*			205	205.039 0
$_{76}$Os	184		$_{82}$Pb	204	204.036 5
	186			206	206.038 1
	187			207	207.039 8
	188	(188.016 7)		208	208.040 9
	189	(189.018 6)	$_{83}$Bi	209	209.045 7
	190	(190.015 2)	$_{90}$Th	232*	232.111 8
	192	(192.023)	$_{92}$U	234*	234.114 7
$_{77}$Ir	191	(191.024 1)		235*	
	193	(193.028 1)		238*	238.124 3

ANSWERS

to odd-numbered numerical problems

CHAPTER 1

1-9 4×10^{42}

1-17 33 weber/m², down

1-23 0.086 microwatt

1-31 2.7×10^3 m/sec

1-33 4.1×10^{-16} joule, 2.6×10^3 **ev**

1-35 0.020 m

1-37 3.2×10^{-16} joule, 2.0×10^3 ev

1-39 (a) 2×10^{-7}; (b) 3m; (c) 6×10^6

CHAPTER 2

2-1 (a) $p_x = 5.3 \times 10^{-23}$ kg-m/sec, $p_y = 0$, $K = 2.7 \times 10^{-20}$ joule; (b) $p_x = p_y = 0$ $K = 1.4 \times 10^{-20}$ joule

2-3 3.33 hr, 2.00 hr, 1.20 hr

2-7 $0.987 c$

2-11 0.72 m, 78.7°

2-13 (a) 2.0 m; (b) first observer: 3.4×10^{-8} sec, second observer: 0.7×10^{-8} sec

2-15 (a) $0.98 c$; (b) 2.45×10^{-18} kg-m/sec

2-17 (a) electron: 8.9×10^{-23} kg, proton: 8.9×10^{-23} kg; (b) electron: 9.8×10^4 **per** cent, proton: 53 per cent; (c) electron: $0.9999995 c$, proton: $0.76 c$

2-19 (a) 0.96 microampere; (b) 3.2×10^{-6} newton

2-21 1836

2-23 (a) $0.99995 c$; (b) 50.6 Mev; (c) 92.6 Bev

2-25 (a) 1.05 cm; (b) $1.79 m_0$

2-29 (a) 8.5 Kev; (b) 4.0 Mev

2-35 (a) 7.2×10^{20} joule; (b) 8000 kg; (c) 8.0×10^5 kg

2-37 0.062 ev/molecule

2-39 1.5 nuclear fuel/10^6 kg oil fuel

CHAPTER 3

3-1 (a) 6×10^{10}; (b) 6×10^{-18} kg

3-3 (a) 2.3 ev; (b) 3.9 ev

3-5 (a) 4.85 Å; (b) 0.0061 Å

3-7 (a) 12.4 Kev, 6.4×10^7 m/sec; (b) 5.1×10^{-23} kg-m/sec, **5.7 ev**

3-9 7.1 ev

3-13 (a) 51 Mev; (b) γ-ray

3-15 (a) 2.0×10^2; (b) 1.5×10^{11}; (c) 5.0×10^{-6}

3-21 (a) 0.022 Å; (b) 0.056 Å; (c) 0.27 Mev; (d) 0.61 **Mev**

3-23 1.87 Mev

3-25 (a) 8.3×10^{-19} newton; (b) 16.6×10^{-19} newton

3-29 (a) 1.0 Mev; (b) 5.4×10^{-22} kg-m/sec; (c) 2.8 ev

3-31 127

3-35 115 Kev

3-37 (a) 1.99 Mev; (b) $0.98 c$; (c) $0.9998 c$

3-39 16.1 m⁻¹

3-41 (a) one; (b) 7.6×10^{-3}

CHAPTER 4

4-1 (a) 3.3×10^4 m/sec; (b) 5.7 volt
4-3 7.4 Å
4-9 (a) 15 Kev, 6.6×10^{-23} kg-m/sec; (b) 0.12 Mev, 6.6×10^{-23} kg-m/sec
4-11 3.62 Å
4-13 (a) 27.4°; (b) 54.8°
4-15 2.3 cm versus 1.6 cm
4-17 6.0 Å
4-19 2890 mc/sec
4-23 4.80×10^8
4-25 5.9×10^9
4-27 X-component, 7.3×10^3 m; Y-component, 0
4-31 (a) 6.6×10^{-20} kg-m/sec; (b) 8.2 Mev, 4×10^7 m/sec
4-33 2.1 Mev

CHAPTER 5

5-1 5.9×10^{22} cm^{-3}
5-3 7.8 per cent
5-5 (a) 2.8×10^{-14} m; (b) 8.0 Mev; (c) 1.4×10^{-14} m
5-7 (a) 2.6×10^{-13} m; (b) 0.50 per cent
5-11 (a) kinetic energy: 2.7 Mev; (b) potential energy: 2.3 Mev
5-17 (a) 13.4 ev, 7.2×10^{-27} kg-m/sec, 920 Å; (b) 4.3 m/sec
5-21 (a) 912 Å, 1216 Å; (b) 3646 Å, 6563 Å
5-23 First line of Lyman series, $(2 \rightarrow 1)$
5-31 (a) $4 \rightarrow 2$; (b) 0.49 Å
5-33 (a) 52.1 ev; (b) 2.2 ev
5-37 (a) 0.5487×10^{-3} Å^{-1}; (b) 13,130 Å; (c) 6.79 ev
5-39 (a) 6.68×10^9 sec^{-1}; (b) 6.58×10^9 sec^{-1}; (c) 1.5 per cent
5-41 $\sim$0.3 Å

CHAPTER 6

6-5 $5D \rightarrow 4P$
6-7 $1.85\ e$
6-13 (b) 3β
6-17 (a) 5×10^5; (b) 5×10^{-29} joule-sec versus 10^{-9} joule-sec
6-19 2.90×10^{-5} ev
6-23 6.95×10^{-5} ev
6-27 5.18×10^{11} sec^{-1}
6-29 18.7°
6-35 $_{15}P$
6-39 $_{12}Mg$

CHAPTER 7

7-1 4.15×10^{-15} volt-sec
7-3 4.90 volt
7-7 (a) 16 Kev
7-11 4.13×10^{-15} volt-sec
7-13 Less
7-15 The selection rule $\Delta l = \pm 1$ precludes the transition for which $l = 0$ in both the K and L shells.
7-19 9.0 kilovolt
7-21 0.071 Å
7-23 (a) 35.5 Kev; (b) yes

7-25 K_β

7-27 3.84 cm

CHAPTER 8

8-1 Greater

8-3 3×10^{-11} amp

8-5 1.67×10^{-13} amp

8-7 10^{10} counts

8-9 2.4×10^{-8} sec

8-11 0.8×10^{-5} °C

8-13 4.0 cm

8-15 (a) 0.46 cm; (b) 0.15 m; (c) 5.7 m

8-17 (a) parabola; (b) circle or helix

8-19 0.18 cm

8-21 (a) 69 per cent Cu^{63}, 31 per cent Cu^{65}; (b) 63.6

8-23 1.33 amp

8-25 (b) 0.24 m; (c) 0.8 m

8-27 (a) 4.9 cm, 5.0 cm; (b) 208; (c) 20 m

8-29 36 ev

8-31 (a) 62 Mev; (b) 31 Mev; (c) 62 Mev

8-33 (a) 31 mc/sec; (b) 16 mc/sec; (c) 16 mc/sec; (d) 48 Mev; (e) 24 Mev; (f) 48 Mev

8-35 (a) $\sim$75 m; (b) $\sim 4.4 \times 10^{-6}$ sec

8-37 (a) 10^3 km; (b) 0.003 sec versus 0.017 sec

8-39 (a) 6.0 mc/sec; (b) 0.17 sec; (c) 50×10^4 km

CHAPTER 9

9-1 (b) 12.8 mc/sec; (c) radio

9-3 (a) 2.79×10^{-15} m; (b) 2.48×10^{-15} m; (c) 9.05×10^{-15} m

9-5 $_{14}Si^{30}$

9-7 10^{14}

9-11 (a) $_{12}Mg^{24}$, 11.7 Mev; $_{11}Na^{23}$, 8.8 Mev; $_{10}Ne^{22}$, 10.3 Mev; $_{10}Ne^{21}$, 6.8 Mev; $_{10}Ne^{20}$, 12.8 Mev; $_9F^{19}$, 8.0 Mev; (b) generally larger; (c) proton and neutron levels are filled with two nucleons alternately.

9-13 (E_b/A) for $_1H^1$ is farthest from that nuclide $(_8O^{16})$ for which the whole-number rule is exact.

9-17 3.05×10^{-14} kg

9-19 13 min

9-21 93 yr

9-23 1.0 min^{-1}

9-25 2.74 Mev

9-27 0.000020 amu

9-29 14.0130 amu

9-31 (a) 99.9994 per cent; (b) $_{19}K^{41}$

9-33 (a) 980 ev; (b) 3.37×10^4 km

9-35 11 per cent

9-37 β^-

9-39 (a) zero rest mass, $v = c$, $E = p/c$; (b) angular momentum: neutrino, $\frac{1}{2}\,\hbar$, photon $\hbar$; neutrino interacts weakly with charged particles, photon is electromagnetic and interacts strongly with charged particles.

9-41 (a) $C^{14}/C^{12} \simeq 10^{-12}$; (b) 8660 yr

9-43 10^{21}

9-47 (a) 6.28 Mev and 6.63 Mev; (b) 6.72 Mev; (c) 0.36 Mev

Chapter 10

10-5   9.95 Mev

10-7 $C^{12}(p, \gamma)N^{13}$, $C^{12}(\alpha, \gamma)O^{16}$, $C^{12}(d, p)C^{13}$, $C^{12}(d, \alpha)B^{10}$, $C^{12}(d, n)N^{13}$, $C^{12}(d, \gamma)N^{14}$

10-9   0.00083 amu

10-13   (a) 3.45 Mev;   (b) 0.35 Mev

10-19   (a) 19 barn;   (b) 0.89 mm

10-21 $_{13}Al^{27}(p; {}_{12}Mg^{24}(\alpha; {}_{14}Si^{28}(\gamma$

10-23   (a) 2.13×10^3 m/sec;   (b) 0.024 ev

10-25   5.6°

10-27   (a) 1.1×10^{-11} kg/sec;   (b) 5×10^4 watt;   (c) 1 gm;   (d) 1.1×10^{-11} kg/sec

10-29   (a) 0.72 Mev;   (b) 5.5×10^9 °K

10-31   (a) $\sim\!10^{24}$ kw-hr;   (b) 10^{13} ton

10-33   (a) 0.5 inch;   (b) 3.3×10^{-11} sec

10-35   (a) 5.1 cm

10-37   (a) 9.8×10^{-15} m;   (b) 1.9 Mev;   (c) 5.3×10^{-15} m

10-41   The final stable particles are: 1 proton, 2 electrons, 2 neutrinos, and 4 anti-neutrinos.

Chapter 11

11-1   11 Å

11-3   (a) 2.80×10^{-5} ev, 4.41 cm; 5.60×10^{-5} ev, 2.20 cm;   (b) microwave radio

11-9   (a) 1.0×10^{-15} ev;   (b) $\sim\!10^{28}$

11-11 $\sim$1 Å

11-15 6.3×10^2 joule

11-17   43, 46, 10, and 1 per cent

11-21 1.2×10^{15}

11-23   (a) 3.6×10^{-9} joule/m³;   (b) 1.7×10^{-32} joule/m³

11-27   0.5 R

11-29   0.030 cal/gm °K

11-33   (a) 4.05×10^5 joule;   (b) no

11-37   2

11-39   The $3s$ and $3p$ bands overlap

11-41   (a) 8.5 Å;   (b) 0.05 ev

INDEX

PHYSICAL
CONSTANTS
COMMONLY
NEEDED IN
COMPUTATIONS